Complex Made Simple

Complex Made Simple

David C. Ullrich

Graduate Studies
in Mathematics
Volume 97

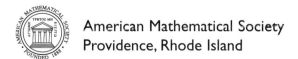

American Mathematical Society
Providence, Rhode Island

Editorial Board
David Cox (Chair)
Steven G. Krantz
Rafe Mazzeo
Martin Scharlemann

2000 *Mathematics Subject Classification*. Primary 30–01.

For additional information and updates on this book, visit
www.ams.org/bookpages/gsm-97

Library of Congress Cataloging-in-Publication Data
Ullrich, David C., 1954–
  Complex made simple / David C. Ullrich.
    p. cm. — (Graduate studies in mathematics ; v. 97)
  Includes bibliographical references and index.
  ISBN 978-0-8218-4479-3 (alk. paper)
  1. Functions of complex variables. I. Title.

QA331.7.U55 2008
515'.9—dc22
                                                                    2008017911

Copying and reprinting. Individual readers of this publication, and nonprofit libraries acting for them, are permitted to make fair use of the material, such as to copy a chapter for use in teaching or research. Permission is granted to quote brief passages from this publication in reviews, provided the customary acknowledgment of the source is given.

Republication, systematic copying, or multiple reproduction of any material in this publication is permitted only under license from the American Mathematical Society. Requests for such permission should be addressed to the Acquisitions Department, American Mathematical Society, 201 Charles Street, Providence, Rhode Island 02904-2294, USA. Requests can also be made by e-mail to reprint-permission@ams.org.

© 2008 by the American Mathematical Society. All rights reserved.
The American Mathematical Society retains all rights
except those granted to the United States Government.
Printed in the United States of America.

∞ The paper used in this book is acid-free and falls within the guidelines
established to ensure permanence and durability.
Visit the AMS home page at http://www.ams.org/

10 9 8 7 6 5 4 3 2 1   13 12 11 10 09 08

Contents

Introduction	ix

Part 1. Complex Made Simple

Chapter 0.	Differentiability and the Cauchy-Riemann Equations	3
Chapter 1.	Power Series	9
Chapter 2.	Preliminary Results on Holomorphic Functions	15
Chapter 3.	Elementary Results on Holomorphic Functions	33
Chapter 4.	Logarithms, Winding Numbers and Cauchy's Theorem	51
Chapter 5.	Counting Zeroes and the Open Mapping Theorem	79
Chapter 6.	Euler's Formula for $\sin(z)$	87
	6.0. Motivation	87
	6.1. Proof by the Residue Theorem	89
	6.2. Estimating Sums Using Integrals	96
	6.3. Proof Using Liouville's Theorem	99
Chapter 7.	Inverses of Holomorphic Maps	105
Chapter 8.	Conformal Mappings	113
	8.0. Meromorphic Functions and the Riemann Sphere	113

8.1. Linear-Fractional Transformations, Part I	117
8.2. Linear-Fractional Transformations, Part II	120
8.3. Linear-Fractional Transformations, Part III	128
8.4. Linear-Fractional Transformations, Part IV: The Schwarz Lemma and Automorphisms of the Disk	130
8.5. More on the Schwarz Lemma	135

Chapter 9. Normal Families and the Riemann Mapping Theorem — 141
   9.0. Introduction — 141
   9.1. Quasi-Metrics — 143
   9.2. Convergence and Compactness in $C(D)$ — 149
   9.3. Montel's Theorem — 155
   9.4. The Riemann Mapping Theorem — 159
   9.5. Montel's Theorem Again — 164

Chapter 10. Harmonic Functions — 167
   10.0. Introduction — 167
   10.1. Poisson Integrals and the Dirichlet Problem — 171
   10.2. Poisson Integrals and Aut($\mathbb{D}$) — 182
   10.3. Poisson Integrals and Cauchy Integrals — 183
   10.4. Series Representations for Harmonic Functions in the Disk — 184
   10.5. Green's Functions and Conformal Mappings — 189
   10.6. Intermission: Harmonic Functions and Brownian Motion — 199
   10.7. The Schwarz Reflection Principle and Harnack's Theorem — 215

Chapter 11. Simply Connected Open Sets — 225

Chapter 12. Runge's Theorem and the Mittag-Leffler Theorem — 229

Chapter 13. The Weierstrass Factorization Theorem — 245

Chapter 14. Carathéodory's Theorem — 257

Chapter 15. More on Aut($\mathbb{D}$) — 267
   15.0. Classification of Elements of Aut($\mathbb{D}$) — 267
   15.1. Functions Invariant under Group Elements — 271

Chapter 16. Analytic Continuation — 277
   16.0. Introduction — 277

16.1. Continuation along Curves	280
16.2. The Complete Analytic Function	287
16.3. Unrestricted Continuation — the Monodromy Theorem	288
16.4. Another Point of View	297
Chapter 17. Orientation	307
Chapter 18. The Modular Function	319
Chapter 19. Preliminaries for the Picard Theorems	337
19.0. Holomorphic Covering Maps	337
19.1. Examples of Holomorphic Covering Maps	343
19.2. Two More Automorphism Groups	345
19.3. Normalizers of Covering Groups	349
19.4. Covering Groups and Conformal Equivalence	353
Chapter 20. The Picard Theorems	357

Part 2. Further Results

Chapter 21. Abel's Theorem	367
Chapter 22. More on Brownian Motion	375
Chapter 23. More on the Maximum Modulus Theorem	385
23.0. Theorems of Hadamard and Phragmén-Lindelöf	385
23.1. An Application: The Hausdorff-Young Inequality	390
Chapter 24. The Gamma Function	399
Chapter 25. Universal Covering Spaces	421
Chapter 26. Cauchy's Theorem for Nonholomorphic Functions	435
Chapter 27. Harmonic Conjugates	441

Part 3. Appendices

Appendix 1. Complex Numbers	445
Appendix 2. Complex Numbers, Continued	449
Appendix 3. Sin, Cos and Exp	455

Appendix 4. Metric Spaces   461

Appendix 5. Convexity   473

Appendix 6. Four Counterexamples   475

Appendix 7. The Cauchy-Riemann Equations Revisited   479

References   483

Index of Notations   485

Index   487

Introduction

Complex Made Simple is intended as a text on complex analysis at the beginning graduate level — students who have already taken a course on this topic may nonetheless be interested in the results in the second half of the book, beginning somewhere around Chapter 16, and experts in the field may be amused by the proof of the Big Picard Theorem in Chapter 20.

The main prerequisite is a course typically called "Advanced Calculus" or "Analysis", including topics such as uniform convergence, continuity and compactness in Euclidean spaces. A hypothetical student who has never heard of complex numbers should begin with Appendices 1 and 2 (students who are familiar with basic manipulations with complex numbers on an informal level can skip Appendix 2, although many such students should probably read Appendix 1). Definitions and results concerning metric spaces are summarized in Appendix 4, with most proofs left as exercises. We decided not to include a similar summary of elementary point-set topology: General topological spaces occur in only a few sections, dealing with Riemann surfaces (students unfamiliar with general topology can skip those sections or pretend that a topological space is just a metric space). The only abstract algebra required is a rudimentary bit of group theory (normal subgroups and homomorphisms), while the deepest fact from linear algebra used in the text is that similar matrices have the same eigenvalues.

Of course the analysis here is really no simpler than that in any other text on the topic at the same level (although we hope we have made it simple to understand). A more accurate title might be *Complex Explained in Excruciating Detail*: Since our main intent is pedagogical, we place great emphasis on motivation, attempting to distinguish clearly between clever ideas and routine calculations, to explain what various results "really mean", to show

how one might have *found* a certain argument, etc. In several places we give two versions of a proof, one more "abstract" than the other; the reader who wishes to attain a clear understanding of the difference between the forest and the trees is encouraged to contemplate both proofs until he or she sees how they are really just different expositions of the same underlying idea.

Many results in elementary complex analysis (pointwise differentiability implies smoothness, a uniform limit of holomorphic functions is holomorphic, etc.) are really quite surprising. Or at least they *should* be surprising; we include examples from real analysis for the benefit of readers who might not otherwise see what the big deal is.

There are a few ways in which the content differs from that of the typical text. First, the reader will notice an emphasis on (holomorphic) automorphism groups and an explicit mention of the notion of covering spaces. These concepts are used in incidental ways in the first half of the book; for example linear-fractional transformations arise naturally as the automorphisms of the Riemann sphere instead of being introduced as an *ad hoc* class of conformal maps in which it just happens that various calculations are easy, covering maps serve to unify various results on analytic continuation, etc.; then it turns out that some not-quite-trivial results on automorphisms and covering maps are crucial to the proof of the Big Picard Theorem.

Probably the most unusual aspect of the content is the inclusion of a section on the relation between Brownian motion and the Dirichlet problem. In most of the text we have tried to achieve a fairly high standard of rigor, but in this section the notion of rigor simply flies out the window: We do not even include precise definitions of the things we're talking about! We decided to include a discussion of this topic even though we could not possibly do so rigorously (considering the prerequisites we assume) because Brownian motion gives the clearest possible intuition concerning the Dirichlet problem. Readers who are offended by the informal nature of the exposition in this section are encouraged to think of it not so much as a lecture but rather a conversation in the departmental lounge or over a few beers on a Friday afternoon.

Finally, the proof of the Big Picard Theorem will probably be new to most readers, possibly including many experts. The proof is certainly not simpler or shorter than the proofs found in typical texts, but it seems very interesting, at least to me: It proceeds by essentially a direct generalization of the standard "one-line" proof of the Little Picard Theorem. (See the discussion of Theorem A and Theorem B in Chapter 20.)

It will be clear to many readers that I first learned much of this material from [**R**]. Very few references are given; all of the results are quite standard, and I doubt that any of the proofs are new. Indeed, for some time I thought that the proof of the Big Picard Theorem was original with me — Anthony Kable discovered that it is essentially the same as the original proof [**J**]. (This raises the question of why the original proof is not so well known; I conjecture that it fell out of favor because various concepts and techniques were much newer and fuzzier in Picard's time than they are at present.) The list of references at the end of the book should not be construed as a guide to the literature or even as a list of suggestions for further reading; it is simply a list of the references that happened to come up in the text. (I decided that including a "Further Reading" section would border on arrogance — further reading here could include topics in almost any area of mathematics.)

It is a pleasure to thank various students and past and present colleagues for mathematical and moral support through the years, including Benny Evans, Alan Noell, Wade Ramey, David Wright, and in particular Robert Myers, who gave a very careful reading of the sections on topology, and especially Anthony Kable, who made various valuable comments at every stage of the project. We enjoyed working with the people at the AMS: Barbara Beeton provided staggeringly competent and often witty TEXnical advice, and Edward Dunne was a very enthusiastic and helpful senior editor.

Any errors or omissions are the responsibility of the author. However, readers who feel that the whole book is just one big mistake need to discuss the matter with Walter Rudin: Before reading *Real and Complex Analysis* I had no idea I was interested in the subject. It seems presumptuous to publish another book in a field where there already exists a text so beautiful it makes your eyes hurt, but several people kept bugging me to write up my lecture notes — this seemed like the only way to shut them up.

Part 1

Complex Made Simple

Chapter 0

Differentiability and the Cauchy-Riemann Equations

A few preliminaries are in order. The letter $\mathbb{C}$ will denote the set of all complex numbers.

This raises the question of exactly what a complex number *is*. In most places in this book (and indeed in most situations in mathematics) one thinks of a complex number as "a quantity of the form $x + iy$, where $x, y \in \mathbb{R}$ and $i^2 = -1$"; one thinks of the ordered pair (x, y) as a way of "representing" the complex number $x + iy$ as a point in the plane. The present chapter is one of the few places where we need to keep in mind that in fact $x + iy$ is literally *equal* to (x, y), by definition. (There are many possible ways to define the complex numbers, but saying that $x + iy = (x, y)$ is the simplest and most common; if the idea that $x + iy = (x, y)$ is not familiar then you should read Appendix 1 before proceeding further.)

Now, since $\mathbb{C} = \mathbb{R}^2$, so in particular $\mathbb{C}$ is a real vector space as well as being a field in its own right, there are at least two things that one might mean by saying that $T : \mathbb{C} \to \mathbb{C}$ is "linear". Recall that T is *linear* if for every $z, w \in \mathbb{C}$ and every scalar c we have

$$T(z + w) = Tz + Tw, \quad T(cz) = cTz.$$

If this holds for all real scalars c we will say that T is $\mathbb{R}$-*linear*, while if it holds for all complex scalars c we will say that T is $\mathbb{C}$-*linear*.

3

Suppose that $f : V \to \mathbb{C}$, where V is an open subset of $\mathbb{C}$. Suppose that $z \in V$. We say that f is *differentiable* at z if the limit

$$f'(z) = \lim_{h \to 0} \frac{f(z+h) - f(z)}{h}$$

exists. This is what the word "differentiable" will mean throughout the book; in the present chapter, since we are comparing this notion with another sort of differentiability, we will refer to functions differentiable in this sense as "complex-differentiable".

Since $\mathbb{C} = \mathbb{R}^2$, there is something else that the word "differentiable" could mean: Recall from advanced calculus that if V is an open subset of $\mathbb{R}^2$ then a function $f : V \to \mathbb{R}^2$ is said to be differentiable at the point $z \in V$ if there exists an $\mathbb{R}$-linear map $T : \mathbb{R}^2 \to \mathbb{R}^2$ such that

$$f(z+h) = f(z) + Th + o(h)$$

as $h \to 0$; in this case the operator T is known as the Fréchet derivative of f at z, or $Df(z)$. Here we are using the Landau "little-oh" notation[1]: The equation above means that

$$f(z+h) = f(z) + Th + E(h),$$

where the "error term" $E(h)$ satisfies $E(0) = 0$ and

$$\lim_{h \to 0} \frac{||E(h)||}{||h||} = 0.$$

We will say that a function differentiable in this sense is *real-differentiable*.

Now if V is an open subset of $\mathbb{C}$ and $f : V \to \mathbb{C}$ then it is also true that V is an open subset of $\mathbb{R}^2$ and $f : V \to \mathbb{R}^2$, which raises the question of how the two notions of differentiability are related. If we rearrange the definition of $f'(z)$ above we see that f is complex-differentiable at z, with derivative $f'(z) = a$, if and only if

$$f(z+h) = f(z) + ah + o(h).$$

Since the mapping from $\mathbb{C}$ to $\mathbb{C}$ defined by $h \mapsto ah$ is certainly $\mathbb{R}$-linear, this shows that complex differentiability is a stronger condition than real differentiability. Indeed, it is clear that the $\mathbb{C}$-linear maps from $\mathbb{C}$ to $\mathbb{C}$ are precisely the maps of the form $h \mapsto ah$ for some complex number a.

---

[1] Elsewhere we will use the similar "big-oh" notation: $O(F(x))$ means "some function $G(x)$ such that $|G(x)| \leq cF(x)$ for all x".

0. Differentiability and the Cauchy-Riemann Equations

Thus we have proved the following proposition:

Proposition 0.0. *Suppose that V is an open subset of $\mathbb{C}$, that $f : V \to \mathbb{C}$, and $z \in V$. Then f is complex-differentiable at z if and only if it is real-differentiable at z and the real derivative $Df(z)$ is complex-linear.*

Now suppose that f is real-differentiable at z. Let u and v denote the real and imaginary parts of f, respectively. The entries in the 2×2 matrix representing the real derivative $Df(z)$ are given by the partial derivatives of u and v:

$$Df \sim \begin{bmatrix} u_x & u_y \\ v_x & v_y \end{bmatrix}.$$

(Of course the existence of the partials u_x, u_y, v_x and v_y at a point does not imply that f is real-differentiable at that point.)

On the other hand, it is easy to see that the real 2×2 matrix

$$\begin{bmatrix} a & b \\ c & d \end{bmatrix}$$

represents a complex-linear map from $\mathbb{C}$ to $\mathbb{C}$ if and only if $a = d$ and $c = -b$, and so we finally deduce

Proposition 0.1. *Suppose that V is an open subset of $\mathbb{C}$, that $f : V \to \mathbb{C}$, and $z \in V$. Then $f = u + iv$ is complex-differentiable at z if and only if it is real-differentiable at z and the real and imaginary parts satisfy the "Cauchy-Riemann equations"*

$$u_x(z) = v_y(z), \quad u_y(z) = -v_x(z).$$

This is the reason we introduced Fréchet differentiability: Proposition 0.1 gives a *characterization* of complex differentiability at a single point in terms of the Cauchy-Riemann equations, with no extra hypotheses. If we try to characterize complex differentiability in terms of the Cauchy-Riemann equations, but using only partial derivatives, the best we can do is to state the following two facts; note that the converse of each of the following corollaries is false (see the exercises):

Corollary 0.2. *Suppose that V is an open subset of $\mathbb{C}$, that $f : V \to \mathbb{C}$, and $z \in V$; write $f = u + iv$. If f is complex-differentiable at $z \in V$ then u and v satisfy the Cauchy-Riemann equations at z.*

Corollary 0.3. *Suppose that V is an open subset of $\mathbb{C}$, that $f : V \to \mathbb{C}$, and $z \in V$; write $f = u + iv$. If the first-order partial derivatives of u and v are continuous at $z \in V$ and satisfy the Cauchy-Riemann equations there then f is complex-differentiable at z.*

Proof. We recall from advanced calculus that continuity of the partial derivatives at z implies that f is real-differentiable at z. □

Note. We remind the reader that in the rest of the book the word "differentiable" will mean "complex-differentiable".

Some of the following exercises ask you to prove things that were stated above. The results of several of the exercises will be used later without comment.

Exercises

0.1. Suppose that f is defined in a neighboorhood of the complex number z. Show that f is complex-differentiable at z if and only if there exists a complex number a such that $f(z+h) = f(z) + ah + o(h)$ as $h \to 0$.

0.2. Suppose that f is complex-differentiable at z. Show that f is continuous at z.

0.3. Suppose that $T : \mathbb{C} \to \mathbb{C}$ is $\mathbb{R}$-linear. Show that T is $\mathbb{C}$-linear if and only if $T(iz) = iTz$ for all z.

0.4. Suppose that $T : \mathbb{R}^2 \to \mathbb{R}^2$ is the $\mathbb{R}$-linear mapping defined by the matrix $\begin{bmatrix} a & b \\ c & d \end{bmatrix}$ (that is, $T(x,y) = (ax+by, cx+dy)$). Show that T is $\mathbb{C}$-linear if and only if $a = d$ and $b = -c$.

Hint: Use the fact that $i(x,y) = (-y,x)$. (If "$i(x,y) = (-y,x)$" makes no sense to you then you should read Appendix 1 first.)

0.5. Suppose that f and g are complex-differentiable at z.

(i) Show that $f+g$ and fg are differentiable at z (and that their derivatives are given by the same formulas as in elementary calculus).

(ii) Suppose in addition that $g(z) \ne 0$ and show that f/g is differentiable at z (with derivative given by the same formula as in elementary calculus).

(iii) Suppose that n is a positive integer and define $f(z) = z^n$. Show that f is differentiable and $f'(z) = nz^{n-1}$.

(iv) Same as part (iii), but for negative integers n, with the restriction $z \ne 0$.

0. Differentiability and the Cauchy-Riemann Equations

0.6. Suppose that f is differentiable at z and that g is differentiable at $f(z)$. Show that the composition $g \circ f$ is differentiable at z, with
$$(g \circ f)'(z) = g'(f(z))f'(z).$$

Note. An obvious approach to Exercise 6 is to write
$$\frac{g \circ f(z) - g \circ f(w)}{z - w} = \left(\frac{g(f(z)) - g(f(w))}{f(z) - f(w)}\right)\left(\frac{f(z) - f(w)}{z - w}\right)$$
and use Exercise 2, which shows that $f(w)$ tends to $f(z)$ as w approaches z. There is a slight problem with this approach: It may happen that $f(w) = f(z)$ for w near z, in which case we are dividing by 0.

If you don't see how to fix this problem, you can find the solution in any calculus book. Better is to simply avoid the problem by using Exercise 0.1 instead of the definition of the derivative via quotients!

A slightly informal version of a solution using Exercise 0.1 would use the exercise twice to show that
$$\begin{aligned} g \circ f(z + h) &= g(f(z) + f'(z)h + o(h)) \\ &= g(f(z)) + g'(f(z))(f'(z)h + o(h)) + o(f'(z)h + o(h)) \\ &= g(f(z)) + g'(f(z))f'(z)h + g'(f(z))o(h) + o(f'(z)h + o(h)) \\ &= g(f(z)) + g'(f(z))f'(z)h + o(h), \end{aligned}$$
and then invoke the exercise once more to conclude that $(g \circ f)'(z) = g'(f(z))f'(z)$. The whole point of this "o" notation is to allow one to perform this sort of manipulation with error terms. However, if we're new to all this we probably want to give a more careful argument; in this regard you might note that $f(z + h) = f(z) + f'(z)h + o(h)$ holds if and only if $f(z + h) = f(z) + f'(z)h + E(h)$, where E satisfies this condition: For every $\epsilon > 0$ there exists $\delta > 0$ such that $|E(h)| \leq \epsilon |h|$ for all h with $|h| < \delta$. (The point to the rephrasing is to avoid the division in the definition of "$o(h)$".)

The next two exercises illustrate the problems that arise in attempting to give a version of Proposition 0.1 involving only partial derivatives (without extra hypotheses like continuity of the partials).

0.7. Define $f : \mathbb{C} \to \mathbb{C}$ by
$$f(x + iy) = \begin{cases} 0 & (x = 0), \\ 0 & (y = 0), \\ 1 & (\text{otherwise}). \end{cases}$$
Show that f satisfies the Cauchy-Riemann equations at the origin although f is not complex-differentiable at the origin.

0.8. Define $f : \mathbb{C} \to \mathbb{C}$ by
$$f(z) = \begin{cases} |z|^2 \sin(1/|z|) & (z \neq 0), \\ 0 & (z = 0). \end{cases}$$

Show that f is complex-differentiable at the origin although the partial derivative u_x is not continuous at the origin. (Here u is the real part of f, as above.)

Chapter 1

Power Series

One of the amazing results in elementary complex analysis is that the mere existence of the derivative of f at each point of an open set implies that f is infinitely differentiable, and indeed is locally represented by a convergent power series; that is, a function complex-differentiable in an open set must be *analytic*. (By way of contrast, recall that differentiability of a function on the line does not imply that the derivative is even continuous, much less differentiable, and that an infinitely differentiable function on the line need not be given by a power series.) This will be proved in the next chapter; here we begin with a brief review of how power series work.

If $z \in \mathbb{C}$ and $r > 0$ then the open and closed disks with center z and radius r are defined by

$$D(z,r) = \{\, w \in \mathbb{C} : |z - w| < r \,\}, \quad \overline{D}(z,r) = \{\, w \in \mathbb{C} : |z - w| \leq r \,\}.$$

(We will usually use this notation only for finite r, but in the next lemma the case $r = \infty$ may arise; note that $D(z_0, \infty) = \overline{D}(z_0, \infty) = \mathbb{C}$.)

Lemma 1.0. *Suppose* $(c_n)_{n=0}^\infty$ *is a sequence of complex numbers, and define* $R \in [0, \infty]$ *by*

$$R = \sup\{\, r \geq 0 : \text{the sequence } (c_n r^n) \text{ is bounded}\,\}.$$

Then the power series $\sum_{n=0}^\infty c_n (z - z_0)^n$ *converges absolutely and uniformly on every compact subset of the disk* $D(z_0, R)$ *and diverges at every point z with* $|z - z_0| > R$.

We should make a few comments about the statement of the lemma.

First a comment on the notation: If $z = z_0$ we have

$$\sum_{n=0}^{\infty} c_n(z - z_0)^n = c_0 0^0 + c_1 0^1 + \cdots.$$

Now, "0^0" is known as an "indeterminate form" in elementary calculus (see Exercise 1.1). As of now we *define* $0^0 = 1$, so that if $f(z) = z^0$ then f is continuous at the origin; now if $z = z_0$ we have

$$\sum_{n=0}^{\infty} c_n(z - z_0)^n = c_0.$$

The number R in the lemma is known as the *radius of convergence* of the power series, while $D(z_0, R)$ is the *disk of convergence*. You may have noticed that the lemma says nothing about convergence or divergence on the boundary of the disk of convergence. There is a reason for that: Roughly speaking, anything can happen on the boundary. For example, the three series $\sum_{n=1}^{\infty} z^n$, $\sum_{n=1}^{\infty} z^n/n$ and $\sum_{n=1}^{\infty} z^n/n^2$ all have radius of convergence 1; the first series converges at no point of the boundary, the third converges at every boundary point, and it follows by a technique known as "summation by parts", or from a result sometimes called Dirichlet's test, that the second series diverges at $z = 1$ and converges at every other boundary point. The set of boundary points where a power series converges can be much more complicated than these examples indicate. (See Chapter 21 for summation by parts, and also for some interesting results regarding the relation between convergence of a power series at a boundary point and the behavior of the corresponding function near that point.)

The lemma says that the radius of convergence of a power series is the same as what might be called the "radius of boundedness" (note that it is not actually called that). We defined R the way we did because it can be easier to check whether a sequence is bounded than whether a series converges; although of course in general a series with bounded terms need not converge, one of the nice facts about power series is that the radius of convergence is in fact the same as the "radius of boundedness".

The terminology "converges absolutely and uniformly" is standard but unfortunately does not say quite what is meant; recall that the series $\sum f_n$ is said to converge absolutely and uniformly on the set K if the series $\sum |f_n|$ converges uniformly on K (which is actually a stronger condition than saying just that the series converges uniformly and also absolutely).

Finally, it is possible to misunderstand the definition of R. By definition, R is the supremum of the set of $r \geq 0$ with the property that there exists $M = M(r)$ such that $|c_n r^n| \leq M(r)$ for all n. That is, $M(r)$ may depend on

r, so in particular it is not necessarily the case that $\{\, c_n r^n : 0 \leq r < R,\, n = 0, 1, \ldots \,\}$ is bounded.

Proof (Lemma 1.0). The second part of the conclusion is clear; indeed, if $|z - z_0| > R$ then the terms in the series $\sum_{n=0}^{\infty} c_n (z - z_0)^n$ are not even bounded, so the series certainly diverges. Suppose now that $0 \leq r < R$. We will show that the series converges absolutely and uniformly on the closed disk $\overline{D}(z_0, r)$; this is sufficient, since any compact subset of $D(z_0, R)$ is contained in $\overline{D}(z_0, r)$ for some $r < R$.

Choose a number ρ with $r < \rho < R$. The definition of R shows that $(c_n \rho^n)$ is bounded; say $|c_n| \rho^n \leq A$ for all n. Then for $z \in \overline{D}(z_0, r)$ we have $|c_n (z - z_0)^n| \leq |c_n| \rho^n (r/\rho)^n \leq A(r/\rho)^n$; since $\sum (r/\rho)^n < \infty$, this shows that the power series converges absolutely and uniformly on $\overline{D}(z_0, r)$. $\square$

A traditional formula for the radius of convergence of the power series $\sum_{n=0}^{\infty} c_n (z - z_0)^n$ is given by $R' = \liminf_{n \to \infty} |c_n|^{-1/n}$. We will have no need for this formula; you can verify for yourself that $R' = R$, where R is as above. (If you find yourself stuck on this, you might note that it is enough to show that $r < R$ if and only if $r < R'$. Now if $r < R$ then as above there exists $\rho > r$ such that $c_n \rho^n$ is bounded, and hence ... On the other hand, if $r < R'$ then $|c_n|^{-1/n} > r$ for all but finitely many values of n, so that ...)

In real variables the uniform convergence of an infinite sum of differentiable functions does not imply that the sum is differentiable, unless we add additional hypotheses such as the uniform convergence of the differentiated series. We shall see later that in fact no extra hypotheses are needed in the present context: If each f_n is differentiable in an open set V and the series $f = \sum f_n$ converges uniformly on every compact subset of V then f is automatically differentiable, with $f' = \sum f_n'$. We will prove this amazing and important fact soon, once we know a little bit about complex analysis; for now we use a little trick to prove an important special case:

Proposition 1.1. *Suppose that the power series $\sum_{n=0}^{\infty} c_n (z - z_0)^n$ has radius of convergence $R > 0$. Then the function $f(z) = \sum_{n=0}^{\infty} c_n (z - z_0)^n$ is differentiable in the disk $D(z_0, R)$, with derivative*

$$f'(z) = \sum_{n=1}^{\infty} n c_n (z - z_0)^{n-1} = \sum_{n=0}^{\infty} (n+1) c_{n+1} (z - z_0)^n.$$

Corollary 1.2. *If the power series $\sum_{n=0}^{\infty} c_n (z - z_0)^n$ has positive radius of convergence R then the function*

$$f(z) = \sum_{n=0}^{\infty} c_n (z - z_0)^n$$

is in fact infinitely differentiable in the disk $D(z_0, R)$; further, the coefficients are given by
$$c_n = \frac{f^{(n)}(z_0)}{n!}.$$

The corollary follows from the proposition by induction. It has a further corollary:

(*Uniqueness*)
Corollary 1.3. *If the power series $\sum_{n=0}^{\infty} c_n(z-z_0)^n$ has positive radius of convergence R and there exists $r \in (0, R)$ such that $\sum_{n=0}^{\infty} c_n(z-z_0)^n = 0$ for all $z \in D(z_0, r)$ then $c_n = 0$ for all n.*

Proof (Proposition 1.1). We may assume without loss of generality that $z_0 = 0$. (In other words: We will prove the proposition under the additional hypothesis that $z_0 = 0$, and we leave it to the reader to show why the general case follows.)

Define a function of two complex variables by
$$F(z, w) = \sum_{n=1}^{\infty} c_n \left(\sum_{k=0}^{n-1} z^k w^{n-k-1} \right).$$

It is easy to see that the series defining F converges uniformly on compact subsets of $D(0, R) \times D(0, R)$: Fix $r \in (0, R)$, choose $\rho \in (r, R)$, choose A so that $|c_n|\rho^n \leq A$ for all n, and note that for $(z, w) \in \overline{D}(0, r) \times \overline{D}(0, r)$ we have
$$\left| c_n \sum_{k=0}^{n-1} z^k w^{n-k-1} \right| \leq n|c_n|r^{n-1} \leq n\frac{A}{r} \left(\frac{r}{\rho} \right)^n,$$
while $\sum n(r/\rho)^n < \infty$, as before.

Thus F is continuous on $D(0, R) \times D(0, R)$. Note however that $F(z, z) = \sum_{n=1}^{\infty} nc_n z^{n-1}$, while the factorization of $z^n - w^n$ shows that $F(z, w) = (f(z) - f(w))/(z - w)$ for $z \neq w$. Since F is continuous, it follows that $\lim_{w \to z}(f(z) - f(w))/(z - w) = \sum_{n=1}^{\infty} nc_n z^{n-1}$. In other words, f is differentiable, and f' is given by a term-by-term differentiation of the power series defining f. □

We finish by noting for future reference that if $f(z)$ and $g(z)$ are both defined by a convergent power series in the disk $D(z_0, r)$ then the same is true of the sum $f(z) + g(z)$ and the product $f(z)g(z)$. This is quite clear for the sum; for the product it holds because convergent power series converge *absolutely* in the interior of the disk of convergence: If $f(z) = \sum_{n=0}^{\infty} a_n(z-z_0)^n$ and $g(z) = \sum_{n=0}^{\infty} b_n(z-z_0)^n$ both converge absolutely then we may rearrange the terms in the product in any way we like, in particular

1. Power Series

obtaining $f(z)g(z) = \sum_{n=0}^{\infty} c_n(z - z_0)^n$, where $c_n = \sum_{k=0}^{n} a_k b_{n-k}$. (We do not need to worry about the radius of convergence of that last series; we know that it does converge for $z \in D(z_0, r)$, and *hence* the radius of convergence is at least r.)

Note. At this point you should probably read Appendix 3, where the functions exp, sin and cos are defined and Proposition 1.1 is used in deriving some of their basic properties.

Exercises

1.1. Suppose $a \in [0, 1]$. Show that there exist sequences $x_n > 0$ and $y_n > 0$ such that $\lim_{n \to \infty} x_n = 0$, $\lim_{n \to \infty} y_n = 0$, and
$$\lim_{n \to \infty} x_n^{y_n} = a.$$

Hint: You might show first that if $0 < x \leq \alpha \leq 1$ then there exists $y \in [0, 1]$ such that $x^y = \alpha$.

1.2. Suppose the power series $\sum_{n=0}^{\infty} c_n z^n$ has radius of convergence $R > 0$, and define $f : D(0, R) \to \mathbb{C}$ by
$$f(z) = \sum_{n=0}^{\infty} c_n z^n.$$

We have shown that
$$f'(z) = \sum_{n=1}^{\infty} n c_n z^{n-1} \quad (z \in D(0, R));$$

explain why it would not be quite correct to write
$$f'(z) = \sum_{n=0}^{\infty} n c_n z^{n-1} \quad (z \in D(0, R))$$

instead, although it *is* perfectly correct to say $f' = \sum_{n=0}^{\infty} f_n'$, where $f_n(z) = c_n z^n$.

Chapter 2

Preliminary Results on Holomorphic Functions

Now we are going to show that a function differentiable in an open set in the plane is in fact represented by a power series (locally — that is, in any disk contained in the domain of the function); in particular a differentiable function is automatically infinitely differentiable, by Corollary 1.2. Along the way we will get preliminary versions of some of the basic results of complex analysis (Cauchy's Theorem, the Cauchy Integral Formula).

It is really not clear how just pointwise differentiability can possibly give any sort of uniform smoothness condition, especially because in real analysis it does not. And indeed the pioneers did not know that differentiability was enough — the original definition of "analytic" *assumed* that the derivative was continuous. But in fact the derivative of a complex-differentiable function must be continuous; the crucial step is the Cauchy-Goursat Theorem below.

We should give an example to show why this is all so amazing. Define $f : \mathbb{R} \to \mathbb{R}$ by

$$f(x) = \begin{cases} x^2 \sin(1/x) & (x \neq 0), \\ 0 & (x = 0). \end{cases}$$

Then f is differentiable everywhere but f' is not continuous:

If $x \neq 0$ then one simply calculates that $f'(x) = 2x \sin(1/x) - \cos(1/x)$; hence $\lim_{x \to 0} f'(x)$ does not exist, so f' is certainly not continuous at the origin. This would not be all that interesting except that f *is* differentiable at the origin. In fact if $h \neq 0$ then

$$\left| \frac{f(h) - f(0)}{h - 0} \right| = \left| \frac{h^2 \sin(1/h)}{h} \right| \leq |h|,$$

which shows that $f'(0) = 0$.

So in this example f is differentiable but f' is not continuous; that cannot happen with a complex-differentiable function. Or can it?

Exercise 2.1. Define $f : \mathbb{C} \to \mathbb{C}$ by $f(z) = z^2 \sin(1/z)$ for $z \neq 0$, $f(0) = 0$. The last few paragraphs almost seem to give a proof that f is differentiable everywhere but f' is not continuous at the origin. This is impossible — where does the "proof" fail?

Note. If you have not yet read Appendix 3 you should consult it for the definition of $\sin(z)$ for complex z.

We now begin the proof of various facts about complex-differentiable functions:

First we need to define complex line integrals. We will restrict the definition to integration over fairly smooth curves to save time — in fact one can integrate over curves which are merely "rectifiable", but that is one of those things that rarely comes up in practice. (Also the basic concepts involved in integration over rectifiable curves are best understood in terms of some concepts from real analysis that we do not know yet: For future reference, a curve is rectifiable if and only if it (is continuous and) has bounded variation. Nobody knows why the same concept is called "rectifiable" in one context and "bounded variation" in another — in any case, someday when you learn about functions of bounded variation in reals you should stop and consider what this concept could have to do with integrals over curves.)

We will be talking about various *metric spaces* throughout the text; if you are not familiar with the definition and basic properties of metric spaces, you should consult Appendix 4 as needed. (In a few places we will talk about a more general notion, *topological* spaces. We decided not to include the defintion; if you have never studied basic point-set topology, you can either skip those sections or just pretend that we are talking about metric spaces instead. Results explicitly mentioning topological spaces will not be used except in other such results.)

Definition. If X is a metric space (or a topological space), a *curve* in X is a continuous map $\gamma : [a, b] \to X$; γ is a *closed* curve if $\gamma(a) = \gamma(b)$. The interval $[a, b]$ will be referred to as the "parameter interval" of γ. The *image* (or *trace*) of γ will be denoted $\gamma^* = \gamma([a, b])$.

2. Preliminary Results on Holomorphic Functions

Note that people often write "γ" when strictly speaking they are referring to γ^*; for example we may say something about "the curve $\gamma \subset V$".

Definition. A *piecewise continuously differentiable* (or "piecewise C^1") *curve* in $\mathbb{C}$ is a curve $\gamma : [a,b] \to \mathbb{C}$ which is continuously differentiable except at finitely many exceptional points, at each of which the derivative γ' has one-sided limits.

To be very explicit, the requirement is that $a = a_0 < \cdots < a_n = b$ in such a way that the derivative γ' is continuous on each of the open intervals (a_{j-1}, a_j) and has one-sided limits at each a_j (a limit from the right at a, a limit from the left at b, and both a limit from the right and a limit from the left at a_j, $1 \leq j \leq n-1$).

We will be referring to piecewise continuously differentiable curves quite a bit, hence the following definition:

Definition. A *smooth curve* is a piecewise continuously differentiable curve.

(You should note that the word "smooth" means various different things in various contexts. Possibly its most common meaning is "infinitely differentiable", but it often means more or less than this — any time you see the word you should realize that it may mean something other than what you think it means.)

Definition. Suppose that $\gamma : [a,b] \to \mathbb{C}$ is smooth and f is continuous on γ^*. Then
$$\int_\gamma f(z)\,dz = \int_a^b f(\gamma(t))\gamma'(t)\,dt.$$

We should note that one may also regard the integral as a limit of "Riemann-Stieltjes sums"
$$\sum_{j=1}^n f(\gamma(t_j^*))(\gamma(t_j) - \gamma(t_{j-1})),$$
where $a = t_0 < \cdots < t_n = b$ and $t_{j-1} \leq t_j^* \leq t_j$. We will not worry about this much — you may wish to regard the formula above as sort of an explanation of what the integral "really means". As long as we restrict to smooth curves the definition above will suffice. (See Lemma 12.6 below.)

We start with two easy lemmas:

Lemma 2.0 (Cauchy's Theorem for derivatives). *Suppose V is an open subset of the plane, $f : V \to \mathbb{C}$ is continuous, and there exists an*

$F : V \to \mathbb{C}$ such that $f = F'$ in V. Then
$$\int_\gamma f(z)\,dz = 0$$
for any smooth closed curve γ in V.

Note for future reference that the hypotheses are redundant: The hypothesis $f = F'$ in V actually implies that f is continuous, although we do not know that yet.

Note. The lemma may be regarded as a version of Cauchy's Theorem. We will be seeing many versions of Cauchy's Theorem; most of them will be special cases of the "big" one (Theorem 4.10 below), but this version is actually not contained in Theorem 4.10. So although most of the stuff we are doing now is just preliminary, the lemma is worth remembering for that reason — it can be applied in situations where Theorem 4.10 does not work!

Proof. Suppose that $\gamma : [a,b] \to V$ is a smooth closed curve. It follows from the chain rule and the fundamental theorem of calculus that
$$\int_\gamma f(z)\,dz = \int_a^b F'(\gamma(t))\gamma'(t)\,dt = \int_a^b (F \circ \gamma)'\,dt = F(\gamma(b)) - F(\gamma(a)) = 0.$$
$\square$

The next lemma is sometimes called the *ML inequality*. (You should note that that is not quite standard terminology; many people who certainly know the inequality will not recognize the phrase "ML inequality".) Recall that the *length* of a smooth curve $\gamma : [a,b] \to \mathbb{C}$ is given by
$$L = \int_a^b |\gamma'(t)|\,dt.$$

Lemma 2.1 (ML Inequality). *Suppose that $\gamma : [a,b] \to \mathbb{C}$ is a smooth curve and f is continuous on γ^*. If $|f| \le M$ on γ^* and L is the length of γ then*
$$\left| \int_\gamma f(z)\,dz \right| \le ML.$$

Proof.
$$\left| \int_\gamma f(z)\,dz \right| = \left| \int_a^b f(\gamma(t))\gamma'(t)\,dt \right|$$
$$\le \int_a^b |f(\gamma(t))\gamma'(t)|\,dt \le M \int_a^b |\gamma'(t)|\,dt = ML. \quad \square$$

2. Preliminary Results on Holomorphic Functions

Note. We used the fact that if $a < b$ and $f : [a, b] \to \mathbb{C}$ is continuous then

$$\left| \int_a^b f(t)\, dt \right| \leq \int_a^b |f(t)|\, dt.$$

This is easy to prove: First choose a complex number α such that $|\alpha| = 1$ and

$$\left| \int_a^b f(t)\, dt \right| = \alpha \int_a^b f(t)\, dt.$$

It follows that

$$\left| \int_a^b f(t)\, dt \right| = \int_a^b \alpha f(t)\, dt = \operatorname{Re} \int_a^b \alpha f(t)\, dt$$
$$= \int_a^b \operatorname{Re}(\alpha f(t))\, dt \leq \int_a^b |\alpha f(t)|\, dt = \int_a^b |f(t)|\, dt.$$

Note. You should not jump to the conclusion that

$$\left| \int_\gamma f(z)\, dz \right| \leq \int_\gamma |f(z)|\, dz\,!$$

That inequality is clearly nonsense; the right side is not even non-negative in general. What is true is

$$\left| \int_\gamma f(z)\, dz \right| \leq \int_\gamma |f(z)|\, |dz|,$$

where $|dz|$ is a medium-standard notation for "with respect to arclength", which is to say that

$$\int_\gamma f(z)\, |dz| = \int_a^b f(\gamma(t)) |\gamma'(t)|\, dt,$$

by definition.

All of the various versions of Cauchy's Theorem we will be seeing say that under certain hypotheses the integral of an analytic function over a closed curve must vanish. The following example shows that you cannot omit the extra hypotheses — it is also a very important example for various other reasons:

Example. Define $f : \mathbb{C} \setminus \{0\} \to \mathbb{C}$ by $f(z) = 1/z$. Fix $r > 0$ and define $\gamma : [0, 2\pi] \to \mathbb{C}$ by $\gamma(t) = re^{it}$ (so γ is the circle of radius r about the origin, traversed counterclockwise). Then

$$\int_\gamma f(z)\,dz = 2\pi i.$$

You should work out the integral and verify this. It really is an important example; we will see later that it is in some sense just about the only way to get something nonzero by integrating an analytic function with an "isolated singularity" over a closed curve.

To speak very informally, one version of Cauchy's Theorem says that if f is differentiable at every point on and inside the smooth closed curve γ then $\int_\gamma f(z)\,dz = 0$; the reason this does not contradict the example above is that the function $1/z$ is not even *defined* at every point inside the unit circle, much less differentiable at every such point.

The reason we call that a "very informal" statement of Cauchy's Theorem is that we have not given a definition of what it means for a point to lie "inside" a closed curve. In fact we will not be giving any such definition; doing so in a rigorous manner is harder than you might think (this actually involves the Jordan Curve Theorem from topology, which is far from trivial). In the versions of Cauchy's Theorem that we prove the so-called "topological" condition regarding points lying "inside" a curve will be replaced by various analytic conditions, involving integrals, convexity, and so on.

More definitions: If $\gamma_1, \ldots, \gamma_n$ are smooth curves and we write

$$\Gamma = \gamma_1 \dot{+} \cdots \dot{+} \gamma_n,$$

that means "Γ is something over which we can integrate a function, and by definition

$$\int_\Gamma f(z)\,dz = \int_{\gamma_1} f(z)\,dz + \cdots + \int_{\gamma_n} f(z)\,dz$$

for every f which is continuous on the union of the γ_j^*". We will refer to Γ as a *chain* (or a *cycle* if the curves are closed). The *trace* of Γ is $\Gamma^* = \bigcup_{j=1}^n \gamma_j^*$.

(If you are bothered by the informality in saying that "Γ is something over which we can integrate a function": A more formal definition would say that Γ is a "linear functional" on $C(\Gamma^*)$, that is, a linear operator from the space of continuous functions on Γ^* to $\mathbb{C}$. The value of this functional at $f \in C(\Gamma^*)$ just happens to be written using an integral sign.)

Similarly, if Γ, Γ_1 and Γ_2 are chains then $\Gamma_1 = \Gamma_2$ means that

$$\int_{\Gamma_1} f(z)\,dz = \int_{\Gamma_2} f(z)\,dz$$

2. Preliminary Results on Holomorphic Functions

for all continuous f, while $\Gamma = \Gamma_1 \dotplus \Gamma_2$ means that

$$\int_\Gamma f(z)\,dz = \int_{\Gamma_1} f(z)\,dz + \int_{\Gamma_2} f(z)\,dz$$

for all continuous f; similarly $\Gamma = 0$ means that

$$\int_\Gamma f(z)\,dz = 0$$

for all continuous f, and $\Gamma_1 = \dotminus \Gamma_2$ means that

$$\int_{\Gamma_1} f(z)\,dz = -\int_{\Gamma_2} f(z)\,dz.$$

Note. Various versions of Cauchy's Theorem say that if f is differentiable in a certain region and Γ satisfies some hypothesis then

$$\int_\Gamma f(z)\,dz = 0.$$

This conclusion does *not* say that $\Gamma = 0$! The reason is this: $\Gamma = 0$ says that the integral of every *continuous* function vanishes, while Cauchy's Theorem deals only with *differentiable* functions.

If a and b are complex numbers then the notation $[a,b]$ will denote the line segment from a to b. That is, $[a,b]$ is the curve $\gamma : [0,1] \to \mathbb{C}$ defined by $\gamma(t) = (1-t)a + tb$. We will use the same notation for the trace of this curve: $[a,b] = \{\,(1-t)a + tb : 0 \le t \le 1\,\}$.

Exercise 2.2.

(i) Show that $[b,a] = \dotminus [a,b]$.

(ii) Generalize this: Suppose that γ_1 is a smooth curve in the plane. How can you define another smooth curve γ_2 so that $\gamma_2 = \dotminus \gamma_1$?

Exercise 2.3. Suppose that $p \in [a,b]$ and show that $[a,b] = [a,p] \dotplus [p,b]$.

The set S is *convex* if $[a,b] \subset S$ for all $a, b \in S$. The *convex hull* of a set S is the smallest convex set containing S. (To show that there *is* such a thing, consider the intersection of all the convex sets containing S. See Appendix 5.)

A *triangle* is the convex hull of three points, called the *vertices* of the triangle; the triangle with vertices a, b, c will be denoted $[a,b,c]$. (Note that this is consistent with our notation for intervals; $[a,b]$ is the convex hull of $\{a,b\}$.) If $T = [a,b,c]$ then the *boundary* of T is the chain defined by

$$\partial T = [a,b] \dotplus [b,c] \dotplus [c,a].$$

Note. We are really talking about "oriented triangles" (or *simplices*), not just triangles; in particular the value of the boundary depends on the order in which the vertices are listed, for example

$$\partial[a,b,c] = \partial[b,c,a] = \dot{-}\partial[b,a,c].$$

If you have studied permutation groups then you should consider the following question: Which permutations of $\{a,b,c\}$ switch the sign of $\partial[a,b,c]$?

We now introduce notation for a decomposition of a triangle into four smaller triangles, which will be used in the proof of the Cauchy-Goursat Theorem: Suppose $T = [a,b,c]$. Let $p = (a+b)/2$, $q = (b+c)/2$, $r = (c+a)/2$, and define

$$T^1 = [a,p,r],$$
$$T^2 = [b,q,p],$$
$$T^3 = [c,r,q],$$
$$T^4 = [p,q,r],$$

as in Figure 2.0:

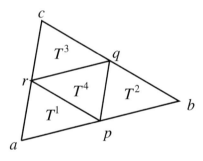

Figure 2.0

Exercise 2.4. Show that $\partial T = \partial T^1 \dot{+} \partial T^2 \dot{+} \partial T^3 \dot{+} \partial T^4$ for any triangle T. (See the preceding two exercises.)

We are now ready for a major result:

Theorem 2.2 (Cauchy-Goursat: Cauchy's Theorem for triangles). *Suppose that $V \subset \mathbb{C}$ is open and $f : V \to \mathbb{C}$ is differentiable at every point of V. Then*

$$\int_{\partial T} f(z)\, dz = 0$$

for every triangle $T \subset V$.

2. Preliminary Results on Holomorphic Functions

Note. The hypothesis is $T \subset V$, not just $\partial T \subset V$! Indeed, if $V = \mathbb{C} \setminus \{0\}$ and $T = [i, -1-i, 1-i]$ then $\partial T \subset V$ although (as is clear from things we have not yet covered) $\int_{\partial T} f(z)\, dz = 2\pi i$ for $f(z) = 1/z$.

Also note that the differentiability of f implies that f is continuous, so that the integral exists.

In case it is not clear yet why the theorem is amazing: The hypothesis that f is differentiable is only a pointwise thing. It says a certain limit exists at every point, and one would expect to need to assume that the limit exists uniformly (or something) to be able to deduce conclusions like this. In fact, if you assume that $\partial f/\partial x$ and $\partial f/\partial y$ are continuous then you can get the conclusion from Green's theorem (you should try this — it is instructive to see how the Cauchy-Riemann equations come in).

The fiendishly clever aspect of the proof below is that it only uses the differentiability of f at one point of V, so the question of what sort of uniformity is required does not arise (alas we do not know in advance which point it is going to be ...).

Proof. Suppose not: Then there exists a triangle $T_0 \subset V$ such that

$$\left| \int_{\partial T_0} f(z)\, dz \right| = \delta > 0.$$

Now, using the notation defined above for the four "quarters" of a triangle, Exercise 2.4 above shows that

$$\left| \sum_{j=1}^{4} \int_{\partial T_0^j} f(z)\, dz \right| = \delta.$$

The triangle inequality shows that we must have

$$\left| \int_{\partial T_0^j} f(z)\, dz \right| \geq \delta/4$$

for at least one value of j; we define $T_1 = T_0^j$, where j is chosen so that

$$\left| \int_{\partial T_1} f(z)\, dz \right| \geq \delta/4.$$

And we do it again: Having chosen T_n, we set T_{n+1} equal to one of the "quarters" of T_n, in such a way that $\left| \int_{\partial T_{n+1}} f(z)\, dz \right| \geq \left| \int_{\partial T_n} f(z)\, dz \right| \big/ 4$. It follows that

$$\left| \int_{\partial T_n} f(z)\, dz \right| \geq 4^{-n} \delta$$

and
$$\text{diam}(T_n) = 2^{-n}\text{diam}(T_0)$$
for $n = 0, 1, \ldots$.

We have a nested sequence of closed sets with diameter tending to 0; hence their intersection consists of exactly one point: There exists $z_0 \in V$ with
$$\{z_0\} = \bigcap_{n=0}^{\infty} T_n$$
(see Exercise A4.14 in Appendix 4).

The fact that f is differentiable at z_0 is going to lead to a contradiction. If we set $\alpha = f(z_0)$ and $\beta = f'(z_0)$ then we have
$$f(z) = \alpha + \beta(z - z_0) + g(z),$$
where
$$\lim_{z \to z_0} \frac{g(z)}{z - z_0} = 0.$$

Now $\alpha + \beta(z - z_0)$ is the derivative of the function $\alpha(z - z_0) + \beta(z - z_0)^2/2$, so Lemma 2.0 shows that $\int_{\partial T_n} (\alpha + \beta(z - z_0))\, dz = 0$, and hence
$$\left| \int_{\partial T_n} g(z)\, dz \right| = \left| \int_{\partial T_n} f(z)\, dz \right| \geq 4^{-n}\delta.$$

If $M_n = \sup_{z \in \partial T_n} |g(z)|$ then the fact that $g(z)/(z - z_0) \to 0$ shows that $M_n = o(\text{diam}(T_n)) = o(2^{-n})$ (that is, $M_n/2^{-n} \to 0$). Since the length of ∂T_n is $c\, 2^{-n}$, it follows from Lemma 2.1 that
$$\left| \int_{\partial T_n} g(z)\, dz \right| \leq c\, 2^{-n} M_n = c\, 2^{-n} o(2^{-n}) = o(4^{-n}),$$
contradicting the fact that $\left| \int_{\partial T_n} g(z)\, dz \right| \geq 4^{-n}\delta$. □

So far we have two versions of Cauchy's Theorem, both of which are somewhat lame: To apply Lemma 2.0 we need to know somehow that our function is a derivative, while Theorem 2.2 only works for triangles. It is interesting that we can combine these two weak versions of Cauchy's Theorem to get Theorem 2.4 below: While Theorem 2.4 should still be regarded as a preliminary version of Cauchy's Theorem, it turns out to be powerful enough to derive many significant results in complex analysis (as we shall see in this and the next chapter).

The following proposition will be used both in Cauchy's Theorem and its converse, Morera's Theorem:

2. Preliminary Results on Holomorphic Functions

Proposition 2.3. *Suppose that V is a convex open subset of the plane, and suppose that $f : V \to \mathbb{C}$ is a continuous function with the property that $\int_{\partial T} f(z)\, dz = 0$ for every triangle $T \subset V$. Then there exists a function $F : V \to \mathbb{C}$ such that $F' = f$ in V.*

Proof. Fix a point $z_0 \in V$ and define F by the formula

$$F(z) = \int_{[z_0, z]} f(w)\, dw \quad (z \in V).$$

(Note first that the convexity of V shows that $[z_0, z] \subset V$; now the continuity of f shows that the integral exists.) If $z \in V$ and $0 \neq h \in \mathbb{C}$ is small enough that also $z + h \in V$ then the convexity of V shows that the triangle $T = [z_0, z, z + h] \subset V$; hence the integral of f over ∂T vanishes. If we use the definition of F and the fact that $[b, a] = -[a, b]$, we deduce that

$$\frac{F(z+h) - F(z)}{h} = \frac{1}{h} \int_{[z, z+h]} f(w)\, dw,$$

so we need to show that

$$\lim_{h \to 0} \frac{1}{h} \int_{[z, z+h]} f(w)\, dw = f(z).$$

But $\int_{[z, z+h]} dw = h$. (You can verify this directly from the definition of the integral, or you can note that $g'(w) = 1$ if $g(w) = w$ and argue as in the proof of Lemma 2.0 above; probably you should do both, for practice.) Hence the above is equivalent to

$$\lim_{h \to 0} \frac{1}{h} \int_{[z, z+h]} (f(w) - f(z))\, dw = 0,$$

which is easily verified using Lemma 2.1: Given $\epsilon > 0$ we choose $\delta > 0$ so that $|f(w) - f(z)| < \epsilon$ whenever $|w - z| < \delta$, we note that $|w - z| < \delta$ whenever $w \in [z+h, z]$ with $0 < |h| < \delta$, and we deduce that for all such h we have

$$\left| \frac{1}{h} \int_{[z, z+h]} (f(w) - f(z))\, dw \right| \leq \left| \frac{1}{h} \right| \epsilon |h| = \epsilon. \quad \square$$

The next exercise shows that the convexity of V cannot be omitted in the previous result:

Exercise 2.5. Let $V = \mathbb{C} \setminus \{0\}$ and define $f : V \to \mathbb{C}$ by $f(z) = 1/z$. Then $\int_{\partial T} f(z)\, dz = 0$ for every triangle $T \subset V$ but there does not exist $F : V \to \mathbb{C}$ such that $F' = f$ in V. (Note that both assertions follow from results proved above.)

(We should perhaps point out again that $T \subset V$ is a stronger condition than $\partial T \subset V$; in fact there do exist triangles T such that $\partial T \subset V$ but $\int_{\partial T} f(z)\, dz \neq 0$.)

We will see later that an open set V has the property that every differentiable function in V is a derivative if and only if V is "simply connected"; thus the exercise shows that $\mathbb{C} \setminus \{0\}$ is not simply connected.

We can now give a usable version of Cauchy's Theorem:

Theorem 2.4 (Cauchy's Theorem in a convex set). *Suppose that V is a convex open subset of the plane and $f : V \to \mathbb{C}$ is differentiable at every point of V. Then*

$$\int_\gamma f(z)\, dz = 0$$

for every smooth closed curve $\gamma \subset V$.

Proof. This is immediate from results we have proved: The Cauchy-Goursat Theorem shows that the integral of f over the boundary of any triangle must be zero, and thus Proposition 2.3 shows that f is the derivative of something, and the conclusion follows from Lemma 2.0. $\square$

We can now prove a first version of the Cauchy Integral Formula, which says (again, under some technical hypotheses that vary from version to version) that we can recover the values of an analytic function at points "inside" a closed curve by an integral involving the values on the curve. This preliminary version is restricted to disks — we shall see that it is nonetheless very useful.

Theorem 2.5 (Cauchy's Integral Formula for disks). *Suppose V is an open subset of the plane and $f : V \to \mathbb{C}$ is differentiable at every point of V. Suppose that $\overline{D(z_0, r)} \subset V$, and define $\gamma : [0, 2\pi] \to \mathbb{C}$ by $\gamma(t) = z_0 + re^{it}$ (so that γ is the boundary of the disk $D(z_0, r)$, traversed counterclockwise). Then*

$$f(z) = \frac{1}{2\pi i} \int_\gamma \frac{f(w)\, dw}{w - z}$$

for all z in the open disk $D(z_0, r)$.

2. Preliminary Results on Holomorphic Functions

Proof. Fix a point $z \in D(z_0, r)$, and for $0 < \rho < r - |z - z_0|$ define $\gamma_\rho : [0, 2\pi] \to V$ by $\gamma_\rho(t) = z + \rho e^{it}$; thus γ_ρ is the boundary of a small disk centered at z.

Let $V' = V \setminus \{z\}$ and define $g : V' \to \mathbb{C}$ by the formula

$$g(w) = \frac{f(w)}{w - z};$$

note that g is differentiable at every point of V', by the quotient rule.

An argument like Exercise 2.4 above shows that

$$\gamma \dotminus \gamma_\rho = \gamma^1 \dotplus \gamma^2 \dotplus \gamma^3,$$

where γ^1, γ^2 and γ^3 are "defined" as in Figure 2.1:

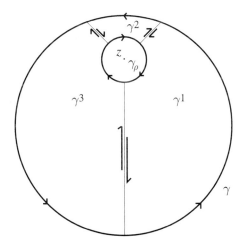

Figure 2.1

(Each γ^j consists of a counterclockwise arc on the outer circle, followed by a line segment joining the outer circle to the inner circle, a clockwise arc on the inner circle, and then a line segment joining the inner circle to the outer circle.)

Now although V' is certainly not convex, in fact for each j we can find a convex open set W_j such that $\gamma^j \subset W_j \subset V'$. Theorem 2.4 shows therefore that

$$\int_{\gamma^j} g(w)\, dw = 0$$

for $j = 1, 2, 3$, and, since $\gamma \dotminus \gamma_\rho = \gamma^1 \dotplus \gamma^2 \dotplus \gamma^3$, this shows that

$$\int_\gamma g(w)\, dw = \int_{\gamma_\rho} g(w)\, dw.$$

We are trying to show that $\int_\gamma g(w)\,dw = 2\pi i f(z)$, and so it is enough to show that
$$\lim_{\rho \to 0^+} \int_{\gamma_\rho} g(w)\,dw = 2\pi i f(z).$$
But simply inserting various definitions shows that
$$\int_{\gamma_\rho} g(w)\,dw = i\int_0^{2\pi} f(z + \rho e^{it})\,dt,$$
and, since $f(z + \rho e^{it}) \to f(z)$ uniformly for $t \in [0, 2\pi]$ as $\rho \to 0$, it follows that
$$\lim_{\rho \to 0^+} i\int_0^{2\pi} f(z + \rho e^{it})\,dt = i\int_0^{2\pi} f(z)\,dt = 2\pi i f(z). \qquad \square$$

It follows immediately that differentiable functions satisfy the Mean Value Property, which is to say the value at the center of a circle is the average of the values on the boundary:

Exercise 2.6. Assuming the hypotheses of Theorem 2.5, show that
$$f(z_0) = \frac{1}{2\pi} \int_0^{2\pi} f(z_0 + re^{it})\,dt.$$

We will see more sophisticated versions of Cauchy's Theorem and the Cauchy Integral Formula a little later. This version of the Cauchy Integral Formula is enough to show that differentiable functions are given by power series:

Theorem 2.6. *Suppose V is an open subset of the plane and $f : V \to \mathbb{C}$ is differentiable at every point of V. Suppose that $\overline{D(z_0, r)} \subset V$. Then there exists a sequence of complex numbers (c_n) such that*
$$f(z) = \sum_{n=0}^{\infty} c_n (z - z_0)^n$$
for every $z \in D(z_0, r)$ (and hence the series converges uniformly on every compact subset of $D(z_0, r)$). Further, the coefficients are given by the formula
$$c_n = \frac{1}{2\pi i} \int_\gamma \frac{f(w)\,dw}{(w - z_0)^{n+1}},$$
where $\gamma : [0, 2\pi] \to V$ is defined by $\gamma(t) = z_0 + re^{it}$.

Note. If we compare the preceding with Corollary 1.2 above we see that the derivatives of f at z_0 are given by

$$f^{(n)}(z_0) = \frac{n!}{2\pi i} \int_\gamma \frac{f(w)\,dw}{(w-z_0)^{n+1}}$$

for $n = 0, 1, \ldots$. We will see a more general version of this formula in Corollary 3.1 below.

Proof. Theorem 2.5 shows that

$$f(z) = \frac{1}{2\pi i} \int_\gamma \frac{f(w)\,dw}{w-z}$$

for $z \in D(z_0, r)$. Now for $z \in D(z_0, r)$ and $w \in \gamma$ we have $\left|\frac{z-z_0}{w-z_0}\right| = \frac{|z-z_0|}{r} < 1$, so that

$$\frac{1}{w-z} = \frac{1}{(w-z_0)\left(1 - \frac{z-z_0}{w-z_0}\right)} = \frac{1}{w-z_0} \sum_{n=0}^\infty \left(\frac{z-z_0}{w-z_0}\right)^n.$$

And in fact if $z \in D(z_0, r)$ is fixed then this series converges uniformly for $w \in \gamma$; this shows that the integral of the sum is the sum of the integrals in the following:

$$f(z) = \frac{1}{2\pi i} \int_\gamma \frac{f(w)\,dw}{w-z}$$
$$= \frac{1}{2\pi i} \int_\gamma \frac{f(w)}{w-z_0} \sum_{n=0}^\infty \left(\frac{z-z_0}{w-z_0}\right)^n dw = \sum_{n=0}^\infty c_n(z-z_0)^n,$$

where c_n is as above. $\square$

Time for a few more definitions:

Definition. Suppose that $E \subset \mathbb{C}$ and $f : E \to \mathbb{C}$. We say that f is <u>analytic</u> on E if f is (or can be extended so as to be) differentiable at every point of some neighborhood of E.

So, by results proved above, f is analytic if and only if it can be represented locally by a power series (that is, if and only if for every z in the domain of f there is a power series centered at z which converges to f at every point of the domain of f near z). We should note that in real analysis local representability by power series is the *definition* of "analytic" (often written "real-analytic" to avoid confusion); for functions of a real variable this is a much stronger condition than differentiability.

Definition. Suppose that $V \subset \mathbb{C}$ is open and $f : V \to \mathbb{C}$. We say that f is *holomorphic* in V if f is differentiable at every point of V. The class of all functions holomorphic in V is denoted $H(V)$.

Note. The words "analytic" and "holomorphic" are not quite synonymous — we only speak of functions holomorphic in an *open* set, while a function can be analytic on any sort of set. (Another difference is that "analytic" has many other meanings in mathematics that have essentially nothing to do with complex analysis; on the other hand, while the word "holomorphic" is used in somewhat different ways in various contexts, as far as I am aware it always means something very closely related to its present meaning.)

We can now do a lot of basic complex analysis. We close this chapter with a few results that are essentially just restatements of things we have already proved — in the next chapter we will see a large number of other consequences of what we have done so far.

Corollary 2.7 (Cauchy's Theorem in a convex set). *Suppose that $V \subset \mathbb{C}$ is a convex open set and $f \in H(V)$. Then*
$$\int_\gamma f(z)\,dz = 0$$
for any smooth closed curve $\gamma \subset V$.

Proof. This is literally just a restatement of Theorem 2.4, repeated in slightly different terminology for emphasis in case you missed it the first time. □

Cauchy's Theorem has a very useful converse, Morera's Theorem:

Corollary 2.8 (Morera's Theorem). *Suppose that V is an open subset of the plane and $f : V \to \mathbb{C}$ is continuous. If $\int_{\partial T} f(z)\,dz = 0$ for every triangle $T \subset V$ then $f \in H(V)$.*

Proof. Note first that we can assume that V is convex (since in general for any $z \in V$ we can find a convex open set W with $z \in W \subset V$; now if we know the result for convex open sets then we know that f is differentiable at z, which is what we are trying to prove). Proposition 2.3 shows that $f = F'$ for some F. In particular F is differentiable, and now Theorem 2.6 shows that f must be differentiable as well (recall Proposition 1.1). □

Let's record that last bit for future reference:

Corollary 2.9. *If $f \in H(V)$ then $f' \in H(V)$; hence by induction f is infinitely differentiable.*

Proof. This follows from Theorem 2.6, by Proposition 1.1. □

Corollary 2.9 is our second illustration of the simply magical nature of complex analysis: How in the world can the existence of one derivative imply that the derivative is itself differentiable? Totally amazing.

(Some would say that the "real" answer to the question of how Corollary 2.9 can hold is that the Cauchy-Riemann equations form an "elliptic" system of partial differential equations; alas we cannot say exactly what that means here.)

Chapter 3

Elementary Results on Holomorphic Functions

So far we have preliminary versions of Cauchy's Theorem and the Cauchy Integral Formula, and we have used the second to obtain the power series representation for a function holomorphic in a disk. The usefulness of the power series representation is a little limited because it is a local thing; it only works in disks. But it is nonetheless a very useful tool, making many important results almost obvious. In the present chapter we will see that the results in the previous chapter can be used to prove many basic results in complex analysis; we will get back to work on a "final" version of Cauchy's Theorem in the next chapter.

Corollary 3.0. *Suppose that $f_n \in H(V)$ for $n = 1, 2, \ldots$ and $f_n \to f$ uniformly on compact subsets of V. Then $f \in H(V)$.*

Note. The hypothesis means that if K is any compact subset of V then $f_n \to f$ uniformly on K; it may be appropriate to point out that it does not follow that $f_n \to f$ uniformly on V (for example if $f_n(z) = z/n$ then $f_n \to 0$ uniformly on compact subsets of $\mathbb{C}$ but not uniformly on $\mathbb{C}$).

The conclusion is another version of the fact that holomorphic functions are magic: All you would expect *a priori* is that the limit must be continuous, not differentiable (and indeed any continuous function on the interval [0, 1] is a uniform limit of differentiable functions, which is to say the corollary is far from true in the context of real analysis). One would expect to deduce that the limit was differentiable if one knew that the *derivatives* converged uniformly on compact sets; we shall see (Proposition 3.5 below) that in fact the hypotheses of Corollary 3.0 do imply that $f'_n \to f'$ uniformly on compact subsets of V.[1]

Proof. Suppose that $T \subset V$ is a triangle. Theorem 2.2 above shows that $\int_{\partial T} f_n(z)\, dz = 0$ for $n = 1, 2, \ldots$. Since $f_n \to f$ uniformly on ∂T, it follows that $\int_{\partial T} f(z)\, dz = 0$, and now Morera's Theorem (Corollary 2.8) shows that $f \in H(V)$. □

Example. Define $f_n : \mathbb{R} \to \mathbb{R}$ $(n = 1, 2, \ldots)$ by

$$f_n(x) = \left(x^2 + \frac{1}{n}\right)^{1/2}.$$

Then each f_n is differentiable (in fact each f_n is real-analytic). But it is easy to see that $f_n(x) \to |x|$ uniformly on compact subsets of $\mathbb{R}$ (and in fact it is only slightly harder to show that the convergence is uniform on $\mathbb{R}$). So in real analysis a uniform limit of differentiable functions need not be differentiable. (As with previous examples this raises the question of why defining $f_n(z)$ for $z \in \mathbb{C}$ by the same formula would not give a counterexample to Corollary 3.0. The reader may wish to return to this question after we have discussed various issues regarding the square root in the complex domain.)

The Cauchy Integral Formula (Theorem 2.5) shows that if f is holomorphic in a neighborhood of a closed disk then we can recover the values of f in the interior of the disk from the values on the boundary. Since f' is also holomorphic, we can get the values of f' at points inside the disk from the values of f' on the boundary by the same formula. Somewhat more interesting is the fact that we can get the values of f' in the interior from the values of f on the boundary:

Corollary 3.1 (Cauchy Integral Formula for derivatives). *Suppose* $f \in H(V)$ *and* $\overline{D(z_0, r)} \subset V$. *Define* $\gamma : [0, 2\pi] \to V$ *by* $\gamma(t) = z_0 + re^{it}$ *(so that γ is the boundary of the disk $D(z_0, r)$, traversed counterclockwise). Then*

$$f^{(n)}(z) = \frac{n!}{2\pi i} \int_\gamma \frac{f(w)\, dw}{(w-z)^{n+1}} \quad (n = 0, 1, 2, \ldots)$$

for all $z \in D(z_0, r)$.

Note that we have already proved Corollary 3.1 in the special case $z_0 = z$. We will use Cauchy's Theorem to derive the general case from this special case. (Alert readers are beginning to get the idea that Cauchy's Theorem is important.)

---

[1] Thus replacing one mystery by another ...

3. Elementary Results on Holomorphic Functions

Proof. The proof begins just like the proof of Theorem 2.5 above: We fix a point $z \in D(z_0, r)$, and for $0 < \rho < r - |z - z_0|$ we define $\gamma_\rho : [0, 2\pi] \to V$ by $\gamma_\rho(t) = z + \rho e^{it}$. We set $V' = V \setminus \{z\}$, define $g : V' \to \mathbb{C}$ by the formula

$$g(w) = \frac{f(w)}{(w-z)^{n+1}},$$

and note that $g \in H(V')$. Exactly the same argument as in the proof of Theorem 2.5 shows that

$$\int_\gamma g(w)\,dw = \int_{\gamma_\rho} g(w)\,dw,$$

and the note following Theorem 2.6 (with $D(z, \rho)$ in place of $D(z_0, r)$) shows that

$$\frac{n!}{2\pi i} \int_{\gamma_\rho} g(w)\,dw = f^{(n)}(z). \qquad \square$$

Corollary 3.2 (Cauchy's Estimates). *Suppose that f is holomorphic in a neighborhood of the closed disk $\overline{D(z_0, r)}$ and $|f| \leq M$ in $\overline{D(z_0, r)}$. Then*

$$|f^{(n)}(z_0)| \leq \frac{Mn!}{r^n} \quad (n = 0, 1, 2, \dots).$$

Proof. This is immediate from Corollary 3.1 and the ML inequality: If γ is the (positively oriented) boundary of the disk then

$$|f^{(n)}(z_0)| = \left| \frac{n!}{2\pi i} \int_\gamma \frac{f(w)\,dw}{(w-z_0)^{n+1}} \right| \leq \frac{n!}{2\pi} \frac{M}{r^{n+1}} \operatorname{length}(\gamma) = \frac{Mn!}{r^n}. \qquad \square$$

We can use Cauchy's Estimates to prove an important result: Convergence of a sequence of holomorphic functions (i.e., uniform convergence on compact subsets of an open set) implies convergence of the derivatives. First we need a few basic facts about compact subsets of the plane:

Lemma 3.3. *Suppose that $V \subset \mathbb{C}$ is open and $K \subset V$ is compact. Then K is at positive distance from the complement of V: There exists $\rho > 0$ such that*

$$|z - w| > \rho$$

for all $z \in K$ and $w \in \mathbb{C} \setminus V$.

Proof. Suppose not. Then there exist sequences $(z_n) \subset K$ and $(w_n) \subset \mathbb{C}\setminus V$ such that
$$|z_n - w_n| < 1/n \quad (n = 1, 2\ldots).$$
Since K is compact, the sequence (z_n) has a convergent subsequence: $z_{n_j} \to z \in K$. But, since $|z_n - w_n| \to 0$, this shows that we also have $w_{n_j} \to z$; since $\mathbb{C} \setminus V$ is closed, this shows that $z \in \mathbb{C} \setminus V$, contradicting the fact that $K \subset V$. $\square$

Lemma 3.4. *Suppose that $K \subset \mathbb{C}$ is compact, $\rho > 0$, and let*
$$K' = \bigcup_{z \in K} \overline{D(z, \rho)}.$$

Then K' is compact.

Proof. Suppose that $(z_n) \subset K'$. We need to show that (z_n) has a subsequence convergent to a point of K'. For every n there exist $\alpha_n \in K$ and $\beta_n \in \overline{D(z, \rho)}$ such that $z_n = \alpha_n + \beta_n$. Since K is compact, there exists a subsequence $\alpha_{n_j} \to \alpha \in K$; now, since $\overline{D(z, \rho)}$ is compact, the sequence (β_{n_j}) has a subsequence $\beta_{n_{j_k}} \to \beta \in \overline{D(z, \rho)}$. So
$$z_{n_{j_k}} = \alpha_{n_{j_k}} + \beta_{n_{j_k}} \to \alpha + \beta \in K'.$$
$\square$

Proposition 3.5. *Suppose that $f_n \in H(V)$ for $n = 1, 2, \ldots$ and $f_n \to f$ uniformly on compact subsets of V. Then $f_n' \to f'$ uniformly on compact subsets of V.*

Proof. Suppose that K is a compact subset of V; we need to show that $f_n' \to f'$ uniformly on K. Let $\epsilon > 0$.

The previous two lemmas show that we can choose $\rho > 0$ such that if
$$K' = \bigcup_{z \in K} \overline{D(z, \rho)}$$
then K' is a compact subset of V. Hence $f_n \to f$ uniformly on K', which implies that there exists N such that
$$|f_n(z) - f(z)| < \epsilon\rho \quad (z \in K', n > N).$$
This says that if $n > N$ and $z \in K$ then $|f_n - f| < \epsilon\rho$ on $\overline{D(z, \rho)}$, and now Cauchy's Estimates show that
$$|f_n'(z) - f'(z)| < \epsilon \quad (z \in K, n > N).$$
$\square$

3. Elementary Results on Holomorphic Functions

As mentioned above, Corollary 3.0 makes much more sense now that we know Proposition 3.5 (of course the latter proposition depends on Cauchy's Estimates, which are themselves somewhat mysterious).

Liouville's Theorem on bounded entire functions is another immediate corollary of Cauchy's Estimates:

Definition. An *entire function* is a function holomorphic in the entire complex plane.

The definition should really say that a function is entire if it is represented by a single power series convergent in the whole plane. We know that in the present context this is equivalent to the definition above, but the next exercise shows that the two notions are different for real-analytic functions:

- **Exercise 3.1.** Define $f : \mathbb{R} \to \mathbb{R}$ by $f(x) = 1/(1+x^2)$. Show that for every $x \in \mathbb{R}$ there exists $\delta > 0$ such that f is represented by a power series convergent in $(x - \delta, x + \delta)$, although the power series at the origin does not converge on the entire line.

  Hint: This is an example of an application of complex analysis to real analysis.

The function in the exercise is thus real-analytic on all of $\mathbb{R}$ but should not really be called real-entire (at least in our opinion; some authors differ on this point).

Corollary 3.6 (Liouville's Theorem). *A bounded entire function must be constant.*

Proof. Suppose that $f \in H(\mathbb{C})$ and $|f(z)| \leq M$ for all $z \in \mathbb{C}$. It follows from Cauchy's Estimates that for each complex number z and each $r > 0$ we have $|f'(z)| \leq M/r$. Thus $f'(z) = 0$ for all z, and hence f must be constant. (For example we can write $f(z) - f(w) = \int_{[w,z]} f'(s)\, ds = 0$.) $\square$

Liouville's Theorem has various applications — for example it can be used to prove the Fundamental Theorem of Algebra:

Corollary 3.7 (Fundamental Theorem of Algebra). *A complex polynomial of positive degree has a (complex) root.*

Before giving the proof we should say something about what "tends to infinity" means in the present context. Roughly speaking, in complex analysis "$z \to \infty$" is by definition the same as "$|z| \to \infty$". More precisely, if we say

$$\lim_{z \to \infty} f(z) = \infty$$

then we mean that for every $A > 0$ there exists R such that $|f(z)| > A$ for all z with $|z| > R$. Similarly, if $z_0 \in \mathbb{C}$ then

$$\lim_{z \to z_0} f(z) = \infty$$

means that for every $A > 0$ there exists $\delta > 0$ such that $|f(z)| > A$ for all z with $0 < |z - z_0| < \delta$, and

$$\lim_{z \to \infty} f(z) = z_0 \in \mathbb{C}$$

means that for every $\epsilon > 0$ there exists R such that $|f(z) - z_0| < \epsilon$ for every $z \in \mathbb{C}$ with $|z| > R$.

Proof. Suppose that n is a positive integer, $a_0, \ldots, a_n$ are complex numbers with $a_n \neq 0$, and set

$$P(z) = \sum_{j=0}^{n} a_j z^j.$$

We need to show that $P(z) = 0$ for at least one $z \in \mathbb{C}$. Suppose not. Then we can define an entire function f by $f(z) = 1/P(z)$.

Now, it is easy to see that $P(z) \to \infty$ as $z \to \infty$: For $z \neq 0$ we have

$$P(z) = a_n z^n \left(1 + \sum_{j=0}^{n-1} \frac{a_j}{a_n} z^{j-n}\right),$$

which tends to infinity at infinity since $a_n \neq 0$ and $\sum_{j=0}^{n-1}(a_j/a_n)z^{j-n} \to 0$. This shows that $\lim_{z \to \infty} f(z) = 0$, so that f is a bounded entire function. Thus f is constant, and hence in fact $f(z) = 0$ for all z, which is clearly impossible since $f = 1/P$. □

(The theorem says that $\mathbb{C}$ is "algebraically closed"; it follows by elementary algebra that if P is as in the proof above then in fact

$$P(z) = a_n \prod_{j=1}^{n} (z - z_j)$$

for some $z_1, \ldots, z_n \in \mathbb{C}$.)

The Fundamental Theorem of Algebra is indeed fundamental in various fields (for example algebra), but for us it's just an amusing digression, at least for now. Back to work:

Corollary 3.8. *Suppose that $f \in H(V)$ and V is connected. If all the derivatives of f vanish at some point of V then f is constant.*

3. Elementary Results on Holomorphic Functions

Proof. Let $A = \{z \in V : f^{(n)}(z) = 0, n = 1, 2, \ldots\}$. We are supposing that A is nonempty, and the fact that each derivative $f^{(n)}$ is continuous shows that A is closed (i.e. relatively closed in V). If we can show that A is open then it follows that $A = V$ (because V is connected; see Appendix 4); in particular it follows that $f'(z) = 0$ for all $z \in V$ and hence f is constant (see Exercise 3.9 below).

But the fact that A is open follows from the power-series representation: Suppose that $z \in A$. Choose $r > 0$ so that $D(z, r) \subset V$. Then we have

$$f(w) = \sum_{n=0}^{\infty} \frac{f^{(n)}(z)}{n!}(w - z)^n = f(z)$$

for all $w \in D(z, r)$. Hence $f^{(n)}(w) = 0$ for all $w \in D(z, r)$ and $n \geq 1$, which says precisely that $D(z, r) \subset A$. $\square$

So an analytic function must be constant if all the derivatives vanish at some point. This shows that a zero of a nonconstant analytic function must have a finite order: Suppose f is analytic at z (i.e., holomorphic in a neighborhood of z), $f(z) = 0$, but f is not constant near z. Then there exists a positive integer N such that $f^{(n)}(z) = 0$ for $0 \leq n < N$ but $f^{(N)}(z) \neq 0$; in such a case f is said to have a *zero of order N* at z. A zero of order 1 is known as a *simple zero*. (It is sometimes convenient to extend this terminology to the case where $f(z) \neq 0$: If f is analytic at z and $f(z) \neq 0$ then we may say that f has a zero of order 0 at z.) The power-series representation shows that this terminology makes sense:

Lemma 3.9. *Suppose that $f \in H(V)$ and f has a zero of order N at $z \in V$. Then there exists $g \in H(V)$ with $g(z) \neq 0$ such that*

$$f(w) = (w - z)^N g(w)$$

for all $w \in V$.

Proof. Choose $r > 0$ so that $D(z, r) \subset V$. The first N coefficients in the power series centered at z vanish; hence

$$f(w) = \sum_{n=N}^{\infty} c_n(w - z)^n = (w - z)^N \sum_{n=0}^{\infty} c_{n+N}(w - z)^n$$

in $D(z, r)$. This shows that

$$f(w)/(w - z)^N = \sum_{n=0}^{\infty} c_{n+N}(w - z)^n$$

in $D(z,r) \cap (V \setminus \{z\})$, and so there exists a function $g \in H(V)$ which equals $\sum_{n=0}^{\infty} c_{n+N}(w-z)^n$ in $D(z,r)$ and $f(w)/(w-z)^N$ in $V \setminus \{z\}$. Hence $f(w) = (w-z)^N g(w)$ in all of V, and $g(z) = c_N \neq 0$. □

We used Exercise 3.2 in the preceding proof without comment, since the exercise is extremely trivial. Nonetheless, the reader who thinks that there is nothing at all to be done in Exercise 3.2 should note Exercise 3.3, in particular noting that a correct solution to Exercise 3.2 must use the fact that A and B are open.

- **Exercise 3.2.** Suppose that A and B are open sets, $f \in H(A)$, $g \in H(B)$, and $f = g$ in $A \cap B$. Show that there exists $F \in H(A \cup B)$ such that $F = f$ in A and $F = g$ in B.

- **Exercise 3.3.** Give an example of two sets $A, B \subset \mathbb{C}$, a function f analytic on A and a function g analytic on B such that $A \cap B = \emptyset$ but there does *not* exist a function F analytic on $A \cup B$ with $F = f$ on A and $F = g$ on B.

It follows that the zeros of a nonconstant holomorphic function f defined in a connected open set cannot have an accumulation point in the interior of the domain of f; each zero is *isolated*. In general, if $f \in H(V)$ we will let $Z(f) = \{z \in V : f(z) = 0\}$ denote the zero set of f.

Corollary 3.10. *Suppose that $f \in H(V)$ and V is connected. If $Z(f)$ has a limit point in V then f is constant.*

Note that the zero set of a nonconstant holomorphic function *can* have limit points on the boundary of the domain of the function; for example let $V = \mathbb{C} \setminus \{0\}$ and define $f \in H(V)$ by $f(z) = \sin(1/z)$.

Proof. Suppose that f is not constant and $z \in V$ is a limit point of $Z(f)$; then $f(z) = 0$ since f is continuous. Lemma 3.9 shows that there exists $r > 0$ with the property that f has no zero in $D(z,r) \setminus \{z\}$ (if g is as in the lemma then g has no zero near z, since $g(z) \neq 0$ and g is continuous). Thus z cannot be a limit point of $Z(f)$ after all. □

The fact that the zeroes of a nonconstant holomorphic function cannot have a limit point inside the domain of the function has a number of useful consequences. For a trivial example, it can sometimes be used to extend identities from the reals to the complex numbers: Suppose that we had defined the function $\exp \in H(\mathbb{C})$ and we knew somehow that $\exp(s+t) = \exp(s)\exp(t)$ for all real s and t. It follows that $\exp(w+z) = \exp(w)\exp(z)$ for all complex z and w: First fix a number $s \in \mathbb{R}$ and define $f(z) = \exp(s+z) - \exp(s)\exp(z)$. Then f is an entire function and we know that f

vanishes on the real axis. It follows that f vanishes in the entire plane; thus $\exp(s+z) = \exp(s)\exp(z)$ for all real s and complex z. Repeating the same argument shows that $\exp(w+z) = \exp(w)\exp(z)$ for all complex z and w.

The Maximum Modulus Theorem (also known as the "maximum principle") is another very important property of analytic functions that is immediate from the power series representation:

Theorem 3.11 (Maximum Modulus Theorem). *Suppose that $f \in H(V)$ and V is connected. Then $|f|$ cannot achieve a (local) maximum in V unless f is constant: If f is nonconstant then for every $a \in V$ and $\delta > 0$ there exists $z \in V$ with $|f(z)| > |f(a)|$ and $|z - a| < \delta$.*

This is such an important result that we give two proofs: The first shows exactly how to find a point z near a with $|f(z)| > |f(a)|$, while the second shows that the average of $|f|^2$ over small circles centered at a is larger than $|f(a)|^2$, so that there must be a large number of nearby points where $|f| > |f(z)|$ (although the second proof does not show exactly how to find one).

Proof 1. Suppose that f is nonconstant and $a \in V$. Arguing as in the proof of Lemma 3.9 we see that

$$f(z) = f(a) + c(z-a)^N + (z-a)^{N+1}h(z),$$

where N is a positive integer, c is a nonzero complex number, and $h \in H(V)$. The plan is this: If we first specify the argument of $z - a$ appropriately we will have

$$|f(a) + c(z-a)^N| = |f(a)| + |c(z-a)^N| > |f(a)|,$$

and then if the modulus of $z - a$ is small enough the remaining term $(z-a)^{N+1}h(z)$ will not matter much. In detail:

First choose $\alpha, \beta \in \mathbb{R}$ so that $f(a) = |f(a)|e^{i\alpha}$ and $c = |c|e^{i\beta}$. Choose $\theta \in \mathbb{R}$ so that $\beta + N\theta = \alpha$ (this is possible since $N \neq 0$). Now if $r > 0$ and $z = a + re^{i\theta}$ it follows that

$$f(a) + c(z-a)^N = (|f(a)| + |c|r^N)e^{i(\beta + N\theta)},$$

so that

$$|f(a) + c(z-a)^N| = |f(a)| + |c|r^N.$$

Now, since h is continuous in V, it is bounded near a; say $|h(z)| \leq M$ for $|z-a| \leq \rho$, where $\rho > 0$ and $M < \infty$. The triangle inequality shows that if $0 < r < \rho$ we have

$$|f(z)| \geq |f(a) + c(z-a)^N| - |(z-a)^{N+1}h(z)|$$
$$\geq |f(a)| + |c|r^N - Mr^{N+1} = |f(a)| + (|c| - Mr)r^N.$$

Choose $\delta \in (0, \rho)$ so that $M\delta < |c|/2$; then if $0 < r < \delta$ we have

$$|f(z)| \geq |f(a)| + \frac{|c|r^N}{2} > |f(a)|. \qquad \square$$

The second proof of the Maximum Modulus Theorem uses a thing called the Parseval Formula; this is important enough that we make it a separate proposition. (If you have studied Fourier series the Parseval Formula should be familiar, and you should see that the second proof of the Maximum Modulus Theorem simply uses the fact that the restriction of a power series to a circle is a Fourier series.)

Proposition 3.12 (Parseval Formula). *Suppose that the power series*

$$f(z) = \sum_{n=0}^{\infty} c_n (z-a)^n$$

converges for $|z - a| < r$. Then

$$\frac{1}{2\pi} \int_0^{2\pi} |f(a + \rho e^{it})|^2 \, dt = \sum_{n=0}^{\infty} |c_n|^2 \rho^{2n}$$

for every $\rho \in [0, r)$.

Proof. Let

$$s_N(z) = \sum_{n=0}^{N} c_n (z-a)^n.$$

Now, one can simply calculate that if $k \in \mathbb{Z}$ then

$$\frac{1}{2\pi} \int_0^{2\pi} e^{ikt} \, dt = \begin{cases} 0 & (k \neq 0), \\ 1 & (k = 0) \end{cases}$$

(or one can show that the integral above vanishes for $k \neq 0$ by symmetry — see Exercise 3.16 below). Hence

$$\frac{1}{2\pi} \int_0^{2\pi} |s_N(a + \rho e^{it})|^2 \, dt = \frac{1}{2\pi} \int_0^{2\pi} \sum_{n=0}^{N} c_n \rho^n e^{int} \overline{\sum_{m=0}^{N} c_m \rho^m e^{imt}} \, dt$$

$$= \sum_{n,m=0}^{N} c_n \overline{c_m} \rho^{n+m} \frac{1}{2\pi} \int_0^{2\pi} e^{i(n-m)t} \, dt$$

$$= \sum_{n=0}^{N} |c_n|^2 \rho^{2n}.$$

3. Elementary Results on Holomorphic Functions

Now if $\rho < r$ then $s_N \to f$ uniformly on $\partial D(a, \rho)$, and hence

$$\frac{1}{2\pi} \int_0^{2\pi} |f(a + \rho e^{it})|^2 \, dt = \lim_{N \to \infty} \frac{1}{2\pi} \int_0^{2\pi} |s_N(a + \rho e^{it})|^2 \, dt$$

$$= \lim_{N \to \infty} \sum_{n=0}^{N} |c_n|^2 \rho^{2n} = \sum_{n=0}^{\infty} |c_n|^2 \rho^{2n}. \qquad \square$$

The Parseval Formula makes the Maximum Modulus Theorem more or less obvious:

Proof 2 (Theorem 3.11). Suppose that f is nonconstant and $D(a, r) \subset V$. Then f has a power series representation

$$f(z) = \sum_{n=0}^{\infty} c_n (z - a)^n$$

valid for $|z - a| < r$. Hence for $0 < \rho < r$ we have

$$\frac{1}{2\pi} \int_0^{2\pi} |f(a + \rho e^{it})|^2 \, dt = \sum_{n=0}^{\infty} |c_n|^2 \rho^{2n} = |f(a)|^2 + \sum_{n=1}^{\infty} |c_n|^2 \rho^{2n}.$$

Now the fact that f is not constant shows that $c_n \neq 0$ for at least one $n \geq 1$, and so

$$\frac{1}{2\pi} \int_0^{2\pi} |f(a + \rho e^{it})|^2 \, dt > |f(a)|^2$$

for every $\rho \in (0, r)$, which shows that for every such ρ there exists t with $|f(a + \rho e^{it})| > |f(a)|$. (If on the other hand $|f(a + \rho e^{it})| \leq |f(a)|$ for all t then it would follow that

$$\frac{1}{2\pi} \int_0^{2\pi} |f(a + \rho e^{it})|^2 \, dt \leq \frac{1}{2\pi} \int_0^{2\pi} |f(a)|^2 \, dt = |f(a)|^2.) \qquad \square$$

Our next topic is the classification of isolated singularities.

Definition. For $z \in \mathbb{C}$ and $r > 0$ the *punctured disk* (or *deleted disk*) with center z and radius r is

$$D'(z, r) = D(z, r) \setminus \{z\} = \{w \in \mathbb{C} : 0 < |w - z| < r\}.$$

Definition. The function $f \in H(V)$ has an *isolated singularity* at z if $D'(z, r) \subset V$ for some $r > 0$ but $z \notin V$.

Definition. Suppose that f has an isolated singularity at z. The singularity is *removable* if f can be defined at z so as to make f holomorphic in a neighborhood of z. The singularity is a *pole* if $\lim_{w \to z} f(w) = \infty$. Finally, an isolated singularity which is neither a pole nor a removable singularity is an *essential singularity*.

At first it may appear that poles are bad and essential singularities are maybe not so bad, since a function actually tends to infinity at a pole. But it turns out that if f has a pole at z then f blows up at z in a very nice way. Indeed, if we let S denote the "Riemann sphere" (to be defined later) then there is a natural notion of "S-valued holomorphic function", and a function holomorphic except for poles (i.e. a *meromorphic* function) is the same thing as an S-valued holomorphic function; in particular a function with a pole is really a very smooth sort of thing if you look at it from the right point of view. On the other hand a function must be very, very bad near an essential singularity, as Proposition 3.14 shows. First an important lemma:

Lemma 3.13. *If a holomorphic function is bounded near an isolated singularity then the singularity is removable.*

Proof. Suppose that $f \in H(D'(z,r))$ and f is bounded in $D'(z,r)$. Define a function $g : D(z,r) \to \mathbb{C}$ by

$$g(w) = \begin{cases} (w-z)^2 f(w) & (w \ne z), \\ 0 & (w = z). \end{cases}$$

The boundedness of f makes it clear that g is continuous at z; in fact it is easy to see from the definition that g is differentiable at z and $g'(z) = 0$. Hence $g \in H(D(z,r))$ and g has a zero of order at least 2 at z, which shows that there exists $h \in H(D(z,r))$ with $g(w) = (w-z)^2 h(w)$ in $D(z,r)$.

In particular $h \in H(D(z,r))$ and $h = f$ in $D'(z,r)$, which says precisely that f has a removable singularity at z. □

Proposition 3.14. *Suppose that $f \in H(D'(z,r))$.*

(i) f has a pole at z if and only if there exist a positive integer N and complex numbers $(c_n)_{n \ge -N}$ such that $c_{-N} \ne 0$ and

$$f(w) = \sum_{n=-N}^{\infty} c_n (w-z)^n$$

for $w \in D'(z,r)$.

3. Elementary Results on Holomorphic Functions

(ii) f has an essential singularity at z if and only if $f(D'(z,\rho))$ is dense in $\mathbb{C}$ for all $\rho \in (0, r)$.

If f has a pole at z then the integer N in part (i) of the proposition is known as the *order* of the pole; a pole of order 1 is a *simple* pole.

Proof. For part (i), note first that if $f(w) = \sum_{n=-N}^{\infty} c_n(w-x)^n$ as above then it is clear that $f(w) \to \infty$ as $w \to z$, so f has a pole at z. Suppose now that f has a pole at z.

There exists $\rho \in (0, r)$ such that $|f| > 1$ in $D'(z, \rho)$. Hence $1/f$ is bounded in $D'(z, \rho)$ and so the previous lemma shows that there exists $g \in H(D(z, \rho))$ with $g = 1/f$ in $D'(z, \rho)$. The fact that f has a pole at z shows that $g(z) = 0$; since g is certainly not identically zero, there exist a positive integer N and a function $h \in H(D(z, \rho))$, with $h(z) \neq 0$, such that

$$g(w) = (w-z)^N h(w)$$

in $D(z, \rho)$. Now g has no zero in $D'(z, \rho)$, since it equals $1/f$ there, and so h has no zero in $D(z, \rho)$, which shows that $1/h$ is given by a power series:

$$\frac{1}{h(w)} = \sum_{n=0}^{\infty} a_n(w-z)^n \quad (w \in D(z, \rho));$$

note that $a_0 = 1/h(z) \neq 0$. It follows that

$$f(w) = (w-z)^{-N} \sum_{n=0}^{\infty} a_n(w-z)^n$$

in $D'(z, \rho)$, which can be rewritten as

$$f(w) = \sum_{n=-N}^{\infty} c_n(w-z)^n,$$

where $c_{-N} = a_0 \neq 0$. (We have shown that $f(w) = \sum_{n=-N}^{\infty} c_n(w-z)^n$ in $D'(z, \rho)$; now it follows from Corollary 3.10 that in fact $f(w) = \sum_{n=-N}^{\infty} c_n(w-z)^n$ in $D'(z, r)$.)

For part (ii), note first that if f has a pole or a removable singularity at z then $f(w)$ has a (finite or infinite) limit as $w \to z$, so it is certainly not true that $f(D'(z, \rho))$ is dense in $\mathbb{C}$ for all $\rho \in (0, r)$. Suppose now that $\rho \in (0, r)$ and $f(D'(z, \rho))$ is not dense in $\mathbb{C}$; we need to show that f has a pole or a removable singularity at z.

We can choose $\alpha \in \mathbb{C}$ and $\delta > 0$ so that

$$f(D'(z, \rho)) \cap D(\alpha, \delta) = \emptyset.$$

This says that
$$|\alpha - f(w)| \geq \delta$$
in $D'(z,\rho)$; in particular $\alpha - f$ has no zero in $D'(z,\rho)$, so
$$\frac{1}{\alpha - f} \in H(D'(z,\rho)).$$

Now $|1/(\alpha - f)| \leq 1/\delta$ in $D'(z,\rho)$, so the previous lemma shows that there exists $g \in H(D(z,\rho))$ with $g = 1/(\alpha - f)$ in $D'(z,\rho)$.

Hence $f = \alpha - 1/g$ in $D'(z,\rho)$. If $g(z) \neq 0$ then $\alpha - 1/g \in H(D(z,\rho))$, which shows that f has a removable singularity at z. On the other hand if $g(z) = 0$ then f tends to infinity at z and so has a pole there. □

Definition. If f has a pole at z and hence is given by a series as in part (i) of the previous proposition then the *principal part* of f at z is the sum
$$\sum_{n=-N}^{-1} c_n (w-z)^n.$$

The series appearing in part (i) of Proposition 3.14 is a special case of a "Laurent series". The Laurent series representation of a function holomorphic in an annulus is analogous to the power series for a function holomorphic in a disk: For $z \in \mathbb{C}$ and $0 \leq r < R \leq \infty$ we define the annulus
$$A(z,r,R) = \{w \in \mathbb{C} : r < |w-z| < R\}.$$
(Note that $D'(z,r) = A(z,0,r)$.) It turns out that if $f \in H(A(z,r,R))$ then f has a series expansion of the form
$$f(w) = \sum_{n=-\infty}^{\infty} c_n (w-z)^n,$$
where the series converges uniformly and absolutely on compact subsets of the annulus; this is a *Laurent series*. We will prove that functions holomorphic in annuli are given by Laurent series in the next chapter; this follows from the "global" Cauchy Integral Formula in much the same way as the power series representation for functions holomorphic in a disk follows from the Cauchy Integral Formula for disks. Let us assume this for now, as well as the fact that the coefficients in the series are uniquely determined by f. It follows that a function with an isolated singularity must have a Laurent series valid near the singularity, since punctured disks are annuli. You can read off the nature of the singularity from the Laurent series as follows:

The uniqueness of the coefficients in the Laurent series shows that f has a removable singularity if and only if the Laurent series is actually a power series, which is to say that $c_n = 0$ for all $n < 0$. On the other hand part (i) of the previous proposition shows that f has a pole at z if and only if $c_n \neq 0$ for at least one but only finitely many $n < 0$. Since an essential singularity is precisely an isolated singularity which is neither removable nor a pole, we see that f has an essential singularity if and only if $c_n \neq 0$ for infinitely many $n < 0$.

We should record that for future reference:

Proposition 3.15. *Suppose that $f \in H(D'(z_0, r))$ has the Laurent series expansion*

$$f(z) = \sum_{n=-\infty}^{\infty} c_n (z - z_0)^n.$$

Then

(i) f has a removable singularity at z_0 if and only if $c_n = 0$ for all $n < 0$.

(ii) f has a pole of order $N > 0$ at z_0 if and only if $c_{-N} \neq 0$ and $c_n = 0$ for all $n < -N$.

(iii) f has an essential singularity at z_0 if and only if there exist infinitely many values of $n < 0$ such that $c_n \neq 0$.

Exercises

3.4. Suppose that $f \in H(\mathbb{C})$ and $|f(z)| \leq e^{\operatorname{Re} z}$ for all z. Show that $f(z) = ce^z$ for some constant c.

3.5. Suppose that $f \in H(\mathbb{C})$, n is a positive integer and $|f(z)| \leq (1 + |z|)^n$ for all z. Show that f is a polynomial.

3.6. Suppose that $f, g \in H(D(z, r))$, $1 \leq m \leq n$, f has a zero of order n at z and g has a zero of order m at z. Show that f/g has a removable singularity at z.

3.7. Suppose that $f \in H(\mathbb{C})$ and $f(n) = 0$ for all $n \in \mathbb{Z}$. Show that all the singularities of $f(z)/\sin(\pi z)$ are removable.

3.8. Suppose that $f \in H(\mathbb{C})$, $f(z+1) = -f(z)$ for all z, $f(0) = 0$, and $|f(z)| \leq e^{\pi|\operatorname{Im} z|}$ for all z. Show that $f(z) = c\sin(\pi z)$ for some constant c.

Hint: Liouville's Theorem and the previous exercises.

3.9. Suppose that V is a connected open set, $f \in H(V)$, and $f'(z) = 0$ for all $z \in V$. Show that f is constant.

Hint: Choose z_0 and $r > 0$ such that $D(z_0, r) \subset V$. If $a, b \in D(z_0, r)$ then
$$f(b) - f(a) = \int_{[a,b]} f'(z)\,dx = 0,$$
as in the proof of Lemma 2.0. So f is constant in $D(z_0, r)$, and now it follows easily from one of the results in the chapter that f is constant in V. For another solution, see the comment following Exercise A4.35 in Appendix 4.

3.10. Give an example of an open set V and a function $f \in H(V)$ such that $f'(z) = 0$ for all $z \in V$ but f is not constant.

3.11. Let $V = \mathbb{C} \setminus \{0\}$. Show that there does not exist $f \in H(V)$ such that $e^{f(z)} = z$ for all $z \in V$.

Hint: Suppose that such an f exists. First determine what $f'(z)$ must be, and then show that this contradicts Lemma 2.0.

Note that it's cheating in Exercise 3.11 to start with the fact that $f(z) = \log(z)$, because we don't know anything about complex logarithms; that's a long story coming up later. (In fact the exercise *shows* that there must be some subtleties involved with complex logarithms, otherwise $f(z) = \log(z)$ would be a counterexample to the statement of the exercise!) Pretending for a second that $f(z) = \log(z)$ might be a reasonable way to guess what $f'(z)$ should be, but your actual *proof* that $f'(z)$ is what it is should not mention logarithms.

Recall that in general $C(X)$ is the space of all continuous functions $f : X \to \mathbb{C}$. Note that strictly speaking $C(\overline{V}) \cap H(V)$ is always empty (except when $V = \mathbb{C}$) because a function cannot have domain V and also domain $\overline{V}$; thus when we say $f \in C(\overline{V}) \cap H(V)$, as people do, we mean that $f \in C(\overline{V})$ and $f|_V \in H(V)$.

3.12. Suppose that V is a bounded open subset of the plane and $f \in C(\overline{V}) \cap H(V)$. Show that if $M \geq 0$ and $|f(z)| \leq M$ for all $z \in \partial V$ then $|f(z)| \leq M$ for all $z \in V$.

3.13. Let $V = D(0,1)$. Suppose that $f \in C(\overline{V}) \cap H(V)$, $f(0) = 0$ and $|f(z)| \leq 1$ for all $z \in \overline{V}$. Show that $|f(z)| \leq |z|$ for all $z \in V$.

Hint: Apply two previous exercises to the function $g(z) = f(z)/z$.

3. Elementary Results on Holomorphic Functions

3.14. Let $V = D(0,1)$. Suppose that $f \in H(V)$, $f(0) = 0$ and $|f(z)| \leq 1$ for all $z \in V$. Show that $|f(z)| \leq |z|$ for all $z \in V$.

Hint: For $0 < r < 1$ define $f_r(z) = f(rz)$ and apply the previous exercise to f_r.

3.15. Suppose that $g \in C([0,1])$, let $V = \mathbb{C} \setminus [0,1]$, and define $f : V \to \mathbb{C}$ by
$$f(z) = \int_0^1 \frac{g(t)\,dt}{t-z}.$$
Show that $f \in H(V)$. (You should give two proofs of this, one directly from the definitions and one using one of the theorems above.)

3.16. Suppose k is a nonzero integer and define $f : \mathbb{R} \to \mathbb{C}$ by $f(t) = e^{ikt}$. Let $\alpha = \pi/k$ and set
$$I = \frac{1}{2\pi}\int_0^{2\pi} f(t)\,dt.$$
Using the fact that f has period 2π show that
$$\frac{1}{2\pi}\int_\alpha^{2\pi+\alpha} f(t)\,dt = I.$$
Using the fact that $f(s+t) = f(s)f(t)$ show that
$$\frac{1}{2\pi}\int_\alpha^{2\pi+\alpha} f(t)\,dt = -I;$$
conclude that $I = 0$.

3.17. Suppose that $\alpha < 1$ and $f \in H(D'(z_0, r))$ satisfies
$$|f(z)| \leq c|z - z_0|^{-\alpha}.$$
Show that f has a removable singularity at z_0. *Two proofs.*

3.18. Suppose that N is a positive integer, $\alpha < N+1$, and $f \in H(D'(z_0, r))$ satisfies *prop 3.14 / 3.9*
$$|f(z)| \leq c|z - z_0|^{-\alpha}.$$
Show that f has either a removable singularity or a pole of order no larger than N at z_0.

3.19. Suppose that $f \in H(D'(z_0, r))$. Show that f has a pole of order N at z_0 if and only if $1/f$ has a zero of order N at z_0 (or, more precisely, if and only if $1/f$ can be extended to a function holomorphic in $D(z_0, r)$ which has a zero of order N at z_0).

The statement of the next exercise is somewhat informal. You should give a more precise statement (see the previous exercise) and then prove it.

3.20. Suppose we agree that a pole of order N is a zero of order $-N$, and that if f is holomorphic near z_0 and $f(z_0) \neq 0$ then f has a zero of order 0 at z_0. Suppose that f and g are holomorphic in $D'(z_0, r)$ and have zeroes of order n, $m \in \mathbb{Z}$ at z_0, respectively. Show that fg has a zero of order $n + m$ at z_0.

Chapter 4

Logarithms, Winding Numbers and Cauchy's Theorem

We now turn to the "global" forms of Cauchy's Theorem and the Cauchy Integral Formula. We will give a relatively recent proof, due to Dixon [**D**]. Among many other things this will allow us to show that a function holomorphic in an annulus has a Laurent series expansion, just as the "local" Cauchy Integral Formula (Theorem 2.4) gave us the power series representation for a function holomorphic in a disk. (We could easily concoct an *ad hoc* version of the Cauchy Integral Formula for this purpose, but, since we are going to do the global Cauchy's Theorem and the Cauchy Integral Formula anyway, we may as well prove them first and then apply them.)

Roughly speaking, Cauchy's Theorem says that if Γ is a cycle in V that "encloses" no point of the complement of V then

$$\int_\Gamma f(z)\,dz = 0$$

for all $f \in H(V)$. (See Theorem 4.10 below for a precise statement.) One might imagine proving this by showing that such a Γ must be a sum of curves, each of which satisfies the hypotheses of some more elementary version of Cauchy's Theorem. We saw an example of this technique in the proof of Theorem 2.5 above: The cycle $\gamma \dot{-} \gamma_\rho$ does not enclose any point of the complement of $V' = V \setminus \{z\}$ (it might appear to "enclose" the point z itself, but it does not, in what turns out to be the relevant sense, because of cancellation: The two circles wind around z in opposite directions), and a key part of the proof was noting that

$$\gamma \dot{-} \gamma_\rho = \gamma^1 \dot{+} \gamma^2 \dot{+} \gamma^3,$$

where each γ^j was a closed curve contained in some convex subset of V'.

That was not too hard, but there we had one specific cycle in mind; showing that an argument like this can be made to work for *any* cycle which encloses no point of the complement would be much harder. Dixon's proof replaces all this "topology" with a very clever application of a little bit of complex analysis, most of which we already know (and the rest of which we need to know for other reasons anyway).

We need to say a bit about "winding numbers" and the complex logarithm before giving the proof or even the statement of Cauchy's Theorem. By way of introduction to all that, we mention another result that might be regarded as another version of Cauchy's Theorem, looking at it from a different point of view. The global Cauchy's Theorem below will provide an answer to the following question:

Question. Given an open set $V \subset \mathbb{C}$ and a cycle Γ in V, when is it true that

$$\int_\Gamma f(z)\,dz = 0$$

for all $f \in H(V)$? (Recall that a *cycle* is a chain composed of *closed* curves.)

There is a similar question that we can answer right now:

Question. Given an open set $V \subset \mathbb{C}$ and $f \in H(V)$, when is it true that

$$\int_\Gamma f(z)\,dz = 0$$

for all cycles Γ in V?

The answer is given by

Theorem 4.0. *Suppose that $V \subset \mathbb{C}$ is open and $f \in H(V)$. We have*

$$\int_\gamma f(z)\,dz = 0$$

for every smooth closed curve γ in V if and only if $f = F'$ for some $F \in H(V)$.

4. Logarithms, Winding Numbers and Cauchy's Theorem

Proof. We already know one direction: If $f = F'$ then Lemma 2.0 shows that $\int_\gamma f(z)\,dz = 0$ for every smooth closed curve γ in V.

Suppose now that $\int_\gamma f(z)\,dz = 0$ for every smooth closed curve γ in V. Since any such curve is in fact contained in one of the components of V (why is that?), we may suppose that V is connected. Since V is open, it follows that there is a smooth path joining any two points of V. Fix a point $z_0 \in V$ and define a function $F : V \to \mathbb{C}$ as follows:

$$F(z) = \int_{\gamma_z} f(w)\,dw,$$

where γ_z is a smooth path from z_0 to z (that is, $\gamma_z : [a,b] \to V$ is smooth and $\gamma(a) = z_0$, $\gamma(b) = z$).

The first thing to note is that F is "well defined". In general, if there were choices made in a purported definition of something we say the something is *well defined* if it turns out to be independent of the choices that were made. Here we need to show that if γ_z and $\tilde\gamma_z$ are two smooth paths in V from z_0 to z then

$$\int_{\gamma_z} f(w)\,dw = \int_{\tilde\gamma_z} f(w)\,dw.$$

But this is clear: In fact $\gamma_z \dot- \tilde\gamma_z$ is a closed curve, so our hypothesis shows that

$$\int_{\gamma_z \dot- \tilde\gamma_z} f(w)\,dw = 0.$$

(Strictly speaking it is not correct to say that $\gamma_z \dot- \tilde\gamma_z$ is a closed curve, because it is not a curve at all, it is a chain. What is true, strictly speaking, is that there exists a closed curve γ_c such that the two chains $\gamma_z \dot- \tilde\gamma_z$ and γ_c are equal (recall the definition of equality we gave for chains above). We will often speak less strictly than this, simply saying things like "$\gamma_z \dot- \tilde\gamma_z$ is a closed curve" without further comment.)

So the definition of $F(z)$ is independent of the particular path from z_0 to z that is used. Suppose now that $z \in V$. Choose $r > 0$ so that $D(z,r) \subset V$. If the complex number h satisfies $|h| < r$ it follows that

$$F(z+h) = \int_{\gamma_z \dot+ [z,z+h]} f(w)\,dw,$$

since $\gamma_z \dot+ [z, z+h]$ is a curve in V from z_0 to $z+h$. This shows that $F' = f$, exactly as in the proof of Proposition 2.3 above. $\square$

The reader should note that the idea of the proof really is the same as in the proof of Proposition 2.3: There we did not need to worry about $F(z)$

being well defined because there was no "choice" made in the definition; we specified that $\gamma_z = [z_0, z]$, which was possible because V was convex.

Theorem 4.0 is something we should know for various reasons, one of which is that it says something about how logarithms work. Let us first talk about "pointwise" logarithms briefly: We say that w is a logarithm of z if

$$e^w = z.$$

That's "*a* logarithm", not "*the* logarithm", because any nonzero $z \in \mathbb{C}$ has infinitely many logarithms; for example any complex number of the form $2\pi i k$ ($k \in \mathbb{Z}$) is a logarithm of 1.

Soon we will be talking about holomorphic logarithms. It is unfortunate that the notation "log" is used both for *any* so-called "branch of the logarithm" (definition below) and also for *the* logarithm function $\log : (0, \infty) \to \mathbb{R}$; in fact the notation "log" is often used in both senses in one equation! Hence we will often use the notation "ln" for the latter function to avoid ambiguity in the notation: When you see $\ln(x)$ below it follows that we are referring to the *real* logarithm of $x > 0$. (We will also often write "log" in place of "ln" — the ambiguity exists in standard notation so you may as well get used to it.)

It follows from the fact that $e^{x+iy} = e^x e^{iy}$ that w is a logarithm of z if and only if $\operatorname{Re} w = \ln(|z|)$ and $\operatorname{Im} w$ is an argument of z; thus the fact that z has many logarithms is closely related to the fact that it has many arguments. (See Appendix 2 for the definition of "argument".)

Definition. Suppose that $V \subset \mathbb{C}$ is open. A branch of the logarithm in V is a function $L \in H(V)$ such that

$$e^{L(z)} = z$$

for all $z \in V$.

The first thing we should note is that while there certainly exist *discontinuous* functions f such that $e^{f(z)} = z$ (for example $f : \mathbb{C} \setminus \{0\} \to \mathbb{C}$ defined by $f(z) = \ln(|z|) + i\operatorname{Arg}(z)$), a *continuous* logarithm defined in an open set must be holomorphic. (This is closely related to something we will prove later, namely that if a holomorphic function has an inverse then the inverse must be holomorphic.)

Proposition 4.1. *Suppose that V is open and $f \in C(V)$ satisfies $e^{f(z)} = z$ for all $z \in V$. Then $f \in H(V)$ (so that f is a branch of the logarithm in V).*

4. Logarithms, Winding Numbers and Cauchy's Theorem

Proof. Note first that f must be one-to-one in V: If $f(z) = f(w)$ then $z = e^{f(z)} = e^{f(w)} = w$. In particular $f(z+h) - f(z) \ne 0$ for $z \in V$ and small $h \ne 0$. Now the fact that f is continuous and $\exp' = \exp$ shows that

$$\lim_{h \to 0} \frac{h}{f(z+h) - f(z)} = \lim_{h \to 0} \frac{e^{f(z+h)} - e^{f(z)}}{f(z+h) - f(z)} = e^{f(z)} = z$$

for all $z \in V$. Since $z = e^{f(z)} \ne 0$ we can take the reciprocal:

$$\lim_{h \to 0} \frac{f(z+h) - f(z)}{h} = \frac{1}{z}. \quad \Rightarrow \text{differentiable} \quad \square$$

The next proposition gives an important characterization of branches of the logarithm:

Proposition 4.2. *Suppose that V is a connected open subset of the plane, $z_0 \in V$ and $L \in H(V)$. The following are equivalent:*

(i) *L is a branch of the logarithm in V.*

(ii) *$e^{L(z_0)} = z_0$ and $L'(z) = 1/z$ for all $z \in V$.*

Proof. Suppose (i). Then $e^{L(z_0)} = z_0$ by definition, while the chain rule shows that $L'(z)e^{L(z)} = 1$ and hence $L'(z)z = 1$.

Now suppose (ii). Let $f(z) = ze^{-L(z)}$. Then $f \in H(V)$ and the chain rule shows that in fact $f'(z) = 0$ for all $z \in V$, so f must be constant (exercise); now $f(z_0) = 1$ and hence $f(z) = 1$ for all $z \in V$ so L is a branch of the logarithm. $\square$

Another way to say more or less the same thing:

Proposition 4.3. *Suppose that V is an open subset of the plane. There exists a branch of the logarithm in V if and only if there exists $f \in H(V)$ with $f'(z) = 1/z$ for all $z \in V$.*

Proof. One direction is trivial and the other is very easy: Suppose that $f \in H(V)$ and $f'(z) = 1/z$ for all $z \in V$. Choose $z_0 \in V$. Note that $z_0 \ne 0$, so z_0 has a logarithm. Hence we can choose a constant $c \in \mathbb{C}$ such that $e^{c+f(z_0)} = z_0$, and now L is a branch of the logarithm in V, if $L(z) = c + f(z)$ for all $z \in V$. $\square$

Corollary 4.4. *If V is a convex open subset of $\mathbb{C} \setminus \{0\}$ then there exists a branch of the logarithm in V.*

Proof. This is clear from the previous proposition and Proposition 2.3 above. □

Corollary 4.4 should be regarded as a preliminary result; we will see soon that the same statement is true in any "simply connected" open set, and it will be clear that a convex set is simply connected.

Combining Proposition 4.3 with Theorem 4.0 yields a geometric characterization of the open subsets of the plane in which there exists a branch of the logarithm:

Proposition 4.5. *Suppose that V is an open subset of $\mathbb{C}$. There exists a branch of the logarithm in V if and only if*

$$\int_\gamma \frac{dz}{z} = 0$$

for all smooth closed curves γ in V.

(Note that the second condition implies in particular that $0 \notin V$, else there would be a closed curve γ for which the integral was undefined.)

It is clear that Proposition 4.5 follows from Theorem 4.0 and Proposition 4.3; on the other hand it may not be so clear why we called it a "geometric" characterization of the open sets in which there exists a branch of the logarithm. This is because the integral $\int_\gamma \frac{dz}{z}$ is equal to $2\pi i$ times "the number of times the curve γ winds around the origin". Our next task is to make that notion precise.

If γ is a smooth closed curve in $\mathbb{C}$ and $a \in \mathbb{C} \setminus \gamma^*$ then the "index" of γ about a will turn out to be given by the formula

$$\mathrm{Ind}(\gamma, a) = \frac{1}{2\pi i} \int_\gamma \frac{dz}{z-a}.$$

For many purposes one could think of that integral as the *definition* of the index. We are not going to take it as the official definition, however, for two reasons: First, the integral only makes sense for smooth curves, and we want a definition that applies to closed curves that are just continuous, both for aesthetic reasons and because there is a place coming up where we will actually need to deal with the index of a nonsmooth continuous closed curve about a point. Second, defining the index as that integral obscures the geometric meaning of the index:

$\mathrm{Ind}(\gamma, a)$ is also known as the "winding number" of γ about a. This is because it really is the number of times the curve γ winds around the point a (where counterclockwise rotations count as positive, clockwise negative).

4. Logarithms, Winding Numbers and Cauchy's Theorem

That's something that needs definition — a first try might be "$1/2\pi$ times the net change in the argument of $z - a$ as z traverses the curve γ". That last needs definition as well ...

The following lemma is something we can actually *prove*, because we know the definitions of all the terms involved. The lemma will allow us to give a precise definition of "the net change in the argument of $z - a$" (we take $a = 0$ in the lemma to make the notation more transparent):

Lemma 4.6. *Suppose that $\gamma : [0, 1] \to \mathbb{C}$ is continuous and $0 \in \mathbb{C} \setminus \gamma^*$.*

(i) If $\theta_0 \in \mathbb{R}$ and $e^{i\theta_0} = \gamma(0)/|\gamma(0)|$ then there exists a unique continuous function $\theta : [0, 1] \to \mathbb{R}$ such that $\theta(0) = \theta_0$ and

$$\gamma(t) = |\gamma(t)| e^{i\theta(t)}$$

for all $t \in [0, 1]$.

(ii) For every $\epsilon > 0$ there exists $\delta > 0$ such that if $\tilde{\gamma} : [0, 1] \to \mathbb{C}$ is continuous and $|\tilde{\gamma}(t) - \gamma(t)| < \delta$ for all $t \in [0, 1]$ then there exists a continuous function $\tilde{\theta} : [0, 1] \to \mathbb{R}$ such that $\tilde{\gamma}(t) = |\tilde{\gamma}(t)| e^{i\tilde{\theta}(t)}$ and also $|\tilde{\theta}(t) - \theta(t)| < \epsilon$ for all $t \in [0, 1]$.

(iii) If γ is a closed curve then $(\theta(1) - \theta(0))/2\pi$ is an integer; if γ is a smooth closed curve then

$$\frac{\theta(1) - \theta(0)}{2\pi} = \frac{1}{2\pi i} \int_\gamma \frac{dz}{z}.$$

The lemma allows us to make the following definition:

Definition. If $\gamma : [0, 1] \to \mathbb{C}$ is a continuous closed curve and $a \in \mathbb{C} \setminus \gamma^*$ then the *index* of γ about a is

$$\text{Ind}(\gamma, a) = \frac{1}{2\pi}(\theta(1) - \theta(0)),$$

where $\theta : [0, 1] \to \mathbb{R}$ is a continuous function such that

$$\gamma(t) - a = |\gamma(t) - a| e^{i\theta(t)}.$$

Note that it follows that $\text{Ind}(\gamma, a)$ is always an integer, and also that if γ is a smooth closed curve then

$$\text{Ind}(\gamma, a) = \frac{1}{2\pi i} \int_\gamma \frac{dz}{z - a}.$$

We extend the definition of the index to smooth cycles Γ by defining

$$\operatorname{Ind}(\Gamma, a) = \frac{1}{2\pi i} \int_\Gamma \frac{dz}{z-a}.$$

We should note that the first and second parts of Lemma 4.6 follow immediately from basic facts about "covering spaces". We will be saying a lot about covering spaces later; one might regard the proof of the lemma as a warmup for that material:

Proof (Lemma 4.6). (i) Note that for any $t \in [0,1]$ we have $\gamma(t) \ne 0$, and hence by continuity and compactness we may choose numbers $0 = t_0 < t_1 < \cdots < t_n = 1$ so that if $I_j = [t_{j-1}, t_j]$ then $\gamma(I_j) \subset D_j$, where D_j is an open disk not containing the origin.

Each D_j is convex and $1/z \in H(D_j)$ since $0 \notin D_j$; hence it follows from Corollary 4.4 that there is a branch of the logarithm in each D_j. So there exists $L_j \in H(D_j)$ such that $e^{iL_j(z)} = z$ for all $z \in D_j$. Note that

$$e^{iL_1(\gamma(0))} = \gamma(0) = |\gamma(0)|e^{i\theta_0};$$

hence $\operatorname{Re}(L_1(0))$ and θ_0 differ by a multiple of 2π, and by adding a constant to L_1 we may obtain $\operatorname{Re}(L_1(0)) = \theta_0$.

Note that $\gamma(t_j) \in D_j \cap D_{j+1}$ for $j \ge 2$, so both $L_j(\gamma(t_j))$ and $L_{j+1}(\gamma(t_j))$ exist. They are both logarithms of $\gamma(t_j)$ divided by i and hence they must differ by a multiple of 2π. Adding a multiple of $2\pi i$ to a branch of the logarithm results in another branch of the logarithm, so if we replace L_2 with $L_2 + 2\pi k$ for an appropriate integer k we obtain

$$L_2(\gamma(t_1)) = L_1(\gamma(t_1)).$$

Having adjusted L_2 in this way we can then add a constant to L_3 so that

$$L_3(\gamma(t_2)) = L_2(\gamma(t_2));$$

proceeding by induction we get L_j satisfying all the properties above and also

$$L_j(\gamma(t_{j-1})) = L_{j-1}(\gamma(t_{j-1})) \quad (j = 2, \ldots, n).$$

Now we have $L_j \circ \gamma = L_k \circ \gamma$ on $I_j \cap I_k$ for all j and k, so there exists a continuous function $\theta : [0,1] \to \mathbb{R}$ with

$$\theta(t) = \operatorname{Re}(L_j(\gamma(t))) \quad (t \in I_j).$$

Since the imaginary part of any logarithm of z is an argument of z, it follows that $\theta(t)$ is an argument of $\gamma(t)$, which is to say that $\gamma(t) = |\gamma(t)|e^{i\theta(t)}$ for all t.

This finishes the proof of the existence of the function θ. Uniqueness is easy: Any two θ's must differ by a multiple of 2π at every point; since they are continuous and agree when $t = 0$, they must agree for every $t \in [0, 1]$.

(ii) Choose I_j, D_j and L_j as above. Note that $\gamma(I_j)$ is a compact subset of D_j. Hence if δ is small enough then we must have

$$\tilde\gamma(I_j) \subset D_j$$

for $j = 1, \ldots, n$.

Recall that $L_j(\gamma(t_{j-1})) = L_{j-1}(\gamma(t_{j-1}))$ $(j = 2, \ldots, n)$. It follows that $L_j = L_{j-1}$ in all of $D_j \cap D_{j-1}$, since $(L_j - L_{j-1})/(2\pi)$ is a continuous integer-valued function on the connected set $D_j \cap D_{j-1}$. In particular we have

$$L_j(\tilde\gamma(t_{j-1})) = L_{j-1}(\tilde\gamma(t_{j-1})) \quad (j = 2, \ldots, n),$$

and hence we may define $\tilde\theta$ by requiring that

$$\tilde\theta(t) = \operatorname{Re}(L_j(\tilde\gamma(t))) \quad (t \in I_j)$$

as before.

Hence

$$\tilde\theta(t) - \theta(t) = \operatorname{Re}(L_j(\tilde\gamma(t)) - L_j(\gamma(t)))$$

for $t \in I_j$. It follows that $|\tilde\theta(t) - \theta(t)| < \epsilon$ for all t if δ is small enough, by continuity of L_j.

(iii) Suppose now that γ is closed (still just continuous, not smooth). Then $\theta(0)$ is an argument of $\gamma(0)$ and $\theta(1)$ is an argument of $\gamma(1)$, but $\gamma(1) = \gamma(0)$; hence $\theta(1) - \theta(0)$ must be a multiple of 2π.

Suppose finally that γ is a *smooth* closed curve. Define "subcurves" $\gamma_j = \gamma|_{I_j}$. Note that γ_j is a curve in D_j and $iL_j' = 1/z$ in D_j; thus we have

$$\frac{1}{2\pi i}\int_\gamma \frac{dz}{z} = \sum_{j=1}^n \frac{1}{2\pi i}\int_{\gamma_j} \frac{dz}{z}$$

$$= \sum_{j=1}^n \frac{1}{2\pi i}(iL_j(\gamma_j(t_j)) - iL_j(\gamma_j(t_{j-1})))$$

$$= \sum_{j=1}^n \frac{1}{2\pi}(L_j(\gamma(t_j)) - L_j(\gamma(t_{j-1})))$$

$$= \frac{1}{2\pi}(L_n(\gamma(1)) - L_1(\gamma(0)))$$

(the sum collapses because $L_j(\gamma(t_{j-1})) = L_{j-1}(\gamma(t_{j-1}))$).

But $L_n(\gamma(1)) = L_1(\gamma(0)) + 2\pi k$ for some integer k, since $e^{iL_n(\gamma(1))} = \gamma(1) = \gamma(0) = e^{iL_1(\gamma(0))}$. In particular $L_n(\gamma(1)) - L_1(\gamma(0))$ is real, so we have

$$\frac{1}{2\pi}(L_n(\gamma(1)) - L_1(\gamma(0))) = \frac{1}{2\pi}\text{Re}\,(L_n(\gamma(1)) - L_1(\gamma(0)))$$
$$= \frac{1}{2\pi}(\theta(1) - \theta(0)). \qquad \square$$

If that proof seems complicated you should probably draw some pictures and think about it. It really is just one simple idea: You define θ "locally", that is, near each point. You adjust your local versions of θ so each one is consistent with the previous one, and you're done. See also Exercise 4.22 below.

Before stating the next result we need to note that if K is a compact subset of $\mathbb{C}$ then the complement $\mathbb{C}\setminus K$ has exactly one unbounded component. (Choose R so that $|z| < R$ for all $z \in K$. Let $A = \{z \in \mathbb{C} : |z| > R\}$. Then A is a connected subset of $\mathbb{C}\setminus K$ and hence is contained in some component of $\mathbb{C}\setminus K$, which must therefore be unbounded. On the other hand all the other components of $\mathbb{C}\setminus K$ are subsets of $\mathbb{C}\setminus A$ and are therefore bounded.)

Proposition 4.7. *Suppose that Γ is a cycle in $\mathbb{C}$. The function $I(a) = \text{Ind}(\Gamma, a)$ is an integer-valued function on $\mathbb{C}\setminus\Gamma^*$ which is constant on the components of $\mathbb{C}\setminus\Gamma^*$ and vanishes on the unbounded component.*

Proof. The previous lemma shows that $\text{Ind}(\Gamma, a)$ is an integer for all $a \in \mathbb{C}\setminus\Gamma^*$. The definition

$$\text{Ind}(\Gamma, a) = \frac{1}{2\pi i}\int_\Gamma \frac{dz}{z-a}$$

shows that I is continuous (if $a_n \to a \in \mathbb{C}\setminus\Gamma^*$ then $1/(z-a_n) \to 1/(z-a)$ uniformly on Γ^*), and hence I must be constant on connected subsets of $\mathbb{C}\setminus\Gamma^*$. The definition also shows that $I(a) \to 0$ as $a \to \infty$; since I is constant on the unbounded component of $\mathbb{C}\setminus\Gamma^*$, it must vanish identically there. $\qquad \square$

Sure enough, Proposition 4.5 *is* a "geometric" characterization of the open sets in which there exist a branch of the logarithm: We can restate it as

4. Logarithms, Winding Numbers and Cauchy's Theorem

Proposition 4.5.1. *Suppose that V is an open subset of $\mathbb{C}$. There exists a branch of the logarithm in V if and only if $0 \notin V$ and*

$$\mathrm{Ind}(\gamma, 0) = 0$$

for all closed curves γ in V.

That is more or less what we need to know about logarithms and winding numbers for the proof of Cauchy's Theorem below. While we are on the topic of branches of the logarithm being antiderivatives of $1/z$ we should say a few things about the logarithm in the punctured plane $A = \mathbb{C} \setminus \{0\}$:

On the one hand the results above show that there is no branch of the logarithm in A, since there is a closed curve γ in A such that $\int_\gamma (dz/z) \neq 0$. On the other hand, one could take the proof of the results as an explanation of where the "multi-valued logarithm function" in A comes from: For $z \in A$ define

$$\log(z) = \int_{\gamma_z} \frac{dw}{w},$$

where γ_z is a curve in A from 1 to z. Of course this is *not* well defined, since we can get different integrals depending on our choice of γ_z. A person can regard this as a way to get various different $\log(z)$'s. (Another point of view says that we actually do get a well defined logarithm function this way, but the domain of this function is not A; instead the domain is a certain "Riemann surface". More on this later, when we talk about analytic continuation.)

The next lemma is a technical detail that will be used in the proof of the Cauchy Integral Formula.

Lemma 4.8. *Suppose that V is an open subset of the plane and $f \in H(V)$. Define $g : V \times V \to \mathbb{C}$ by*

$$g(z, w) = \begin{cases} \dfrac{f(z) - f(w)}{z - w} & (z \neq w), \\ f'(z) & (z = w). \end{cases}$$

Then g is continuous in $V \times V$, and if we define $g_w(z) = g(z, w)$ then $g_w \in H(V)$ for every $w \in V$.

Proof. It is clear that g is continuous at (z, w) and that g_w is differentiable at z if $z \neq w$, so we only need to show that g is continuous near points of the form (z_0, z_0) and that g_{z_0} is holomorphic near z_0. But this follows from the power-series representation for f in $D(z_0, r) \subset V$, as in the proof of Proposition 1.1 above. □

In Chapter 2 we proved Cauchy's Theorem for convex sets and deduced the Cauchy Integral Formula for disks. The proof of the global versions is the other way around — first we will prove the Cauchy Integral Formula and then deduce Cauchy's Theorem. Here is the statement of the Cauchy Integral Formula we are going to prove:

Theorem 4.9 (Cauchy's Integral Formula; homology version). *Suppose that V is an open subset of the plane and Γ is a cycle in V. If Γ has the property that*
$$\text{Ind}(\Gamma, a) = 0$$
for all $a \in \mathbb{C} \setminus V$ then
$$\frac{1}{2\pi i} \int_\Gamma \frac{f(w)\,dw}{w - z} = \text{Ind}(\Gamma, z) f(z)$$
for all $f \in H(V)$ and all $z \in V \setminus \Gamma^*$.

Cauchy's Theorem will follow easily:

Theorem 4.10 (Cauchy's Theorem; homology version). *Suppose that V is an open subset of the plane and Γ is a cycle in V. If Γ has the property that*
$$\text{Ind}(\Gamma, a) = 0$$
for all $a \in \mathbb{C} \setminus V$ then
$$\int_\Gamma f(z)\,dz = 0$$
for all $f \in H(V)$.

Cauchy's Theorem answers the question we said it would answer:

Question. Given an open set $V \subset \mathbb{C}$ and a cycle Γ in V, when is it true that
$$\int_\Gamma f(z)\,dz = 0$$
for all $f \in H(V)$?

Answer. If and only if
$$\text{Ind}(\Gamma, a) = 0$$
for all $a \in \mathbb{C} \setminus V$.

Indeed, it is clear that Cauchy's Theorem gives the "if" part of this statement. What about the "only if"? That half is trivial: Suppose that there exists $a \in \mathbb{C} \setminus V$ such that $\text{Ind}(\Gamma, a) \neq 0$. Then the function $f(z) = 1/(z - a)$ is in $H(V)$ and satisfies $\int_\Gamma f(z)\,dz \neq 0$. $= \text{Ind}(\Gamma, a)$

This allows us to give an interesting rephrasing of Cauchy's Theorem:

4. Logarithms, Winding Numbers and Cauchy's Theorem

Theorem 4.10.1 (Cauchy's Theorem). *Suppose that V is an open subset of the plane and Γ is a cycle in V. If*

$$\int_\Gamma f(z)\,dz = 0$$

for all $f \in H(V)$ of the form $f(z) = 1/(z-a)$ then

$$\int_\Gamma f(z)\,dz = 0$$

for all $f \in H(V)$.

(This is one of the things I had in mind when I said in Chapter 2 that integrating $1/z$ about a circle was in some sense the only way to obtain a counterexample to Cauchy's Theorem: For a given V and Γ, if there is no counterexample of the form $1/(z-a)$ then there is no counterexample at all.)

The plan for the proof of the Cauchy Integral Formula (Theorem 4.9) is fairly simple: We define $g(z, w)$ as in Lemma 4.8 above, and then we set

$$F(z) = \frac{1}{2\pi i} \int_\Gamma g(z, w)\,dw.$$

We show that F is actually an entire function that tends to 0 at infinity; hence F vanishes identically, and it follows from the definition of g that

$$\frac{1}{2\pi i} \int_\Gamma \frac{f(w)\,dw}{w-z} = \frac{1}{2\pi i} \int_\Gamma \frac{f(z)\,dw}{w-z} = f(z) \frac{1}{2\pi i} \int_\Gamma \frac{dw}{w-z} = \mathrm{Ind}(\Gamma, z) f(z)$$

for $z \in V \setminus \Gamma^*$.

That's the plan — you should read the previous paragraph a few times to convince yourself that, for example, $F(z) = 0$ really does imply what it is claimed to imply. The proof made a big splash when it appeared; it is very unexpected that an entire function should arise in the proof when the theorem itself talks only about functions in $H(V)$.

Unfortunately some parts of the plan don't quite make sense, at least at first. Notably the idea that we can define $F(z) = \frac{1}{2\pi i} \int_\Gamma g(z, w)\,dw$ for all $z \in \mathbb{C}$; in fact $g(z, w)$ is undefined for $z \in \mathbb{C} \setminus V$ so there is a slight problem with this. But this is where the hypothesis $\mathrm{Ind}(\Gamma, a) = 0$ for all $a \in \mathbb{C} \setminus V$ gets used: Roughly speaking, that hypothesis shows that the parts of g that are undefined for $z \in \mathbb{C} \setminus V$ don't matter because they get multiplied by 0! Various details in the proof might be regarded as justification for the idea that we can say that "undefined" times 0 equals 0.

Here's the actual proof:

Proof (Theorem 4.9). We are going to be defining an entire function F by two different formulas on two different sets. Let

$$\Omega = \{\, z \in \mathbb{C} \setminus \Gamma^* : \operatorname{Ind}(\Gamma, z) = 0 \,\}.$$

Proposition 4.7 shows that Ω is a union of components of $\mathbb{C} \setminus \Gamma^*$, so that in particular Ω is an open subset of the plane. The fact that $\operatorname{Ind}(\Gamma, a) = 0$ for all $a \in \mathbb{C} \setminus V$ shows that

$$\mathbb{C} \setminus V \subset \Omega,$$

which is to say that

$$V \cup \Omega = \mathbb{C}.$$

Define $g : V \times V \to \mathbb{C}$ as in Lemma 4.8, and now define $F_1 : V \to \mathbb{C}$ by

$$F_1(z) = \frac{1}{2\pi i} \int_\Gamma g(z, w)\, dw.$$

The fact that g is continuous on $V \times V$ shows that F_1 is continuous on V (because if $z_n \to z$ in V then $g(w, z_n) \to g(w, z)$ uniformly for $w \in \Gamma^*$). Now we can use Morera's Theorem (Corollary 2.8) to show that $F_1 \in H(V)$. It is enough to show that F_1 is holomorphic in any convex open subset of V, but if T is a triangle in such a set then

$$\int_{\partial T} F_1(z)\, dz = \int_{\partial T} \left(\int_\Gamma g(z, w)\, dw \right) dz$$

$$= \int_\Gamma \left(\int_{\partial T} g_w(z)\, dz \right) dw = \int_\Gamma 0\, dw = 0,$$

by Theorem 2.2.

If you are wondering why $\int_{\partial T} \int_\Gamma g(z, w)\, dw\, dz = \int_\Gamma \int_{\partial T} g(z, w)\, dz\, dw$ in the previous paragraph, that's good: This is an exchange of limits, and it is not automatic that you get the same result when you swap two limits this way (if interchanging limits always worked then half the theorems in analysis would be trivial). This is exactly the sort of thing you need to be careful about in analysis; it is ok here because it can be reduced (using the definition of the line integral and the fact that our curves have piecewise-continuous derivatives) to what is sometimes called "Fubini's Theorem" in advanced calculus: If $\phi : [0, 1] \times [0, 1] \to \mathbb{R}$ is continuous then

$$\int_0^1 \int_0^1 \phi(x, y)\, dx\, dy = \int_0^1 \int_0^1 \phi(x, y)\, dy\, dx.$$

We define $F_2 \in H(\Omega)$ by a different integral formula:

$$F_2(z) = \frac{1}{2\pi i} \int_\Gamma \frac{f(w)\, dw}{w - z} \quad (z \in \Omega).$$

4. Logarithms, Winding Numbers and Cauchy's Theorem

Note that $\Omega \subset \mathbb{C} \setminus \Gamma^*$ by definition; in particular if $z \in \Omega$ then $w \mapsto 1/(w-z)$ defines a continuous function on Γ^*, so the integral in the definition of $F_2(z)$ exists. The proof that $F_2 \in H(\Omega)$ is essentially identical to the proof that $F_1 \in H(V)$. (This happens a lot, Morera's Theorem being used with Fubini's Theorem to show that a function defined by an integral is holomorphic.)

Next we need to show that $F_1 = F_2$ in $V \cap \Omega$. Note first that

$$V \cap \Omega = \{ z \in V \setminus \Gamma^* : \operatorname{Ind}(\Gamma, z) = 0 \}.$$

In particular if $z \in V \cap \Omega$ then $z \notin \Gamma^*$, so that $g(z,w) = (f(z)-f(w))/(z-w)$ for all $w \in \Gamma^*$, hence

$$\begin{aligned}
F_1(z) &= \frac{1}{2\pi i} \int_\Gamma g(z,w)\, dw \\
&= \frac{1}{2\pi i} \int_\Gamma \frac{f(w) - f(z)}{w - z}\, dw \\
&= \frac{1}{2\pi i} \int_\Gamma \frac{f(w)}{w-z}\, dw - \frac{1}{2\pi i} \int_\Gamma \frac{f(z)}{w-z}\, dw \\
&= F_2(z) - \operatorname{Ind}(\Gamma, z) f(z) \\
&= F_2(z).
\end{aligned}$$

Now we're in business. Since $F_1 \in H(V)$ and $F_2 \in H(\Omega)$ agree in $V \cap \Omega$, there exists a function F holomorphic in $V \cup \Omega = \mathbb{C}$ which equals F_1 in V and equals F_2 in Ω. It follows from Proposition 4.7 that the unbounded component of $\mathbb{C} \setminus \Gamma^*$ is contained in Ω; hence there exists $R > 0$ such that $F(z) = F_2(z)$ whenever $|z| > R$. This shows that

$$\lim_{z \to \infty} F(z) = 0,$$

since if $z_n \to \infty$ then $f(w)/(w - z_n) \to 0$ uniformly for $w \in \Gamma^*$. But F is an entire function, and hence $F(z) = 0$ for all $z \in \mathbb{C}$. (The fact that an entire function which tends to 0 at infinity must vanish identically follows from either of two results in Chapter 3; which ones?)

Suppose now that $z \in V \setminus \Gamma^*$. Then $g(z,w) = (f(z) - f(w))/(z-w)$ for all $w \in \Gamma^*$, so that

$$\frac{1}{2\pi i} \int_\Gamma \frac{f(w)\, dw}{w - z} - \operatorname{Ind}(\Gamma, z) f(z) = \frac{1}{2\pi i} \int_\Gamma \frac{f(w) - f(z)}{w - z}\, dw = F_1(z) = 0. \quad \square$$

Cauchy's Theorem follows easily:

Proof (Theorem 4.10). Suppose that V and Γ are as in the statement of the theorem and $f \in H(V)$. Fix $z_0 \in V \setminus \Gamma^*$ and define $g \in H(V)$ by

$$g(z) = (z - z_0)f(z).$$

Now Theorem 4.9 shows that

$$\frac{1}{2\pi i} \int_\Gamma f(z)\,dz = \frac{1}{2\pi i} \int_\Gamma \frac{g(z)\,dz}{z - z_0} = \operatorname{Ind}(\Gamma, z_0) g(z_0) = 0. \qquad \square$$

So now we know Cauchy's Theorem and Cauchy's Integral Formula. These will be more useful if we know the indexes of some curves about some points:

Definition. If $a \in \mathbb{C}$ and $r > 0$ then the boundary of the disk $D(a, r)$ is the curve $\partial D(a, r) = \gamma$, where $\gamma : [0, 1] \to \mathbb{C}$ is defined by

$$\gamma(t) = a + re^{2\pi i t}.$$

Exercise 4.1. Show directly from the definition that $\operatorname{Ind}(\partial D(a, r), a) = 1$ for any $a \in \mathbb{C}$ and $r > 0$.

Exercise 4.2. Suppose $a \in \mathbb{C}$ and $r > 0$. Show that that

$$\operatorname{Ind}(\partial D(a, r), z) = \begin{cases} 1 & (|z - a| < r), \\ 0 & (|z - a| > r). \end{cases}$$

(What is $\operatorname{Ind}(\partial D(a, r), z)$ when $|z - a| = r$?)

Exercise 4.3. Show that Theorem 4.9 follows from Theorem 4.10, by an argument similar to the proof of Theorem 2.5 above.

That was easy. Now we know Cauchy's Theorem, the Cauchy Integral Formula, and we also know the index of a few curves about a few points. Our first application is the fact that functions holomorphic in annuli have Laurent series:

As before we define

$$A(a, r, R) = \{z \in \mathbb{C} : r < |z - a| < R\}.$$

4. Logarithms, Winding Numbers and Cauchy's Theorem

Theorem 4.11. *Suppose that $a \in \mathbb{C}$ and $0 \leq r < R \leq \infty$. If $f \in H(A(a,r,R))$ then there exists a sequence of complex numbers $(c_n)_{n=-\infty}^{\infty}$ such that*

$$f(z) = \sum_{n=-\infty}^{\infty} c_n(z-a)^n \qquad (z \in A(a,r,R)).$$

The series converges uniformly on compact subsets of $A(a,r,R)$ and the coefficients are given by

$$c_n = \frac{1}{2\pi i} \int_{\partial D(a,\rho)} \frac{f(w)\,dw}{(w-a)^{n+1}}$$

for any $\rho \in (r,R)$.

Proof. To begin, choose numbers r' and R' with $r < r' < R' < R$ and define

$$\Gamma = \partial D(a,R') \dotdiv \partial D(a,r').$$

It follows from Exercise 4.2 that $\operatorname{Ind}(\Gamma,p) = 0$ for all $p \in \mathbb{C} \setminus A(a,r,R)$. (If $|p-a| > R$ then $\operatorname{Ind}(\Gamma,p) = 0 - 0 = 0$, while if $|p-a| < r$ then $\operatorname{Ind}(\Gamma,p) = 1 - 1 = 0$.) Thus we may apply the Cauchy Integral Formula (Theorem 4.9). Suppose that $r' < |z-a| < R'$. Then $\operatorname{Ind}(\Gamma,z) = 1 - 0 = 1$, so the theorem shows that

$$f(z) = \frac{1}{2\pi i} \int_{\Gamma} \frac{f(w)\,dw}{(w-z)} = \frac{1}{2\pi i} \int_{\partial D(a,R')} \frac{f(w)\,dw}{(w-z)} - \frac{1}{2\pi i} \int_{\partial D(a,r')} \frac{f(w)\,dw}{(w-z)}.$$

Inside the first integral we have $|(z-a)/(w-a)| < 1$, so we rewrite the integrand as in the proof of Theorem 2.6. On the other hand we have $|(w-a)/(z-a)| < 1$ inside the second integral, so we rewrite the second integrand somewhat differently:

$$f(z) = \frac{1}{2\pi i} \int_{\partial D(a,R')} \frac{f(w)\,dw}{(w-z)} - \frac{1}{2\pi i} \int_{\partial D(a,r')} \frac{f(w)\,dw}{(w-z)}$$

|w-a|= R' ≥ |z-a| ≥ r'

$$= \frac{1}{2\pi i} \int_{\partial D(a,R')} \frac{f(w)}{(w-a)} \frac{dw}{1 - \frac{z-a}{w-a}} + \frac{1}{2\pi i} \int_{\partial D(a,r')} \frac{f(w)}{(z-a)} \frac{dw}{1 - \frac{w-a}{z-a}}$$

$$= \frac{1}{2\pi i} \int_{\partial D(a,R')} \frac{f(w)}{(w-a)} \sum_{n=0}^{\infty} \left(\frac{z-a}{w-a}\right)^n dw$$

$$+ \frac{1}{2\pi i} \int_{\partial D(a,r')} \frac{f(w)}{(z-u)} \sum_{n=0}^{\infty} \left(\frac{w-a}{z-a}\right)^n dw$$

$$= \sum_{n=0}^{\infty} (z-a)^n \frac{1}{2\pi i} \int_{\partial D(a,R')} \frac{f(w)\,dw}{(w-a)^{n+1}}$$

$$+ \sum_{n=0}^{\infty} (z-a)^{-n-1} \frac{1}{2\pi i} \int_{\partial D(a,r')} f(w)(w-a)^n\,dw$$

$$= \sum_{n=0}^{\infty} (z-a)^n \frac{1}{2\pi i} \int_{\partial D(a,R')} \frac{f(w)\,dw}{(w-a)^{n+1}}$$

$$+ \sum_{n=-\infty}^{-1} (z-a)^n \frac{1}{2\pi i} \int_{\partial D(a,r')} \frac{f(w)\,dw}{(w-a)^{n+1}}$$

$$= \sum_{n=-\infty}^{\infty} c_n(z-a)^n,$$

where

$$c_n = \begin{cases} \dfrac{1}{2\pi i} \int_{\partial D(a,R')} \dfrac{f(w)\,dw}{(w-a)^{n+1}} & (n \geq 0), \\ \dfrac{1}{2\pi i} \int_{\partial D(a,r')} \dfrac{f(w)\,dw}{(w-a)^{n+1}} & (n < 0). \end{cases}$$

This is the representation we want, except that so far it appears that the coefficients may depend on z (we derived the sum assuming that $r' < |z - a| < R'$, and the numbers r' and R' appear in the formulas for the coefficients). But an application of Cauchy's Theorem shows that the quantity $\frac{1}{2\pi i} \int_{\partial D(a,\rho)} \frac{f(w)\,dw}{(w-a)^{n+1}}$ is independent of $\rho \in (r, R)$ (if $r < \rho_1 < \rho_2 < R$ then Theorem 4.10 shows that the integral of $f(w)/(w-a)^{n+1}$ over $\partial D(a, \rho_1) \dotdiv \partial D(a, \rho_2)$ vanishes). In particular the coefficients do not depend on our choice of r' and R', so that $f(z)$ is given by the series above for any $z \in A(a, r, R)$.

Hence the terms are bounded for $z \in A(a, r, R)$, and hence the power series $\sum_{n=0}^{\infty} c_n(z-a)^n$ has radius of convergence at least R; thus it converges uniformly on compact subsets of $D(a, R)$. On the other hand the boundedness of the terms in the other half of the series, $\sum_{n=-\infty}^{0} c_n(z-a)^n$, for $z \in A(a, r, R)$ shows that this latter series must converge uniformly on compact subsets of $\{z \in \mathbb{C} : |z - a| > r\}$ (one could imitate the proof of Lemma 1.0(i) or one could derive this fact from Lemma 1.0 itself by a change of variables $w = 1/(z-a)$). Hence the entire series converges uniformly on compact subsets of $A(a, r, R)$. □

4. Logarithms, Winding Numbers and Cauchy's Theorem

So far the only winding numbers we know are those of the form $\text{Ind}(\partial D(a,r), z)$. Our next result shows that if one closed curve can be continuously deformed into another (in $\mathbb{C} \setminus \{a\}$) then they have the same index about a. This will allow us to "see at a glance" what a lot of winding numbers are; it also allows us to prove the homotopy version of Cauchy's Theorem.

We say that two closed curves in V are *homotopic* (or "homotopic in V") if one can be continuously deformed to the other (without leaving V). More precisely:

Definition. Suppose that V is an open subset of the plane and γ_0, γ_1 are closed curves in V, both with parameter interval $[0,1]$. We say that γ_0 is *homotopic* to γ_1 (in V) if there exists a continuous map $\Gamma : [0,1] \times [0,1] \to V$ such that $\Gamma(0,t) = \gamma_0(t)$ and $\Gamma(1,t) = \gamma_1(t)$ for all $t \in [0,1]$ while $\Gamma(s,0) = \Gamma(s,1)$ for all $s \in [0,1]$. In this case we write

$$\gamma_0 \sim \gamma_1 \quad \text{or} \quad \gamma_0 \underset{V}{\sim} \gamma_1.$$

Proposition 4.12. *Suppose that γ_0 and γ_1 are closed curves in the open set V and $\gamma_0 \underset{V}{\sim} \gamma_1$. Then*

$$\text{Ind}(\gamma_0, a) = \text{Ind}(\gamma_1, a)$$

for all $a \in \mathbb{C} \setminus V$.

Proof. We fix $a \in \mathbb{C} \setminus V$, and we assume without loss of generality that $a = 0$.

Suppose that $\Gamma : [0,1] \times [0,1] \to V$ is a map showing that $\gamma_0 \underset{V}{\sim} \gamma_1$, and define $\gamma_s(t) = \Gamma(s,t)$ for $s, t \in [0,1]$. (Note this does not change the meaning of the notations "γ_0" and "γ_1".) Now part (ii) of Lemma 4.6 shows that $\text{Ind}(\gamma_s, 0)$ is a continuous function of s; since it takes only integer values it must be constant. So $\text{Ind}(\gamma_0, 0) = \text{Ind}(\gamma_1, 0)$. $\square$

As an application we give another version of Cauchy's Theorem:

Theorem 4.13 (Cauchy's Theorem; homotopy version). *Suppose V is an open set in the plane, γ_1 and γ_2 are smooth closed curves in V, and $\gamma_1 \underset{V}{\sim} \gamma_2$. Then*

$$\frac{1}{2\pi i} \int_{\gamma_1} f(z)\, dz = \frac{1}{2\pi i} \int_{\gamma_2} f(z)\, dz$$

for all $f \in H(V)$.

Proof. Proposition 4.12 shows that
$$\text{Ind}(\gamma_1 \dotdiv \gamma_2, a) = 0$$
for all $a \in \mathbb{C} \setminus V$; hence the theorem follows from Theorem 4.10. □

We conclude this section by introducing a class of open sets such that Cauchy's Theorem "always works", that is, such that the index of any cycle about any point of the complement is zero:

Definition. The connected open set $V \subset \mathbb{C}$ is *simply connected* if any closed curve in V is homotopic to a point (that is, if any closed curve γ in V satisfies $\gamma \underset{V}{\sim} \gamma_c$, where $\gamma_c(t)$ is independent of t).

- **Exercise 4.4.** Suppose that V is simply connected, γ is a smooth closed curve in V and $a \in \mathbb{C} \setminus V$. Show that
$$\text{Ind}(\gamma, a) = 0.$$

- **Exercise 4.5.** Show that any convex open set is simply connected.

Theorem 4.14 (Cauchy's Theorem for simply connected sets). *If V is simply connected and Γ is a cycle in V then*
$$\int_\Gamma f(z)\, dz = 0$$
for all $f \in H(V)$.

Proof. This is immediate from Theorem 4.10 and Exercise 4.4. □

Corollary 4.15. *Any nonvanishing holomorphic function in a simply connected set has a holomorphic logarithm. That is, if V is simply connected, $f \in H(V)$ and f has no zero in V then there exists $L \in H(V)$ with*
$$e^L = f$$
in V.

Proof. It follows from Theorem 4.0 and Theorem 4.14 that there exists $F \in H(V)$ such that
$$F' = \frac{f'}{f}.$$
Now the chain rule shows that $(fe^{-F})' = 0$ so fe^{-F} is constant. Setting $L = F + c$ for a suitable $c \in \mathbb{C}$ we obtain $fe^{-L} = 1$. □

4. Logarithms, Winding Numbers and Cauchy's Theorem

- **Exercise 4.6.** Show that any nonvanishing holomorphic function in a simply connected set has a holomorphic square root.

We will see later that a connected open set V such that $\int_\gamma f(z)\, dz = 0$ for all $f \in H(V)$ and all smooth closed curves $\gamma \subset V$ must be simply connected; in fact any connected open set satisfying the conclusion of Exercise 4.6 must be simply connected.

Our next big topic is the Residue Theorem.

Definition. Suppose that f has an isolated singularity at z_0. Let

$$f(z) = \sum_{n=-\infty}^{\infty} c_n(z-z_0)^n \qquad (0 < |z - z_0| < R)$$

be the Laurent series for f near z_0. The *residue* of f at z_0 is

$$\operatorname{Res}(f, z_0) = c_{-1},$$

the coefficient of $(z-z_0)^{-1}$.

If f has a pole at z_0 then there are tricks one can use to calculate $\operatorname{Res}(f, z_0)$: If f has a pole of order N (or less) then in fact we have

$$f(z) = \sum_{n=-N}^{\infty} c_n(z-z_0)^n$$

near z_0, so that if we let $g(z) = (z-z_0)^N f(z)$ then g has a removable singularity at z_0:

$$g(z) = \sum_{n=0}^{\infty} c_{n-N}(z-z_0)^n.$$

Now the residue of f is c_{-1}, which is the coefficient of $(z-z_0)^{N-1}$ in the power series expansion of g about z_0; it follows that the residue is given by the formula

$$\operatorname{Res}(f, z_0) = \frac{g^{(N-1)}(z_0)}{(N-1)!}$$

in this case. Note in particular that if f has a simple pole at z_0 then the residue of f is just $g(z_0)$, or

$$\lim_{z \to z_0} (z-z_0)f(z).$$

See Exercise 4.7 below for more on calculating residues of various functions.

These things do not work if f has an essential singularity, but regardless of what sort of singularity f has at z_0 it follows from Theorem 4.11 that the residue is given by

$$\text{Res}(f, z_0) = \frac{1}{2\pi i} \int_{\partial D(z_0, \rho)} f(w)\, dw$$

if $\rho > 0$ is small enough that f is holomorphic in the punctured disk $A(z_0, 0, r)$ for some $r > \rho$. One can regard the residue as measuring the extent to which Cauchy's Theorem fails when f is integrated over small closed curves enclosing z_0: You would expect the integral to vanish (or you would if you forgot that Cauchy's Theorem had hypotheses that are not always satisfied) — instead all of the integral except for the "residue" of f vanishes.

- **Exercise 4.7.**

    (i) Suppose that f and g are holomorphic near z_0 and f has a simple zero at z_0. Find an expression for the residue of g/f at z_0 (and prove that it is correct).

    (ii) Suppose that f has a simple pole at z_0 and g is holomorphic near z_0. Show that

    $$\text{Res}(fg, z_0) = g(z_0) \text{Res}(f, z_0).$$

    (iii) Suppose that f is holomorphic in a neighborhood of z_0, set $g(z) = f(z)/(z - z_0)^n$, and show that

    $$\text{Res}(g, z_0) = f^{(n-1)}(z_0)/(n-1)!.$$

- **Exercise 4.8.** Suppose that f has an isolated singularity at z_0. Fix $r > 0$ and $n \in \mathbb{Z}$, and define $\gamma : [0, 2\pi] \to \mathbb{C}$ by

$$\gamma(t) = z_0 + re^{int}.$$

Show that $\text{Ind}(\gamma, z_0) = n$ and that

$$\frac{1}{2\pi i} \int_\gamma f(z)\, dz = \text{Ind}(\gamma, z_0) \text{Res}(f, z_0)$$

if $r > 0$ is small enough (be explicit about how small r must be.)

We need to note a fact about discrete subsets of open sets in the plane before we can state the Residue Theorem:

4. Logarithms, Winding Numbers and Cauchy's Theorem

Lemma 4.16. *Suppose that $V \subset \mathbb{C}$ is open, $S \subset V$ is (relatively) closed in V, and every point of S is isolated (that is, for every $z \in S$ there exists $r > 0$ such that $S \cap D(z,r) = \{z\}$). Suppose that Γ is a cycle in $V \setminus S$ such that $\mathrm{Ind}(\Gamma, a) = 0$ for all $a \in \mathbb{C} \setminus V$. Then*

$$\mathrm{Ind}(\Gamma, z) = 0$$

for all but finitely many $z \in S$.

Proof. Let $C = \{\, z \in \mathbb{C} \setminus \Gamma^* : \mathrm{Ind}(\Gamma, z) = 0 \,\}$. Note that C is open since it is a union of components of the open set $\mathbb{C} \setminus \Gamma^*$, and that C contains the unbounded component of $\mathbb{C} \setminus \Gamma^*$. It follows that

$$\Gamma^* \cup \{\, z \in \mathbb{C} \setminus \Gamma^* : \mathrm{Ind}(\Gamma, z) \neq 0 \,\} = \mathbb{C} \setminus C = K,$$

where K is compact. The hypothesis $\mathrm{Ind}(\Gamma, a) = 0$ for all $a \in \mathbb{C} \setminus V$ shows that $K \subset V$. It follows that $K \cap S$ must be finite: For every $z \in K$ we can find $r = r_z > 0$ such that $S \cap D(z, r_z) \subset \{z\}$ (if $z \notin S$ then $S \cap D(z, r_z) = \emptyset$ for some $r_z > 0$ because S is closed in V). Now K is covered by finitely many such disks $D(z_j, r_{z_j})$, $j = 1, \ldots, n$. $\square$

Theorem 4.17 (The Residue Theorem). *Suppose that $V \subset \mathbb{C}$ is open, $S \subset V$ is (relatively) closed in V, and every point of S is <u>isolated</u>. Suppose that $f \in H(V \setminus S)$. (In other words, f is "holomorphic in V except for isolated singularities at the points of S".) If Γ is a cycle in V such that $\Gamma^* \subset V \setminus S$ and $\mathrm{Ind}(\Gamma, a) = 0$ for all $a \in \mathbb{C} \setminus V$ then*

$$\frac{1}{2\pi i} \int_\Gamma f(z)\, dz = \sum_{z \in S} \mathrm{Ind}(\Gamma, z) \mathrm{Res}(f, z).$$

(Note that all but finitely many terms in the sum vanish, by Lemma 4.16.)

Proof. Let $z_1, \ldots, z_n$ be the elements of S about which Γ has nonvanishing index. For $j = 1, \ldots, n$ choose $r_j > 0$ small enough that $\overline{D}(z_j, r_j) \subset V \setminus \Gamma^*$ and also $S \cap \overline{D}(z_j, r_j) = \{z_j\}$. Define curves $\gamma_j : [0, 2\pi] \to V \setminus \Gamma^*$ by

$$\gamma_j(t) = z_j + r_j e^{i d_j t},$$

where $d_j = \mathrm{Ind}(\Gamma, z_j)$.

The curve γ_j is simply the boundary of $D(z_j, r_j)$ traversed d_j times (counterclockwise). In particular

$$\mathrm{Ind}(\gamma_j, z_j) = \mathrm{Ind}(\Gamma, z_j)$$

for $j = 1, \ldots, n$. We define a new cycle $\tilde{\Gamma}$ by

$$\tilde{\Gamma} = \Gamma \dotdiv (\gamma_1 \dotplus \cdots \dotplus \gamma_n).$$

Let $\Omega = V \setminus S$. It follows from our construction that $f \in H(\Omega)$ and $\operatorname{Ind}(\tilde{\Gamma}, a) = 0$ for all $a \in \mathbb{C} \setminus \Omega$. To see this, note first that

$$\mathbb{C} \setminus \Omega = (\mathbb{C} \setminus V) \cup S = (\mathbb{C} \setminus V) \cup (S \setminus \{z_1, \ldots, z_n\}) \cup (\{z_1, \ldots, z_n\}).$$

If $a \in \mathbb{C} \setminus V$ then $\operatorname{Ind}(\Gamma, a) = 0$ by hypothesis and $\operatorname{Ind}(\gamma_j, a) = 0$ for all j by construction, and hence

$$\operatorname{Ind}(\tilde{\Gamma}, a) = 0 - (0 + \cdots + 0) = 0.$$

Similarly $\operatorname{Ind}(\tilde{\Gamma}, a) = 0 - (0 + \cdots + 0) = 0$ for $a \in S \setminus \{z_1, \ldots, z_n\}$. On the other hand, if $a = z_k$ for some k then $\operatorname{Ind}(\gamma_j, a) = 0$ for $j \neq k$ while $\operatorname{Ind}(\gamma_k, a) = \operatorname{Ind}(\gamma_k, z_k) = \operatorname{Ind}(\Gamma, z_k)$, so that $\operatorname{Ind}(\tilde{\Gamma}, a) = \operatorname{Ind}(\Gamma, z_k) - \operatorname{Ind}(\Gamma, z_k) = 0$.

So Cauchy's Theorem implies that

$$\frac{1}{2\pi i} \int_{\tilde{\Gamma}} f(z) \, dz = 0.$$

Since $\frac{1}{2\pi i} \int_{\gamma_j} f(z) \, dz = \operatorname{Ind}(\gamma_j, z_j) \operatorname{Res}(f, z_j) = \operatorname{Ind}(\Gamma, z_j) \operatorname{Res}(f, z_j)$ (Exercise 4.8), the result follows. $\square$

The Residue Theorem has many applications, both concrete and abstract; the exercises below include several examples.

Exercises

4.9. Show that the Cauchy Integral Formula (Theorem 4.9) is an immediate consequence of the Residue Theorem.

4.10. Show that $\int_{-\infty}^{\infty} \frac{dx}{1+x^2} = \pi$, using the Residue Theorem. (For $R > 1$ let $\Gamma = \gamma_1 \dotplus \gamma_2$, where $\gamma_1 = [-R, R]$ and $\gamma_2 : [0, \pi] \to \mathbb{C}$ is defined by $\gamma_2(t) = R e^{it}$. Let $f(z) = \frac{1}{1+z^2}$. Use the Residue Theorem to calculate $\int_{\Gamma} f(z) \, dz$, then show that $\int_{\gamma_1} f(z) \, dz$ tends to $\int_{-\infty}^{\infty} \frac{dx}{1+x^2}$ as $R \to \infty$, while $\int_{\gamma_2} f(z) \, dz$ tends to 0.)

4.11. Suppose that P and Q are polynomials with $\text{degree}(Q) \geq \text{degree}(P) + 2$ and let $f(z) = P(z)/Q(z)$. Suppose that Q has no zero on the real axis and show that $\int_{-\infty}^{\infty} f(x)\, dx$ is equal to $2\pi i$ times the sum of the residues of f in the upper half-plane.

The next exercise gives a version of Parseval's Formula for Laurent series:

4.12. Suppose that $f \in H(A(z_0, r, R))$ has Laurent series $f(z) = \sum_{n=-\infty}^{\infty} c_n (z - z_0)^n$. Show that

$$\frac{1}{2\pi} \int_0^{2\pi} |f(z_0 + \rho e^{it})|^2\, dt = \sum_{n=-\infty}^{\infty} \rho^{2n} |c_n|^2$$

for $r < \rho < R$

4.13. Prove the homology version of the "Cauchy Integral Formula for derivatives": Under the same hypotheses as Theorem 4.9 we have

$$\frac{n!}{2\pi i} \int_\Gamma \frac{f(w)\, dw}{(w - z)^{n+1}} = \text{Ind}(\Gamma, z) f^{(n)}(z)$$

for $n = 0, 1, \ldots$. (Note that this follows in just "one line" from one of our theorems together with one of the previous exercises.)

4.14. Show that Exercise 4.4 follows immediately from Proposition 4.12.

The next few exercises give an example of using the Residue Theorem (actually just Cauchy's Theorem here) to evaluate an improper integral that is a little more elaborate than Exercise 4.10 above.

4.15. Suppose that f has a simple pole (a pole of order 1) at z_0 and $a < b$. For $r > 0$ define $\gamma_r : [a, b] \to \mathbb{C}$ by $\gamma_r(t) = z_0 + re^{it}$, so that γ_r is an arc on the circle with center z_0 and radius r. Show that

$$\lim_{r \to 0^+} \int_{\gamma_r} f(z)\, dz = (b - a) i \text{Res}(f, z_0).$$

4.16. Suppose that $f_n : [a, b] \to \mathbb{C}$ is continuous for $n = 1, 2, \ldots$, $|f_n(t)| \leq M$ for all n and all t, and suppose that $f_n \to 0$ uniformly on $[a + \delta, b - \delta]$ for every $\delta \in (0, (b-a)/2)$. Show that $\int_a^b f_n(t)\, dt \to 0$ as $n \to \infty$.

Hint: Let $\epsilon > 0$. First choose δ so that something is less than $\epsilon/2$ for all n and then show that there exists N such that something else is less than $\epsilon/2$ for all $n > N$.

4.17. Find
$$\lim_{\delta \to 0^+, R \to \infty} \left(\int_{-R}^{-\delta} + \int_{\delta}^{R} \right) \frac{e^{it}}{t} \, dt.$$

Hint: For $0 < \delta < R < \infty$ let $\gamma = \gamma_1 \dotplus \gamma_2 \dotplus \gamma_3 \dotplus \gamma_4$, where $\gamma_1 = [-R, -\delta]$, $\gamma_2 : [0, \pi] \to \mathbb{C}$ is defined by $\gamma_2(t) = \delta e^{i(\pi-t)}$, $\gamma_3 = [\delta, R]$, and $\gamma_4 : [0, \pi] \to \mathbb{C}$ is defined by $\gamma_4(t) = Re^{it}$. (So γ is the same curve as we used in Exercise 4.10 above, except with a little detour around the origin added.)

Let $f(z) = e^{iz}/z$. Find $\int_\gamma f(z) \, dz$ and apply the previous two exercises to figure out what happens to two pieces of the integral as $\delta \to 0$ and $R \to \infty$.

4.18. Consider the integral
$$\lim_{R \to \infty} \int_{-R}^{R} \frac{\sin(t)}{t} \, dt.$$

(i) Explain why you cannot evaluate this integral using the method we used in the previous exercise.

(ii) Use the *result* of the previous exercise to evaluate the integral.

The next two exercises allow one to give a rigorous proof that various winding numbers are what they are without doing any work at all, in most of the cases where the value of the winding number is supposed to be "obvious":

4.19. Suppose that $\gamma_1 : [a, b] \to \mathbb{C} \setminus \{z_0\}$ is a smooth closed curve and let $\gamma_2(t) = \gamma_1(t) - z_0$. Show that $\text{Ind}(\gamma_1, z_0) = \text{Ind}(\gamma_2, 0)$.

4.20. Suppose that $\gamma : [a, b] \to \mathbb{C} \setminus \{z_0\}$ is a smooth closed curve. Suppose that $a < c < b$. Show that if $\text{Re}\,(\gamma(t)) \geq \text{Re}\,(z_0)$ for all $t \in [a, c]$, $\text{Re}\,(\gamma(t)) \leq \text{Re}\,(z_0)$ for all $t \in [c, b]$, $\text{Im}\,(\gamma(a)) < \text{Im}\,(z_0)$ and $\text{Im}\,(\gamma(c)) > \text{Im}\,(z_0)$ then $\text{Ind}(\gamma, z_0) = 1$. (The previous exercise shows that you can assume that $z_0 = 0$. Now use the existence of a branch of the logarithm in the right half-plane and a different branch of the logarithm in the left half-plane.)

4.21. Suppose that $P(z)$ is a polynomial of degree $n \geq 2$ with n distinct zeroes $z_1, \ldots, z_n$. Explain why it follows that every zero of P is simple, and show that
$$\sum_{j=1}^{n} \frac{1}{P'(z_j)} = 0.$$

(Exercise 4.7(i) shows that $1/P'(z_j)$ is the residue at z_j of ...)

In the next exercise we give a sketch of an alternate proof of Lemma 4.6(i). You should study both proofs and convince yourself that they are really the same proof, or at least based on the same idea, but expressed very differently. The notation in the proof in the exercise is simpler than that in the proof in the text; on the other hand the proof in the exercise is somewhat more abstract, so it is not clear to me which proof you will find easier to understand. (I suspect this may vary from student to student!)

So you should learn both proofs, and then the next time you see an argument that reminds you of the version of the proof you find harder to understand, you should consider rephrasing it in the style of the proof you found more transparent:

4.22. Prove part (i) of Lemma 4.6 as follows: Let A be the set of all $x \in [0,1]$ with the property that there exists a continuous function $\theta : [0,x] \to \mathbb{R}$ such that $\theta(0) = \theta_0$ and

$$\gamma(t) = |\gamma(t)|e^{i\theta(t)} \quad (t \in [0,x]).$$

Show that A is nonempty, closed, and relatively open in $[0,1]$, and conclude that $A = [0,1]$ since $[0,1]$ is connected.

4.23. Derive Lemma 3.13 from Theorem 4.9 as follows: Suppose that $f \in H(A(z_0, 0, R))$ is bounded. Fix $r \in (0, R)$, and show that

$$f(z) = \frac{1}{2\pi i} \int_{\partial D(z_0, r)} \frac{f(w)\,dw}{w-z} - \frac{1}{2\pi i} \int_{\partial D(z_0, \rho)} \frac{f(w)\,dw}{w-z}$$

whenever $0 < \rho < |z - z_0| < r$. Now fix z and consider what happens when ρ tends to 0.

Chapter 5

Counting Zeroes and the Open Mapping Theorem

The Residue Theorem can be used to count the zeroes of a holomorphic function.

Note. In complex analysis the "number of zeroes" of a holomorphic function always means the number of zeroes "with multiplicity", which is to say that a zero of order N counts as N zeroes. (For example, the function $f(z) = z^2$ has two zeroes in the plane, even though $f(z) \ne 0$ for $z \ne 0$.)

Theorem 5.0. *Suppose that $f \in H(V)$, and f is not constant on any component of V. Suppose that γ is a smooth curve in V such that $\mathrm{Ind}(\gamma, z)$ is either 0 or 1 for all $z \in \mathbb{C} \setminus \gamma^*$, and equals 0 for all $z \in \mathbb{C} \setminus V$. Suppose that f has no zero on γ^*, and let $\Omega = \{\, z \in V : \mathrm{Ind}(\gamma, z) = 1 \,\}$. Then the number of zeroes of f in Ω is given by*

$$\frac{1}{2\pi i} \int_\gamma \frac{f'(z)}{f(z)} \, dz.$$

Proof. Let $S = \{\, z \in V : f(z) = 0 \,\}$ and define $g \in H(V \setminus S)$ by $g(z) = f'(z)/f(z)$. We know that S is a relatively closed subset of V with no limit point in V, so the Residue Theorem shows that

$$\frac{1}{2\pi i} \int_\gamma g(z) \, dz = \sum_{z \in \Omega \cap S} \mathrm{Res}(g, z),$$

79

and we are done if we can show that $\text{Res}(g, z) = n$ whenever f has a zero of order n at z. But this is clear: If f has a zero of order n at z then $f(w) = (w-z)^n h(z)$, where $h \in H(V)$ is nonzero at z. Hence

$$g(w) = \frac{f'(w)}{f(w)} = \frac{n}{w-z} + \frac{h'(w)}{h(w)},$$

and, since h is nonzero near z, it follows that h'/h is holomorphic near z, so that the first term above, $\frac{n}{w-z}$, must be the principal part of g at z. □

This result has a fascinating geometric interpretation:

Theorem 5.1 (The Argument Principle). *Suppose that f, γ, etc. are as in Theorem 5.0. Define a curve $\tilde{\gamma}$ by $\tilde{\gamma}(t) = f(\gamma(t))$. Then the number of zeroes of f in Ω is equal to*

$$\text{Ind}(\tilde{\gamma}, 0).$$

Proof. This is immediate from the integral formula for $\text{Ind}(\tilde{\gamma}, 0)$: If $\gamma : [a, b] \to \mathbb{C}$ then

$$\text{Ind}(\tilde{\gamma}, 0) = \frac{1}{2\pi i} \int_{\tilde{\gamma}} \frac{dz}{z} = \frac{1}{2\pi i} \int_a^b \frac{\tilde{\gamma}'(t)}{\tilde{\gamma}(t)} dt$$

$$= \frac{1}{2\pi i} \int_a^b \frac{f'(\gamma(t))\gamma'(t)}{f(\gamma(t))} dt = \frac{1}{2\pi i} \int_\gamma \frac{f'(z)}{f(z)} dz,$$

and Theorem 5.0 shows that the last integral is equal to the number of zeroes of f in Ω. □

In other words, the number of zeroes of f "inside" γ is the same as the number of times the curve $f \circ \gamma$ winds around the origin. (This is called "the argument principle" because $\text{Ind}(f \circ \gamma, 0)$ is the net increase in the argument of $f(z)$ as z traverses γ.)

We know that two curves which are homotopic in $\mathbb{C} \setminus \{0\}$ have the same winding number about the origin. It seems clear that two curves which are very close to each other will in fact be homotopic — we now consider the question of how close is close here:

Lemma 5.2. *If $z, w \in \mathbb{C}$ and $|z - w| < |z| + |w|$ then 0 does not lie on the line segment joining z and w.*

5. Counting Zeroes and the Open Mapping Theorem 81

Proof. Suppose that $0 \in [z, w]^*$. This says that

$$0 = tz + (1-t)w$$

for some $t \in [0,1]$. We will show that $|z - w| = |z| + |w|$.

If $t = 0$ then $w = 0$ and hence $|z - w| = |z| + |w|$, so we may assume that $t \neq 0$. It follows that $z = ((t-1)/t)w$, so that

$$|z| = \left|\frac{t-1}{t}\right| |w| = \frac{1-t}{t}|w|.$$

Since $z - w = -w/t$, we see that

$$|z - w| = \frac{1}{t}|w| = |w| + \frac{1-t}{t}|w| = |w| + |z|. \qquad \square$$

Proposition 5.3. *Suppose γ_0 and γ_1 are two closed curves in $\mathbb{C} \setminus \{0\}$ with the same parameter interval. If*

$$|\gamma_1(t) - \gamma_0(t)| < |\gamma_1(t)| + |\gamma_0(t)|$$

for all t then γ_0 and γ_1 are homotopic in $\mathbb{C} \setminus \{0\}$, and in particular we have

$$\mathrm{Ind}(\gamma_0, 0) = \mathrm{Ind}(\gamma_1, 0).$$

Proof. Define a family of curves $(\gamma_s)_{0 \leq s \leq 1}$ by

$$\gamma_s(t) = s\gamma_1(t) + (1-s)\gamma_0(t).$$

Lemma 5.2 shows that $\gamma_s(t) \neq 0$; hence γ_0 and γ_1 are homotopic in $\mathbb{C} \setminus \{0\}$. The fact that γ_0 and γ_1 have the same index about the origin now follows from Proposition 4.12. $\qquad \square$

Proposition 5.3 and the Argument Principle immediately imply Rouché's Theorem, which says that two holomorphic functions which are sufficiently close to each other on the boundary of a region must have the same number of zeroes in the interior:

Theorem 5.4 (Rouché's Theorem). *Suppose that γ is a smooth closed curve in the open set V such that $\mathrm{Ind}(\gamma, z)$ is either 0 or 1 for all $z \in \mathbb{C} \setminus \gamma^*$ and equals 0 for all $z \in \mathbb{C} \setminus V$, and let $\Omega = \{z \in V : \mathrm{Ind}(\gamma, z) = 1\}$. If $f, g \in H(V)$ and*

$$|f(z) - g(z)| < |f(z)| + |g(z)|$$

for all $z \in \gamma^*$ then f and g have the same number of zeroes in Ω.

Proof. Define two curves $\tilde\gamma_f$ and $\tilde\gamma_g$ by $\tilde\gamma_f = f\circ\gamma$ and $\tilde\gamma_g = g\circ\gamma$. Proposition 5.3 shows that
$$\mathrm{Ind}(\tilde\gamma_f, 0) = \mathrm{Ind}(\tilde\gamma_g, 0),$$
and now Theorem 5.1 shows that f and g have the same number of zeroes in Ω. □

See Exercise 5.4 below for another proof of the theorem.

Traditional versions of Rouché's Theorem assume the stronger hypothesis
$$|f(z) - g(z)| < |f(z)|$$
for all $z \in \gamma^*$; as far as I am aware the version above first appeared in [**E**].

The traditional version of Rouché's Theorem is sufficient for most traditional applications of the theorem (which is no surprise if you think about it). This raises the question of whether Theorem 5.4 is good for anything that cannot be done using the traditional version. The following exercise gives an example of something which is at least much simpler using the new version:

- **Exercise 5.1.** Suppose that n is a positive integer and show that the polynomial $z^n(z-2) - 1$ has exactly n roots in the open unit disk $D(0, 1)$.

Just to demonstrate that the example in the exercise was not contrived solely for the sake of illustrating the superiority of the newer version of Rouché's Theorem, we mention that the exercise shows that there exist "Pisot numbers" larger than 2 but arbitrarily close to 2. We are not going to say what that means here; if you are curious see Salem [**S**].

Salem says that the exercise follows from (the traditional version of) Rouché's Theorem "with a little care". It really is much simpler using Theorem 5.4: You have to show only that
$$1 < |z^n(z-2)| + |z^n(z-2) - 1|$$
for all z with $|z| = 1$, which is not so hard (note that in fact $1 < |z - 2|$ at most points of the unit circle ...). The obvious way to do it using the traditional Rouché's Theorem would be to show that
$$1 < |z^n(z-2)|$$
for all z with $|z| = 1$, but that is clearly false (consider $z = 1$). Or you could try to show that
$$1 < |z^n(z-2) - 1|$$

for all z with $|z|=1$, but that is false as well, at least for $n \geq 3$ (this is not quite so obvious; let $z = e^{i\pi/n}$).

In addition to concrete applications to specific functions, as in the exercise above and Exercise 5.3 below, Rouché's Theorem is also very important theoretically; for example it immediately implies a result known as the Open Mapping Theorem. In general a function f is an *open mapping* if $f(S)$ is open for every open subset S of the domain of f.

Theorem 5.5 (The Open Mapping Theorem). *If V is a connected open subset of the plane and $f \in H(V)$ is nonconstant then f is an open mapping.*

Proof. Suppose that $S \subset V$ is open and $z_0 \in S$. Let $g(z) = f(z) - f(z_0)$. Choose $r > 0$ so that $\overline{D}(z_0, r) \subset S$ and such that g has no zero on the boundary of $D(z_0, r)$. Let $\gamma = \partial D(z_0, r)$, and set $\delta = \inf\{|g(z)| : z \in \gamma^*\}$ (note that $\delta > 0$).

Suppose that $h \in \mathbb{C}$ and $|h| < \delta$. Then

$$|(g(z) - h) - g(z)| < \delta \leq |g(z)|$$

for all $z \in \gamma^*$, so Rouché's Theorem implies that g and $g - h$ have the same number of zeroes in $D(z_0, r)$. In particular $g - h$ has at least one zero there, which says that there exists $z \in D(z_0, r)$ with $f(z) = f(z_0) + h$. Since this holds for every $h \in D(0, \delta)$, it follows that

$$D(f(z_0), \delta) \subset f(D(z_0, r)) \subset S;$$

hence $f(S)$ is open. □

Saying that f is an open mapping says that the range of f must contain various disks. We should note for the record that the proof says something about the *size* of these disks under certain circumstances:

Theorem 5.6 (The Open Mapping Theorem, explicit version). *Suppose that f is holomorphic in a neighborhood of $\overline{D}(z_0, r)$ and*

$$|f(z_0 + re^{it}) - f(z_0)| \geq \delta > 0$$

for all $t \in \mathbb{R}$. Then

$$D(f(z_0), \delta) \subset f(D(z_0, r)).$$

Proof. This follows from the proof of Theorem 5.5. □

This raises the question of how one might know that $|f(z_0 + re^{it}) - f(z_0)| \geq \delta > 0$ for all $t \in \mathbb{R}$. One can sometimes use calculus to obtain this sort of information:

Theorem 5.7 (The Open Mapping Theorem, with bounds). *Suppose that $f \in H(V)$ and $|f''(z)| \leq B$ for all $z \in V$. Suppose that $z_0 \in V$ and $f'(z_0) \neq 0$. If $r > 0$ is small enough that $\overline{D}(z_0, r) \subset V$ and also $r < \frac{2|f'(z_0)|}{B}$ then*
$$D(f(z_0), \delta) \subset f(D(z_0, r))$$
for
$$\delta = r|f'(z_0)| - \frac{Br^2}{2}.$$

The idea of the proof is quite simple: An upper bound on $|f''|$ tells us how close to constant f' must be, and hence how close $f(z)$ must be to $f(z_0) + f'(z_0)(z - z_0)$.

Proof. Since $|f''(z)| \leq B$, we must have
$$|f'(z) - f'(z_0)| \leq |z - z_0|B$$
for all $z \in D(z_0, r)$. But
$$f(z) - f(z_0) - f'(z_0)(z - z_0) = \int_{[z_0, z]} (f'(w) - f'(z_0))\, dw,$$
so that
$$|f(z) - f(z_0) - f'(z_0)(z - z_0)| \leq \int_0^{|z-z_0|} tB\, dt = B\frac{|z - z_0|^2}{2},$$
and hence
$$|f(z_0 + re^{it}) - f(z_0)| \geq r|f'(z_0)| - \frac{Br^2}{2} = \delta.$$
Now Theorem 5.6 shows that
$$D(f(z_0), \delta) \subset f(D(z_0, r))$$
(note that the hypothesis $r < \frac{2|f'(z_0)|}{B}$ implies $\delta > 0$). □

You should note that the formula for δ in the statement of the theorem "looks right": If f'' were identically 0 then we would have $f(z) = f(z_0) + f'(z_0)(z - z_0)$, and hence we would have $D(f(z_0), d) = f(D(z_0, r))$ for $d = r|f'(z_0)|$. Now when f'' does not vanish identically we should expect to get a result somewhat weaker than this, but it should be asymptotically the same as $r \to 0$. And in fact our δ is a little smaller than d, but not by much — δ is the same as d plus a higher-order term, controlled by the size of f''.

Exercises

Recall that $\mathbb{D} = D(0,1)$, the open unit disk.

5.2. Suppose that $f \in H(\mathbb{D})$ and let $u = \operatorname{Re} f$, $v = \operatorname{Im} f$.

(i) Show that if $|u| + |v| = 1$ at every point of $\mathbb{D}$ then f is constant.

(ii) Show that if $|u| + |v| \leq 1$ at every point of $\mathbb{D}$ then either f is constant or $|u| + |v| < 1$ everywhere.

5.3. Determine the number of zeroes of $f(z) = 1 + 6z^3 + 3z^{10} + z^{11}$ in the annulus $A(0, 1, 2)$.

Hint: $A(0, 1, 2) = D(0, 2) \setminus (D(0, 1) \cup \partial D(0, 1))$.

5.4. Give another proof of Rouché's Theorem (Theorem 5.4) as follows: Let $W = \mathbb{C} \setminus (-\infty, 0]$ be the plane with the non-negative real axis removed. Show that there exists a branch of the logarithm $L \in H(W)$. (You should write down an explicit formula for L, and also note that the existence of a branch of the logarithm in W follows from one of the general results in Chapter 4.) Now show that there exists an open set O with $\gamma^* \subset O$ such that $f(z)/g(z) \in W$ for all $z \in O$; hence $F = L \circ (f/g) \in H(O)$. Theorem 4.0 shows that $\int_\gamma F'(z)\, dz = 0$; explain how the result follows.

Give an analogous proof for Proposition 5.3.

5.5. Suppose that $f : \overline{\mathbb{D}} \to \overline{\mathbb{D}}$ is continuous and f is holomorphic in $\mathbb{D}$. Show that f has a fixed point in $\overline{\mathbb{D}}$.

Hint: If there exists $r < 1$ such that $|f(z)| \leq r$ for all $z \in \overline{\mathbb{D}}$ then it is easy to show that $f(z) - z$ has a zero in $\mathbb{D}$. In general, let $f_r(z) = rf(z)$; then f_r has a fixed point for every $r \in [0, 1)$. Now let $r \to 1$.

5.6. Prove Hurwitz's Theorem: Suppose that D is an open set, $f_n \in H(D)$, $f_n \to f$ uniformly on compact subsets of D, $\overline{D(z, r)} \subset D$, and f has no zero on $\partial D(z, r)$. Then there exists N such that f_n and f have the same number of zeroes in $D(z, r)$ for all $n > N$.

In particular, if D is connected, $f_n \in H(D)$, $f_n \to f$ uniformly on compact sets, and f_n has no zero then either f has no zero or f vanishes identically.

5.7. Suppose that D is a connected open set, $f_n \in H(D)$, and $f_n \to f$ uniformly on compact subsets of D. If f is nonconstant and $z \in D$ then there exists N and a sequence $z_n \to z$ such that $f_n(z_n) = f(z)$ for all $n > N$.

Hint: Assume that $f(z) = 0$. Apply the previous exercise in the disk $D(z, r_j)$ for a suitable sequence $r_j \to 0$.

Chapter 6

Euler's Formula for sin(z)

6.0. Motivation

In this chapter we take a little break and show how the complex analysis we have learned so far can be used to prove Euler's infinite-product representation of the sine function:

$$\sin(\pi z) = \pi z \prod_{n=1}^{\infty} \left(1 - \frac{z^2}{n^2}\right).$$

We should note that this is a special case of general results about factorizations of entire functions to be discussed later; here we give two *ad hoc* proofs, one using the Residue Theorem and one by Liouville's Theorem.

How could someone come up with a formula like that in the first place? A person might start with the observation that $\sin(\pi z)$ vanishes precisely at the integers, so it could be that a formula something like

$$\sin(\pi z) = \prod_{n=-\infty}^{\infty} (z - n)$$

works; the product on the right apparently vanishes at the integers, so if sine were a polynomial this would be at worst off by a constant factor.

Unfortunately a person realizes very quickly that the product $\prod_{n=-\infty}^{\infty} (z - n)$ simply does not converge; the factors do not even tend to 1, which seems a likely prerequisite for convergence of an infinite product. We could try to fix this this by multiplying each factor by some constant to make the product converge *somewhere*. The simplest approach would be to multiply each factor (or each factor with $n \neq 0$) by a constant so as to make the factor equal 1 at the origin. So we want to replace the factor $(z - n)$ with $(z - n)/(0 - n)$, or $1 - z/n$. That gives us a tentative

$$\sin(\pi z) = \left(\prod_{n=-\infty}^{-1} \left(1 - \frac{z}{n}\right) \right)(z) \left(\prod_{n=1}^{\infty} \left(1 - \frac{z}{n}\right) \right).$$

It is *possible* that that is within a constant factor of being correct. If so then what would the constant be? If you divide both sides by z and then let $z \to 0$, you get $\pi = 1$. So we fix that:

$$\sin(\pi z) = \left(\prod_{n=-\infty}^{-1} \left(1 - \frac{z}{n}\right) \right)(\pi z) \left(\prod_{n=1}^{\infty} \left(1 - \frac{z}{n}\right) \right).$$

This is actually correct. Almost. The products still do not converge. Whether an infinite product converges has to do with how fast the factors tend to 1, and (as we will see below) the fact that the series $\sum_{n=1}^{\infty} 1/n$ diverges implies that the product $\prod_{n=1}^{\infty}(1 - z/n)$ diverges. But it turns out that this product converges if we modify it by grouping certain pairs of factors together: $(1 - z/n)$ and $(1 + z/n)$ are both too far from 1 for convergence, but their product is $(1 - z^2/n^2)$, which is much closer to 1. And so we combine pairs of factors to get

$$\sin(\pi z) = \pi z \prod_{n=1}^{\infty} \left(1 - \frac{z^2}{n^2}\right),$$

which turns out to be correct.

Of course we have not yet given much indication why this product should actually equal $\sin(\pi z)$. It is not hard to see that the product converges to an entire function, and if we define

$$P(z) = \pi z \prod_{n=1}^{\infty} \left(1 - \frac{z^2}{n^2}\right)$$

then it is not hard to show that $P(z) = 0$ if and only if $z \in \mathbb{Z}$, and one can even show without much trouble that $P(z+1) = -P(z)$, but $\sin(\pi z)$ is not the only function with all these properties (for example if g is entire then the function $\sin(\pi z) e^{g(\sin^2(\pi z))}$ has all the properties listed). But it seems that it could be so; now we just have to prove it. There are various ways to do that.

6.1. Proof by the Residue Theorem

One might imagine proving that Euler's infinite product equals $\sin(\pi z)$ by taking the logarithm of both sides — we need only show that

$$\log(\sin(\pi z)) = \log(\pi z) + \sum_{n=1}^{\infty} \log\left(1 - \frac{z^2}{n^2}\right).$$

Of course this cannot quite work because these functions do not quite *have* ("single-valued") logarithms; at the very least we would need to be very careful about saying exactly what branches of the logarithm we are using (note that the factors in the product have zeroes, and we are not going to be able to define branches of the logarithm holomorphic near those zeroes ...). This begins to seem somewhat tricky.

Before abandoning this idea we might note that although most holomorphic functions do not have (holomorphic) logarithms, and in particular these ones do not, the nonexistent logarithm of a holomorphic function *does* have a derivative! Seriously: Two branches of $\log(f)$ differ by a constant, and that constant goes away when we differentiate. So, in spite of our problems with interpreting the previous formula, if we differentiate it we get something that might make sense, might be true, and might then lead to Euler's formula for $\sin(\pi z)$. Noting that the derivative of $\log(f)$ is f'/f, we take the derivative of all the terms in the last formula and we get

$$\pi \cot(\pi z) = \frac{1}{z} + \sum_{n=1}^{\infty} \frac{-2z/n^2}{1 - z^2/n^2} = \frac{1}{z} + \sum_{n=1}^{\infty} \left(\frac{1}{z-n} + \frac{1}{z+n}\right).$$

Could this be right? Yes: Note that the sum on the right has poles at the integers and residue 1 at each pole, just like the function $\pi \cot(\pi z)$.

The actual proof starts here: We will show using the Residue Theorem that the function $\pi \cot(\pi z)$ is given by the infinite sum above, and we will deduce Euler's formula for $\sin(\pi z)$ (after a few preliminaries about infinite products and the derivatives of nonexistent logarithms).

The infinite-sum representation for the cotangent follows from a Cauchy's Integral Formula–Residue Theorem hybrid:

Theorem 6.1.0 (Cauchy Integral Formula for functions with simple poles). *Suppose that $V \subset \mathbb{C}$ is open and Γ is a cycle in V such that $\mathrm{Ind}(\Gamma, a) = 0$ for all $a \in \mathbb{C} \setminus V$. Suppose that S is a (relatively) closed subset of V, $S \cap \Gamma^* = \emptyset$, and every point of S is isolated. Suppose that $f \in H(V \setminus S)$ and f has a simple pole or removable singularity at every point of S. Then for $z \in V \setminus (S \cup \Gamma^*)$ we have*

$$\mathrm{Ind}(\Gamma, z) f(z) = \frac{1}{2\pi i} \int_{\Gamma} \frac{f(w)}{w - z} \, dw + \sum_{p \in S} \frac{\mathrm{Ind}(\Gamma, p) \mathrm{Res}(f, p)}{z - p}.$$

(Note that the sum has only finitely many nonzero terms, as in Theorem 4.17. Note also that the validity of the theorem depends on the fact that f has at worst a *simple* pole at each point of S.)

Proof. Fix $z \in V \setminus (S \cup \Gamma^*)$ and let

$$g(w) = \frac{f(w)}{w - z}.$$

Now g is holomorphic in V except for (possible) simple poles at z and the points of S. Exercise 4.7(ii) shows that $\operatorname{Res}(g, z) = f(z)$ and $\operatorname{Res}(g, p) = \operatorname{Res}(f, p)/(p - z)$ for $p \in S$. So the Residue Theorem (Theorem 4.17) shows that

$$\frac{1}{2\pi i} \int_\Gamma g(w)\, dw = \operatorname{Ind}(\Gamma, z)\operatorname{Res}(g, z) + \sum_{p \in S} \operatorname{Ind}(\Gamma, p)\operatorname{Res}(g, p)$$

$$= \operatorname{Ind}(\Gamma, z) f(z) + \sum_{p \in S} \frac{\operatorname{Ind}(\Gamma, p)\operatorname{Res}(f, p)}{p - z}. \qquad \square$$

For our application we need to know that the cotangent function remains bounded if we stay away from its poles:

Lemma 6.1.1. *There exists a constant M such that*

$$|\cot(\pi z)| \leq M$$

whenever $|\operatorname{Im}(z)| \geq 1$ or $\operatorname{Re}(z) = n + 1/2$ $(n \in \mathbb{Z})$.

Proof. If $z = x + iy$ then

$$\cot(\pi z) = i\frac{e^{-2\pi y}e^{2\pi i x} + 1}{e^{-2\pi y}e^{2\pi i x} - 1} = i\frac{1 + e^{2\pi y}e^{-2\pi i x}}{1 - e^{2\pi y}e^{-2\pi i x}}.$$

The first expression shows that $\cot(\pi z) \to -i$ uniformly in x as $y \to \infty$, while the second shows that $\cot(\pi z) \to i$ uniformly in x as $y \to -\infty$. So there exists a number A such that

$$|\cot(\pi z)| \leq 2 \quad (|y| \geq A).$$

Let

$$K = \{\, x + iy : 0 \leq x \leq 1, 1 \leq |y| \leq A \,\} \cup \{\, \tfrac{1}{2} + iy : |y| \leq 1 \,\}.$$

Since K is compact and the function $\cot(\pi z)$ is continuous on K, there exists $M \geq 2$ such that $|\cot(\pi z)| \leq M$ for $z \in K$. Since $\cot(\pi(z+1)) = \cot(\pi z)$, it follows that $|\cot(\pi z)| \leq M$ for $z \in \bigcup_{n \in \mathbb{Z}}(n + K)$ and hence for $z \in \{\, x + iy : |y| \geq A \,\} \cup \bigcup_{n \in \mathbb{Z}}(n + K)$. $\qquad \square$

6.1. Proof by the Residue Theorem

Definition. If $z \in \mathbb{C}$ and $s > 0$ then the boundary of the square with lower-left corner z and side length s is the cycle
$$[z, z+s] \dot{+} [z+s, z+s+is] \dot{+} [z+s+is, z+is] \dot{+} [z+is, z].$$

Theorem 6.1.2. *If $z \in \mathbb{C} \setminus \mathbb{Z}$ then*
$$\pi \cot(\pi z) = \frac{1}{z} + \sum_{n=1}^{\infty} \left(\frac{1}{z-n} + \frac{1}{z+n} \right).$$

Proof. For N a positive integer let Q_N be the boundary of the square with lower-left corner $-(N+\frac{1}{2}) - i(N+\frac{1}{2})$ and side length $2N+1$. An application of Theorem 6.1.0 shows that if $z \in \mathbb{C} \setminus \mathbb{Z}$, $|\operatorname{Re}(z)| < N + \frac{1}{2}$ and $|\operatorname{Im}(z)| < N + \frac{1}{2}$ then
$$\pi \cot(\pi z) = \sum_{n=-N}^{N} \frac{1}{z-n} + \frac{1}{2\pi i} \int_{Q_N} \frac{\pi \cot(\pi w)}{w-z} \, dw,$$
so we need only show that
$$\lim_{N \to \infty} \int_{Q_N} \frac{\pi \cot(\pi w)}{w-z} \, dw = 0.$$

Now Lemma 6.1.1 shows that $|\cot(\pi z)| \leq M$ on Q_N^*, but this is not quite enough; a direct application of the ML inequality shows only that the integral is bounded, not that it tends to 0. But the ML inequality is a rather crude device for estimating the size of integrals, because it ignores the fact that the absolute value of the integrand may not be constant and also ignores any possible cancellation. It turns out that our integral is much smaller than the ML inequality would appear to indicate, because of cancellation:

We notice that the integral of an *even* function over Q_N must equal 0. Now, $\cot(\pi w)/(w-z)$ is not an even function of w, but if z is fixed and w is large then it is very close to the even function $\cot(\pi w)/w$. So we estimate our integral by comparing it to the integral of $\cot(\pi w)/w$, as follows:
$$\int_{Q_N} \frac{\cot(\pi w)}{w-z} \, dw = \int_{Q_N} \cot(\pi w) \left(\frac{1}{w-z} - \frac{1}{w} \right) dw = z \int_{Q_N} \frac{\cot(\pi w)}{w(w-z)} \, dw.$$

The ML inequality suffices to show that this last integral tends to 0:

Note that $|w| > N$ for $w \in Q_N^*$. Hence if $N > 2|z|$ we have $|w-z| > N/2$ and hence
$$\left| \frac{1}{w(w-z)} \right| < \frac{2}{N^2}$$

for $w \in Q_N^*$. So the ML inequality and Lemma 6.1.1 show that

$$\left| \int_{Q_N} \frac{\cot(\pi w)}{w(w-z)} \, dw \right| < \frac{2M}{N^2}(8N+4)$$

for $N > 2|z|$; hence

$$\lim_{N \to \infty} \int_{Q_N} \frac{\cot(\pi w)}{w(w-z)} \, dw = 0. \qquad \square$$

Note that the proof shows that

$$\pi \cot(\pi z) = \lim_{N \to \infty} \sum_{n=-N}^{N} \frac{1}{z-n}.$$

It would be a bad idea to rewrite this as

$$\pi \cot(\pi z) = \sum_{n=-\infty}^{\infty} \frac{1}{z-n},$$

because that might be taken to mean

$$\pi \cot(\pi z) = \lim_{N,M \to \infty} \sum_{n=-M}^{N} \frac{1}{z-n},$$

and this last equation is not correct, because the indicated limit does not exist!

- **Exercise 6.1.** Show that the series

$$\sum_{n=0}^{\infty} \frac{1}{z-n}$$

diverges for all $z \in \mathbb{C} \setminus \mathbb{Z}$.

Hint: If $\sum a_n$ diverges and $\sum (a_n - b_n)$ converges then $\sum b_n$ diverges.

- **Exercise 6.2.** Show that the series

$$\sum_{n=1}^{\infty} \left(\frac{1}{z-n} + \frac{1}{z+n} \right)$$

converges absolutely for all $z \in \mathbb{C} \setminus \mathbb{Z}$.

We need to say a little bit about infinite products. The main technical detail follows:

6.1. Proof by the Residue Theorem

Lemma 6.1.3. *If* $z_1, \ldots, z_n \in \mathbb{C}$ *and* $\sum_{j=1}^{n-1} |1 - z_j| < 1/2$ *then*

$$\left| 1 - \prod_{j=1}^{n} z_j \right| \leq 2 \sum_{j=1}^{n} |1 - z_j|.$$

Proof. The proof is by induction on n; the case $n = 1$ is clear. (By convention a sum of the form $\sum_{j=1}^{0}$ is equal to 0, being the sum of no terms.)

Suppose we know that $\left| 1 - \prod_{j=1}^{n} z_j \right| \leq 2 \sum_{j=1}^{n} |1 - z_j|$ whenever $\sum_{j=1}^{n-1} |1 - z_j| < 1/2$ and suppose as well that $\sum_{j=1}^{n} |1 - z_j| < 1/2$. Then

$$\left| \prod_{j=1}^{n} z_j \right| \leq 1 + \left| 1 - \prod_{j=1}^{n} z_j \right| \leq 1 + 2 \sum_{j=1}^{n} |1 - z_j| \leq 2,$$

and so

$$\left| 1 - \prod_{j=1}^{n+1} z_j \right| = \left| 1 - \prod_{j=1}^{n} z_j + \prod_{j=1}^{n} z_j - \prod_{j=1}^{n+1} z_j \right|$$

$$\leq \left| 1 - \prod_{j=1}^{n} z_j \right| + \left| \prod_{j=1}^{n} z_j \right| |1 - z_{n+1}|$$

$$\leq 2 \sum_{j=1}^{n} |1 - z_j| + 2|1 - z_{n+1}|$$

$$= 2 \sum_{j=1}^{n+1} |1 - z_j|. \qquad \square$$

Lemma 6.1.4. (i) *If* $z_1, \ldots \in \mathbb{C}$ *and* $\sum_{j=1}^{\infty} |1 - z_j| < \infty$ *then*

$$\prod_{j=1}^{\infty} z_j = \lim_{n \to \infty} \prod_{j=1}^{n} z_j$$

exists; furthermore, $\prod_{j=1}^{\infty} z_j \neq 0$ *unless* $z_j = 0$ *for some* j.

(ii) *If* (f_j) *is a sequence of complex-valued functions on some set* S *and the sum* $\sum_{j=1}^{\infty} |1 - f_j|$ *converges uniformly on* S *then* $P_n = \prod_{j=1}^{n} f_j$ *tends to* $P = \prod_{j=1}^{\infty} f_j$ *uniformly on* S; *if* $z \in S$ *then* $P(z) \neq 0$ *unless* $f_j(z) = 0$ *for some* j.

Note. It is possible for an infinite product to equal 0 even if none of the factors vanish (consider $\prod_{j=1}^{\infty} 1/2$). Some authors say that such a sum "diverges to 0". (And other authors say it converges to 0; for this reason we will try to avoid saying that an infinite product converges or diverges.)

Proof. (i) Choose N so that $\sum_{j=N+1}^{\infty} |1 - z_j| < 1/2$. To show that $\lim_{n \to \infty} \prod_{j=1}^{n} z_j$ exists it is sufficient to show that (p_n) is a Cauchy sequence, where

$$p_n = \prod_{j=N+1}^{N+n} z_j.$$

But Lemma 6.1.3 shows that $|p_n| \leq 2$, and hence another application of the lemma shows that if $n > m$ then

$$|p_n - p_m| = |p_m| \left| 1 - \prod_{j=N+m+1}^{N+n} z_j \right| \leq 4 \sum_{j=N+m+1}^{N+n} |1 - z_j|;$$

hence $|p_n - p_m| \to 0$ as $n, m \to \infty$.

If none of the z_j vanish then $\prod_{j=1}^{N} z_j \neq 0$, and Lemma 6.1.4 shows that $|1 - \prod_{j=N+1}^{\infty} z_j| < 1$, so $\prod_{j=N+1}^{\infty} z_j \neq 0$; hence $\prod_{j=1}^{\infty} z_j \neq 0$.

(ii) The proof of the second part of the lemma is the same as the proof of the first part. □

Now the sum

$$\sum_{n=1}^{\infty} \left| \frac{z^2}{n^2} \right|$$

converges uniformly on compact subsets of the plane; thus Lemma 6.1.4(ii) shows that the partial products

$$P_N(z) = \pi z \prod_{n=1}^{N} \left(1 - \frac{z^2}{n^2} \right)$$

converge uniformly on compact subsets of the plane to

$$P(z) = \pi z \prod_{n=1}^{\infty} \left(1 - \frac{z^2}{n^2} \right),$$

an entire function which vanishes only at the integers.

Note that we also have $P'_N \to P'$, by Proposition 3.5.

6.1. Proof by the Residue Theorem

In general, if f is a differentiable function, we define the *logarithmic derivative* $L(f)$ by

$$L(f) = \frac{f'}{f},$$

at least at points where f is nonzero. The logarithmic derivative of f is the derivative of the logarithm of f when f *has* a logarithm; this makes the formula

$$L(fg) = L(f) + L(g)$$

plausible. In fact the product rule shows that $L(fg) = L(f) + L(g)$ for any differentiable nonvanishing functions f and g.

It follows that

$$L\left(\prod_{n=1}^{N} f_n\right) = \sum_{n=1}^{N} L(f_n)$$

on the set where none of the f_n vanish. (This sometimes gives a convenient and efficient way to calculate derivatives of products of several functions, by the way.)

We will use the following continuity property of L:

Proposition 6.1.5. *Suppose that V is a connected open set, $f_1, f_2, \ldots \in H(V)$ and $f_n \to f$ uniformly on compact subsets of V. Suppose that f is not identically zero. Then*

$$L(f_n) \to L(f)$$

uniformly on K, if K is any compact subset of V on which f has no zero.

Proof. This is immediate from Proposition 3.5. □

And we will use the fact that $L(f)$ determines f, at least up to a constant factor:

Lemma 6.1.6. *Suppose that V is a connected open subset of $\mathbb{C}$, f and g are holomorphic functions in V neither of which vanishes identically, and*

$$L(f) = L(g)$$

on the set where neither f nor g vanishes. Then $f = cg$ for some constant c.

Proof. We may suppose that neither f nor g has a zero in V (because the set where they are both nonzero is in general a dense connected open subset of V). Now the fact that $L(f) = L(g)$ shows that $L(f/g) = 0$, and hence $(f/g)' = 0$; thus f/g is constant. □

We can finally prove Euler's formula. Assume first that $z \in \mathbb{C} \setminus \mathbb{Z}$. Note that
$$L\left(1 - \frac{z^2}{n^2}\right) = L\left(1 - \frac{z}{n}\right) + L\left(1 + \frac{z}{n}\right) = \frac{1}{z-n} + \frac{1}{z+n}.$$

If P_N and P are as above then it follows that
$$L(P_N)(z) = \sum_{n=-N}^{N} \frac{1}{z-n},$$
and now Proposition 6.1.5 shows that
$$L(P) = \lim_{N \to \infty} \sum_{n=-N}^{N} \frac{1}{z-n} = \frac{1}{z} + \sum_{n=1}^{\infty} \left(\frac{1}{z-n} + \frac{1}{z+n}\right).$$

But now Theorem 6.1.2 shows that
$$L(P)(z) = \pi \cot(\pi z) = L(\sin(\pi z)),$$
so that
$$P(z) = c \sin(\pi z)$$
by Lemma 6.1.6. Dividing by πz we see that
$$\frac{c \sin(\pi z)}{\pi z} = \prod_{n=1}^{\infty} \left(1 - \frac{z^2}{n^2}\right) \quad (z \neq 0),$$
and letting $z \to 0$ here shows that $c = 1$.

This proves Euler's formula for $z \in \mathbb{C} \setminus \mathbb{Z}$; the case $z \in \mathbb{Z}$ follows by continuity. □

6.2. Estimating Sums Using Integrals

In the second proof of Euler's formula we will need to use an integral to estimate a certain infinite sum. Let us say a little bit about various ways one might do that. In this section we will assume that $\phi : (0, \infty) \to (0, \infty)$ is continuous, and we will consider the question of what we can say about the sum
$$\sum_{n=1}^{\infty} \phi(n)$$
in terms of integrals.

6.2. Estimating Sums Using Integrals

The simplest approach is something found in a lot of calculus books under the heading "Integral Test". Suppose that ϕ is nonincreasing. This shows that
$$\phi(n) = \int_{n-1}^{n} \phi(n)\, dt \leq \int_{n-1}^{n} \phi(t)\, dt,$$
and hence that
$$\sum_{n=1}^{\infty} \phi(n) \leq \int_{0}^{\infty} \phi(t)\, dt.$$
This can sometimes be useful even if it does not appear that way at first. For example, if we want to know that
$$\sum_{n=1}^{\infty} \frac{1}{n^2}$$
converges then a direct application of the inequality above does not help, because $\int_0^\infty dt/t^2 = \infty$. But of course the same argument shows that
$$\sum_{n=1}^{\infty} \phi(n) = \phi(1) + \sum_{n=1}^{\infty} \phi(n) \leq \phi(1) + \int_{1}^{n} \phi(t)\, dt,$$
which shows that $\sum_{n=1}^{\infty} 1/n^2$ converges. Noting that a similar argument shows that
$$\int_{1}^{\infty} \phi(t)\, dt \leq \sum_{n=1}^{\infty} \phi(n)$$
we obtain the "integral test": If $\phi : [1, \infty) \to (0, \infty)$ is continuous and nonincreasing then $\sum_{n=1}^{\infty} \phi(n)$ converges if and only if $\int_1^\infty \phi(t)\, dt < \infty$.

If we just want to check whether a sum converges then that will often suffice. However, sometimes we need more precise information about the *size* of our sum than is given by the integral test. If you think about it for a second, you decide that $\int_{n-1}^{n} \phi(t)\, dt$ is probably a fairly poor approximation to $\phi(n)$; taking the integral from $n - \frac{1}{2}$ to $n + \frac{1}{2}$ should often give a better approximation. Of course now the problem of estimating the error in this approximation arises.

- **Exercise 6.3.** Suppose that ϕ'' is continuous on $[-1/2, 1/2]$.

    (i) Show that
    $$\int_{-1/2}^{1/2} \phi(t)\, dt - \phi(0) = \frac{1}{2} \int_{-1/2}^{1/2} \left(|t| - \frac{1}{2}\right)^2 \phi''(t)\, dt.$$

(ii) Deduce that

$$\left| \int_{-1/2}^{1/2} \phi(t)\, dt - \phi(0) \right| \le \frac{1}{8} \int_{-1/2}^{1/2} |\phi''(t)|\, dt.$$

Hint: For the first part, write

$$\frac{1}{2} \int_{-1/2}^{1/2} \left(|t| - \frac{1}{2}\right)^2 \phi''(t)\, dt = \frac{1}{2} \int_{0}^{1/2} \left(t - \frac{1}{2}\right)^2 (\phi''(t) + \phi''(-t))\, dt$$

and integrate by parts until done.

The exercise shows that

$$\left| \int_{1/2}^{\infty} \phi(t)\, dt - \sum_{n=1}^{\infty} \phi(n) \right| \le \frac{1}{8} \int_{1/2}^{\infty} |\phi''(t)|\, dt.$$

This can be useful, although dealing with the error term $\frac{1}{8} \int_{1/2}^{\infty} |\phi''|$ can be inconvenient. It is too bad that we do not have a nice simple inequality like $\phi(n) \le \int_{n-1/2}^{n+1/2} \phi$, so we could get a bound for our sum in terms of an integral with no error term, as in our first integral test above ...

But wait. If $\phi'' \ge 0$ then the first part of the exercise above shows that we *do* have

$$\phi(n) \le \int_{n-\frac{1}{2}}^{n+\frac{1}{2}} \phi(t)\, dt;$$

we have been forgetting about the basic principle that convexity is where inequalities come from!

Or in any case *I* forgot this in the first version of this proof; maybe some of us have never seen it stated explicitly. Here it is:

Convexity is the source of many useful inequalities.

Now you've seen it; try to remember this the next time you have an inequality you wish you could prove.

We should mention a technicality: By definition the continuous function ϕ is *convex* if

$$\phi(x) \le \frac{\phi(x+t) + \phi(x-t)}{2}$$

for all x, t such that the interval $[x-t, x+t]$ is contained in the domain of ϕ. If it happens that ϕ'' is continuous, *then* ϕ is convex if and only if $\phi'' \ge 0$.

But convexity is actually a weaker condition: The function $\phi(x) = |x|$ shows that a convex function need not have a second derivative.

And in fact convexity is exactly the condition we need here, not $\phi'' \geq 0$; if ϕ is convex then

$$\phi(n) = 2\int_0^{\frac{1}{2}} \phi(n)\,dt \leq 2\int_0^{\frac{1}{2}} \frac{\phi(n+t) + \phi(n-t)}{2}\,dt = \int_{n-\frac{1}{2}}^{n+\frac{1}{2}} \phi(t)\,dt,$$

a much simpler argument than the previous exercise.

Let us summarize the things we have proved here for future reference; they can all be useful in various places:

Theorem 6.2.0. *Suppose that* $\phi : (0, \infty) \to (0, \infty)$ *is continuous.*

(i) *If ϕ is nonincreasing then*

$$\int_1^\infty \phi(t)\,dt \leq \sum_{n=1}^\infty \phi(n) \leq \int_0^\infty \phi(t)\,dt.$$

(ii) *If ϕ'' is continuous then*

$$\left| \int_{1/2}^\infty \phi(t)\,dt - \sum_{n=1}^\infty \phi(n) \right| \leq \frac{1}{8} \int_{1/2}^\infty |\phi''(t)|\,dt.$$

(iii) *If ϕ is convex then*

$$\sum_{n=1}^\infty \phi(n) \leq \int_{1/2}^\infty \phi(t)\,dt.$$

We will see in the next section that the difference between the $\int_0^\infty \phi$ in part (i) and the $\int_{1/2}^\infty \phi$ in part (iii) can make a big difference.

6.3. Proof Using Liouville's Theorem

One can use Liouville's Theorem to prove Euler's formula; we define

$$P(z) = \pi z \prod_{n=1}^\infty \left(1 - \frac{z^2}{n^2}\right),$$

and it follows from Exercise 3.8 in Chapter 3 that we need only show that P is an entire function such that $P(z+1) = -P(z)$ and $|P(z)| \leq ce^{\pi |\operatorname{Im}(z)|}$.

(The exercise then shows that $P(z) = \alpha \sin(\pi z)$ for some constant α; as noted above we can then divide both sides by z and then consider the limit as $z \to 0$ to show that $\alpha = 1$.)

The fact that P is an entire function is proved as in Section 6.1; this follows from Lemma 6.1.4, since the sum $\sum_{n=1}^{\infty} |z^2/n^2|$ converges uniformly for z in any compact subset of the plane. (The fact that Lemma 6.1.4 is used in both proofs is not a surprise; the lemma is simply the basic tool one uses in checking convergence of infinite products.)

The proof that $P(z+1) = -P(z)$ is not hard. Let

$$P_N(z) = z \prod_{n=1}^{N} \left(1 - \frac{z^2}{n^2}\right).$$

A little rearrangement shows that

$$P_N(z) = \frac{(-1)^N}{(N!)^2} \prod_{n=-N}^{N} (z - n).$$

Hence

$$P_N(z+1) = \frac{(-1)^N}{(N!)^2} \prod_{n=-N}^{N} (z - (n-1)) = \frac{(-1)^N}{(N!)^2} \prod_{n=-N-1}^{N-1} (z - n),$$

so that if $z \notin \mathbb{Z}$ we have

$$\frac{P_N(z+1)}{P_N(z)} = \frac{z - (-N-1)}{z - N}.$$

This shows that $P_N(z+1)/P_N(z) \to -1$ and hence $P(z+1) = -P(z)$ for $z \notin \mathbb{Z}$; the fact that $P(z+1) = -P(z)$ for all z follows by continuity.

The interesting part is the proof that $|P(z)| \leq c e^{\pi |\operatorname{Im}(z)|}$. It seems natural to take the logarithm, since sums can be easier to deal with than products. So we begin by noting that

$$\log\left(\left|\prod_{n=1}^{\infty}\left(1 - \frac{z^2}{n^2}\right)\right|\right) = \sum_{n=1}^{\infty} \log\left(\left|1 - \frac{z^2}{n^2}\right|\right) \leq \sum_{n=1}^{\infty} \log\left(1 + \frac{|z|^2}{n^2}\right).$$

(At first it looks like we threw away too much in that last inequality, so that we could not hope to get a good enough bound starting here. But we are only going to be using this inequality for z close to the imaginary axis; for such z the inequality is close to equality.) Now we estimate that sum by comparing it to an integral, as in the previous section:

6.3. Proof Using Liouville's Theorem

Fix $z \neq 0$ and let $\phi(t) = \log(1 + |z|^2/t^2)$. In the first version of the proof I used part (i) of Theorem 6.1.7:

$$\sum_{n=1}^{\infty} \log\left(1 + \frac{|z|^2}{n^2}\right) \leq \int_0^{\infty} \log\left(1 + \frac{|z|^2}{t^2}\right) dt = \pi|z|.$$

(The integral can be evaluated by elementary calculus; see below for details.) This shows that

$$\prod_{n=1}^{\infty} \left(1 + \frac{|z|^2}{n^2}\right) \leq e^{\pi|z|},$$

which looks like what we want. Unfortunately there is a missing factor on the left-hand side; all we have proved here is that

$$|P(z)| \leq \pi|z|e^{\pi|z|}.$$

Now, this last inequality can in fact be used to prove Euler's formula, but the simple Exercise 3.8 does not quite apply, we need a somewhat clumsier argument.

A miracle happened when I realized I should be applying the convexity of ϕ: The improvement in the estimate of the sum by the integral was exactly enough to take care of that extra factor of z!

You can easily verify that $\phi'' \geq 0$. Hence part (iii) of Theorem 6.1.7 shows that

$$\sum_{n=1}^{\infty} \log\left(1 + \frac{|z|^2}{n^2}\right) \leq \int_{1/2}^{\infty} \log\left(1 + \frac{|z|^2}{t^2}\right) dt.$$

Now a simple integration by parts shows that

$$\int \log\left(1 + \frac{|z|^2}{t^2}\right) dt = t\log\left(1 + \frac{|z|^2}{t^2}\right) + 2|z|\arctan\left(\frac{t}{|z|}\right),$$

at least for $t \in (0, \infty)$. We need to figure out how this antiderivative behaves as $t \to +\infty$. Since $\log(1) = 0$ and $\log'(1) = 1$, it follows that $\log(1 + |z|^2/t^2)$ is approximately $|z|^2/t^2$ for large t, so that $t\log(1 + |z|^2/t^2) \to 0$. It is clear that $\arctan(t/|z|) \to \pi/2$, so we obtain

$$\int_{1/2}^{\infty} \log\left(1 + \frac{|z|^2}{t^2}\right) dt = \pi|z| - (\log(1 + 4|z|^2)/2 + 2|z|\arctan(1/(2|z|)))$$

$$\leq \pi|z| - \log(4|z|^2)/2$$
$$= \pi|z| - \log(|z|) - \log(2)$$
$$< \pi|z| - \log(|z|).$$

That $-\log(|z|)$ saves the day, showing that
$$|P(z)| \le \pi e^{\pi|z|}.$$
Since $P(z+2) = P(z)$, it follows (see Exercise 6.4 below) that
$$|P(z)| \le c e^{\pi|\operatorname{Im}(z)|},$$
and this proves Euler's formula, as noted at the start of this section.

Exercises

6.4. Suppose that $P : \mathbb{C} \to \mathbb{C}$ is continuous, $P(z+2) = P(z)$, and $|P(z)| \le e^{\pi|z|}$. Show that there exists c such that $|P(z)| \le c e^{\pi|\operatorname{Im}(z)|}$ for all z.

6.5. Prove Wallis' formula:
$$\frac{\pi}{2} = \prod_{n=1}^{\infty} \left(\frac{4n^2}{(2n-1)(2n+1)} \right) = \left(\frac{2}{1}\right)\left(\frac{2}{3}\right)\left(\frac{4}{3}\right)\left(\frac{4}{5}\right)\cdots.$$

Note. After you prove that $\frac{\pi}{2} = \prod_{n=1}^{\infty} \left(\frac{4n^2}{(2n-1)(2n+1)}\right)$ there is still a *little* bit of work to do in showing that $\prod_{n=1}^{\infty} \left(\frac{4n^2}{(2n-1)(2n+1)}\right) = \left(\frac{2}{1}\right)\left(\frac{2}{3}\right)\left(\frac{4}{3}\right)\left(\frac{4}{5}\right)\cdots$! Yes, it is clear that
$$\frac{4n^2}{(2n-1)(2n+1)} = \frac{2n}{2n-1}\frac{2n}{2n+1},$$
but that is not *quite* enough by itself, there is a little argument required. For example, it is also true that
$$\frac{4n^2}{(2n-1)(2n+1)} = \frac{4n}{2n-1}\frac{n}{2n+1},$$
but it does not follow from that that
$$\frac{\pi}{2} = \left(\frac{4}{1}\right)\left(\frac{1}{3}\right)\left(\frac{8}{3}\right)\left(\frac{2}{5}\right)\cdots;$$
in fact *that* product does not converge.

6.6. Find $\sum_{n=1}^{\infty} \frac{1}{n^2}$, using the infinite series for $\cot(\pi z)$.

The next exercise shows that, at least for continuous functions, the definition of convexity we gave above is equivalent to an apparently stronger condition. (We take the domain to be all of $\mathbb{R}$ just for convenience.)

6.7. Suppose that $\phi : \mathbb{R} \to \mathbb{R}$ is continuous and satisfies
$$\phi(x) \leq \frac{\phi(x+y) + \phi(x-y)}{2}$$
for all $x, y \in \mathbb{R}$. Show that
$$\phi(tx + (1-t)y) \leq t\phi(x) + (1-t)\phi(y)$$
whenever $0 \leq t \leq 1$.

Hint: First note that the hypothesis is equivalent to
$$\phi\left(\frac{x+y}{2}\right) \leq \frac{\phi(x) + \phi(y)}{2},$$
which gives the conclusion for $t = 1/2$. Since ϕ is continuous, you may assume that $t = k/2^n$ for non-negative integers n, k. Now the result follows by induction on n; for example, the identity
$$\frac{\frac{x+y}{2} + y}{2} = \frac{1}{4}x + \frac{3}{4}y$$
shows how the case $t = 1/4$ follows from the case $t = 1/2$:
$$\phi\left(\frac{1}{4}x + \frac{3}{4}y\right) = \phi\left(\frac{\frac{x+y}{2} + y}{2}\right)$$
$$\leq \frac{\phi\left(\frac{x+y}{2}\right) + \phi(y)}{2} \leq \frac{\frac{\phi(x)+\phi(y)}{2} + \phi(y)}{2} = \frac{1}{4}\phi(x) + \frac{3}{4}\phi(y).$$

Chapter 7

Inverses of Holomorphic Maps

Our first result concerning inverses of holomorphic mappings is that an invertible holomorphic mapping must have a nonvanishing derivative. (Note that there is indeed something to be proved here: Consider the map $f : \mathbb{R} \to \mathbb{R}$ defined by $f(x) = x^3$.)

Theorem 7.0. *Suppose that $f \in H(V)$ and f is one-to-one in V. Then f' has no zero in V.*

Proof. Suppose on the other hand that $z_0 \in V$ and $f'(z_0) = 0$. If f' vanishes identically in a neighborhood of z_0 then f is not one-to-one, so we may assume that f' does not vanish identically near z_0. Now, since the zeroes of holomorphic functions are isolated (Theorem 3.10), we may choose $r > 0$ such that $D(z_0, r) \subset V$ and $f'(z) \neq 0$ for $0 < |z - z_0| < r$.

The function $f - f(z_0)$ has at least two zeroes in $D(z_0, r)$, since it has a zero of order at least 2 at z_0. As in the proof of the Open Mapping Theorem (Theorem 5.5), Rouché's Theorem shows that $f - w$ must have at least two zeroes in $D(z_0, r)$ if $|w - f(z_0)| > 0$ is small enough. But all the zeroes of $f - w$ in $D(z_0, r)$ must be simple, since f' has no zero in $D(z_0, r)$ except at z_0; thus $f - w$ actually has at least two *distinct* zeroes and so f is not one-to-one. □

The converse of Theorem 7.0 fails; in fact the exponential function gives an example of a function with nonvanishing derivative which is not one-to-one. But we shall see in Proposition 7.4 below that a holomorphic function with a nonvanishing derivative must be *locally* one-to-one.

7. Inverses of Holomorphic Maps

The simplest ways to show that a holomorphic function is one-to-one in an open set involve convexity:

Lemma 7.1. *Suppose that V is convex and $f \in H(V)$. If K is a closed convex subset of the plane and $f'(z) \in K$ for all $z \in V$ then*
$$\frac{f(z) - f(w)}{z - w} \in K$$
for all $z, w \in V$ with $z \neq w$.

Proof. Since V is convex, we can calculate $f(z) - f(w)$ by integrating the derivative over the line segment $[w, z]$; parametrizing this curve in the obvious way we see that
$$\frac{f(z) - f(w)}{z - w} = \frac{1}{z - w} \int_{[w,z]} f'(\alpha) \, d\alpha = \int_0^1 f'(w + t(z - w)) \, dt.$$

This integral can be approximated by Riemann sums:
$$\int_0^1 f'(w + t(z - w)) \, dt = \lim_{n \to \infty} \frac{1}{n} \sum_{k=1}^n f'\left(w + \frac{k}{n}(z - w)\right).$$

The convexity of K shows that each sum is an element of K; now the fact that K is closed shows that
$$\int_0^1 f'(w + t(z - w)) \, dt \in K. \qquad \square$$

(Another way to state the lemma is this: If V is convex and $f \in H(V)$ then the values of $(f(z) - f(w))/(z - w)$ must lie in the *closed convex hull* of the set $f'(V)$.)

It follows that while a nonvanishing derivative is not enough to make a function invertible, it *is* enough to assume that the values of the derivative lie in a closed convex set not containing the origin:

Proposition 7.2. *Suppose that V is convex and $f \in H(V)$ is nonconstant; let $W = f(V)$ (note that W is open by Theorem 5.5). Suppose that K is a closed convex subset of the plane and*
$$0 \notin K.$$
If $f'(z) \in K$ for all $z \in K$ then it follows that f is one-to-one in V and the inverse $f^{-1} : W \to V$ is holomorphic.

Note that this is a "temporary" proposition; soon we will know that the inverse of *any* invertible holomorphic mapping is holomorphic, making the last line of the conclusion of the lemma appear a bit curious.

7. Inverses of Holomorphic Maps

Proof. We may choose $\delta > 0$ such that $|z| \geq \delta$ for all $z \in K$. It follows from Lemma 7.1 that
$$\left| \frac{f(z) - f(w)}{z - w} \right| \geq \delta$$
and hence
$$|f(z) - f(w)| \geq \delta |z - w|$$
for all $z, w \in V$ with $z \neq w$.

In particular f is one-to-one in V, so that the inverse $f^{-1} : W \to V$ exists, and we need only show that f^{-1} is holomorphic in W.

Fix $\alpha \in W$ and suppose that (α_n) is a sequence in W converging to α. Then $\alpha = f(z)$ and $\alpha_n = f(z_n)$ for unique $z, z_n \in V$. The inequality above shows that
$$|\alpha - \alpha_n| \geq \delta |z - z_n|$$
so that in particular $z_n \to z$. (Showing that $z_n \to z$ is the hard part of the proof in other situations!) It follows that
$$\frac{f^{-1}(\alpha_n) - f^{-1}(\alpha)}{\alpha_n - \alpha} = \frac{z_n - z}{f(z_n) - f(z)} \to \frac{1}{f'(z)}$$
as $n \to \infty$, so that f^{-1} is differentiable at α, and in fact
$$(f^{-1})'(\alpha) = \frac{1}{f'(z)}. \qquad \square$$

The continuity of the inverse function is indeed a crucial step in the proof; see Exercise 7.3 below.

Proposition 7.2 will suffice for our application, but we should note in passing that it is not really necessary to assume that $0 \notin K$ in the proposition; it suffices to assume that f is nonconstant and $0 \notin K^\circ$.

Proposition 7.3. *Suppose V is convex and $f \in H(V)$ is nonconstant. Suppose K is a closed convex set, $0 \notin K^\circ$, and $f'(z) \in K$ for all $z \in V$. Then f is one-to-one in V.*

Proof. If f' is constant then f is one-to-one, so we may assume that f' is nonconstant. Define
$$\Omega = \left\{ \frac{f(z) - f(w)}{z - w} : z, w \in V, z \neq w \right\}.$$

The Open Mapping Theorem shows that Ω is open. (How? Note that this depends on the fact that f' is nonconstant.) Now Lemma 7.1 shows that $\Omega \subset K$. Thus $\Omega \subset K^\circ$; hence $0 \notin \Omega$, so f is one-to-one. $\qquad \square$

For example, if the nonconstant function f is holomorphic in a convex open set and $\operatorname{Re}(f(z)) \geq 0$ then f is one-to-one.

It follows from Proposition 7.2 that a holomorphic function with a nonvanishing derivative is locally one-to-one:

Proposition 7.4. *Suppose that $f \in H(V)$, $z_0 \in V$, and $f'(z_0) \neq 0$. Then there is an open set Ω containing z_0 in which f is one-to-one; furthermore, $(f|_\Omega)^{-1}$ is holomorphic in $f(\Omega)$.*

Proof. Let
$$K = \overline{D}(f'(z_0), |f'(z_0)|/2).$$
Then K is a closed convex set and $0 \notin K$. Since f' is continuous, we can find a convex open set Ω with $z_0 \in \Omega$ and such that $f'(z) \in K$ for all $z \in \Omega$; now Proposition 7.2 shows that f is one-to-one in Ω and $(f|_\Omega)^{-1}$ is holomorphic in $f(\Omega)$. $\square$

If we put the pieces together it turns out that we have shown that the inverse of an invertible holomorphic mapping is holomorphic:

Theorem 7.5. *Suppose that $f \in H(V)$ and f is one-to-one in V; let $W = f(V)$. Then $f^{-1} \in H(W)$ and*
$$(f^{-1})'(w) = \frac{1}{f'(f^{-1}(w))}.$$

Proof. Fix $w_0 \in W$ and let $z_0 = f^{-1}(w_0)$. Theorem 7.0 shows that $f'(z_0) \neq 0$, and now Proposition 7.4 shows that f^{-1} is holomorphic in a neighborhood of w_0. Hence $f^{-1} \in H(W)$; now the formula for the derivative of f^{-1} follows from the chain rule. $\square$

To summarize our story so far: Inverses of invertible holomorphic functions are holomorphic. An invertible holomorphic function must have a nonvanishing derivative; while a holomorphic function with nonvanishing derivative need not be one-to-one, it is nonetheless locally one-to-one (and conditions slightly stronger than saying the derivative does not vanish do suffice to show the function is one-to-one).

It is curious that one can actually give a "formula" for the inverse of an invertible holomorphic function:

Theorem 7.6. *Suppose that $f \in H(V)$ and f is one-to-one in V. Suppose that $\overline{D}(a,r) \subset V$, and let $W = f(D(a,r))$. Then*
$$f^{-1}(\alpha) = \frac{1}{2\pi i} \int_{\partial D(a,r)} \frac{f'(w)w}{f(w) - \alpha}\, dw$$
for all $\alpha \in W$.

7. Inverses of Holomorphic Maps

Proof. Fix $\alpha \in W$ and let $f^{-1}(\alpha) = p$. Let $g(w) = f'(w)w/(f(w) - \alpha)$. Now g is holomorphic in $D(a, r)$ except for a pole at p. Theorem 7.0 shows that $f - \alpha$ has a simple zero at p, so Exercise 4.7 shows that $\text{Res}(g, p) = f'(p)p/f'(p) = p$. Thus

$$p = \text{Ind}(\partial D(a, r), p) \text{Res}(g, p) = \frac{1}{2\pi i} \int_{\partial D(a,r)} g(w)\, dw,$$

by the Residue Theorem. $\square$

- **Exercise 7.1.** Show that Theorem 7.5 follows "immediately" from Theorem 7.6.

- **Exercise 7.2.** Fill in the details in the following alternate (perhaps more intuitive) proof of Theorem 7.6: Let $\Omega = f(V)$. Note that $f^{-1} \in H(\Omega)$. Define a curve $\Gamma : [0, 2\pi] \to \Omega$ by $\Gamma(t) = f(a + re^{it})$. If you apply the definition of the contour integral, you see that

$$\frac{1}{2\pi i} \int_{\partial D(a,r)} \frac{f'(z)z}{f(z) - \alpha}\, dz = \frac{1}{2\pi i} \int_{\Gamma} \frac{f^{-1}(w)}{w - \alpha}\, dw$$

(you should explain this in a little more detail), and the Cauchy Integral Formula shows that

$$\frac{1}{2\pi i} \int_{\Gamma} \frac{f^{-1}(w)}{w - \alpha}\, dw = f^{-1}(\alpha).$$

(What do we need to know about the curve Γ to make this argument correct, and why does Γ satisfy those conditions? There is a very simple answer based on something we proved earlier.)

We finish with a description of the behavior of an analytic function near points where the derivative vanishes (as well as near points where the derivative does not vanish). Theorem 7.7 below may be regarded as a detailed version of the Open Mapping Theorem.

We saw in Chapter 3 that if $f - f(z_0)$ has a zero of order m at z_0 then there exists a sequence (c_n) with $c_m \neq 0$ such that

$$f(z) = f(z_0) + c_m(z - z_0)^m + \sum_{n=m+1}^{\infty} c_n(z - z_0)^n$$

near z_0. This shows that f behaves very much like the function $P(z) = f(z_0) + c_m(z - z_0)^m$ near z_0; for example

$$\lim_{z \to z_0} \frac{|f(z) - P(z)|}{|z - z_0|^m} = 0.$$

This approximation is of course very useful, but it says nothing *a priori* about, for example, the existence of points where f actually *assumes* various values. On the other hand Theorem 7.7, or rather Corollary 7.8, says that there exists a "change of variables" ϕ near z_0 which actually makes $f(\phi(z))$ equal to
$$f(z_0) + z^m.$$

Theorem 7.7. *Suppose that $f \in H(V)$, $z_0 \in V$, and $f - f(z_0)$ has a zero of finite order m at z_0. Then there is an open set Ω with $z_0 \in \Omega$ and a one-to-one function $g \in H(\Omega)$ such that $g(z_0) = 0$, $g(\Omega) = D(0, r)$ for some $r > 0$, and such that*
$$f(z) = f(z_0) + (g(z))^m$$
for all $z \in \Omega$.

(In particular f is exactly m-to-one in $\Omega \setminus \{z_0\}$.)

For example, suppose that
$$f(z) = z^3 + \sum_{n=4}^{\infty} c_n z^n$$
near the origin. Let us compare this with the function
$$F(z) = z^3.$$
It is clear that
$$\{\, z : F(z) \geq 0 \,\} = L_0 \cup L_1 \cup L_2,$$
where $L_k = \{\, re^{2\pi i k/3} : r \geq 0 \,\}$. Arguments as in Chapter 3 show that $f(z)$ is "almost positive" as z tends to the origin along one of the rays L_k; more precisely,
$$f(re^{2\pi i k/3}) = \rho_k(r) e^{i\theta_k(r)},$$
where $\rho_k(r) \geq 0$ and $\theta_k(r) \to 0$ as $r \to 0$. But arguments of this sort say nothing about the existence of points where f is actually non-negative.

On the other hand Corollary 7.8 gives the much stronger result that there are three infinitely differentiable curves $\gamma_1, \gamma_2, \gamma_3$, each with the origin as one endpoint, such that γ_k is *tangent* to L_k at the origin, and such that $f \geq 0$ on γ_k.

Proof of Theorem 7.7. We know that
$$f(z) = f(z_0) + (z - z_0)^m h(z),$$
where $h \in H(V)$ and $h(z_0) \neq 0$. Hence there is an open set Ω containing z_0 in which h has no zero. Replacing Ω by a subset we may assume that

7. Inverses of Holomorphic Maps

Ω is simply connected, so Corollary 4.15 shows that h has a holomorphic logarithm in Ω: There exists $L \in H(\Omega)$ such that $h = e^L$ in Ω.

For $z \in \Omega$ set $g(z) = (z - z_0)e^{L(z)/m}$; then $g \in H(\Omega)$ and $f(z) = f(z_0) + (g(z))^m$ in Ω, as required. One checks that $g'(z_0) \ne 0$; now Proposition 7.4 shows that there is an open set $\tilde{\Omega}$ with $z_0 \in \tilde{\Omega} \subset \Omega$ and such that g is one-to-one in $\tilde{\Omega}$, and again replacing $\tilde{\Omega}$ by a smaller set if necessary we may assume that $g(\tilde{\Omega})$ is a disk centered at the origin. □

And now if we define $\phi : D(0,r) \to \Omega$ by $\phi = g^{-1}$ we obtain the following:

Corollary 7.8. *Suppose that $f \in H(V)$, $z_0 \in V$, and $f - f(z_0)$ has a zero of finite order m at z_0. Then there exist an open set Ω with $z_0 \in \Omega$, a number $r > 0$ and a function $\phi \in H(D(0,r))$ such that ϕ maps $D(0,r)$ bijectively onto Ω, and such that*
$$f \circ \phi(z) = f(z_0) + z^m$$
for all $z \in D(0,r)$.

- **Exercise 7.3.** Suppose that $f \in H(V)$, W is an open set in the plane, and $h : W \to V$ satisfies
$$f(h(z)) = z$$
for all $z \in W$.

  (i) Give an example showing that it does not follow that $h \in H(W)$.

  Hint: You can take W to be an annulus $A(0, 0, 1)$.

  (ii) Suppose in addition that h is continuous and show that $h \in H(W)$.

Comment. Part (ii) is slightly tricky. If $z \in W$ satisfies $f'(h(z)) \ne 0$ then we have seen several times how to show that h is differentiable at z; the problem is at points where $f'(h(z)) = 0$. But it turns out that there are no such points. If we knew that the range $h(W)$ was open then the fact that there are no points where $f'(h(z)) = 0$ would be immediate from Lemma 7.0. (How? And why will this argument fail if $h(W)$ is not open?) But we do not know that $h(W)$ is open. There are at least two ways to take care of this problem:

One way to handle it is to first prove a lemma: If γ_1 is a continuous closed curve in $\mathbb{C} \setminus \{0\}$, m is a positive integer and $\gamma_2(t) = \gamma_1(t)^m$ for all t then $\text{Ind}(\gamma_2, 0) = m \text{Ind}(\gamma_1, 0)$. It follows from this and Theorem 7.7 that in fact $f'(h(z)) \ne 0$ for all $z \in W$. (Of course you need to explain *how* it follows ... Perhaps as a hint we should point out that you will need to prove the suggested lemma for closed curves γ_1 which are merely continuous, possibly not smooth.)

There is another way to solve the problem which is simpler, although I like it less because it relies more on magic so it does not explain as much about what is really going on: Suppose that $f'(h(z)) = 0$. Show carefully that there exists $\delta > 0$ such that $f'(h(w)) \neq 0$ for all $w \in D'(z, \delta)$. Suppose that we have shown that h is differentiable at w for any w with $f'(h(w)) \neq 0$. Now h is continuous in $D(z, \delta)$ and differentiable in the deleted disk $D'(z, \delta)$, and hence h is differentiable at z, because ... (And at this point we have given a solution to the exercise, but probably we should go on to note that, since we have shown that h is differentiable at z, the chain rule implies that $0 = 1$; hence $f'(h(z)) = 0$ was impossible in the first place.)

Chapter 8

Conformal Mappings

8.0. Meromorphic Functions and the Riemann Sphere

Suppose that $V \subset \mathbb{C}$ is open. A *meromorphic function* in V is a function which is "holomorphic in V except for poles". More carefully, we say that f is meromorphic in V if there exists a relatively closed set $S \subset V$ such that every point of S is isolated, $f \in H(V \setminus S)$, and f has a pole or a removable singularity at every point of S.

If V is connected then (with one exception) a meromorphic function in V is the same thing as a holomorphic mapping of V into the so-called "extended plane" $\mathbb{C}_\infty$ (also known as the "Riemann sphere"), which we define as follows:

Set
$$\mathbb{C}_\infty = \mathbb{C} \cup \{\infty\},$$
where "∞" denotes some object which does not happen to be a complex number. Let
$$S_2 = \{\, x \in \mathbb{R}^3 : x_1^2 + x_2^2 + x_3^2 = 1 \,\}$$
be the unit sphere in $\mathbb{R}^3$. We define a map $P : \mathbb{C}_\infty \to S_2$ by declaring that $P(\infty) = (0, 0, 1)$, while if $z \in \mathbb{C}$ then $P(z)$ is the unique element of S_2 which lies on the straight line through $(0, 0, 1)$ and the point $(z, 0) \in \mathbb{R}^3$ (see Figure 8.0). If you do the math it turns out that

$$P(z) = \left(\frac{2x}{|z|^2+1}, \frac{2y}{|z|^2+1}, \frac{|z|^2-1}{|z|^2+1} \right) \qquad (z = x + iy \in \mathbb{C}).$$

(P is known as a "stereographic projection".)

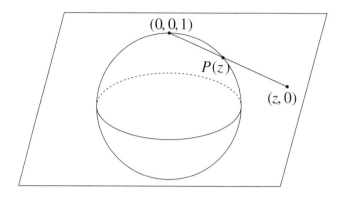

Figure 8.0

The only thing we will use P for is to define a certain metric on $\mathbb{C}_\infty$: We set

$$d_\infty(z,w) = ||P(z) - P(w)|| \qquad (z, w \in \mathbb{C}_\infty).$$

It follows that $(\mathbb{C}_\infty, d_\infty)$ is a compact metric space, since S_2 is a compact subset of $\mathbb{R}^3$ with the usual metric. It is easy to verify that the identity map on $\mathbb{C}$ is continuous when regarded as a map from $\mathbb{C}$ with the usual metric to $(\mathbb{C}, d_\infty)$; the inverse of this map is also continuous. (This says that $(\mathbb{C}_\infty, d_\infty)$ is a "compactification" of $\mathbb{C}$; in fact it is the so-called "one-point compactification".)

We take d_∞ to be the default metric on $\mathbb{C}_\infty$: Statements about convergence in $\mathbb{C}_\infty$ are to be interpreted as statements about convergence with respect to the metric d_∞ unless otherwise specified. We note that the convergent sequences in $\mathbb{C}_\infty$ are as they "should" be: Given a sequence $(z_n)_{n=1}^\infty$ in $\mathbb{C}_\infty$, we have $z_n \to \infty$ if and only if $|z_n| \to \infty$ in the usual sense (with the convention that $|\infty| = \infty$), and we have $z_n \to z \in \mathbb{C}$ if and only if $z_n \in \mathbb{C}$ for all but finitely many n and $|z_n - z| \to 0$.

(In fact we may never mention the metric d_∞ again; the only reason we care about it is that it shows that the natural notion of "$z_n \to z$" for $z_n, z \in \mathbb{C}_\infty$ is equivalent to convergence in some metric, and that $\mathbb{C}_\infty$ is compact in this metric.)

We define $1/0 = \infty$ and $1/\infty = 0$; now the function $z \mapsto 1/z$ is a bijection from $\mathbb{C}_\infty$ to itself.

In fact $\mathbb{C}_\infty$ is a complex manifold in a natural (and standard) way. We are not going to define the term "complex manifold" here; we just note that defining a "structure" making $\mathbb{C}_\infty$ into a complex manifold is equivalent to stating which functions from (subsets of) $\mathbb{C}_\infty$ to $\mathbb{C}_\infty$ are holomorphic, and we *will* do that:

Suppose that V is an open subset of $\mathbb{C}_\infty$ and $f : V \to \mathbb{C}_\infty$. We say that f is holomorphic if f is continuous and each of the four functions $f(z)$, $f(1/z)$, $1/f(z)$, and $1/f(1/z)$ is holomorphic in the usual sense in the (open) subset of $\mathbb{C}$ where it is finite.

(That is, f is holomorphic in $\{z \in \mathbb{C} \cap V : f(z) \in \mathbb{C}\}$, $f(1/z)$ is holomorphic in $\{z \in \mathbb{C} : 1/z \in V, f(1/z) \in \mathbb{C}\}$, $1/f$ is holomorphic in $\{z \in \mathbb{C} \cap V : f(z) \neq 0\}$ (note that this set includes the points $z \in \mathbb{C} \cap V$ where $f(z) = \infty$), and $1/f(1/z)$ is holomorphic in $\{z \subset \mathbb{C} : 1/z \in V, f(1/z) \neq 0\}$.)

A few examples should suffice to make the intent of the definition clear: (i) If in fact $V \subset \mathbb{C}$ and $f(V) \subset \mathbb{C}$ then f is holomorphic as above if and only if it is holomorphic in the usual sense. (ii) If $\infty \in V$ and $f(\infty) \in \mathbb{C}$ then f is holomorphic in a neighborhood of ∞ if and only if $f(1/z)$ is holomorphic in the usual sense in some neighborhood of the origin. (iii) If $z_0 \in \mathbb{C}$ and $f(z_0) = \infty$ then f is holomorphic near z_0 if and only if $1/f$ is holomorphic in the usual sense near z_0, etc. (Similarly: If $f \in H(V)$, where $V \subset \mathbb{C}$ contains a deleted neighborhood of infinity (that is, $\{z \in \mathbb{C} : |z| > R\} \subset V$ for some $R > 0$), then we say f has a removable singularity, pole, or essential singularity at ∞ depending on the sort of singularity the function $f(1/z)$ has at the origin.)

We have extended the meaning of the term "holomorphic". To prevent confusion we will reserve the term "holomorphic function" (and the notation "$f \in H(V)$") for the previously defined concept (differentiable complex-valued functions defined in open subsets of $\mathbb{C}$), while if V is an open subset of $\mathbb{C}_\infty$ and $f : V \to \mathbb{C}_\infty$ is holomorphic in the sense defined above then we will call f a "holomorphic mapping" or "$\mathbb{C}_\infty$-valued holomorphic mapping".

Suppose now that V is an open subset of the plane and f is meromorphic in V; that is, f is a holomorphic function in V except for poles at isolated points. The poles of f are not, strictly speaking, in the domain of f; if we extend f by stating that $f(p) = \infty$ when p is a pole of f then the extended function is a holomorphic mapping of V into $\mathbb{C}_\infty$ — this follows from the fact that $1/f$ is holomorphic in V (except for poles at the zeroes of f) and has removable singularities at the poles of f. Conversely, if V is a connected subset of the plane, f is a holomorphic mapping of V into $\mathbb{C}_\infty$ and f is not identically ∞ then the points where $f = \infty$ are isolated (because the zeroes of $1/f$ are isolated) and it follows that f may be regarded as a meromorphic function in V, with poles at the points where it takes the value ∞. Thus (for connected $V \subset \mathbb{C}$) there is a natural equivalence between the meromorphic functions in V and the holomorphic maps from V into $\mathbb{C}_\infty$

(with one exception, namely that if $f(z) = \infty$ for all $z \in V$ then f does not correspond to a meromorphic function in V).

It is interesting to note that there are no "transcendental" holomorphic mappings from $\mathbb{C}_\infty$ to itself:

Theorem 8.0.0. *If $f : \mathbb{C}_\infty \to \mathbb{C}_\infty$ is holomorphic then there exist polynomials P and Q such that*
$$f(z) = \frac{P(z)}{Q(z)}$$
for all $z \in \mathbb{C}_\infty$.

Proof. If $f(z) = \infty$ for all $z \in \mathbb{C}_\infty$ then we can take $P = 1$ and $Q = 0$. (This special case explains why we did not simply say that f is a rational function in the statement of the theorem!) So we may assume that f is not identically equal to ∞; now the remarks above about the correspondence between $\mathbb{C}_\infty$-valued holomorphic mappings and meromorphic functions show that we may assume that f is meromorphic in $\mathbb{C}$ with a pole or removable singularity at ∞, and we need to show that f is a rational function.

A constant is certainly a rational function, so we may suppose further that f is nonconstant. Since f has an *isolated* singularity at ∞, there exists $R > 0$ such that f is holomorphic in $\{z \in \mathbb{C} : |z| > R\}$. It follows that f has only finitely many poles in $\mathbb{C}$. Similarly the fact that the zeroes of f are isolated (including the zero at ∞ if there is one) shows that f has only finitely many zeroes in $\mathbb{C}$.

Suppose that the zeroes of f in $\mathbb{C}$ are $\{z_1, \ldots, z_n\}$ and the poles are $\{p_1, \ldots, p_m\}$ (both repeated according to multiplicity). Let
$$g(z) = f(z) \frac{\prod_{j=1}^m (z - p_j)}{\prod_{k=1}^n (z - z_j)}.$$
Then g is (or can be extended at removable singularities so as to become) an entire function with no zero in $\mathbb{C}$. And g has a (possibly infinite) limit at ∞: The fact that f has a pole or removable singularity at ∞ shows that
$$\lim_{z \to \infty} \frac{f(z)}{cz^N} = 1$$
for some $N \in \mathbb{Z}$ and nonzero $c \in \mathbb{C}$; hence
$$\lim_{z \to \infty} \frac{g(z)}{cz^{N+m-n}} = 1.$$
So g has a pole or a removable singularity at ∞. In fact g cannot have a pole at ∞: That would imply that $1/g$ was an entire function which tended to 0 at ∞, which would imply that $1/g$ was identically 0. So g has a finite limit at ∞ (i.e. $N + m - n \geq 0$). Thus g is a bounded entire function and hence constant. □

8.1. Linear-Fractional Transformations, Part I

Suppose that $f \in H(V)$. If f' has no zero in V then f is said to be a _conformal mapping_. This is because f "preserves angles"; what we mean by that is contained in the following proposition:

Proposition 8.1.0. *Suppose that $f \in H(V)$, $z_0 \in V$, and $f'(z_0) \neq 0$. Suppose that $\gamma_0, \gamma_1 : [0,1] \to V$ are curves with nonvanishing (right) derivatives at 0, with $\gamma_j(0) = z_0$. Let $\tilde{\gamma}_j(t) = f(\gamma_j(t))$. Then*

$$\arg\left(\frac{\tilde{\gamma}_1'(0)}{\tilde{\gamma}_0'(0)}\right) = \arg\left(\frac{\gamma_1'(0)}{\gamma_0'(0)}\right)$$

(by which we mean that every argument of $\tilde{\gamma}_1'(0)/\tilde{\gamma}_0'(0)$ is an argument of $\gamma_1'(0)/\gamma_0'(0)$ and conversely).

Proof. Let

$$\gamma_j'(0) = r_j e^{i\theta_j} \quad (j = 0, 1)$$

with $r_j > 0$ and $\theta_j \in \mathbb{R}$. Similarly let

$$f'(z_0) = re^{i\theta}.$$

Then

$$\tilde{\gamma}_j'(0) = f'(z_0)\gamma_j'(0) = rr_j e^{i(\theta_j + \theta)},$$

so

$$\arg\left(\frac{\tilde{\gamma}_1'(0)}{\tilde{\gamma}_0'(0)}\right) = (\theta_1 + \theta) - (\theta_0 + \theta) = \theta_1 - \theta_0 = \arg\left(\frac{\gamma_1'(0)}{\gamma_0'(0)}\right). \quad \square$$

The conclusion of the proposition could also be stated thus:

$$\arg(\tilde{\gamma}_1'(0)) - \arg(\tilde{\gamma}_0'(0)) = \arg(\gamma_1'(0)) - \arg(\gamma_0'(0)).$$

We leave it to the reader to say exactly what that equation means, given the "multi-valued" nature of $\arg(z)$. (Things would be simpler if $\arg(z)$ had been defined to be an element of $\mathbb{R}/2\pi\mathbb{Z}$ instead of one of various real numbers.)

Suppose now that $f \in H(V)$ and f is one-to-one in V; let $W = f(V)$. In this case we say that f is a _conformal equivalence_, and that W is _conformally equivalent_ to V.

Note that results in the previous chapter show that the derivative of a conformal equivalence cannot vanish, so that a conformal equivalence is in fact a conformal map. Note as well that the inverse of a conformal equivalence is a conformal equivalence, so that "is conformally equivalent to" defines an equivalence relation. (Some caution is required here: Many authors define "conformal map" and "conformal equivalence" as above but then speak as though the two phrases meant the same thing. In fact the exponential function is a conformal map by our definitions, although it is not an example of what is usually intended when people say "conformal mapping", since it is not one-to-one.)

Note. In the present context the phrase "biholomorphic map" is a synonym for "conformal equivalence" or "invertible holomorphic mapping". People also use the term "biholomorphic map" to mean "invertible holomorphic mapping" in several complex variables, where such maps should not be called conformal, since they do not preserve angles.

If V is an open subset of the plane (or if $V = \mathbb{C}_\infty$), we will use the notation Aut(V) to denote the group of invertible holomorphic mappings from V to itself; Aut(V) is the (holomorphic) "automorphism group" of V. For most open sets Aut(V) consists of just the identity map; we will calculate the automorphism groups in a few interesting special cases.

Definition. A *linear-fractional transformation* (or *Möbius transformation*) is a mapping of the form
$$\phi(z) = \frac{az+b}{cz+d}$$
where a, b, c, $d \in \mathbb{C}$ satisfy
$$ad - bc \neq 0.$$

Note that the condition $ad - bc \neq 0$ implies (among other things) that c and d cannot both vanish; thus $cz + d$ is not the zero polynomial, and a linear-fractional transformation is thus a rational function. So we can regard a linear-fractional transformation as a meromorphic function in the plane (with at most one pole) or as a holomorphic mapping from $\mathbb{C}_\infty$ to itself.

Theorem 8.1.1. Aut($\mathbb{C}_\infty$) *is the set of all linear-fractional transformations.*

Proof. Suppose first that ϕ is a linear-fractional transformation:
$$\phi(z) = \frac{az+b}{cz+d}$$
with $ad - bc \neq 0$. As noted above it follows that ϕ is a holomorphic mapping from $\mathbb{C}_\infty$ to itself; we need only show that ϕ is a bijection. But one can calculate the inverse of ϕ explicitly: If
$$\psi(z) = \frac{dz-b}{-cz+a}$$

8.1. Linear-Fractional Transformations, Part I

then
$$\phi(\psi(z)) = \psi(\phi(z)) = z$$
for all $z \in \mathbb{C}_\infty$ (in checking this it might be convenient to begin by verifying the identity for $z \in \mathbb{C}$ such that neither denominator vanishes, and then note that the identity follows for all $z \in \mathbb{C}_\infty$ by continuity).

So every linear-fractional transformation defines an element of $\mathrm{Aut}(\mathbb{C}_\infty)$. Suppose on the other hand that $\phi \in \mathrm{Aut}(\mathbb{C}_\infty)$. In particular ϕ is a holomorphic mapping of $\mathbb{C}_\infty$ to itself, so Theorem 8.0.0 shows that ϕ is rational:
$$\phi(z) = c \frac{\prod_{k=1}^n (z - z_k)}{\prod_{j=1}^m (z - p_j)},$$
where $z_1, \ldots, z_n$ are the zeroes of ϕ in $\mathbb{C}$ and $p_1, \ldots, p_m$ are the poles in $\mathbb{C}$. (Note that $\{z_k\} \cap \{p_j\} = \emptyset$.)

Now the fact that $\phi : \mathbb{C}_\infty \to \mathbb{C}_\infty$ is one-to-one shows that ϕ has at most one zero in $\mathbb{C}$, so that $n \leq 1$. Similarly $m \leq 1$. We cannot have $n = m = 0$, because in that case $\phi(z) = c$ and $\phi \notin \mathrm{Aut}(\mathbb{C}_\infty)$. This leaves three possibilities:
$$\phi(z) = c \frac{z - z_1}{z - p_1} \quad (z_1 \neq p_1),$$
$$\phi(z) = c \frac{1}{z - p_1}$$
and
$$\phi(z) = c(z - z_1).$$

In each case ϕ is a linear-fractional transformation (note that the condition $z_1 \neq p_1$ implies that $1(-p_1) - (-z_1)(1) \neq 0$, so that $(z - z_1)/(z - p_1)$ is in fact a linear-fractional transformation.) $\square$

We can use this to find all the holomorphic automorphisms of the plane. Recall that an *affine map* is a map of the form $z \mapsto az + b$.

Theorem 8.1.2. $\mathrm{Aut}(\mathbb{C})$ *is equal to the set of nonconstant affine maps.*

Proof. It is clear that a nonconstant affine map is an automorphism of the plane. Suppose on the other hand that $\phi \in \mathrm{Aut}(\mathbb{C})$. Now the fact that ϕ^{-1} is continuous shows that $\phi^{-1}(K)$ is compact for every compact $K \subset \mathbb{C}$ (which is to say that ϕ is a "proper" map from the plane to itself). It follows that
$$\lim_{z \to \infty} \phi(z) = \infty,$$
so that if we extend ϕ to a map $\tilde{\phi} : \mathbb{C}_\infty \to \mathbb{C}_\infty$ by defining $\tilde{\phi}(\infty) = \infty$ then $\tilde{\phi}$ is continuous. The fact that ϕ is a bijection from $\mathbb{C}$ to $\mathbb{C}$ shows that $\tilde{\phi}$

is a bijection from $\mathbb{C}_\infty$ to $\mathbb{C}_\infty$. And $\tilde\phi$ is a holomorphic mapping from $\mathbb{C}_\infty$ to $\mathbb{C}_\infty$ (by definition we need to verify that $1/\tilde\phi(1/z)$ is holomorphic in a neighborhood of 0, but this is clear: Since $\lim_{z\to 0} 1/\phi(1/z) = 0$, $1/\phi(1/z)$ has a removable singularity at the origin.)

Thus $\tilde\phi \in \mathrm{Aut}(\mathbb{C}_\infty)$, and so the previous theorem shows that ϕ is a linear-fractional transformation. Since ϕ has no poles in $\mathbb{C}$, it follows that $\phi(z) = az + b$. □

If we had not already proved Theorem 8.0.0, we could have easily proved Theorem 8.1.2 directly and then deduced Theorem 8.1.1:

- **Exercise 8.1.** Show directly (without using any results in Chapters 7 or 8) that an entire function with a pole at ∞ must be a polynomial.

- **Exercise 8.2.** Give a direct proof of Theorem 8.1.2, without using Theorem 8.1.1.

  Hint: You can just use the proof in the text to show that $\lim_{z\to\infty} \phi(z) = \infty$; now use something other than Theorem 8.1.1 to show that ϕ is affine.

- **Exercise 8.3.** Show that Theorem 8.1.1 follows from Theorem 8.1.2.

  Hint: Take the fact that a linear-fractional transformation lies in $\mathrm{Aut}(\mathbb{C}_\infty)$ as obvious. Now suppose that $\phi \in \mathrm{Aut}(\mathbb{C}_\infty)$. If in addition $\phi(\infty) = \infty$ then it follows from Theorem 8.1.2 that ϕ is a linear-fractional transformation (explain). The main point to the exercise is to figure out how the general case follows from the case $\phi(\infty) = \infty$; this has something to do with the fact that $\mathrm{Aut}(\mathbb{C}_\infty)$ is a group ... .

8.2. Linear-Fractional Transformations, Part II

Linear-fractional transfomations are important for many different reasons. We saw in the last section that they arise naturally as the holomorphic automorphisms of $\mathbb{C}_\infty$. We will see later that the automorphism groups of disks and half-planes in $\mathbb{C}$ are in fact groups of linear-fractional transformations; they are also often very useful in constructing conformal equivalences between pairs of open sets in the plane (in particular between the unit disk and some other open set).

We will let the symbol $\mathcal{M}$ denote the group of all linear-fractional transformations. We will be discussing three topics here: The natural correspondence between linear-fractional transformations and invertible 2×2 complex matrices, how lines and circles transform under linear-fractional transformations, and linear transformations with specified values at three points.

8.2. Linear-Fractional Transformations, Part II

The set of all invertible 2×2 complex matrices forms a group under matrix multiplication, commonly known as $GL_2(\mathbb{C})$, the 2×2 complex "general linear group". Recalling that the complex 2×2 matrix $A = \begin{bmatrix} a & b \\ c & d \end{bmatrix}$ is invertible if and only if $ad - bc \neq 0$, we see that

$$\phi_A(z) = \frac{az+b}{cz+d}$$

defines a linear-fractional transformation $\phi_A \in \mathcal{M}$ if and only if $A \in GL_2(\mathbb{C})$. One may verify directly that

$$\phi_A \circ \phi_B = \phi_{AB}$$

for all $A, B \in GL_2(\mathbb{C})$. That's not hard, but it does not seem to explain much; since matrix multiplication is equivalent to composition of the associated linear maps, it seems that there should be a more interesting proof that $\phi_A \circ \phi_B = \phi_{AB}$, somehow relating the composition of the two linear-fractional transformations to the composition of the linear maps associated with the matrices A and B. It turns out that the fact that $\phi_A \circ \phi_B = \phi_{AB}$ *is* an "immediate" consequence of the fact that matrix multiplication corresponds to composition of linear maps, if you look at things from the right point of view:

We will take $\mathbb{C}^2$ to be a space of "column vectors":

$$\mathbb{C}^2 = \left\{ \begin{bmatrix} z_1 \\ z_2 \end{bmatrix} : z_1, z_2 \in \mathbb{C} \right\}.$$

Let $\mathbb{P}_1$ be the set of all one-dimensional complex-linear subspaces of the vector space $\mathbb{C}^2$. If $A \in GL_2(\mathbb{C})$ then multiplication by A defines an invertible linear operator $T_A : \mathbb{C}^2 \to \mathbb{C}^2$:

$$T_A(z) = Az \quad (z \in \mathbb{C}^2).$$

Now if S is a one-dimensional subspace of $\mathbb{C}^2$ then the invertibility of T_A shows that the image $T_A(S)$ is also a one-dimensional subspace of $\mathbb{C}^2$; hence A induces a map

$$P_A : \mathbb{P}_1 \to \mathbb{P}_1$$

in a natural way, via the formula

$$P_A(S) = T_A(S).$$

Note. Unfortunately, the formula $P_A(S) = T_A(S)$ makes it look as though $P_A = T_A$, which is not so. The reason for the problem is an ambiguity in traditional notation: The S on the left is an element of the domain of P_A, while the S on the right is a *subset* of the domain of T_A. One might clarify things by rewriting the definition of P_A as follows:

$$P_A(S) = \{T_A(z) : z \in S\} \quad (S \in \mathbb{P}_1).$$

Now, there is a natural (although not quite canonical) way to identify elements of $\mathbb{P}_1$ with elements of $\mathbb{C}_\infty$: If $S \in \mathbb{P}_1$ then either

$$S = \left\{ \begin{bmatrix} z \\ 0 \end{bmatrix} : z \in \mathbb{C} \right\}$$

or

$$S = \left\{ \begin{bmatrix} \alpha z \\ z \end{bmatrix} : z \in \mathbb{C} \right\}$$

for some $\alpha \in \mathbb{C}$. If we let

$$L(\infty) = \left\{ \begin{bmatrix} z \\ 0 \end{bmatrix} : z \in \mathbb{C} \right\}$$

and

$$L(\alpha) = \left\{ \begin{bmatrix} \alpha z \\ z \end{bmatrix} : z \in \mathbb{C} \right\} \quad (\alpha \in \mathbb{C})$$

then $L : \mathbb{C}_\infty \to \mathbb{P}_1$ is a bijection; note that if $S \in \mathbb{P}_1$ then

$$L^{-1}(S) = z_1/z_2$$

for any nonzero $z \in S$.

So if $A \in GL_2(\mathbb{C})$ then

$$M_A : \mathbb{C}_\infty \to \mathbb{C}_\infty,$$

where

$$M_A = L^{-1} \circ P_A \circ L.$$

We know that $T_A \circ T_B = T_{AB}$, it follows that $P_A \circ P_B = P_{AB}$, and now the definition of M shows that $M_A \circ M_B = M_{AB}$. Simply tracing through the definitions shows that in fact $M_A = \phi_A$, so that all this makes it "obvious" that $\phi_A \circ \phi_B = \phi_{AB}$.

- **Exercise 8.4.** Give the details of both proofs of the identity $\phi_A \circ \phi_B = \phi_{AB}$ sketched above: Write out a "direct" proof and also verify carefully that $M_A = \phi_A$.

Of course simply verifying that $\phi_A \circ \phi_B = \phi_{AB}$ directly by working out both sides and seeing they come to the same thing takes less space than

8.2. Linear-Fractional Transformations, Part II

the second proof above. But an identity involving matrix multiplication "should" have something to do with composition of linear maps, and the last few paragraphs explain the connection. (Note also that linear-fractional transformations are called "linear" in the older literature on the subject; one suspects that the story above is part of the reason for that.)

One way or another we see that $\phi_A \circ \phi_B = \phi_{AB}$, which says that $A \mapsto \phi_A$ defines a homomorphism of $GL_2(\mathbb{C})$ onto $\mathcal{M}$. The obvious next question is to find the *kernel* of this homomorphism:

Suppose that ϕ_A is the identity: $(az+b)/(cz+d) = z$ for all $z \in \mathbb{C}_\infty$. Setting $z = 0$ shows that $b = 0$. Setting $z = \infty$ shows that $a/c = \infty$, so $c = 0$. Now setting $z = 1$ shows that $a/d = 1$, so $a = d$; thus A is a scalar multiple of the 2×2 identity matrix.

So we see that ϕ_A is the identity if and only if A is a (non-zero) scalar multiple of the 2×2 identity matrix. Now if $\phi_A = \phi_B$ then $\phi_{AB^{-1}}$ must be the identity transformation, so that AB^{-1} is a scalar multiple of the identity:

$\phi_A = \phi_B$ if and only if A is a scalar multiple of B.

Let N denote the class of nonzero scalar multiples of the 2×2 identity matrix. The reader should verify directly that N is a normal subgroup of $GL_2(\mathbb{C})$; we already know this must be so, since N is the kernel of a homomorphism. We have shown that $\mathcal{M}$ is isomorphic to the quotient group $GL_2(\mathbb{C})/N$; fascinating. (One might suspect that writing $\mathcal{M}$ as a quotient of matrix groups this way would be useful in studying $\mathcal{M}$; it might be a surprise to hear that in fact it often happens that people studying matrix groups obtain information about them by looking at them as subgroups of $\mathcal{M}$!)

So much for the question of what linear-fractional transformations have to do with 2×2 complex matrices. The next topic is what they have to do with lines and circles. We will let $\mathcal{C}$ denote the class of all lines and circles in the plane, almost:

$$\mathcal{C} = \{\, C : C \subset \mathbb{C} \text{ is a circle}\,\} \cup \{\, L \cup \{\infty\} : L \subset \mathbb{C} \text{ is a line}\,\}.$$

A very important property of linear-fractional transformations is this: If $\phi \in \mathcal{M}$ and $C \in \mathcal{C}$ then the set $\phi(C)$ is also an element of $\mathcal{C}$; in other words, "linear-fractional transformations take lines and circles to lines and circles". (The elements of $\mathcal{C}$ are sometimes called just circles; a line is regarded as a circle passing through ∞. This makes a lot of sense from the point of view of projective geometry; one might also note that if we identify $\mathbb{C}_\infty$ with the unit sphere in $\mathbb{R}^3$ as in Section 8.0 then elements of $\mathcal{C}$ become actual circles. But lines do not *look* round in $\mathbb{C}_\infty$; we will call the elements of $\mathcal{C}$ "lines and circles" or, more simply and accurately, "elements of $\mathcal{C}$".)

The fact that linear-fractional transformations take lines and circles to lines and circles involves a bit of computation. We note first that $\mathcal{M}$ is generated by a few subgroups:

If $\alpha \in \mathbb{C}$ we define $\tau_\alpha \in \mathcal{M}$ by

$$\tau_\alpha(z) = z + \alpha.$$

(That's τ for "translation". As always in this chapter, formulas that are supposed to define elements of $\mathcal{M}$ are extended by continuity at points where they do not make sense *a priori*; here that means that $\tau_\alpha(\infty) = \infty$.) Not that it matters for our purposes, but it is clear that $\tau_\alpha \circ \tau_\beta = \tau_{\alpha+\beta}$, so that $\{\tau_\alpha : \alpha \in \mathbb{C}\}$ is a subgroup of $\mathcal{M}$ isomorphic to the additive complex numbers. For $t > 0$ we define

$$\delta_t(z) = tz$$

(δ for "dilation") and for $\theta \in \mathbb{R}$ we define

$$\rho_\theta(z) = e^{i\theta} z$$

(ρ for "rotation"). Note that $\{\delta_t : t > 0\}$ is a subgroup of $\mathcal{M}$ (isomorphic to the multiplicative group of positive reals) and $\{\rho_\theta : \theta \in \mathbb{R}\}$ is another subgroup of $\mathcal{M}$ (isomorphic to the multiplicative group of complex numbers of modulus 1). Finally, we set

$$j(z) = \frac{1}{z}$$

(that's "j" for "we would call it 'i' or 'I', meaning either 'inverse' or 'involution', but those letters are taken"). Note that j generates a subgroup of $\mathcal{M}$ of order 2 (isomorphic to any other group of order 2).

Lemma 8.2.0. *The group $\mathcal{M}$ is generated by the subgroups $\{\tau_\alpha : \alpha \in \mathbb{C}\}$, $\{\delta_t : t > 0\}$, $\{\rho_\theta : \theta \in \mathbb{R}\}$, and $\{j, I\}$ (here I is the identity: $I(z) = z$).*

Proof. Let G denote the subgroup of $\mathcal{M}$ generated by the four subgroups listed in the statement of the lemma. Note first that if we extend the definition of δ_α to

$$\delta_\alpha(z) = \alpha z \qquad (\alpha \in \mathbb{C}, \, \alpha \neq 0)$$

then $\delta_\alpha \in G$ for any nonzero complex number α, because if $\alpha = re^{it}$ (with $r > 0$ and $t \in \mathbb{R}$) then

$$\delta_\alpha = \delta_r \circ \rho_t.$$

(The only reason we did not define δ_α this way to begin with is then it is not quite right to call δ_α a dilation.)

8.2. Linear-Fractional Transformations, Part II

Now suppose that $\phi \in \mathcal{M}$; choose $a, b, c, d \in \mathbb{C}$ so that

$$\phi(z) = \frac{az+b}{cz+d}.$$

If $c = 0$ then the nondegeneracy condition $ad - bc \neq 0$ shows that $d \neq 0$, and so

$$\phi(z) = \frac{a}{d}z + \frac{b}{d},$$

which shows that

$$\phi = \tau_{\frac{b}{d}} \circ \delta_{\frac{a}{d}}.$$

Suppose on the other hand that $c \neq 0$. Then

$$\phi(z) = \frac{a}{c} + \frac{b - \frac{ad}{c}}{cz+d},$$

which shows that

$$\phi = \tau_{\frac{a}{c}} \circ \delta_{b - \frac{ad}{c}} \circ j \circ \tau_d \circ \delta_c. \qquad \square$$

Theorem 8.2.1. *If $C \in \mathcal{C}$ and $\phi \in \mathcal{M}$ then $\phi(C) \in \mathcal{C}$.*

(Here $\phi(C) = \{\phi(z) : z \in C\}$ as usual.)

Proof. We may assume that ϕ is an element of one of the four subgroups of $\mathcal{M}$ listed in the statement of Lemma 8.2.0. But the conclusion is clear if ϕ is a translation, rotation, or dilation; hence we may assume that $\phi = j$: $\phi(z) = 1/z$.

If C is a line through the origin (together with the point at infinity) then it is clear that $j(C)$ is another line through the origin, simply because

$$\frac{1}{re^{it}} = \frac{1}{r}e^{-it}$$

(so if C is the line passing through the origin and the point e^{it} then $j(C)$ is the line passing through the origin and the point e^{-it}). Suppose instead that $C = C_0$ is the line defined by $\operatorname{Re}(z) = 1$:

$$C_0 = \{\infty\} \cup \{1 + iy : y \in \mathbb{R}\}.$$

Now *if* $j(C_0)$ is a line or circle it must be a circle, since $\infty \notin j(C_0)$ (because $0 \notin C_0$). It must be a circle passing through the origin because $\infty \in C_0$; now since 1 is the point on C_0 of minimum absolute value, it follows that $1 = 1/1$ must be the point on $j(C_0)$ of maximum absolute value. That is, 0 and 1 must be endpoints of a diameter of this circle, so the center must be

1/2 and the radius must be 1/2. So we try to prove that, by showing that $|j(1+iy) - 1/2| = 1/2$. This is actually easy, once we have decided what to try to prove: For any $y \in \mathbb{R}$ we have

$$\left| j(1+iy) - \frac{1}{2} \right| = \left| \frac{1}{1+iy} - \frac{1}{2} \right| = \frac{1}{2} \left| \frac{1-iy}{1+iy} \right| = \frac{1}{2}.$$

This shows that $j(C_0)$ is a subset of C_1, the circle with center 1/2 and radius 1/2. (It shows that j maps a dense subset of C_0 into C_1; now $j(\infty)$ must also be an element of C_1, by continuity. Or note that $j(\infty) = 0$.) But $j(C_0)$ must be homeomorphic to a circle, since C_0 is homeomorphic to a circle, and no proper subset of a circle is homeomorphic to a circle (Exercise 8.6). So $j(C_0)$ must be exactly C_1.

Suppose now that C is any line not passing through the origin. One could claim that the argument we used for C_0 above must work in this case. Better, one can derive the fact that $j(C)$ is a circle from the fact that $j(C_0)$ is a circle: There exists a nonzero complex number α such that

$$C = \delta_\alpha(C_0).$$

Since

$$j \circ \delta_\alpha = \delta_{1/\alpha} \circ j,$$

it follows that

$$j(C) = j(\delta_\alpha(C_0)) = \delta_{1/\alpha}(j(C_0)) = \delta_{1/\alpha}(C_1),$$

which shows that $j(C)$ is a circle passing through the origin, since C_1 is.

Note now that any circle passing through the origin is equal to $\delta_{1/\alpha}(C_1)$ for a suitable (nonzero) $\alpha \in \mathbb{C}$; since j is its own inverse, it follows that if $C \in \mathcal{C}$ is a circle passing through the origin then $j(C)$ is a line not passing through the origin. So we have shown that $j(C) \in \mathcal{C}$ if $C \in \mathcal{C}$ is either a line (containing the origin or not) or a circle passing through the origin, and all that remains is to show that $j(C) \in \mathcal{C}$ if C is a circle not passing through the origin. We may suppose in addition (just to simplify the notation) that the center of the circle lies on the positive real axis, because any circle not passing through the origin is equal to $\delta_\alpha(C)$ where C is a circle not passing through the origin and with center on the positive real axis. Suppose then that $p > 0$, $r > 0$, $r \neq p$, and

$$C = \{\, z \in \mathbb{C} : |z - p| = r \,\}.$$

As before, one can determine what the center of $j(C)$ must be if $j(C)$ is a circle: The points on C of maximum and minimum absolute value are $p + r$

and $p-r$, respectively, so if $j(C)$ is a circle it must have a diameter with endpoints $1/(p+r)$ and $1/(p-r)$, and so the center must be

$$\frac{1}{2}\left(\frac{1}{p+r}+\frac{1}{p-r}\right) = \frac{p}{p^2-r^2}.$$

So we attempt to show that $j(C)$ is a circle by calculating the distance from $p/(p^2-r^2)$ to the reciprocal of the generic point of C:

$$\left|\frac{1}{p+re^{it}}-\frac{p}{p^2-r^2}\right| = \frac{r}{|p^2-r^2|}\left|\frac{pe^{it}+r}{p+re^{it}}\right|$$

$$= \frac{r}{|p^2-r^2|}\left|\frac{p+re^{-it}}{p+re^{it}}\right| - \frac{r}{|p^2-r^2|}.$$

This shows that $j(C)$ is a subset of the circle with center $p/(p^2-r^2)$ and radius $r/|p^2-r^2|$; as before it follows that $j(C)$ must actually equal this circle. $\square$

The last of the big deals concerning linear-fractional transformations is this:

Theorem 8.2.2. *If z_1, z_2, z_3 are three distinct elements of $\mathbb{C}_\infty$ and w_1, w_2, w_3 are three distinct elements of $\mathbb{C}_\infty$ then there exists a unique $\phi \in \mathcal{M}$ with $\phi(z_j) = w_j$ ($j = 1, 2, 3$).*

Proof. We prove this in a few pieces. First we show that if z_1, z_2, z_3 are three distinct elements of $\mathbb{C}_\infty$ then there exists $\phi \in \mathcal{M}$ such that $\phi(z_1) = 0$, $\phi(z_2) = 1$, and $\phi(z_3) = \infty$. If all three of z_1, z_2, z_3 are finite then we simply let

$$\phi(z) = \frac{(z-z_1)(z_2-z_3)}{(z-z_3)(z_2-z_1)}.$$

This needs to be modified slightly if one of the z_j is infinite: If $z_1 = \infty$ take $\phi(z) = (z_2-z_3)/(z-z_3)$, if $z_2 = \infty$ take $\phi(z) = (z-z_1)/(z-z_3)$, and if $z_3 = \infty$ take $\phi(z) = (z-z_1)/(z_2-z_1)$.

That gives existence in the special case $w_1 = 0$, $w_2 = 1$, $w_3 = \infty$. In general, given three distinct z_1, z_2, z_3 and three distinct w_1, w_2, w_3 we can find a $\phi_1 \in \mathcal{M}$ with $\phi_1(z_1) = 0$, $\phi_1(z_2) = 1$, and $\phi_1(z_3) = \infty$ and we can also find $\phi_2 \in \mathcal{M}$ with $\phi_2(w_1) = 0$, $\phi_2(w_2) = 1$, and $\phi_2(w_3) = \infty$. Now $\phi = \phi_2^{-1} \circ \phi_1$ satisfies $\phi(z_j) = w_j$ for $j = 1, 2, 3$.

For uniqueness we begin with the following observation: If $\phi \in \mathcal{M}$ and ϕ fixes all three of the points $0, 1, \infty$ then ϕ is the identity. (Suppose that

$$\phi(z) = \frac{az+b}{cz+d}.$$

The fact that $\phi(\infty) = \infty$ shows that $a/c = \infty$, so that $c = 0$, and hence $d \neq 0$. Now $\phi(0) = 0$ implies that $b = 0$; thus $\phi(z) = az/d$. Since $\phi(1) = 1$, we must have $a = d$, and hence $\phi(z) = z$.)

Suppose now that z_1, z_2, z_3 are three distinct elements of $\mathbb{C}_\infty$, that w_1, w_2, w_3 are three distinct elements of $\mathbb{C}_\infty$, and that $\phi_1, \phi_2 \in \mathcal{M}$ satisfy

$$\phi_k(z_j) = w_j \quad (k = 1, 2;\ j = 1, 2, 3).$$

We need to show that $\phi_1 = \phi_2$. Choose $\psi \in \mathcal{M}$ with $\psi(z_1) = 0$, $\psi(z_2) = 1$, and $\psi(z_3) = \infty$, and consider

$$\phi = \psi \circ \phi_2^{-1} \circ \phi_1 \circ \psi^{-1}.$$

It is easy to see that ϕ fixes 0, 1, and ∞. Hence ϕ must be the identity; it follows that $\phi_2^{-1} \circ \phi_1$ is the identity, and hence $\phi_2 = \phi_1$. □

- **Exercise 8.5.** The proof of Theorem 8.2.2 contains a proof that if $\phi \in \mathcal{M}$ fixes 0, 1, and ∞ then ϕ is the identity. Show how it follows immediately that if $\phi \in \mathcal{M}$ has three fixed points it must be the identity. (The proof should exploit the fact that $\mathcal{M}$ is a group, as in the proof of Theorem 8.2.2, instead of doing calculations with the coefficients of linear-fractional transformations.)

- **Exercise 8.6.** Suppose that C is a circle. Show that the only subset of C homeomorphic to a circle is C itself.

    Hint: One way is to note that $C \setminus \{z\}$ is connected for all $z \in C$.

8.3. Linear-Fractional Transformations, Part III

One of the big theorems coming up is the Riemann Mapping Theorem, which states that any simply connected open subset of $\mathbb{C}$, other than $\mathbb{C}$ itself, is conformally equivalent to the unit disk. This is a result of fundamental importance in pure complex analysis and also in applied mathematics: It happens that various applied problems in an open set can be transferred to equivalent problems about the unit disk, where they are easier to solve (or where the problem has already been solved). The "Dirichlet problem" is an example of this (see Chapter 10, especially Section 10.5).

Of course successful application of this strategy often depends on knowing the function that maps the open set to the disk (or failing that, knowing something about the mapping function). We give a few examples of conformal mappings in this section, to illustrate a few techniques, and also to show how linear-fractional transformations come into the picture. (Books on applied complex analysis often have large tables of conformal mappings, in the place where you find a table of integrals in a calculus book.)

8.3. Linear-Fractional Transformations, Part III

We start with the Cayley transform; this is a conformal mapping of the disk onto the upper half-plane $\Pi^+ = \{\, z : \operatorname{Im}(z) > 0 \,\}$. Let's see; first it's clear from the results in the previous section that there *is* a linear-fractional transformation mapping the disk onto Π^+. For example, we know that there is a linear-fractional transformation mapping the unit circle onto the real axis; now the complement of the unit circle in $\mathbb{C}_\infty$ has two components, the unit disk and the set of all complex numbers of modulus greater than 1. The complement of $\mathbb{R} \cup \{\infty\}$ in $\mathbb{C}_\infty$ also has two components, the upper half-plane and the lower half-plane. So if we start with a linear-fractional transformation mapping the unit circle to the real line then it must map the unit disk to the upper half-plane or to the lower half-plane, and if we happen to get the lower half-plane on our first try we can multiply by -1. We could obtain a linear-fractional transformation mapping the unit circle to the real axis by, for example, finding $\phi \in \mathcal{M}$ with $\phi(1) = 0$, $\phi(i) = 1$ and $\phi(-1) = \infty$. It's easy to write down a formula for a $\phi \in \mathcal{M}$ doing those three things, as in the proof of Theorem 8.2.2; now the image of the unit circle under ϕ must be a line or circle including the three points 0, 1, and ∞, which is to say it must be the real axis.

But there's an easier and much more fun way to do it, and a way that makes the resulting formula 100% unforgettable. When is a complex number in Π^+ anyway? If you think about it, it's clear that $z \in \Pi^+$ if and only if z is closer to i than to $-i$; that is, if and only if $|z - i| < |z + i|$. And there you are:

$$z \in \Pi^+ \iff \left|\frac{z-i}{z+i}\right| < 1,$$

which says that ψ_C is a conformal map from the upper half-plane onto the unit disk if

$$\psi_C(z) = \frac{z-i}{z+i}.$$

(Thanks to Anthony Kable for pointing this out.) Now $\begin{bmatrix} 1 & -i \\ 1 & i \end{bmatrix} \begin{bmatrix} -i & -i \\ 1 & -1 \end{bmatrix}$
$= \begin{bmatrix} -2i & 0 \\ 0 & -2i \end{bmatrix}$, a multiple of the 2×2 identity matrix, so $\psi_C^{-1} = \phi_C$, where

$$\phi_C(z) = i\frac{1+z}{1-z}.$$

So: ϕ_C is the Caley transform, a conformal equivalence mapping the unit disk onto Π^+; the inverse of the Cayley transform is ψ_C.

The same reasoning applies to other half-planes. Suppose for example that $a, b \in \mathbb{C}$. Let L be the perpendicular bisector of the segment $[a, b]$. Now

L divides the plane into two half-planes; say S is the half-plane containing the point a. It's clear that

$$\phi(z) = \frac{z-a}{z-b}$$

is a conformal map from S onto the unit disk.

I don't know a way to see that ϕ_C maps the disk onto the upper half-plane that's anywhere near as transparent as the reasoning showing that ψ_C maps the upper half-plane onto the disk, but it's possible to make *some* sense of the formula for ϕ_C:

We want to show that the imaginary part of $i\frac{1+z}{1-z}$ is positive if and only if $|z| < 1$. This is the same as showing that the real part of $\frac{1+z}{1-z}$ is positive if and only if $|z| < 1$. But $\operatorname{Re}(w) > 0$ if and only if $|\arg(w)| < \pi/2$ (for some choice of $\arg(w)$). So we need to show that

$$|\arg(1+z) - \arg(1-z)| < \pi/2$$

if and only if $|z| < 1$, and this is clear from the fact that an angle inscribed in a semicircle is a right angle (see Figure 8.1).

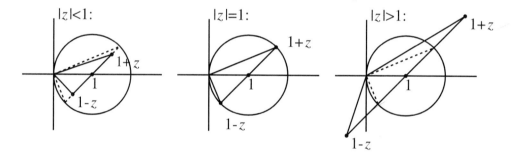

Figure 8.1

8.4. Linear-Fractional Transformations, Part IV: The Schwarz Lemma and Automorphisms of the Disk

In this section $\mathbb{D} = D(0,1)$, the unit disk. Our next goal is a description of $\operatorname{Aut}(\mathbb{D})$. For this we will need the "Schwarz Lemma":

Theorem 8.4.0 (Schwarz Lemma). *Suppose that $f : \mathbb{D} \to \mathbb{D}$ is holomorphic and $f(0) = 0$. Then $|f(z)| \leq |z|$ for all $z \in \mathbb{D}$ and $|f'(0)| \leq 1$. Furthermore, if $|f'(0)| = 1$ or $|f(z)| = |z|$ for some nonzero $z \in \mathbb{D}$ then f is a rotation: $f(z) = \beta z$ for some constant β with $|\beta| = 1$.*

8.4. The Schwarz Lemma and Automorphisms of the Disk

Proof. Since $f(0) = 0$, we can define a function $g \in H(\mathbb{D})$ by

$$g(z) = \begin{cases} \dfrac{f(z)}{z} & (z \neq 0), \\ f'(0) & (z = 0). \end{cases}$$

The result follows from the Maximum Modulus Theorem applied to g. One might argue as follows: It is clear that $|g| \leq 1$ on $\partial \mathbb{D}$, hence the Maximum Modulus Theorem shows that $|g| \leq 1$ in $\mathbb{D}$, and hence $|f(z)| \leq |z|$ in $\mathbb{D}$ and $|f'(0)| \leq 1$.

Of course if one argued that way one would be speaking nonsense: g is not even *defined* on $\partial \mathbb{D}$ so it has little chance of being smaller than 1 in absolute value there. There are various interesting ways to fix this problem: We could give a different version of the Maximum Modulus Theorem with a hypothesis involving the lim sup of a holomorphic function at points of the boundary (see Lemma 10.5.2). An even more interesting approach would be to show that a bounded holomorphic function in the disk must have a boundary value at "almost every" boundary point, and the supremum of the function is equal to the "essential supremum" of the boundary values; for this we would need some real analysis. The simplest thing is to apply the Maximum Modulus Theorem that we already know on smaller disks centered at the origin:

Suppose $r \in (0,1)$. It is clear that $|g| \leq 1/r$ on $\partial D(0,r)$, and so the Maximum Modulus Theorem shows that $|g| \leq 1/r$ in $D(0,r)$. Since this holds for all $r \in (0,1)$, it follows that $|g| \leq 1$ in $\mathbb{D}$ (and hence that $|f(z)| \leq |z|$ and $|f'(0)| \leq 1$).

Finally, if we have $|f'(0)| = 1$ or $|f(z)| = |z|$ for some nonzero $z \in \mathbb{D}$ then we have $|g(z)| = 1$ for some $z \in \mathbb{D}$ and hence the Maximum Modulus Theorem shows that g is constant. $\square$

(We should note that the fact that $|f'(0)| \leq 1$ follows already from Cauchy's Estimates.)

The Schwarz Lemma allows us to determine the subgroup of $\text{Aut}(\mathbb{D})$ fixing the origin:

Theorem 8.4.1. *Suppose that $\phi \in \text{Aut}(\mathbb{D})$ and $\phi(0) = 0$. Then ϕ is a rotation*:

$$\phi(z) = \beta z$$

for some $\beta \in \mathbb{C}$ with $|\beta| = 1$.

Proof. The Schwarz Lemma shows that $|\phi'(0)| \leq 1$. If we let ψ be the inverse of ϕ then we have $|\psi'(0)| \leq 1$ for exactly the same reason. But the chain rule shows that $\psi'(0) = 1/\phi'(0)$, and so we must have $|\phi'(0)| = 1$; now the Schwarz Lemma shows that ϕ is a rotation. □

So we have identified the subgroup of $\mathrm{Aut}(\mathbb{D})$ fixing the origin. If we can find an element of $\mathrm{Aut}(\mathbb{D})$ taking an arbitrary point of $\mathbb{D}$ to the origin then we've got the whole group. And we can do that — below we define $\phi_a \in \mathrm{Aut}(\mathbb{D})$ for each $a \in \mathbb{D}$ such that $\phi_a(a) = 0$.

The definition of ϕ_a below may appear to be somewhat lacking in motivation. We should note that if $b \in \Pi^+$ then it is obvious (why?) that there exists $\psi_b \in \mathrm{Aut}(\Pi^+)$ with $\psi_b(b) = i$, and it is clear from this (why?) that for every $a \in \mathbb{D}$ there exists $\phi_a \in \mathrm{Aut}(\mathbb{D})$ with $\phi_a(a) = 0$.

Definition. If $a \in \mathbb{D}$ then

$$\phi_a(z) = \frac{a - z}{1 - \bar{a}z}.$$

These ϕ_a things will come up many times in the material to follow. We collect a few of their important properties in a lemma:

Lemma 8.4.2. (i) $\phi_a(a) = 0$ and $\phi_a(0) = a$.

(ii) Each ϕ_a is its own inverse. (That is, ϕ_a is an "involution".)

(iii) $\phi_a \in \mathrm{Aut}(\mathbb{D})$ for any $a \in \mathbb{D}$.

(iv) For any $a, z \in \mathbb{D}$ we have

$$1 - |\phi_a(z)|^2 = \frac{(1 - |a|^2)(1 - |z|^2)}{|1 - \bar{a}z|^2}.$$

(v) $|\phi_a'(0)| = 1 - |a|^2$ and $|\phi_a'(a)| = 1/(1 - |a|^2)$.

Proof. The first part is clear and the second part is a trivial calculation. We should note that (iv) shows that $\phi_a : \mathbb{D} \to \mathbb{D}$, and so (iii) follows. But one can give a direct proof that $\phi_a : \mathbb{D} \to \mathbb{D}$ which may be somewhat more transparent than the calculations that will be used in proving (iv):

Suppose first that $|z| = 1$. Then $z\bar{z} = 1$, so that

$$|\phi_a(z)| = \frac{|a - z|}{|1 - \bar{a}z|} = \frac{|a - z|}{|z(1 - a\bar{z})|} = \frac{|a - z|}{|z - a|} = 1.$$

So ϕ_a maps the unit circle to itself. Since ϕ_a is a homeomorphism of $\mathbb{C}_\infty$, it must map components of the complement of the unit circle in $\mathbb{C}_\infty$ to components of the complement of the unit circle, which says that the image of $\mathbb{D}$ must be either $\mathbb{D}$ or the "exterior" of the circle, that is $\{z \in \mathbb{C}_\infty : |z| > 1\}$. But the second case cannot hold since $\phi_a(0) = a$ lies in $\mathbb{D}$. So ϕ_a maps $\mathbb{D}$ to itself. (And now the fact that $\phi_a \in \text{Aut}(\mathbb{D})$ follows either from generalities about elements of $\mathcal{M}$ or from part (ii).)

For (iv) we just do the math:

$$\begin{aligned}
1 - |\phi_a(z)|^2 &= 1 - \frac{(a-z)(\bar{a}-\bar{z})}{(1-\bar{a}z)(1-a\bar{z})} \\
&= \frac{1 - \bar{a}z - a\bar{z} + |a|^2|z|^2 - (|a|^2 - a\bar{z} - \bar{a}z + |z|^2)}{|1-\bar{a}z|^2} \\
&= \frac{1 + |a|^2|z|^2 - |a|^2 - |z|^2}{|1-\bar{a}z|^2} \\
&= \frac{(1-|a|^2)(1-|z|^2)}{|1-\bar{a}z|^2}.
\end{aligned}$$

We leave (v) to the reader. $\square$

It follows immediately that $\text{Aut}(\mathbb{D})$ is generated by the subgroup of rotations and the subset $\{\phi_a : a \in \mathbb{D}\}$:

Theorem 8.4.3. *For any $\psi \in \text{Aut}(\mathbb{D})$ there exist a unique $a \in \mathbb{D}$ and $\beta \in \mathbb{C}$ with $|\beta| = 1$ such that*

$$\psi(z) = \beta\phi_a(z)$$

for all $z \in \mathbb{D}$.

Proof. Let $a = \psi^{-1}(0)$ and set $\psi_1 = \psi \circ \phi_a$. Now $\psi_1 \in \text{Aut}(\mathbb{D})$ and

$$\psi_1(0) = \psi(a) = 0,$$

so Theorem 8.4.1 shows that ψ_1 is a rotation: There exists β with $|\beta| = 1$ such that
$$\psi_1(z) = \beta z$$
for all z. Since ϕ_a is its own inverse, we have $\psi = \psi_1 \circ \phi_a$, and hence

$$\psi(z) = \psi_1(\phi_a(z)) = \beta\phi_a(z)$$

as claimed.

(The uniqueness is clear: If $\psi(z) = \beta\phi_a(z)$ for all $z \in \mathbb{D}$ then $\psi(a) = 0$ and hence $a = \psi^{-1}(0)$.) $\square$

We should note that by essentially the same argument one may show that any $\psi \in \mathrm{Aut}(\mathbb{D})$ has a unique representation

$$\psi(z) = \phi_\alpha(\gamma z)$$

for some $\alpha \in \mathbb{D}$ and γ with $|\gamma| = 1$.

And that's the automorphism group of the unit disk. This is actually one of the groups that people study a lot for various reasons, some of which we will see later. We know that the unit disk is conformally equivalent to the upper half-plane, so knowing $\mathrm{Aut}(\mathbb{D})$ is in a sense the same thing as knowing $\mathrm{Aut}(\Pi^+)$. It turns out that the elements of $\mathrm{Aut}(\Pi^+)$ have a very interesting explicit description:

- **Exercise 8.7.** Suppose that $\psi \in \mathrm{Aut}(\Pi^+)$. Show that there exist a, b, c, $d \in \mathbb{R}$, with $ad - bc = 1$, such that

$$\psi(z) = \frac{az + b}{cz + d}$$

for all $z \in \Pi^+$. Show that a, b, c, d are not unique but are almost unique (state and prove a precise statement regarding how close to unique they are).

Hint: Show first that ψ must be a linear-fractional transformation — this is immediate from things we have proved. So you know that

$$\psi(z) = \frac{az + b}{cz + d}$$

for some a, b, c, $d \in \mathbb{C}$. Don't try to *show* that a, b, c, $d \in \mathbb{R}$; that's impossible (why?). Show that it is possible to choose another quadruple $\tilde{a}$, $\tilde{b}$, $\tilde{c}$, $\tilde{d} \in \mathbb{R}$ that also gives

$$\psi(z) = \frac{\tilde{a}z + \tilde{b}}{\tilde{c}z + \tilde{d}}.$$

- **Exercise 8.8.** Explain why it is obvious that for any $a \in \Pi^+$ there exists $\psi \in \mathrm{Aut}(\Pi^+)$ with $\psi(a) = i$. (This follows from the corresponding fact in the disk. But an example is much more obvious in the upper half-plane; pretend you don't know the result for the disk and give the obvious argument for Π^+.)

If the coefficients in Exercise 8.7 were unique, this would say that $\mathrm{Aut}(\Pi^+)$ was isomorphic to $SL_2(\mathbb{R})$, the group of all 2×2 real matrices with determinant 1. The fact that the coefficients are actually almost unique shows that $\mathrm{Aut}(\Pi^+)$ is in fact isomorphic to the quotient $SL_2(\mathbb{R})/G$, where

G is a certain small subgroup of $SL_2(\mathbb{R})$ (your solution to the exercise should make it clear what G is).

We used the vanilla version of the Schwarz Lemma in calculating the automorphism group of the disk. Now that we know the automorphisms of the disk we can show how a somewhat more sophisticated version of the Schwarz Lemma follows.

8.5. More on the Schwarz Lemma

The origin is a special point in the Schwarz Lemma (Theorem 8.4.0). Since Aut($\mathbb{D}$) acts transitively on $\mathbb{D}$ (that is, for any $z, w \in \mathbb{D}$ there exists $\psi \in$ Aut($\mathbb{D}$) with $\psi(z) = w$, namely $\psi = \phi_w \circ \phi_z$), every point of $\mathbb{D}$ is really the same as every other point; there should be an "invariant" form of the Schwarz Lemma that reflects this symmetry. And indeed there is.

Since the function f in Theorem 8.4.0 is assumed to fix the origin, the conclusion
$$|f(z)| \leq |z|$$
is the same as
$$|f(z) - f(0)| \leq |z - 0|;$$
the distance from $f(z)$ to $f(0)$ is no larger than the distance from z to 0. The invariant form of the theorem below says that if f is *any* holomorphic map from the disk to itself then the distance from $f(z)$ to $f(w)$ is no larger than the distance from z to w, for *any* $z, w \in \mathbb{D}$. Except not for the euclidean distance; instead this is true for a certain "invariant" distance, that is, for a certain metric d on $\mathbb{D}$ such that $d(\psi(z), \psi(w)) = d(z, w)$ for all $z, w \in \mathbb{D}$ and $\psi \in$ Aut($\mathbb{D}$).

There are various ways one can construct an invariant metric on $\mathbb{D}$. We will take the crudest possible approach, and try to get a metric d which is invariant and which also satisfies
$$d(z, 0) = |z|$$
for $z \in \mathbb{D}$. If you think about it for a second, you see that *if* d is an invariant metric on $\mathbb{D}$ satisfying this condition then we must have
$$d(z, w) = |\phi_z(w)|$$
for all $z, w \in \mathbb{D}$. It is not quite obvious that this formula actually defines a metric on $\mathbb{D}$ but it turns out to be true; the metric so defined is known as the *pseudo-hyperbolic distance* on $\mathbb{D}$. In this section we shall show that the pseudo-hyperbolic distance is in fact an invariant metric on $\mathbb{D}$ and also prove

the Aut($\mathbb{D}$)-invariant form of the Schwarz Lemma for the pseudo-hyperbolic distance, due to Pick.

First the Schwarz Lemma:

Theorem 8.5.0 (Invariant Schwarz Lemma). *Suppose that $f : \mathbb{D} \to \mathbb{D}$ is holomorphic. If $f \notin \mathrm{Aut}(\mathbb{D})$ then*

$$d(f(z), f(w)) < d(z, w)$$

for all $z, w \in \mathbb{D}$ with $z \neq w$ and

$$\frac{|f'(z)|}{1 - |f(z)|^2} < \frac{1}{1 - |z|^2}$$

for all $z \in \mathbb{D}$, while if $f \in \mathrm{Aut}(\mathbb{D})$ then

$$d(f(z), f(w)) = d(z, w)$$

and

$$\frac{|f'(z)|}{1 - |f(z)|^2} = \frac{1}{1 - |z|^2}$$

for all $z, w \in \mathbb{D}$. (Here $d(z, w) = |\phi_z(w)|$, as above.)

Proof. Fix $z \in \mathbb{D}$ and let $a = f(z)$. Let

$$g = \phi_a \circ f \circ \phi_z.$$

Now $g(0) = 0$, and so Theorem 8.4.0 shows that

$$|g(w)| \leq |w|$$

for all $w \in \mathbb{D}$. That is,

$$|\phi_a(f(\phi_z(w)))| \leq |w|$$

for all $w \in \mathbb{D}$; replacing w by $\phi_z(w)$ this shows that

$$|\phi_a(f(w))| \leq |\phi_z(w)|$$

for all w. In other words,

$$|\phi_{f(z)}(f(w))| \leq |\phi_z(w)|,$$

or

$$d(f(z), f(w)) \leq d(z, w).$$

8.5. More on the Schwarz Lemma

Similarly, Theorem 8.4.0 shows that

$$|g'(0)| \leq 1,$$

so that

$$|\phi_a'(a)||f'(z)||\phi_z'(0)| \leq 1$$

by the chain rule. Now Lemma 8.4.2 shows that

$$\frac{|f'(z)|(1-|z|^2)}{1-|a|^2} \leq 1,$$

or

$$\frac{|f'(z)|}{1-|f(z)|^2} \leq \frac{1}{1-|z|^2}.$$

We are done except for showing that if $f \in \text{Aut}(\mathbb{D})$ then we have equality throughout, while if $f \notin \text{Aut}(\mathbb{D})$ we have strict inequality throughout (at least for $z \neq w$).

Suppose first that $f \in \text{Aut}(\mathbb{D})$. Then g is an element of $\text{Aut}(\mathbb{D})$ fixing the origin, and hence g is a rotation, by Theorem 8.4.1. Hence $|g(w)| = |w|$ for all w and $|g'(0)| = 1$, which gives equality in all the inequalities above.

Finally, if we have $d(f(z), f(w)) = d(z, w)$ for some $w \neq z$ then we have $|g(w')| = |w'|$ for $w' = \phi_z(w)$ (note that $w' \neq 0$), while if we have $\frac{|f'(z)|}{1-|f(z)|^2} = \frac{1}{1-|z|^2}$ then we have $|g'(0)| = 1$. In either case Theorem 8.4.0 shows that g is a rotation, and hence $f \in \text{Aut}(\mathbb{D})$. □

Very interesting: Automorphisms of the disk are isometries in the pseudo-hyperbolic distance, while holomorphic maps from the disk to itself other than automorphisms are strictly distance-decreasing in the pseudo-hyperbolic distance. It would be nice to know that d is actually an invariant metric.

Theorem 8.5.0 says in part that $d(\phi(z), \phi(w)) = d(z, w)$ for all $z, w \in \mathbb{D}$ if $\phi \in \text{Aut}(\mathbb{D})$. It is clear that $d(z, w) \geq 0$ for all $z, w \in \mathbb{D}$ and that $d(z, w) = 0$ if and only if $z = w$ (because $\phi_z(w) = 0$ if and only if $z = w$.) It is also clear from the definition of $\phi_z(w)$ that $d(z, w) = d(w, z)$; the only thing that is not obvious is the triangle inequality.

Since d is invariant under the action of $\text{Aut}(\mathbb{D})$, it suffices to show that

$$d(z, w) \leq d(z, 0) + d(0, w)$$

for all $z, w \in \mathbb{D}$ (because $d(\phi_a(z), \phi_a(w)) \leq d(\phi_a(z), 0) + d(0, \phi_a(w))$ implies that $d(z, w) \leq d(z, a) + d(a, w)$.) That is, we need only verify that

$$d(z, w) \leq |z| + |w|.$$

In fact the stronger inequality

$$d(z,w) \le \frac{|z|+|w|}{1+|z||w|}$$

holds:

If one just applies the ordinary triangle directly to the definition of $d(z,w)$ one obtains the *weaker* inequality

$$d(z,w) \le \frac{|z|+|w|}{1-|z||w|}.$$

The problem is we need to apply the triangle inequality in one direction for the numerator and in the other direction for the denominator. But if we use our formula for $1-|\phi_a|^2$, the proof is immediate:

$$1-d(z,w)^2 = \frac{(1-|z|^2)(1-|w|^2)}{|1-\bar{z}w|^2} \ge \frac{(1-|z|^2)(1-|w|^2)}{(1+|z||w|)^2} = 1-d(|z|,-|w|)^2,$$

so

$$d(z,w) \le d(|z|,-|w|) = \left|\frac{|z|-(-|w|)}{1-\overline{|z|}(-|w|)}\right| = \frac{|z|+|w|}{1+|z||w|}.$$

So d is an invariant metric on $\mathbb{D}$. We should note for future reference that a similar argument shows that

$$d(|z|,|w|) \le d(z,w) \le d(|z|,-|w|).$$

The reader may be wondering what if anything the second half of Theorem 8.5.0 says about f decreasing distances between points. The inequality

$$\frac{|f'(z)|}{1-|f(z)|^2} \le \frac{1}{1-|z|^2}$$

says precisely that f does not increase distances in the *hyperbolic* metric. We are not going to say exactly what this means; it would take a little while because for example the hyperbolic metric is not a metric at all in the sense in which we are using the word, instead it is a "riemannian metric". But we can give a hint: If $f : \mathbb{D} \to \mathbb{D}$ is holomorphic then

$$\frac{|f'(z)|(1-|z|^2)}{1-|f(z)|^2} = \lim_{w \to z} \frac{d(f(z),f(w))}{d(z,w)}.$$

In fact a somewhat stronger statement is easy to prove:

8.5. More on the Schwarz Lemma

- **Exercise 8.9.** Suppose that $f : \mathbb{D} \to \mathbb{D}$ is holomorphic and define $\Phi : \mathbb{D} \times \mathbb{D} \to \mathbb{C}$ by

$$\Phi(z,w) = \begin{cases} \dfrac{d(f(z), f(w))}{d(z,w)} & (z \neq w), \\ \dfrac{|f'(z)|(1-|z|^2)}{1-|f(z)|^2} & (z = w). \end{cases}$$

Show that Φ is continuous.

Hint: If $f(z) \neq f(w)$ then

$$\frac{d(f(z), f(w))}{d(z,w)} = \left| \frac{f(z) - f(w)}{z - w} \right| \frac{\left| \dfrac{\phi_{f(z)}(f(w)) - \phi_{f(z)}(f(z))}{f(z) - f(w)} \right|}{\left| \dfrac{\phi_z(w) - \phi_z(z)}{z - w} \right|}.$$

It follows from Exercise 8.9 and Theorem 8.5.0 that if $f : \mathbb{D} \to \mathbb{D}$ is holomorphic, $f \notin \text{Aut}(\mathbb{D})$, and $K \subset \mathbb{D}$ is compact then there exists $c < 1$ such that $|\Phi| \leq c$ on $K \times K$, and hence

$$d(f(z), f(w)) \leq c\, d(z,w) \quad (z, w \in K, z \neq w),$$

which is to say that the inequality $d(f(z), f(w)) < d(z,w)$ holds uniformly (in a sense) on compact sets, as it should.

One final comment regarding the Schwarz Lemma versus hyperbolic geometry (whatever that is) before we get back to work: If $\gamma : [a,b] \to \mathbb{D}$ is a smooth curve in the unit disk then we define the *hyperbolic length* of γ by

$$\mathcal{L}(\gamma) = \int_a^b \frac{|\gamma'(t)|}{1 - |\gamma(t)|^2} \, dt.$$

Just like the ordinary length except for that factor of $1/(1-|\gamma(t)|^2)$ (including this factor means that the parts of the curve near the boundary of the disk are longer than they appear).

- **Exercise 8.10.** Suppose that $f : \mathbb{D} \to \mathbb{D}$ is holomorphic and γ is a smooth curve in $\mathbb{D}$. Show that

$$\mathcal{L}(f \circ \gamma) \leq \mathcal{L}(\gamma),$$

with equality if and only if $f \in \text{Aut}(\mathbb{D})$ or $\mathcal{L}(\gamma) = 0$.

- **Exercise 8.11.** Suppose $z \in \mathbb{D}$ and $K \subset \mathbb{D}$ is compact. Show that there exists $c > 0$ such that

$$1 - |f(w)| \leq c(1 - |f(z)|) \quad (w \in K)$$

for every holomorphic $f : \mathbb{D} \to \mathbb{D}$.

  Hint: Lemma 8.4.2(iv) may be useful, as in the proof of the triangle inequality for the pseudo-hyperbolic metric d. Begin by showing that there exists $r < 1$ such that $d(z, w) \leq r$ for all $w \in K$; now apply Theorem 8.5.0. Note that if $a, b \in [0, 1)$ then

$$\frac{(1-a^2)(1-b^2)}{(1-ab)^2} \leq \frac{(1-a^2)(1-b^2)}{(1-b)^2} = \frac{(1-a)(1+a)(1+b)}{1-b} \leq 4\frac{1-a}{1-b}.$$

Chapter 9

Normal Families and the Riemann Mapping Theorem

9.0. Introduction

Our next major goal is the Riemann Mapping Theorem, which states that if D is a simply connected open subset of the plane (other than the plane itself) then D is conformally equivalent to the unit disk.

This is a remarkable result; one wonders where the conformal mapping could possibly come from. In the proof we shall give the mapping arises as the solution to a certain extremal problem (as often happens in existence proofs in analysis). In very rough outline: We fix a point $z_0 \in D$, and we choose a one-to-one holomorphic function $f : D \to \mathbb{D}$ with $f(z_0) = 0$ which maximizes $|f'(z_0)|$. It turns out that this implies that $f(D) = \mathbb{D}$; if we assume that $p \in \mathbb{D} \setminus f(D)$ we can diddle things to get another such function g with $|g'(z_0)| > |f'(z_0)|$.

It is not obvious how any of that is actually going to work. But we already have the tools we need to show that an f maximizing $|f'(z_0)|$ must be surjective (not that this is supposed to be obvious at present); what we do not have yet is a way to show that there *is* an f maximizing $|f'(z_0)|$. (Simply assuming that extremal problems have solutions is a traditional way to construct incorrect proofs of theorems that then need to be fixed by the next generation of mathematicians.) It turns out that there must exist such an extremal f because the set of all injective holomorphic f mapping D to $\mathbb{D}$ with $f(z_0) = 0$ and $|f'(z_0)| \geq c > 0$ is compact in a certain metric.

This compactness is given to us by Montel's Theorem on "normal families". We begin the chapter with a fairly large amount of preliminary material which will be used in the proof of Montel's Theorem. One can give an *ad hoc* proof of Montel's Theorem without all the machinery of quasi-metrics, equicontinuity, and so on that we develop in the next two sections. But that material is important in various places in analysis, much of it seems quite interesting (for example, the fact that there is a metric d on $H(D)$ such that $f_n \to f$ uniformly on compact subsets of D if and only if $d(f_n, f) \to 0$, or the characterization of compactness in $C(D)$ in terms of equicontinuity), and we feel it gives the best explanation for why Montel's Theorem is really true. Impatient readers can skip Sections 9.1, 9.2 and 9.3 in this chapter; we give the *ad hoc* proof of Montel's Theorem in section 9.5.

There are other possible approaches to the Riemann Mapping Theorem; for example in Section 10.5 below we will see that it follows from the assumption that the "Dirichlet problem" can be solved for D. (From our point of view that is a much weaker version of the theorem, because the hypotheses are so much stronger. But the version of the theorem in Section 10.5 is perfectly adequate for most of the simply connected open sets D that actually arise in practice. Things were a lot simpler a century or so ago, when an open set in the plane was at worst something bounded by a few piecewise-smooth curves ...)

The fact that maximizing $|f'(z_0)|$ (subject to the restrictions outlined above) gives a function with $f(D) = \mathbb{D}$ may seem mysterious. In order to demonstrate that it *should* seem at least somewhat mysterious, we mention another approach to proving the theorem which makes at least as much sense *a priori* but which does not actually work: Suppose that $f = u + iv$ is an injective holomorphic mapping from D into $\mathbb{D}$. The Cauchy-Riemann equations show that

$$\det \begin{bmatrix} u_x & u_y \\ v_x & v_y \end{bmatrix} = u_x v_y - u_y v_x = u_x^2 + v_x^2 = |f'|^2.$$

That is, $|f'|^2$ is the Jacobian of f, regarded as a mapping from $\mathbb{R}^2$ to $\mathbb{R}^2$. So the change-of-variables formula from advanced calculus shows that

$$\iint_D |f'(x+iy)|^2 \, dx \, dy$$

is just the area of the image $f(D)$, and so it seems at least plausible that maximizing $\iint_D |f'|^2$ should have the effect of making $f(D)$ as large as possible, in particular implying that $f(D) = \mathbb{D}$.

That doesn't quite work; we needn't worry about how we might show that there exists f as above with $\iint_D |f'|^2 = \pi$ because that would still not imply that $f(D) = \mathbb{D}$. For example, the Riemann Mapping Theorem implies that there exists a conformal equivalence mapping D onto the slit disk $\mathbb{D} \setminus [1/2, 1)$, and for this f we have $\iint_D |f'|^2 = \pi$ but $f(D) \neq \mathbb{D}$.

We will see why maximizing $|f'(z_0)|$ *does* imply that $f(D) = \mathbb{D}$ in Section 9.4, after the preliminaries alluded to above.

9.1. Quasi-Metrics

A *quasi-metric* on the set X is a function $d : X \times X \to [0, \infty)$ satisfying the conditions
$$d(x, x) = 0 \quad (x \in X),$$
$$d(x, y) = d(y, x) \quad (x, y \in X),$$
and
$$d(x, z) \leq d(x, y) + d(y, z) \quad (x, y, z \in X).$$

That is, a quasi-metric is a metric except that $d(x, y)$ can equal 0 even when $x \neq y$. (If d is a quasi-metric on X it is easy to see that $d(x, y) = 0$ defines an equivalence relation on X and that d induces a metric on the corresponding space of equivalence classes.)

It is often convenient to know that given a quasi-metric d there exists a bounded quasi-metric $\tilde{d}$ which is equivalent to d in the sense that $d(x_n, y_n) \to 0$ if and only if $\tilde{d}(x_n, y_n) \to 0$. For this we introduce the notion of *concave function*:

Suppose that $I \subset \mathbb{R}$ is convex (in other words, I is an open, half-open or closed, bounded or unbounded interval). The function $\psi : I \to \mathbb{R}$ is said to be *concave* if
$$\psi(tx + (1-t)y) \geq t\psi(x) + (1-t)\psi(y)$$
whenever $x, y \in I$ and $0 \leq t \leq 1$. (Thus ψ is concave if and only if $-\psi$ is convex. Note that if ψ is continuously differentiable then it is easy to see that ψ is concave if and only if ψ' is nonincreasing.)

Lemma 9.1.0. *Suppose that $\psi : I \to \mathbb{R}$ is concave. Suppose that a, b, a', $b' \in I$, $a < b$, $a' < b'$, $a' \geq a$, and $b' \geq b$. Then*
$$\frac{\psi(b') - \psi(a')}{b' - a'} \leq \frac{\psi(b) - \psi(a)}{b - a}.$$

That is, the slope of the segment joining two points on the graph of ψ decreases as the points move to the right.

Proof. Since $a < b \le b'$, there exists $t \in [0,1)$ such that $b = ta + (1-t)b'$. It follows that
$$\psi(b) = \psi(ta + (1-t)b') \ge t\psi(a) + (1-t)\psi(b');$$
rearranging this shows that
$$\psi(b) - \psi(a) \ge (1-t)(\psi(b') - \psi(a)),$$
and noting that $1 - t = (b-a)/(b'-a)$ shows that
$$\frac{\psi(b) - \psi(a)}{b-a} \ge \frac{\psi(b') - \psi(a)}{b'-a}.$$
Similarly
$$\frac{\psi(b') - \psi(a)}{b'-a} \ge \frac{\psi(b') - \psi(a')}{b'-a'},$$
and the result follows. $\square$

Lemma 9.1.1. *Suppose that $\psi : [0,\infty) \to \mathbb{R}$ is concave and*
$$\psi(0) = 0.$$
Then
$$\psi(x+y) \le \psi(x) + \psi(y)$$
for all $x, y \ge 0$.

Proof. We may suppose that $x > 0$. An application of the previous lemma shows that
$$\frac{\psi(x+y) - \psi(y)}{x} \le \frac{\psi(x) - \psi(0)}{x}$$
and then, since $\psi(0) = 0$, the result follows. $\square$

Lemma 9.1.2. *Suppose that $\psi : [0,\infty) \to \mathbb{R}$ is concave and*
$$\psi(0) = 0.$$
Suppose further that ψ is nondecreasing, $\psi(t) > 0$ for all $t > 0$, and ψ is continuous at 0.

If d is a quasi-metric on X then $\tilde{d} = \psi \circ d$ is also a quasi-metric on X, such that

(i) $\tilde{d}(x,y) = 0$ *if and only if* $d(x,y) = 0$,

(ii) $\tilde{d}(x_n, y_n) \to 0$ *if and only if* $d(x_n, y_n) \to 0$.

(Note that if d is a metric then so is $\tilde{d}$. Regarding the hypotheses of the lemma we should note that the concavity of ψ easily implies that ψ must be continuous on $(0, \infty)$, but it does not imply continuity at 0, which is actually the only point where we care about continuity.)

9.1. Quasi-Metrics

Proof. It is clear that $\tilde{d}(x,y) = \tilde{d}(y,x)$ and $\tilde{d}(x,x) = 0$. It is also clear that $\tilde{d}(x,y) = 0$ if and only if $d(x,y) = 0$, since $\psi(t) = 0$ if and only if $t = 0$. The triangle inequality for $\tilde{d}$ follows from the triangle inequality for d, the monotonicity of ψ, and the previous lemma:

$$\tilde{d}(x,z) = \psi(d(x,z)) \le \psi(d(x,y) + d(y,z))$$
$$\le \psi(d(x,y)) + \psi(d(y,z)) = \tilde{d}(x,y) + \tilde{d}(y,z).$$

Finally, if $d(x_n, y_n) \to 0$ then the fact that ψ is continuous at 0 shows that $\tilde{d}(x_n, y_n) \to 0$. Suppose on the other hand that $d(x_n, y_n) \not\to 0$; then there exists $\epsilon > 0$ such that $d(x_n, y_n) > \epsilon$ for infinitely many values of n. Since ψ is nondecreasing, it follows that $\tilde{d}(x_n, y_n) \ge \psi(\epsilon) > 0$ whenever $d(x_n, y_n) \ge \epsilon$, so that $\tilde{d}(x_n, y_n) \not\to 0$. □

There exist bounded functions ψ satisfying the hypotheses of Lemma 9.1.2; indeed, two traditional choices are

$$\psi_1(t) = \frac{t}{1+t}$$

and

$$\psi_2(t) = \begin{cases} t & (0 \le t \le 1), \\ 1 & (t > 1). \end{cases}$$

(One might regard ψ_1 as preferable because it is smooth, or one might prefer ψ_2 because it is easy to show directly that it does what Lemma 9.1.2 says it does.) Hence the lemma shows that given a quasi-metric there exists a bounded quasi-metric which is "equivalent" in the sense described in the lemma. (We should note that the word "equivalent" often means something different in this context, hence the quote marks.) The reason we care is that bounded things are easier to add up. For example, the next lemma, showing that given a countable family of quasi-metrics (d_j) there exists a single quasi-metric d such that $d(x_n, y_n) \to 0$ as $n \to \infty$ if and only if $d_j(x_n, y_n) \to 0$ for every j, begins by converting the original family of quasi-metrics to a family of bounded quasi-metrics.

Lemma 9.1.3. *Suppose that d_j is a quasi-metric on X for $j = 1, 2, \ldots$. Define*

$$d(x,y) = \sum_{j=1}^{\infty} 2^{-j} \frac{d_j(x,y)}{1 + d_j(x,y)} \quad (x, y \in X).$$

Then d is a quasi-metric on X with the property that $d(x_n, y_n) \to 0$ as $n \to \infty$ if and only if $d_j(x_n, y_n) \to 0$ for every j.

Furthermore, d is a metric on X if and only if for every $x, y \in X$ with $x \ne y$ there exists a positive integer j such that $d_j(x,y) > 0$ (that is, if and only if the family (d_j) "separates points" of X).

Proof. Note first that the sum converges for all $x, y \in X$, and in fact $d(x,y) \leq 1$. One may easily verify that if $\psi(t) = t/(1+t)$ then ψ satisfies the hypotheses of Lemma 9.1.2 (the concavity follows from the fact that $\psi'' < 0$); this shows that $\tilde{d}_j = d_j/(1+d_j)$ is a quasi-metric for each j, and hence that d is a quasi-metric on X.

Suppose that (x_n) and (y_n) are two sequences in X, such that for every j we have $d_j(x_n, y_n) \to 0$ as $n \to \infty$. It follows from Lemma 9.1.4 below that $d(x_n, y_n) \to 0$.

If on the other hand there exists a positive integer j such that $d_j(x_n, y_n) \not\to 0$ then it follows that $\tilde{d}_j(x_n, y_n) \not\to 0$, and hence that $d(x_n, y_n) \not\to 0$, since $d(x,y) \geq 2^{-j}\tilde{d}_j(x,y)$.

Finally, it is clear that d is a metric if and only if for every $x, y \in X$ with $x \neq y$ there exists a positive integer j such that $d_j(x,y) > 0$. $\square$

Readers who know some real analysis will recognize Lemma 9.1.4 as a trivial special case of the Dominated Convergence Theorem; Lemma 17.7 below is another. (Readers who have not studied measure theory may wish to look at both lemmas, convince themselves that the basic idea behind the two lemmas is the same, and then contemplate the question of what sort of abstraction could contain both lemmas as a special case; the answer is "measure theory".)

Lemma 9.1.4. *Suppose that $c_j \geq 0$ for $j = 1, 2, \ldots$ and*

$$\sum_{j=1}^{\infty} c_j < \infty.$$

Suppose that $(a_{j,n})$ is a double sequence such that

$$|a_{j,n}| \leq c_j$$

for all j and n, and such that

$$\lim_{n \to \infty} a_{j,n} = a_j$$

for all j. Then

$$\lim_{n \to \infty} \sum_{j=1}^{\infty} a_{j,n} = \sum_{j=1}^{\infty} a_j.$$

9.1. Quasi-Metrics

Proof. Let $\epsilon > 0$. Choose M such that

$$\sum_{j=M+1}^{\infty} c_j < \epsilon/4.$$

Now it is clear that

$$\lim_{n \to \infty} \sum_{j=1}^{M} a_{j,n} = \sum_{j=1}^{M} a_j,$$

so we can find N such that

$$\left| \sum_{j=1}^{M} a_{j,n} - \sum_{j=1}^{M} a_j \right| < \epsilon/2$$

for all $n > N$. For such n we have

$$\left| \sum_{j=1}^{\infty} a_{j,n} - \sum_{j=1}^{\infty} a_j \right| \le \left| \sum_{j=1}^{M} a_{j,n} - \sum_{j=1}^{M} a_j \right| + \left| \sum_{j=M+1}^{\infty} a_{j,n} - \sum_{j=M+1}^{\infty} a_j \right|$$

$$< \epsilon/2 + \sum_{j=M+1}^{\infty} (|a_{j,n}| + |a_j|)$$

$$\le \epsilon/2 + 2 \sum_{j=M+1}^{\infty} c_j$$

$$\le \epsilon/2 + 2\epsilon/4$$

$$= \epsilon. \qquad \square$$

It follows that the open balls in this metric are "equivalent" in a sense to the open balls associated with our quasi-metrics:

Lemma 9.1.5. *Suppose that (d_j) is a sequence of quasi-metrics on X which separates points of X, so that*

$$d(x,y) = \sum_{j=1}^{\infty} 2^{-j} \frac{d_j(x,y)}{1 + d_j(x,y)} \quad (x, y \in X)$$

defines a metric on X, as in Lemma 9.1.3. For $x \in X$ and $r > 0$ define

$$B_d(x, r) = \{ y \in X : d(x,y) < r \},$$

as usual. For each positive integer N, $x \in X$ and $\epsilon > 0$ define

$$B_N(x, \epsilon) = \{ y \in X : d_j(x,y) < \epsilon, \, j = 1, 2, \ldots, N \}.$$

(i) *Each $B_N(x,\epsilon)$ is open in the metric space (X,d).*

(ii) *For every $r > 0$ there exist a positive integer N and a number $\epsilon > 0$ such that*
$$B_N(x,\epsilon) \subset B_d(x,r)$$
for all $x \in X$.

(iii) *For every $\epsilon > 0$ and positive integer N there exists a number $r > 0$ such that*
$$B_d(x,r) \subset B_N(x,\epsilon)$$
for all $x \in X$.

We will not be using (i) so we leave it as an exercise. (It follows from a modification of the proof of (iii) given below; it is possible one could instead deduce it from the statement of (iii) itself, plus a little argument.) We should point out that, appearances to the contrary, (iii) does not quite follow from (i). If we know that $B_N(x,\epsilon)$ is open then it follows that for every $\epsilon > 0$, every positive integer N and every $x \in X$ there exists a number $r > 0$ such that $B_d(x,r) \subset B_N(x,\epsilon)$; (iii) is a stronger statement, because the value of r in (iii) is independent of x. The distinction is important in the application below (Proposition 9.1.6).

Proof. We leave (i) as an exercise. For (ii): Suppose $r > 0$ is given. Choose a positive integer N such that
$$\sum_{j=N+1}^{\infty} 2^{-j} < r/2$$
and let $\epsilon = r/2$.

Suppose that $y \in B_N(x,\epsilon)$. Then $d_j(x,y) < r/2$ for $1 \leq j \leq N$, and so
$$d(x,y) \leq \sum_{j=1}^{N} 2^{-j} d_j(x,y) + \sum_{j=N+1}^{\infty} 2^{-j} < \frac{r}{2}\sum_{j=1}^{\infty} 2^{-j} + \frac{r}{2} = \frac{r}{2} + \frac{r}{2} = r;$$
hence $y \in B_d(x,r)$.

Now for (iii): Suppose that $\epsilon > 0$ and a positive integer N are given. Choose $\delta > 0$ such that $d_j(p,q) < \epsilon$ whenever $\tilde{d}_j(p,q) < \delta$ (here $\tilde{d}_j = d_j/(1+d_j)$ as previously). Let $r = 2^{-N}\delta$.

Suppose that $y \in B_d(x,r)$ for some $x \in X$. Since $2^{-j}\tilde{d}_j \leq d$, it follows that
$$\tilde{d}_j(x,y) < 2^j r \leq 2^N r = \delta$$
for $1 \leq j \leq N$. Our choice of δ shows that $d_j(x,y) < \epsilon$ for $1 \leq j \leq N$, and so $y \in B_N(x,\epsilon)$. □

We finally get to the reason we're doing all this:

9.2. Convergence and Compactness in $C(D)$

Proposition 9.1.6. *Suppose that (d_j) is a sequence of quasi-metrics on X which separates points of X (that is, for any $x, y \in X$ with $x \neq y$ there exists a j with $d_j(x,y) > 0$). There exists a metric d on X with the following two properties:*

(i) *For any sequences (x_n) and (y_n) in X we have $d(x_n, y_n) \to 0$ if and only if $d_j(x_n, y_n) \to 0$ for all j.*

(ii) *A set $S \subset X$ is totally bounded in the metric space (X, d) if and only if for every positive integer N and every $\epsilon > 0$ there exist finitely many points $x_1, \ldots, x_n \in X$ such that*

$$S \subset \bigcup_{k=1}^{n} B_N(x_k, \epsilon).$$

The notation in Proposition 9.1.6 is as in Lemma 9.1.5. We leave the proof as an exercise. (See Appendix 4 for the definition of "totally bounded".)

- **Exercise 9.1.** Prove part (i) of Lemma 9.1.5.

- **Exercise 9.2.** Prove Proposition 9.1.6.

9.2. Convergence and Compactness in $C(D)$

In this section we will give a version of the Arzelà-Ascoli Theorem, a characterization of the compact subsets of $C(D)$. We will restrict attention to a special case; it is clear the results can be generalized.

Throughout this section D will be an open subset of the plane, and $C(D)$ will denote the space of all continuous functions from D to $\mathbb{C}$. Given a sequence $(f_n) \subset C(D)$ and a function $f \in C(D)$ we say that $f_n \to f$ "in $C(D)$" if $f_n \to f$ uniformly on K for every compact $K \subset D$.

Using the standard notation

$$\|f\|_K = \sup_{z \in K} |f(z)|$$

we see that $f_n \to f$ in $C(D)$ if and only if

$$\|f_n - f\|_K \to 0$$

for every compact $K \subset D$.

It is easy to see that if K is a compact subset of D and $d_K(f, g) = \|f - g\|_K$ then d_K is a quasi-metric on $C(D)$. So we have a family of quasi-metrics on $C(D)$ and convergence in $C(D)$ is equivalent to simultaneous convergence with respect to each of these quasi-metrics. We wish to conclude that convergence in $C(D)$ is actually convergence in a certain metric; this is exactly why we proved the results in the previous section. We cannot quite apply Lemma 9.1.3 yet, because D has uncountably many compact subsets. But it turns out that one can find a countable collection of compact sets that suffice:

Lemma 9.2.0. *Suppose that D is an open subset of the plane. There exist compact sets K_j $(j = 1, 2, \dots)$ such that*

$$K_j \subset K_{j+1}^\circ$$

and

$$D = \bigcup_{j=1}^\infty K_j.$$

We will call such a sequence of compact sets an *exhaustion* of D.

Proof. If $D = \mathbb{C}$ we can take K_j to be $\overline{D(0, j)}$; assume that $D \neq \mathbb{C}$. For $z \in D$ let $\rho(z)$ denote the distance from z to the complement of D:

$$\rho(z) = \inf_{w \in \mathbb{C} \setminus D} |z - w|.$$

Define

$$K_j = \{\, z \in D : |z| \leq j,\, \rho(z) \geq 1/j \,\}.$$

Now K_j is certainly a subset of D, and the fact that D is open shows that $D = \bigcup_{j=1}^\infty K_j$: If $z \in D$ then there exists $r > 0$ such that $D(z, r) \subset D$, and now if we choose j so that $1/j < r$ and $j > |z|$ it follows that $z \in K_j$.

Finally, if $z \in K_j$ and $r = 1/j - 1/(j+1)$ then $D(z, r) \subset K_{j+1}$; hence $K_j \subset K_{j+1}^\circ$. $\square$

Note that it follows that $D = \bigcup_{j=1}^\infty K_j^\circ$. This shows that any compact subset of D is contained in one of the K_j, and hence that $f_n \to f$ in $C(D)$ if and only if $\|f_n - f\|_{K_j} \to 0$ for every j. And so we obtain the following:

Theorem 9.2.1. *Suppose that D is an open subset of the plane. Let (K_j) be an exhaustion of D, and define*

$$d(f, g) = \sum_{j=1}^\infty 2^{-j} \frac{\|f - g\|_{K_j}}{1 + \|f - g\|_{K_j}}.$$

Then d is a metric on $C(D)$, and for a sequence $(f_n) \subset C(D)$ and a function $f \in C(D)$ we have $f_n \to f$ in $C(D)$ if and only if $d(f_n, f) \to 0$.

9.2. Convergence and Compactness in $C(D)$

Proof. Lemma 9.1.3 shows that d is a quasi-metric on $C(D)$ and that $d(f_n, f) \to 0$ if and only if $||f_n - f||_{K_j} \to 0$ for every j. But if K is any compact subset of D then there exists j such that $K \subset K_j$, which implies that $||f_n - f||_K \le ||f_n - f||_{K_j}$. So if $||f_n - f||_{K_j} \to 0$ for every j then it follows that $||f_n - f||_K \to 0$ for every compact $K \subset D$.

Lemma 9.1.3 also shows that d is actually a metric on D. For this we need to show that if $f, g \in C(D)$ and $f \ne g$ then there exists j such that $||f - g||_{K_j} > 0$. This is clear: Since f and g are functions on D, by definition if $f \ne g$ then there exists $z \in D$ with $f(z) \ne g(z)$. Now there exists j with $z \in K_j$, and hence

$$||f - g||_{K_j} \ge |f(z) - g(z)| > 0. \qquad \square$$

So our notion of "$f_n \to f$ in $C(D)$" is actually convergence with respect to a certain metric. This is a good thing, because we know something about metric spaces. In particular we know (see Appendix 4) that a subset of a complete metric space is compact if and only if it is closed and totally bounded; we are going to use this to characterize the compact subsets of $C(D)$.

We should note that D has more than one exhaustion, and so the preceding theorem gives various metrics on $C(D)$, one for each exhaustion. Note however that if d and $\tilde{d}$ are two metrics on $C(D)$, obtained from Theorem 9.2.1 using different exhaustions of D, then the theorem shows that $d(f_n, f) \to 0$ if and only if $\tilde{d}(f_n, f) \to 0$.

For the rest of this section and the next we will assume that we have fixed an exhaustion of D and let d denote the metric associated with this exhaustion by the theorem.

Lemma 9.2.2. *The metric space $(C(D), d)$ is complete.*

Proof. Suppose that (f_n) is a Cauchy sequence in $C(D)$. This says that $d(f_n, f_m) \to 0$ as $n, m \to \infty$, which implies that $||f_n - f_m||_{K_j} \to 0$ for every j. Hence for each j there is a continuous function F_j defined on K_j such that $f_n \to F_j$ uniformly on K_j. It follows that $F_j = F_k$ on $K_j \cap K_k$, so we may define a function $F : D \to \mathbb{C}$ by requiring that $F(z) = F_j(z)$ for $z \in K_j$. The fact that (K_j) is an exhaustion of D shows that

$$D = \bigcup_{j=1}^{\infty} K_j^\circ,$$

and it follows that F is continuous on D, since each K_j° is open. (There is a slightly subtle point here; see the example following the proof.) The fact that $f_n \to F_j$ uniformly on K_j shows that

$$\|f_n - F\|_{K_j} \to 0$$

for every j, which implies that $d(f_n, F) \to 0$. □

The reader should contemplate the following example, and then go back and make certain he or she understands exactly why we know that the function F constructed in the proof of Lemma 9.2.2 is continuous:

Example. Let $X = \{0\} \cup \{1/k : k = 1, 2, 3, \ldots\}$ (with the standard metric). For $j = 1, 2, \ldots$ let $K_j = \{0\} \cup \{1/k : 1 \leq k \leq j\}$. Define $F_j : K_j \to \mathbb{C}$ by $F_j(0) = 0$, $F_j(x) = 1$ for $x \in K_j$ with $x \neq 0$.

Now K_j is a compact subset of X, each F_j is continuous on K_j, and we also have $F_j = F_k$ on $K_j \cap K_k$, so that there exists a function F defined on X with $F(x) = F_j(x)$ for all $x \in K_j$. But F is not continuous on X.

The example gives another indication of why we take an exhaustion to satisfy $K_j \subset K_{j+1}^\circ$ instead of just $K_j \subset K_{j+1}$.

We are interested in determining the pre-compact subsets of $C(D)$, that is, the subsets with compact closure. Since $C(D)$ is a complete metric space, these are the same as the totally bounded subsets (again, see Appendix 4). It turns out that a subset of $C(D)$ is totally bounded if and only if it is pointwise bounded and "equicontinuous" at every point:

Definition. Suppose $S \subset C(D)$ and $z \in D$. We say that S is *equicontinuous* at z if for every $\epsilon > 0$ there exists $\delta > 0$ such that $|f(w) - f(z)| < \epsilon$ whenever $f \in S$ and $|w - z| < \delta$.

That is, a family of functions is equicontinuous at z if "the same δ works for every function in the family". One should contrast this with the notion of uniform continuity, which says that the same δ works for every z.

One might guess the definition of "uniformly equicontinuous". People often do not talk explicitly about uniform equicontinuity, because in the contexts where it's likely to come up it often happens that uniform equicontinuity follows from pointwise equicontinuity plus the other hypotheses. (For example a family of functions which is equicontinuous at every point of a compact set is automatically uniformly equicontinuous on that compact set, just as a function continuous on a compact set must be uniformly continuous.)

The first thing to do is to note that Proposition 9.1.6 more or less reduces the problem of determining the totally bounded subsets of $C(D)$ to finding the totally bounded subsets of $C(K)$ for K compact. We have too many

9.2. Convergence and Compactness in $C(D)$

B's already, so we may as well add one more: If K is a compact subset of D we define
$$B_K(f,\epsilon) = \{\, g \in C(D) : \|f-g\|_K < \epsilon \,\}.$$

Lemma 9.2.3. *The set $S \subset C(D)$ is totally bounded if and only if for every compact $K \subset D$ and every $\epsilon > 0$ there exist finitely many functions $f_1,\ldots,f_n \in S$ such that*
$$S \subset \bigcup_{k=1}^{n} B_K(f_k,\epsilon).$$

Proof. Let (K_j) be an exhaustion of D, and define $d_j(f,g) = \|f-g\|_{K_j}$ as previously. The fact that $K_{j+1} \subset K_j$ shows that $d_{j+1} \geq d_j$, and hence in the notation used in Proposition 9.1.6 we have
$$B_N(f,\epsilon) = B_{K_N}(f,\epsilon).$$

So Proposition 9.1.6 shows that S is totally bounded if and only if for every positive integer N and every $\epsilon > 0$ there exist finitely many functions $f_1,\ldots,f_n \in S$ such that
$$S \subset \bigcup_{k=1}^{n} B_{K_N}(f_k,\epsilon).$$

Since every K_N is compact and every compact subset of D is contained in some K_N, this is equivalent to the statement that for every compact $K \subset D$ and every $\epsilon > 0$ there exist finitely many functions $f_1,\ldots,f_n \in S$ such that
$$S \subset \bigcup_{k=1}^{n} B_K(f_k,\epsilon). \qquad \square$$

We can now state and prove a version of the Arzelà-Ascoli Theorem:

Theorem 9.2.4 (Arzelà-Ascoli). *The set $S \subset C(D)$ is <u>totally bounded</u> if and only if it is <u>bounded and equicontinuous at every point of D.</u>*

To be explicit, the first hypothesis means that for every $z \in D$ there exists $M = M(z)$ such that $|f(z)| \leq M$ for all $f \in S$.

Proof. Suppose first that S is totally bounded. Suppose that $z \in D$.

We can show that S is bounded at z by applying Lemma 9.2.3 with $K = \{z\}$ and $\epsilon = 1$: The lemma shows that there exist finitely many functions $f_1,\ldots,f_n \in S$ such that for every $f \in S$ there exists j with

$|f(z) - f_j(z)| < 1$, and so $|f(z)| \le M(z) = 1 + \max(|f_1(z)|, \ldots, |f_n(z)|)$ for all $f \in S$.

Showing that S is equicontinuous at z is almost as easy: Choose $r > 0$ so that $K = \overline{D(z,r)} \subset D$, and suppose that $\epsilon > 0$. By Lemma 9.2.3 there exist finitely many functions $f_1, \ldots, f_n \in S$ such that $S \subset \bigcup_{k=1}^n B_K(f_k, \epsilon/3)$. For each $j = 1, \ldots, n$ we can choose $\delta_j > 0$ so that $\delta_j < r$ and such that $|f_j(w) - f_j(z)| < \epsilon/3$ whenever $|z - w| < \delta_j$; thus if we set $\delta = \min(\delta_1, \ldots, \delta_n)$ then we have $|f_j(w) - f_j(z)| < \epsilon/3$ for $j = 1, \ldots, n$ whenever $|z - w| < \delta$. Now for any $f \in S$ we can choose j such that $|f - f_j| < \epsilon/3$ on all of K, and then if $|z - w| < \delta$ we obtain

$$|f(z) - f(w)| \le |f(z) - f_j(z)| + |f_j(z) - f_j(w)| + |f_j(w) - f(w)|$$
$$< \epsilon/3 + \epsilon/3 + \epsilon/3 = \epsilon.$$

The proof of the converse is slightly more intricate. Suppose that S is bounded and equicontinuous at every point of D. Suppose that $K \subset D$ is compact and $\epsilon > 0$; by Lemma 9.2.3 we need only show that there exist finitely many functions $f_1, \ldots, f_n \in S$ such that

$$S \subset \bigcup_{k=1}^n B_K(f_k, \epsilon).$$

For each $z \in D$ there exists $\delta_z > 0$ such that if $|z - w| < \delta_z$ then $|f(z) - f(w)| < \epsilon/3$ for all $f \in S$. By compactness there exist finitely many points $z_1, \ldots, z_m \in K$ such that

$$K \subset \bigcup_{k=1}^m D(z_k, \delta_{z_k}) = \bigcup_{k=1}^m D_k.$$

Now there exists $M < \infty$ such that for all $f \in S$ we have $|f(z_k)| \le M$ for $k = 1, \ldots, m$.

Choose finitely many disks $Q_1, \ldots, Q_N$ of radius $\epsilon/6$ such that $D(0, M) \subset \bigcup_{j=1}^N Q_j$. Let Ω be the set of all m-tuples α such that each coordinate α_k is a positive integer between 1 and N (that is, let $\Omega = \{1, \ldots, N\}^m$). Now, for each $\alpha \in \Omega$, if there exists $f \in S$ such that $f(z_k) \in Q_{\alpha_k}$ for $k = 1, \ldots, m$ then choose one and call it f_α; if there is no such f then let f_α be any element of S.

We now have a finite set $\{f_\alpha : \alpha \in \Omega\} \subset S$ with the following property: For any $f \in S$ there exists an α such that $|f(z_k) - f_\alpha(z_k)| < \epsilon/3$ for $k = 1, \ldots, m$. But this last condition implies that $|f - f_\alpha| < \epsilon$ on all of K:

If $z \in K$ then there exists k such that $|z - z_k| < \delta_{z_k}$, and hence

$$|f(z) - f_\alpha(z)| \leq |f(z) - f(z_k)| + |f(z_k) - f_\alpha(z_k)| + |f_\alpha(z_k) - f(z_k)|$$
$$< \epsilon/3 + \epsilon/3 + \epsilon/3 = \epsilon.$$

That is, $\|f - f_\alpha\|_K < \epsilon$, so $f \in B_K(f_\alpha, \epsilon)$.

We have shown that for any compact $K \subset D$ and any $\epsilon > 0$ there exists a finite set $\{f_\alpha : \alpha \in \Omega\} \subset S$ such that

$$S \subset \bigcup_{\alpha \in \Omega} B_K(f_\alpha, \epsilon).$$

This shows that S is totally bounded, by Lemma 9.2.3. $\square$

Recalling that a complete metric space is compact if and only if it is totally bounded, that compactness for metric spaces is equivalent to sequential compactness, and that convergence in $C(D)$ is the same as uniform convergence on compact subsets of D, we obtain the following corollary:

Theorem 9.2.5 (Arzelà-Ascoli). *Suppose that D is an open set in the plane and $S \subset C(D)$. The following are equivalent:*

(i) *S is pointwise bounded and pointwise equicontinuous.*

(ii) *Any sequence in S has a subsequence which converges uniformly on compact subsets of D.*

9.3. Montel's Theorem

The last few sections have actually been topics in real analysis, in case you hadn't noticed. In this section we get back to complex variables with Montel's Theorem on "normal families". This is a basic result characterizing the compact subsets of $H(D)$ (not that it's always phrased in those terms). It's useful for a lot of things; we will use this result in the proof of the Riemann Mapping Theorem in the next section.

If (f_n) is a sequence of functions holomorphic in the open set $D \subset \mathbb{C}$ and $f \in H(D)$ then we say that "$f_n \to f$ in $H(D)$" if $f_n \to f$ uniformly on every compact subset of D. This is the same as the definition of convergence in $C(D)$ given in the previous section; thus the results of the previous section show that there is a metric on $H(D)$ such that convergence "in $H(D)$" is equivalent to convergence in this metric.

It's easy to see that $H(D)$ is a closed subset of $C(D)$, and hence a subset of $H(D)$ is compact if and only if it is compact when regarded as a subset of $C(D)$. Thus the Arzelà-Ascoli Theorem (Theorems 9.2.4, 9.2.5) gives a characterization of the compact subsets of $H(D)$. But Montel's Theorem is a little more than just a restatement of the Arzelà-Ascoli Theorem; it turns out that the hypothesis of equicontinuity can be omitted, because it follows from uniform boundedness on compact sets:

Lemma 9.3.0. *If $D \subset \mathbb{C}$ is open then $H(D)$ is a closed subspace of $C(D)$.*

Proof. This is something we already know: Suppose that (f_n) is a sequence in $H(D)$ and $f_n \to f$ in $C(D)$. Then $f_n \to f$ uniformly on compact subsets of D, and hence $f \in H(D)$, by Corollary 3.0. □

We feel compelled to point out again just how amazing Lemma 9.3.0 is: If one knew a lot of real analysis but nothing about complex analysis, one would expect a uniform limit of differentiable functions to be just continuous, not necessarily differentiable.

Recall that $\|f\|_K = \sup_{z \in K} |f(z)|$.

Theorem 9.3.1 (Montel's Theorem). *Suppose $D \subset \mathbb{C}$ is open and $S \subset H(D)$ is closed. Then S is compact if and only if it is uniformly bounded on compact sets (that is, if and only if for every compact $K \subset D$ there exists a number M such that $\|f\|_K \leq M$ for all $f \in S$).*

Proof. Suppose first that S is compact. Since the metric on $H(D)$ is the one it inherits from $C(D)$, it follows that S is also compact when regarded as a subset of $C(D)$, and so Theorem 9.2.4 shows that S is pointwise bounded and equicontinuous. It follows that S is uniformly bounded on compact sets:

For every $z \in D$ there exists $A = A_z$ such that $|f(z)| \leq A$ for all $f \in S$, and there exists $\delta = \delta_z > 0$ such that $D(z, \delta) \subset D$ and $|f(w) - f(z)| < 1$ for all $w \in D(z, \delta)$ and $f \in S$. It follows that $|f(w)| < A_z + 1$ whenever $w \in D(z, \delta_z)$. Now any compact set $K \subset D$ is contained in the union of finitely many of the disks $D(z, \delta_z)$, and it follows that there exists M such that $|f| \leq M$ on K, for all $f \in S$.

That direction did not use anything special about holomorphic functions; the same proof shows that any compact subset of $C(D)$ is uniformly bounded on compact subsets of D. The converse is where things get interesting:

Suppose then that $S \subset H(D)$ is closed and uniformly bounded on compact subsets of D. It follows that S is pointwise bounded, since singletons are compact. So if we can show that S is (pointwise) equicontinuous then the Arzelà-Ascoli Theorem will show that S is compact. But the equicontinuity is immediate from Cauchy's Estimates:

Fix $z \in D$ and choose $r > 0$ such that $K = \overline{D(z, 2r)} \subset D$. Choose M so that $\|f\|_K \leq M$ for all $f \in S$. Suppose that $w \in D(z, r)$. It follows

9.3. Montel's Theorem

that $D(w,r) \subset D(z, 2r)$, so that $|f| \le M$ in $D(w,r)$ for every $f \in S$. Now Cauchy's Estimates show that
$$|f'(w)| \le \frac{M}{r} \quad (f \in S, w \in D(z,r)).$$
This shows that
$$|f(z) - f(w)| \le \frac{|z-w|M}{r} \quad (f \in S, w \in D(z,r)),$$
from which the required equicontinuity follows. $\square$

The traditional statement of Montel's Theorem does not use the word "compact". First a definition: A *normal family* is a set $\mathcal{F} \subset H(D)$ which is uniformly bounded on compact sets (for every compact $K \subset D$ there exists M such that $\|f\|_K \le M$ for all $f \in \mathcal{F}$.)

Note. This is the notion of "normal family" appropriate to the study of holomorphic functions; there are other versions of "normal family" which similarly characterize compactness in spaces of meromorphic functions, for example.

Theorem 9.3.2 (Montel's Theorem). *If $\mathcal{F} \subset H(D)$ is a normal family and (f_n) is a sequence in $\mathcal{F}$ then there exists a subsequence f_{n_j} such that $f_{n_j} \to f \in H(D)$ uniformly on compact subsets of D.*

Proof. Theorem 9.3.1 shows that the closure of $\mathcal{F}$ in $H(D)$ is compact and hence sequentially compact. $\square$

There are many standard exercises on Montel's Theorem. For example:

- **Exercise 9.3.** Suppose that D is a connected open set in the plane. For $\mathcal{F} \subset H(D)$ define
$$\mathcal{F}' = \{ f' : f \in \mathcal{F} \}.$$
  (i) Show that if $\mathcal{F}$ is a normal family then so is $\mathcal{F}'$.
  (ii) Give an example showing that the converse of part (i) is false.
  (iii) Show that the converse of (i) becomes true if you add one more very small hypothesis.

- **Exercise 9.4.** Suppose that D is a connected open set, $\mathcal{F} \subset H(D)$ is a normal family, and (f_n) is a sequence of elements of $\mathcal{F}$. Suppose that $S \subset D$ has a limit point in D and that $f \in H(D)$ has the property that $f_n(z) \to f(z)$ for every $z \in S$. Show that $f_n \to f$ in $H(D)$.

    Hint: You might show first that if (x_n) is a sequence in the compact metric space X, $x \in X$, and $x_n \not\to x$ then there is a subsequence (x_{n_j}) which converges to some element of X other than x.

- **Exercise 9.5.** Suppose that D is an open set in the plane, M is a positive number, and let
$$\mathcal{F} = \left\{ f \in H(D) : \iint_D |f(z)|^2 \, dx dy \leq M \right\}.$$
Show that $\mathcal{F}$ is a normal family.

  Hint: You might begin by looking for a proof that
$$|f(a)|^2 \leq \frac{1}{2\pi} \int_0^{2\pi} |f(a + re^{it})|^2 \, dt$$
whenever $f \in H(D)$ and $\overline{D}(a,r) \subset D$, and then deducing that
$$|f(a)|^2 \leq \frac{1}{\pi r^2} \iint_{D(a,r)} |f(x+iy)|^2 \, dx \, dy.$$

- **Exercise 9.6.** Suppose that D is a connected open set in the plane, $(f_n)_{n=1}^\infty \subset H(D)$ is a normal family, and each f_n has no zero in D. If there exists $z \in D$ with $f_n(z) \to 0$ then $f_n \to 0$ uniformly on compact subsets of D.

  (This is a version of Hurwitz's Theorem; see the comment following Exercise 5.6.)

  Although it's just a curiosity, at least for our present purposes, I cannot resist the temptation to point out that Theorem 9.3.1 shows that the compact subsets of $H(D)$ are exactly the closed and bounded subsets, so that $H(D)$ is what is sometimes called a "Heine-Borel space". *Note* however that the word "bounded" here does not mean what you may think it means, because we are talking about bounded subsets of $H(D)$ regarding $H(D)$ as a topological vector space, not a metric space:

  A topological vector space is a real or complex vector space together with a topology making all the vector-space operations continuous. (If you don't know what a "topology" is then for our purposes here you can assume that there is a metric with respect to which the vector-space operations are continuous.) If S is a subset of a topological vector space then S is said to be *bounded* if for every neighborhood of the origin N there exists a number $\lambda > 0$ such that
$$S \subset \lambda N = \{\lambda x : x \in N\}.$$

  One can easily verify that $S \subset C(D)$ is bounded in this sense if and only if for every compact $K \subset D$ there exists M such that $\|f\|_K \leq M$ for all $f \in S$. So Theorem 9.3.1 says exactly that a subset of $H(D)$ is compact if and only if it is closed and bounded. (This is curious because in "most" infinite-dimensional topological vector spaces closed and bounded sets are not compact.)

We should probably say again that here we are not talking about boundedness in the metric-space sense. (Indeed, $H(D)$ itself is bounded in the metric we gave it, but it is certainly not bounded in the sense above.)

- **Exercise 9.7.** Show that $S \subset C(D)$ is bounded (in the topological-vector-space sense as above) if and only if for every compact $K \subset D$ there exists M such that $\|f\|_K \leq M$ for all $f \in S$.

- **Exercise 9.8.** Let L be the space of all absolutely summable real sequences: $x \in L$ if and only if $x = (x_n)$ where each $x_n \in \mathbb{R}$ and $\sum_{n=1}^{\infty} |x_n| < \infty$. Define a metric d on L by
$$d(x,y) = \sum_{n=1}^{\infty} |x_n - y_n|.$$

   (i) Show that this formula actually defines a complete metric on L.

   (ii) Show that a subset of L is bounded in the topological-vector-space sense if and only if it is bounded as a subset of the metric space (L, d).

   Hint: If $B(0, r)$ is the open ball about the origin in this metric then $tB(0, r) = B(0, tr)$.

   (iii) Show that the closed ball $\overline{B}(0, 1)$ is not compact.

   Hint: Let e_n be the vector with all coordinates equal to 0 except for the n-th coordinate, which equals 1. What is $d(e_n, e_m)$ for $n \neq m$?

9.4. The Riemann Mapping Theorem

In this section we will prove the Riemann Mapping Theorem:

Theorem 9.4.0 (Riemann Mapping Theorem). *Suppose that $D \subset \mathbb{C}$ is a nonempty simply connected open set and $D \neq \mathbb{C}$. Then D is conformally equivalent to the unit disk $\mathbb{D}$.*

Note that Liouville's Theorem shows that $\mathbb{C}$ is not conformally equivalent to $\mathbb{D}$.

The plan is as suggested in the introduction: We will fix $z_0 \in D$ and consider the class of all one-to-one holomorphic functions $f : D \to \mathbb{D}$ with $f(z_0) = 0$. Montel's Theorem will allow us to choose such an f with $|f'(z_0)|$ maximal, and this maximality will be used to show that in fact $f(D) = \mathbb{D}$.

This makes sense intuitively: If $a \in \mathbb{D} \setminus f(D)$ then we should be able to modify f so as to spread its values more uniformly over the disk, so as to include a in the range. Now when we modify f to spread the values farther apart it seems that the size of the derivative should increase, and this contradicts the maximality of $|f'(z_0)|$. The way we are actually going to do this is to note that, since $\phi_a \circ f$ has no zero in D and D is simply connected, the function $\phi_a \circ f$ has a holomorphic square root; we let $g = \phi_b \circ (\phi_a \circ f)^{1/2}$, where b is chosen so that $g(z_0) = 0$, and then we show that $|g'(z_0)| > |f'(z_0)|$.

It is not immediately clear, at least to me, why taking the square root makes sense here: Since

$$|z^2 - w^2| = |z - w||z + w|,$$

taking a square root can either increase distances or decrease distances. But it is clear that the square root *increases* distances in the pseudo-hyperbolic metric: If we (temporarily) define $d(z,w) = |\phi_z(w)|$ as in Section 8.5 then the Invariant Schwarz Lemma (Theorem 8.5.0) shows that

$$d(z^2, w^2) < d(z, w)$$

for all $z, w \in \mathbb{D}$ with $z \ne w$, and it follows that

$$d(z^{1/2}, w^{1/2}) > d(z, w)$$

(for any choice of the square roots). We need an "infinitesimal" version of this inequality:

If $f : D \to \mathbb{D}$ is holomorphic we define

$$H_f(z) = \frac{|f'(z)|}{1 - |f(z)|^2}.$$

Lemma 9.4.1. *Suppose that $f : D \to \mathbb{D}$ and $\psi : \mathbb{D} \to \mathbb{D}$ are holomorphic. If $\psi \in \mathrm{Aut}(\mathbb{D})$ then*

$$H_{\psi \circ f}(z) = H_f(z)$$

for all $z \in D$, while if $\psi \notin \mathrm{Aut}(\mathbb{D})$ then

$$H_{\psi \circ f}(z) < H_f(z)$$

for all $z \in D$ with $f'(z) \ne 0$. (In particular

$$H_{f^2}(z) < H_f(z)$$

for all $z \in D$ with $f'(z) \ne 0$.)

9.4. The Riemann Mapping Theorem

Proof. Theorem 8.5.0 shows that

$$\frac{|\psi'(w)|}{1-|\psi(w)|^2} \le \frac{1}{1-|w|^2}$$

for all $w \in \mathbb{D}$, with equality for all $w \in \mathbb{D}$ if $\psi \in \text{Aut}(\mathbb{D})$ and strict inequality for all $w \in \mathbb{D}$ if $\psi \notin \text{Aut}(\mathbb{D})$. In particular

$$\frac{|\psi'(f(z))|}{1-|\psi(f(z))|^2} \le \frac{1}{1-|f(z)|^2},$$

and so

$$H_{\psi \circ f}(z) = \frac{|\psi'(f(z))||f'(z)|}{1-|\psi(f(z))|^2} \le \frac{|f'(z)|}{1-|f(z)|^2} = H_f(z),$$

with equality if $\psi \in \text{Aut}(\mathbb{D})$ or $f'(z) = 0$ and strict inequality otherwise. $\square$

We recall that any nonvanishing holomorphic function in a simply connected open set has a holomorphic logarithm and hence a holomorphic square root. This is actually the only property of simply connected open sets that will be used in the proof of the Riemann Mapping Theorem; for future reference we give a theorem where the hypothesis involves square roots of holomorphic functions instead of simple connectedness (Theorem 9.4.0 is an immediate consequence):

Theorem 9.4.2 (Riemann Mapping Theorem, restated). *Suppose that $D \subset \mathbb{C}$ is open, connected, nonempty, and has the property that any nonvanishing function holomorphic in D has a holomorphic square root. If $D \ne \mathbb{C}$ then D is conformally equivalent to $\mathbb{D}$.*

If $S \subset \mathbb{C}$ we define $-S = \{-z : z \in \mathbb{C}\}$.

Proof. Fix $z_0 \in D$. Let $\mathcal{F}$ be the family of holomorphic functions $f : D \to \mathbb{D}$ such that f is one-to-one and $f(z_0) = 0$.

To begin, the fact that $D \ne \mathbb{C}$ shows that $\mathcal{F}$ is nonempty: Suppose that $p \in \mathbb{C} \setminus D$. The function $z - p$ has no zero in D and hence by hypothesis there exists $g \in H(D)$ with

$$g(z)^2 = z - p \quad (z \in D).$$

Note that g is one-to-one; in fact g satisfies a stronger condition: If $z, w \in D$ and either $g(z) = g(w)$ or $g(z) = -g(w)$ then it follows that $z = w$, since

$$z - p = g(z)^2 = g(w)^2 = w - p.$$

(One might say here that g was "one-to-two".) It follows that

$$g(D) \cap (-g(D)) = \emptyset.$$

(Indeed, if $\alpha \in g(D) \cap (-g(D))$ then we have $\alpha = g(z)$ and $\alpha = -g(w)$ for some $z, w \in D$; this implies that $z = w$ and so $\alpha = -\alpha$, hence $\alpha = 0$, which is impossible since g has no zero.)

The Open Mapping Theorem shows that $g(D)$ is open; it follows that $g(D)$ is not dense in the plane, since if $D(q, r) \subset g(D)$ for some $q \in \mathbb{C}$ and $r > 0$ then

$$D(-q, r) \cap g(D) = (-D(q, r)) \cap g(D) \subset (-g(D)) \cap g(D) = \emptyset.$$

That is, $|g(z) + q| \geq r$ for all $z \in D$, and hence $h \in \mathcal{F}$, if

$$h(z) = \frac{r/3}{g(z) + q} - \frac{r/3}{g(z_0) + q}.$$

So $\mathcal{F}$ is nonempty. It is clear that there exists a number $M > 0$ such that

$$|f'(z_0)| \leq M$$

for all $f \in \mathcal{F}$; indeed, if $r > 0$ and $D(z_0, r) \subset D$ then Cauchy's Estimates show that

$$|f'(z_0)| \leq \frac{1}{r},$$

since $|f| < 1$.

Let $m = \sup\{|f'(z_0)| : f \in \mathcal{F}\}$ and choose a sequence (f_n) in $\mathcal{F}$ such that $|f_n'(z_0)| \to m$. Montel's Theorem shows that (f_n) has a subsequence converging in $H(D)$; replacing (f_n) by this subsequence we may assume that $F \in H(D)$ and

$$f_n \to F \text{ in } H(D)$$

(recall that this means that $f_n \to F$ uniformly on every compact subset of D).

It follows that $f_n' \to F'$ in $H(D)$, and hence that $|F'(z_0)| = m$. Since $m > 0$, this shows that F is not constant. Now this shows in turn that $F(D) \subset \mathbb{D}$: It is clear that $F(D) \subset \overline{\mathbb{D}}$, and if $|F(z)| = 1$ for some $z \in D$ then the Maximum Modulus Principle (or the Open Mapping Theorem) would show that F was constant.

Rouché's Theorem shows that F is one-to-one in D: Suppose to the contrary that $z, w \in D$, $z \neq w$, but $\alpha = F(z) = F(w)$. Rouché's Theorem (applied in a small disk about z and a small disk about w) shows that if n

9.4. The Riemann Mapping Theorem

is large enough then $f_n - \alpha$ has at least two zeroes, one near z and one near w, contradicting the fact that f_n is one-to-one.

So $F \in \mathcal{F}$. Now for the good part: The maximality of $|F'(z_0)|$ shows that $F(D) = \mathbb{D}$. If, on the other hand, there exists $a \in \mathbb{D} \setminus F(D)$ then $\phi_a \circ F \in H(D)$ has no zero in D. It follows that $\phi_a \circ F$ has a holomorphic square root, which we shall denote using the somewhat risky notation $(\phi_a \circ F)^{1/2}$. As at the start of the proof it follows that $(\phi_a \circ F)^{1/2}$ is one-to-one (in fact one-to-two) and hence

$$G = \phi_b \circ (\phi_a \circ F)^{1/2} \in \mathcal{F},$$

if $b = (\phi_a \circ F)^{1/2}(z_0)$. Since $(\phi_a \circ F)^{1/2}$ is one-to-one, its derivative never vanishes, and so it follows from Lemma 9.4.1 that

$$H_{\phi_a \circ F} < H_{(\phi_a \circ F)^{1/2}}$$

in D. Again by Lemma 9.4.1

$$H_G = H_{\phi_b \circ (\phi_a \circ F)^{1/2}} = H_{(\phi_a \circ F)^{1/2}} > H_{\phi_a \circ F} = H_F$$

in D. In particular, since $F(z_0) = G(z_0) = 0$, we have

$$|G'(z_0)| = H_G(z_0) > H_F(z_0) = |F'(z_0)|,$$

contradicting the maximality of $|F'(z_0)|$. □

If we add a suitable normalization then it is easy to see that the conformal equivalence is unique:

Corollary 9.4.3. *Suppose that D is a simply connected open set, $D \neq \mathbb{C}$, and $z_0 \in D$. Then there exists a unique conformal equivalence $F : D \to \mathbb{D}$ such that $F(z_0) = 0$ and $F'(z_0) > 0$.*

Proof. The proof of the theorem shows that there exists a conformal equivalence F with $F(z_0) = 0$, and multiplying F by a complex number of modulus 1 if necessary we may also obtain $F'(z_0) > 0$. If $\tilde{F} : D \to \mathbb{D}$ is another conformal equivalence satisfying these two conditions then $\phi = \tilde{F} \circ F^{-1} \in \text{Aut}(\mathbb{D})$ satisfies $\phi(0) = 0$ and $\phi'(0) > 0$; now Theorem 8.4.1 shows that ϕ is the identity. □

9.5. Montel's Theorem Again

In this section we give the promised direct proof of Montel's Theorem, without all the technical machinery developed at the start of the chapter. Ambitious readers should study both proofs and convince themselves that they are really in some sense the same argument, expressed in very different ways.

We need a notation for a subsequence of a subsequence of a subsequence of a subsequence ... of a sequence. If S is an infinite set of natural numbers we will write
$$\lim_{j \in S} a_j = a$$
if the restriction of the sequence (a_j) to $j \in S$ tends to a (that is, for every $\epsilon > 0$ there exists N such that $|a_j - a| < \epsilon$ for all $j \in S$ with $j > N$).

Theorem 9.3.2 (Montel's Theorem). *If $\mathcal{F} \subset H(D)$ is a normal family and (f_n) is a sequence in $\mathcal{F}$ then there exists a subsequence f_{n_j} such that $f_{n_j} \to f \in H(D)$ uniformly on compact subsets of D.*

Proof. Let $A = \{z_1, z_2, \ldots\}$ be a countable dense subset of D (for example, take all the points with rational real and imaginary parts). A simple "diagonal argument" shows that the sequence (f_n) has a subsequence which is convergent at every point of A:

Since $f_j(z_1)$ is bounded, there exists an infinite set $S_1 \subset \mathbb{N}$ such that
$$\lim_{j \in S_1} f_j(z_1)$$
exists. Suppose that we have chosen $S_1, \ldots, S_n$. Since $f_j(z_{n+1})$ is bounded, there is an infinite set $S_{n+1} \subset S_n$ such that
$$\lim_{j \in S_{n+1}} f_j(z_{n+1})$$
exists.

We now extract the "diagonal subsequence" from our chain of sub-sub-subsequences. Let n_j be the j-th element of S_j (thus n_1 is the smallest element of S_1, n_2 is the second-smallest element of S_2, etc.). Since $S_{j+1} \subset S_j$, it follows that $n_{j+1} > n_j$. Further, $n_j \in S_k$ for all $j \geq k$, so that
$$\lim_{j \to \infty} f_{n_j}(z_k) = \lim_{j \in S_k} f_j(z_k)$$
exists; the sequence (f_{n_j}) converges at every point of A, as promised.

It follows that there exists $f \in H(D)$ such that $f_{n_j} \to f$ uniformly on compact subsets of D. It is enough to show that the sequence (f_{n_j}) is

9.5. Montel's Theorem Again

"uniformly Cauchy on compact sets"; thus we fix a compact set $K \subset D$ and $\epsilon > 0$, and we need to show that there exists N such that

$$|f_{n_j}(z) - f_{n_k}(z)| < \epsilon$$

for all $z \in K$, whenever $j, k > N$.

For $r > 0$ let K_r be the compact set

$$K_r = \bigcup_{z \in K} \overline{D(z,r)}.$$

Choose $r > 0$ so that $K_{2r} \subset D$, and then choose M so that

$$|f_n(z)| \leq M \quad (z \in K_{2r})$$

for all n. If $z \in K_r$ then $\overline{D(z,r)} \subset K_{2r}$, so Cauchy's Estimates show that

$$|f_n'(z)| \leq \frac{M}{r} \quad (z \in K_r).$$

Now if $z, w \in K$ and $|z - w| < r$ then $[z, w] \subset D(z, r) \subset K_r$, so that

$$|f_n(z) - f_n(w)| \leq \frac{M|z-w|}{r} \quad (z, w \in K, |z-w| < r).$$

Choose $\delta \in (0, r)$ so that

$$\frac{M\delta}{r} < \epsilon/3.$$

Since K is compact and A is dense, there exists a finite set $F \subset A$ such that

$$K \subset \bigcup_{w \in F} D(w, \delta).$$

Since F is finite and the sequence (f_{n_j}) converges at every point of A, there exists N so that

$$|f_{n_j}(w) - f_{n_k}(w)| < \epsilon/3 \quad (w \in F, j, k > N).$$

We are done: Suppose that $z \in K$ and $j, k > N$. Choose $w \in F$ with $|z - w| < \delta$. Then

$$\begin{aligned}
|f_{n_j}(z) &- f_{n_k}(z)| \\
&\leq |f_{n_j}(z) - f_{n_j}(w)| + |f_{n_j}(w) - f_{n_k}(w)| + |f_{n_k}(w) - f_{n_k}(z)| \\
&< \frac{M|z-w|}{r} + \frac{\epsilon}{3} + \frac{M|z-w|}{r} \\
&< \frac{\epsilon}{3} + \frac{\epsilon}{3} + \frac{\epsilon}{3} \\
&= \epsilon.
\end{aligned}$$

□

Chapter 10

Harmonic Functions

10.0. Introduction

Suppose that D is an open subset of $\mathbb{R}^2$ and $u : D \to \mathbb{C}$ is twice continuously differentiable (in the real sense; i.e., the partial derivatives of $u(x,y)$ of order one and two are continuous). We say that u is *harmonic* in D if it satisfies Laplace's equation

$$\Delta u = \frac{\partial^2 u}{\partial x^2} + \frac{\partial^2 u}{\partial y^2} = 0,$$

sometimes written

$$u_{xx} + u_{yy} = 0.$$

The operator $\Delta = \partial^2/\partial x^2 + \partial^2/\partial y^2$ is known as the *Laplacian*.

Probably the first thing to note here is that many authors require a harmonic function to be real-valued, while others allow complex-valued harmonic functions as we do. There is no real difference in the theory, because (as is clear from the definition) a complex-valued function u is harmonic if and only if $\operatorname{Re}(u)$ and $\operatorname{Im}(u)$ are both harmonic; thus our harmonic functions are simply linear combinations of the "harmonic functions" in some other texts. There are advantages to both versions of the definition. (The disadvantage to our convention is that when we want to say something about real-valued harmonic functions we need to include the words "real-valued". On the other hand, we will see below that a holomorphic function is harmonic; with the other convention we could say only that the real and imaginary parts of a holomorphic function are harmonic.)

The theory of harmonic functions can be regarded as a topic in real analysis: There is an obvious definition of "harmonic" for functions defined in an open subset of $\mathbb{R}^n$, and the theory of harmonic functions in $\mathbb{R}^n$ is very much like the theory in $\mathbb{R}^2$. But there are also close connections between the theory of harmonic functions in $\mathbb{R}^2$ and the theory of holomorphic functions in $\mathbb{C}$; the fact that there is no notion of "holomorphic" in $\mathbb{R}^n$ is the reason for most of the differences between the theory of harmonic functions in $\mathbb{R}^2$ and $\mathbb{R}^n$. (These days using "complex methods" in studying harmonic functions and related topics in $\mathbb{R}^2$ is deprecated somewhat, precisely because such methods do not generalize readily to $\mathbb{R}^n$. We are not going to worry about that.)

We know that a function is harmonic if and only if its real and imaginary parts are both harmonic. The corresponding statement for holomorphic functions is about as wrong as it gets: In a connected open set the real and imaginary parts of a nonconstant holomorphic function are *never* holomorphic. We will see that the real and imaginary parts of a holomorphic function are harmonic, however, and that in a simply connected open set any real-valued harmonic function is the real part of some holomorphic function. (We will also see that this fails in any connected open set which is not simply connected.)

The Laplacian can be expressed in terms of the operators $\partial/\partial z$ and $\partial/\partial \bar{z}$ introduced in Appendix 7; this gives an indication of why there should be a connection between harmonic functions and holomorphic functions. Recall the definitions

$$\frac{\partial f}{\partial z} = \frac{1}{2}\left(\frac{\partial f}{\partial x} - i\frac{\partial f}{\partial y}\right)$$

and

$$\frac{\partial f}{\partial \bar{z}} = \frac{1}{2}\left(\frac{\partial f}{\partial x} + i\frac{\partial f}{\partial y}\right).$$

Note that in general we have

$$\overline{\left(\frac{\partial f}{\partial z}\right)} = \frac{\partial \bar{f}}{\partial \bar{z}}$$

and

$$\overline{\left(\frac{\partial f}{\partial \bar{z}}\right)} = \frac{\partial \bar{f}}{\partial z}.$$

If u is twice continuously differentiable then $\partial^2 u/\partial x \partial y = \partial^2 u/\partial y \partial x$, and hence

$$\frac{\partial^2 u}{\partial z \partial \bar{z}} = \frac{1}{4}\left(\frac{\partial^2 u}{\partial x^2} + i\frac{\partial^2 u}{\partial x \partial y} - i\frac{\partial^2 u}{\partial y \partial x} + \frac{\partial^2 u}{\partial y^2}\right)$$

$$= \frac{1}{4}\left(\frac{\partial^2 u}{\partial x^2} + \frac{\partial^2 u}{\partial y^2}\right).$$

10.0. Introduction

This makes the following clear:

Lemma 10.0.0. *Suppose that $D \subset \mathbb{R}^2$ is open and $u : D \to \mathbb{C}$ is twice continuously differentiable (in the real sense). Then u is harmonic if and only if $\partial^2 u / \partial z \partial \bar{z} = 0$.*

Now recall (see Appendix 7) that the Cauchy-Riemann equations can be written simply as
$$\frac{\partial f}{\partial \bar{z}} = 0,$$
and that if f is holomorphic then
$$f' = \frac{\partial f}{\partial z}.$$
It follows that if f is holomorphic we also have
$$f' = 2 \frac{\partial (\operatorname{Re}(f))}{\partial z}.$$
Indeed,
$$2 \frac{\partial (\operatorname{Re}(f))}{\partial z} = \frac{\partial (f + \bar{f})}{\partial z} = \frac{\partial f}{\partial z} + \frac{\partial \bar{f}}{\partial z} = \frac{\partial f}{\partial z} + \overline{\left(\frac{\partial f}{\partial \bar{z}}\right)} = \frac{\partial f}{\partial z} = f'.$$

The basic relations between harmonic and holomorphic functions in the plane follow easily:

Lemma 10.0.1. (i) *A holomorphic function is harmonic (and hence the real and imaginary parts of a holomorphic function are harmonic).*

(ii) *If u is harmonic then $f = \partial u / \partial z$ is holomorphic.*

(iii) *If D is simply connected and u is a real-valued harmonic function in D then there exists $f \in H(D)$ with $u = \operatorname{Re}(f)$.*

Proof. Suppose that f is holomorphic. Then f is harmonic, since it satisfies the Cauchy-Riemann equations:
$$\frac{\partial^2 f}{\partial z \partial \bar{z}} = \frac{\partial}{\partial z} 0 = 0.$$

This gives (i). The proof of (ii) is similar: Suppose u is harmonic and let $f = \partial u / \partial z$. The fact that u has continuous second-order partial derivatives shows that f has continuous first-order partial derivatives; hence f is holomorphic, since it satisfies the Cauchy-Riemann equations:
$$\frac{\partial f}{\partial \bar{z}} = \frac{\partial^2 u}{\partial \bar{z} \partial z} = 0.$$

Now for (iii): Suppose that D is simply connected and u is a real-valued harmonic function in D. Let $g = \partial u/\partial z$. We know from part (ii) that $g \in H(D)$, and, since D is simply connected, it follows that there exists $f \in H(D)$ with $g = f'$ (cf. Theorem 4.14 and Theorem 4.0). Hence

$$\frac{\partial u}{\partial z} = f' = 2\frac{\partial(\operatorname{Re}(f))}{\partial z},$$

which implies that $u - 2\operatorname{Re}(f) \in H(D)$, since

$$\frac{\partial(u - 2\operatorname{Re}(f))}{\partial \overline{z}} = \overline{\left(\frac{\partial \overline{(u - 2\operatorname{Re}(f))}}{\partial z}\right)} = \overline{\left(\frac{\partial(u - 2\operatorname{Re}(f))}{\partial z}\right)} = \overline{0} = 0.$$

Now $u - 2\operatorname{Re}(f)$ must be a real constant, since it is a real-valued holomorphic function; thus $u = c + 2\operatorname{Re}(f) = \operatorname{Re}(c + 2f)$ with $c + 2f \in H(D)$. □

So in a simply connected open set the real-valued harmonic functions are precisely the real parts of holomorphic functions. It turns out this is true only in simply connected open sets:

Theorem 10.0.2. *Suppose that D is a connected open subset of the plane. Then D is simply connected if and only if every real-valued harmonic function in D is the real part of some holomorphic function.*

Proof. One half is contained in Lemma 10.0.1; suppose now that every harmonic function in D is the real part of some holomorphic function.

We want to show that D is simply connected; since $\mathbb{C}$ is simply connected, we may assume that $D \neq \mathbb{C}$. But now it suffices to show that every nonvanishing holomorphic function in D has a holomorphic logarithm, because this implies that every nonvanishing holomorphic function has a holomorphic square root, and then Theorem 9.4.2 shows that D is homeomorphic to the unit disk, which implies that D is simply connected.

So we suppose that $f \in H(D)$ has no zero in D and we need to show that f has a holomorphic logarithm in D. Let $u = \log(|f|)$. Now the fact that f has a holomorphic logarithm will follow from the fact that u is harmonic, which we will prove by using the fact that f has a holomorphic logarithm!

What? The preceding sentence sounds circular only because we did not specify the *domains* of the logarithms in question. In fact we already know that f has a holomorphic logarithm locally; this shows that u is "locally harmonic", which means simply that u is harmonic. In detail: Suppose that $D(z, r) \subset D$ for some $r > 0$. Since $D(z, r)$ is simply connected, we know that there exists $L = L_{z,r} \in H(D(z,r))$ with $e^L = f$ in $D(z, r)$. Hence

$\operatorname{Re}(L)$ is harmonic in $D(z,r)$. But $\operatorname{Re}(L) = \log(|f|) = u$ in $D(z,r)$; hence u is harmonic in every disk contained in D, which says that u is harmonic.

We are assuming that every function harmonic in D is the real part of some element of $H(D)$; thus there exists $L \in H(D)$ with $\operatorname{Re}(L) = u$ in all of D. Now we have

$$|e^L| = e^{\operatorname{Re}(L)} = e^u = e^{\log(|f|)} = |f|$$

in all of D. It follows that $e^L/f \in H(D)$ and $|e^L/f| = 1$ in D, so e^L/f must be constant, and in fact a constant of absolute value 1. Replacing L by $L + ic$ for an appropriate $c \in \mathbb{R}$ gives $e^L/f = 1$, so that L is a logarithm for f. $\square$

If u is a real-valued harmonic function, a *harmonic conjugate* for u is a real-valued harmonic function v such that $u + iv$ is holomorphic. The previous theorem can be rephrased in terms of harmonic conjugates:

Theorem 10.0.2.1. *A connected open subset of the plane is simply connected if and only if every real-valued harmonic function has a harmonic conjugate.*

We should give an explicit statement of a few facts which arose in the proof of Theorem 10.0.2: If $f \in H(D)$ has no zero in D then it is always true that $u = \log(|f|)$ is harmonic. It is also always true that u has a harmonic conjugate *locally*, that is in a neighborhood of any given point; this is equivalent to saying that f always has a holomorphic logarithm defined locally. But u does not always have a global harmonic conjugate; it has a harmonic conjugate defined in all of D precisely when f has a holomorphic logarithm defined in all of D.

10.1. Poisson Integrals and the Dirichlet Problem

In the previous section we saw some of the basic connections between the theory of harmonic functions and the theory of holomorphic functions in the plane. In the current section we give a solution to the so-called "Dirichlet problem" in the unit disk, using a thing called the Poisson integral.

Suppose that D is a bounded open set in the plane and $f \in C(\partial D)$ is a continuous function defined on the boundary of D. To solve the *Dirichlet problem* is to find (or otherwise prove the existence of) a function $u \in C(\overline{D})$ such that $u|_{\partial D} = f$ and such that $u|_D$ is harmonic in D. (That is, to find a harmonic function in D with "boundary values" given by f.)

In this section we will use the Poisson integral to give an explicit solution to the Dirichlet problem in the unit disk. One can prove existence and uniqueness for the Dirichlet problem in open sets other than disks, if the set satisfies quite weak regularity conditions, but this is not so easy; in particular there is no single formula, like the Cauchy Integral Formula for holomorphic functions, that gives the solution to the Dirichlet problem in an arbitrary open set. (There is a fascinating, intuitively plausible hand-waving "solution" to the general Dirichlet problem in Section 6 of this chapter, with many missing details, in fact with many missing definitions.)

The fact that any continuous function on the boundary of the disk can be extended to the closed disk so as to be harmonic in the interior shows that harmonic functions are very different from holomorphic functions: Not just any function on the boundary of the disk can be the boundary value of a holomorphic function. We will see later how one can tell whether or not a given continuous function on the circle can be extended to a function holomorphic in the disk; for now we just give an example:

Define $f \in C(\partial \mathbb{D})$ by $f(e^{it}) = e^{-it}$. There does not exist a function $u \in C(\overline{\mathbb{D}})$ which is equal to f on the boundary and holomorphic in the interior: Suppose u is such a function, and define $v(z) = 1 - zu(z)$. The Maximum Modulus Theorem shows that $v(z) = 0$ for all $z \in \mathbb{D}$, and now if we set $z = 0$ we obtain $0 = 1$, the canonical contradiction.

- **Exercise 10.1.** Explain why the following argument showing that there is no continuous function in the closed unit disk which is holomorphic in the interior and equal to e^{-it} at every boundary point e^{it} is wrong: If u were such a function then $v(z) = u(z) - 1/z$ would define a holomorphic function in $\mathbb{D} \setminus \{0\}$ vanishing at every point of the boundary of the disk. But if the zero set of a holomorphic function has a limit point the function must vanish identically. Thus $u(z) = 1/z$ for $z \neq 0$ and hence u is not continuous at the origin.

Note. This argument really is *wrong*, not just incomplete or badly written: If it were essentially correct, the same argument would show that there is no function continuous in the closed disk and holomorphic in the interior which equals e^{-it} for infinitely many boundary points e^{it}, but this is false; although we will not prove it, in fact there do exist such functions.

We will show shortly that harmonic functions in disks are given by "Poisson integrals". There is a special case of this fact that we already know. Recall that holomorphic functions satisfy the Mean Value Property: If f is holomorphic in a neighborhood of $\overline{D}(z,r)$ then

$$f(z) = \frac{1}{2\pi} \int_0^{2\pi} f(z + re^{it}) \, dt.$$

10.1. Poisson Integrals and the Dirichlet Problem

Note that this is an ordinary integral over an interval on the line, not an integral over some curve in the plane; hence

$$\text{Re}\,(f(z)) = \text{Re}\left(\frac{1}{2\pi}\int_0^{2\pi} f(z+re^{it})\,dt\right) = \frac{1}{2\pi}\int_0^{2\pi}\text{Re}\,(f(z+re^{it}))\,dt.$$

It follows immediately that harmonic functions also enjoy the Mean Value Property:

Proposition 10.1.0. *If u is harmonic in a neighborhood of $\overline{D}(z,r)$ then*

$$u(z) = \frac{1}{2\pi}\int_0^{2\pi} u(z+re^{it})\,dt.$$

Proof. Since u is harmonic and

$$\text{Re}\left(\frac{1}{2\pi}\int_0^{2\pi} u(z+re^{it})\,dt\right) = \frac{1}{2\pi}\int_0^{2\pi}\text{Re}\,(u(z+re^{it}))\,dt,$$

we may assume that u is real-valued. Now, u is harmonic in the disk $D(z,r')$ for some $r' > r$; since this disk is simply connected, there exists a function $f \in H(D(z,r'))$ with $u = \text{Re}\,(f)$ in this disk, and so the result follows from the comments above. $\square$

So holomorphic functions satisfy the Mean Value Property and now it turns out that harmonic functions do as well; one wonders where this stops. It stops here: We will see soon that (given a few technical hypotheses) a function satisfying the Mean Value Property must be harmonic.

We will solve the Dirichlet problem in the unit disk by writing down the answer. This will involve a function known as the *Poisson kernel*, which is defined as follows:

If $0 \leq r < 1$ and $t \in \mathbb{R}$ then

$$P_r(t) = \sum_{n=-\infty}^{\infty} r^{|n|} e^{int}.$$

Note that if $0 \leq R < 1$ then the sum converges absolutely and uniformly for $(r,t) \in [0,R] \times \mathbb{R}$; thus $P_r(t)$ is continuous on $[0,1) \times \mathbb{R}$. We will use various different representations for the Poisson kernel given by the next lemma.

Lemma 10.1.1. *If $0 \leq r < 1$ and $t \in \mathbb{R}$ then*

$$P_r(t) = \sum_{n=-\infty}^{\infty} r^{|n|} e^{int} = \text{Re}\left(\frac{1+re^{it}}{1-re^{it}}\right) = \frac{1-r^2}{1-2r\cos(t)+r^2}$$

$$= \frac{1-r^2}{(1-r\cos(t))^2 + r^2\sin^2(t)}.$$

Proof. We use the fact that $z + \bar{z} = 2\text{Re}\,(z)$ and the formula for the sum of a geometric series; then we rationalize the denominator and finally complete the square in the denominator:

$$P_r(t) = \sum_{n=-\infty}^{\infty} r^{|n|} e^{int} = 1 + \sum_{n=1}^{\infty} r^n (e^{int} + e^{-int}) = 1 + 2\text{Re}\left(\sum_{n=1}^{\infty} r^n e^{int}\right)$$

$$= 1 + 2\text{Re}\left(\frac{re^{it}}{1 - re^{it}}\right) = \text{Re}\left(1 + 2\frac{re^{it}}{1 - re^{it}}\right) = \text{Re}\left(\frac{1 + re^{it}}{1 - re^{it}}\right)$$

$$= \text{Re}\left(\frac{(1 + re^{it})(1 - re^{-it})}{(1 - re^{it})(1 - re^{-it})}\right) = \text{Re}\left(\frac{1 + re^{it} - re^{-it} - r^2}{1 - re^{it} - re^{-it} + r^2}\right)$$

$$= \text{Re}\left(\frac{1 + 2ir\sin(t) - r^2}{1 - 2r\cos(t) + r^2}\right) = \frac{1 - r^2}{1 - 2r\cos(t) + r^2}$$

$$= \frac{1 - r^2}{(1 - r\cos(t))^2 + r^2 \sin^2(t)}. \qquad \square$$

The next lemma says exactly that $P_r(t)$ (regarded as a family of functions of t, indexed by r) forms an *approximate identity*:

Lemma 10.1.2. (i) $P_r(t) > 0 \quad (0 \le r < 1, t \in \mathbb{R})$.

(ii) $\frac{1}{2\pi} \int_0^{2\pi} P_r(t)\, dt = 1 \quad (0 \le r < 1)$.

(iii) If $0 < \delta \le \pi$ then $P_r(t) \to 0$ as $r \to 1$, uniformly for $t \in [-\pi, -\delta] \cup [\delta, \pi]$.

Proof. The fact that $P_r(t) = (1 - r^2)/((1 - r\cos(t))^2 + r^2 \sin^2(t))$ makes it clear that $P_r(t) > 0$. For a fixed $r \in [0, 1)$ the infinite series in the definition converges uniformly on $\mathbb{R}$, and hence the integral of the sum over a bounded interval is the sum of the integrals:

$$\frac{1}{2\pi} \int_0^{2\pi} P_r(t)\, dt = \frac{1}{2\pi} \int_0^{2\pi} \sum_{n=-\infty}^{\infty} r^{|n|} e^{int}\, dt$$

$$= \sum_{n=-\infty}^{\infty} r^{|n|} \frac{1}{2\pi} \int_0^{2\pi} e^{int}\, dt$$

$$= \cdots + 0 + 0 + 1 + 0 + 0 + \cdots = 1.$$

Suppose now that $0 < \delta \le \pi$. There exists $\lambda < 1$ such that $\cos(t) \le \lambda$ for all $t \in I = [-\pi, -\delta] \cup [\delta, \pi]$, and it follows that

$$P_r(t) = \frac{1 - r^2}{(1 - r\cos(t))^2 + r^2 \sin^2(t)} \le \frac{1 - r^2}{(1 - r\cos(t))^2} \le \frac{1 - r^2}{(1 - r\lambda)^2}$$

10.1. Poisson Integrals and the Dirichlet Problem

for all $t \in I$. This shows that $P_r \to 0$ uniformly on I as $r \to 1$. □

We finally get to the point: If $f \in C(\partial \mathbb{D})$ and $re^{i\theta} \in \mathbb{D}$ we define the *Poisson integral* $P[f]$ by

$$P[f](re^{i\theta}) = \frac{1}{2\pi} \int_0^{2\pi} f(e^{it}) P_r(\theta - t)\, dt.$$

Some readers will recognize the Poisson integral as a *convolution*: Using the notation

$$f_r(\theta) = P[f](re^{i\theta})$$

we have

$$f_r = f * P_r.$$

Now the fact that the family (P_r) is an approximate identity says that $f_r = f * P_r \to f$ as $r \to 1$ (in some sense or other, given appropriate hypotheses on f). The proof of the next result is a simple example of an "approximate identity argument":

Theorem 10.1.3. *Suppose that $f \in C(\partial \mathbb{D})$. Define $u : \overline{\mathbb{D}} \to \mathbb{C}$ by*

$$u(re^{i\theta}) = \begin{cases} f(e^{it}) & (r = 1), \\ P[f](re^{i\theta}) & (0 \le r < 1). \end{cases}$$

Then $u|_{\partial \mathbb{D}} = f$, $u \in C(\overline{\mathbb{D}})$, and $u|_{\mathbb{D}}$ is harmonic in $\mathbb{D}$.

That is, $P[f]$ gives a solution to the Dirichlet problem in $\mathbb{D}$ with "boundary data" f.

Proof. The fact that $u|_{\partial \mathbb{D}} = f$ is clear from the definition of u.

Since $\operatorname{Re}(P[f]) = P[\operatorname{Re}(f)]$, we may assume that f is real-valued. Now for $re^{i\theta} \in \mathbb{D}$ we have

$$P[f](re^{i\theta}) = \frac{1}{2\pi} \int_0^{2\pi} f(e^{it}) \operatorname{Re}\left(\frac{1 + re^{i(\theta - t)}}{1 - re^{i(\theta - t)}}\right) dt$$

$$= \operatorname{Re}\left(\frac{1}{2\pi} \int_0^{2\pi} f(e^{it}) \frac{1 + e^{-it} re^{i\theta}}{1 - e^{-it} re^{i\theta}}\, dt\right).$$

In other words, for $z \in \mathbb{D}$ we have

$$u(z) = \operatorname{Re}(F(z)),$$

where

$$F(z) = \frac{1}{2\pi} \int_0^{2\pi} f(e^{it}) \frac{1 + e^{-it} z}{1 - e^{-it} z}\, dt.$$

An argument using Morera's Theorem as in the proof of Theorem 4.9 shows that $F \in H(\mathbb{D})$: If $T \subset \mathbb{D}$ is a triangle then

$$\int_{\partial T} F(z)\,dz = \int_{\partial T} \left(\frac{1}{2\pi} \int_0^{2\pi} f(e^{it}) \frac{1+e^{-it}z}{1-e^{-it}z}\,dt \right) dz$$

$$= \frac{1}{2\pi} \int_0^{2\pi} \left(\int_{\partial T} f(e^{it}) \frac{1+e^{-it}z}{1-e^{-it}z}\,dz \right) dt$$

$$= \frac{1}{2\pi} \int_0^{2\pi} 0\,dt = 0.$$

Hence u is harmonic in $\mathbb{D}$, since u is the real part of a holomorphic function.

This shows that u is continuous in $\mathbb{D}$, and hence to show that u is continuous in $\overline{\mathbb{D}}$ it suffices to show that $u(re^{it}) \to u(e^{it})$ uniformly as $r \to 1$. This is the promised "approximate identity argument": We split the integral into two pieces. One piece is small because of part (iii) of Lemma 10.1.2, while part (ii) of the same lemma shows that the rest of the integral is more or less an average of values of f on an interval small enough that f is close to constant. In detail:

Let $\epsilon > 0$. Now f is a continuous function on a compact set and so is uniformly continuous; thus we may choose $\delta > 0$ such that $|f(e^{it}) - f(e^{i\theta})| < \epsilon/2$ whenever $|e^{it} - e^{i\theta}| < \delta$. (Take $\delta < \pi$ to prevent confusion in the notation below.) Similarly, since f is a continuous function on a compact set, it must be bounded; choose $M > 0$ such that $|f(e^{it})| \leq M$ for all $e^{it} \in \partial \mathbb{D}$. Lemma 10.1.2 shows that there exists $r' < 1$ such that $P_r(t) < \epsilon/(4M)$ whenever $r' < r < 1$ and $t \in I = [-\pi, -\delta] \cup [\delta, \pi]$.

Suppose that $r' < r < 1$ and fix $e^{i\theta} \in \partial \mathbb{D}$. Using part (ii) of Lemma 10.1.2 and noting that everything in sight has period 2π we obtain

$$u(re^{i\theta}) - u(e^{i\theta}) = \frac{1}{2\pi} \int_0^{2\pi} f(e^{it}) P_r(\theta - t)\,dt - f(e^{i\theta})$$

$$= \frac{1}{2\pi} \int_0^{2\pi} f(e^{i(\theta-t)}) P_r(t)\,dt - f(e^{i\theta})$$

$$= \frac{1}{2\pi} \int_0^{2\pi} (f(e^{i(\theta-t)}) - f(e^{i\theta})) P_r(t)\,dt$$

$$= \frac{1}{2\pi} \int_{-\pi}^{\pi} (f(e^{i(\theta-t)}) - f(e^{i\theta})) P_r(t)\,dt$$

10.1. Poisson Integrals and the Dirichlet Problem

$$= \frac{1}{2\pi} \int_I (f(e^{i(\theta-t)}) - f(e^{i\theta})) P_r(t)\, dt$$

$$+ \frac{1}{2\pi} \int_{(-\delta,\delta)} (f(e^{i(\theta-t)}) - f(e^{i\theta})) P_r(t)\, dt.$$

Both pieces are small, for different reasons. First, our choice of r' shows that

$$\left| \frac{1}{2\pi} \int_I (f(e^{i(\theta-t)}) - f(e^{i\theta})) P_r(t)\, dt \right|$$

$$\leq 2M \frac{1}{2\pi} \int_I P_r(t)\, dt \leq 2M \frac{1}{2\pi} \int_I \frac{\epsilon}{4M}\, dt < \frac{\epsilon}{2},$$

since the sum of the lengths of the intervals comprising I is less than 2π. And now our choice of δ shows that

$$\left| \frac{1}{2\pi} \int_{(-\delta,\delta)} (f(e^{i(\theta-t)}) - f(e^{i\theta})) P_r(t)\, dt \right|$$

$$\leq \frac{1}{2\pi} \int_{(-\delta,\delta)} |f(e^{i(\theta-t)}) - f(e^{i\theta})| P_r(t)\, dt$$

$$\leq \frac{\epsilon}{2} \frac{1}{2\pi} \int_{(-\delta,\delta)} P_r(t)\, dt$$

$$\leq \frac{\epsilon}{2} \frac{1}{2\pi} \int_{-\pi}^{\pi} P_r(t)\, dt = \frac{\epsilon}{2}.$$

Combining these inequalities we see that $|u(re^{i\theta}) - u(e^{i\theta})| < \epsilon$ for all θ whenever $r' < r < 1$, which says exactly that $u(re^{i\theta}) \to u(e^{i\theta})$ uniformly as $r \to 1$. □

- **Exercise 10.2.** Suppose that $u : \overline{\mathbb{D}} \to \mathbb{C}$ is continuous in $\mathbb{D}$ and that $u(re^{it}) \to u(e^{it})$ uniformly as $r \to 1$. Show that u is continuous in $\overline{\mathbb{D}}$. (Note that $\epsilon = \epsilon/2 + \epsilon/2$.)

- **Exercise 10.3.** There is actually an application of the inequality

$$\left| e^{it} - e^{is} \right| \leq |t - s| \quad (s, t \in \mathbb{R})$$

hidden in the proof above (not that it was really important).

   (i) Explain where this inequality was used.

   (ii) Prove the inequality.

So the Poisson integral solves the Dirichlet problem in the disk. The previous theorem does not say anything about uniqueness; it does not follow directly from the statement of the theorem that a harmonic function in the unit disk which is continuous up to the boundary must equal the Poisson integral of its boundary values. ("Continuous up to the boundary" is a standard phrase meaning "continuous on the closure of the domain" — we point this out because the phrase *seems* to say much less than that.) This uniqueness will follow from the next lemma, as will the fact that Mean Value Property characterizes harmonic functions.

Note that although we have only proved existence for the Dirichlet problem in the unit disk the proof of uniqueness is no harder in a general open set. Also for an application later we need to derive the uniqueness from a weak form of the Mean Value Property:

Lemma 10.1.4. *Suppose that D is a bounded open set in the plane, and that $u \in C(\overline{D})$ satisfies a weak Mean Value Property in D: For every $z \in D$ there exists $\rho = \rho(z) > 0$ such that $\overline{D}(z, \rho(z)) \subset D$ and*

$$u(z) = \frac{1}{2\pi} \int_0^{2\pi} u(z + re^{it})\, dt \quad (0 < r < \rho(z)).$$

If u vanishes identically on ∂D then u vanishes identically in D.

Proof. We may suppose that u is real-valued. We need only show that $u \leq 0$ in D, because if we can show that then it follows that $-u \leq 0$ as well.

Suppose on the other hand that $u(z) > 0$ for some $z \in D$. Let $M = \sup\{u(z) : z \in \overline{D}\}$ and set

$$K = \{z \in \overline{D} : u(z) = M\}.$$

Then K is a compact subset of $\overline{D}$; since u vanishes on ∂D and $M > 0$, it follows that $K \subset D$.

Choose a point $z \in K$ with the distance from z to ∂D minimal, and choose $r \in (0, \rho(z))$. Note that $u(z + re^{it}) \leq M$ for all t, while in fact $u(z + re^{it}) < M$ for t in some open subinterval of $[0, 2\pi]$ (in particular $u(z + re^{it}) < M$ for all t such that $d(z + re^{it}, \partial D) < d(z, \partial D)$, since $z + re^{it} \notin K$). It follows that $M < M$:

$$M = u(z) = \frac{1}{2\pi} \int_0^{2\pi} u(z + re^{it})\, dt < \frac{1}{2\pi} \int_0^{2\pi} M\, dt = M. \qquad \square$$

Uniqueness for the Dirichlet problem in any bounded open set follows:

10.1. Poisson Integrals and the Dirichlet Problem

Lemma 10.1.4.1. *Suppose that D is a bounded open set in the plane, and that $u_1, u_2 \in C(\overline{D})$ satisfy a weak Mean Value Property in D: For every $z \in D$ there exists $\rho = \rho(z) > 0$ such that $\overline{D}(z, \rho(z)) \subset D$ and*

$$u_j(z) = \frac{1}{2\pi} \int_0^{2\pi} u_j(z + re^{it})\, dt \quad (j = 1, 2;\ 0 < r < \rho(z)).$$

If $u_1 = u_2$ at every point of ∂D then $u_1 = u_2$ everywhere in D.

Proof. Apply the preceding lemma to $u = u_1 - u_2$. □

And now, since we know that the Poisson integral gives a solution to the Dirichlet problem in $\mathbb{D}$, it follows that harmonic functions in $\mathbb{D}$ are given by Poisson integrals (at least if they are continuous up to the boundary):

Theorem 10.1.5. *Suppose that $u \in C(\overline{\mathbb{D}})$ is harmonic in $\mathbb{D}$. Then*

$$u|_{\mathbb{D}} = P[u|_{\partial \mathbb{D}}].$$

Proof. Define $v : \overline{\mathbb{D}} :\to \mathbb{C}$ by

$$v(z) = \begin{cases} 0 & (z \in \partial \mathbb{D}), \\ u(z) - P[u|_{\partial \mathbb{D}}](z) & (u \in \mathbb{D}). \end{cases}$$

Theorem 10.1.3 shows that v is continuous in $\overline{\mathbb{D}}$ and that v is harmonic in $\mathbb{D}$; hence v satisfies the Mean Value Property in $\mathbb{D}$, by Proposition 10.1.0. Since v certainly vanishes on $\partial \mathbb{D}$, the preceding lemma shows that $v = 0$ in $\mathbb{D}$. □

Essentially the same argument shows that the Mean Value Property characterizes harmonic functions. This actually holds for harmonic functions in any open set; we give the result first for functions continuous in the closed unit disk:

Theorem 10.1.6. *Suppose that $u \in C(\overline{\mathbb{D}})$ satisfies a weak Mean Value Property in $\mathbb{D}$: For every $z \in \mathbb{D}$ there exists $\rho = \rho(z) > 0$ such that $\overline{D}(z, \rho(z)) \subset \mathbb{D}$ and*

$$u(z) = \frac{1}{2\pi} \int_0^{2\pi} u(z + re^{it})\, dt \quad (0 < r < \rho(z)).$$

Then u is harmonic in $\mathbb{D}$.

Proof. As before we define $v : \overline{\mathbb{D}} \to \mathbb{C}$ by

$$v(z) = \begin{cases} 0 & (z \in \partial \mathbb{D}), \\ u(z) - P[u|_{\partial \mathbb{D}}](z) & (u \in \mathbb{D}). \end{cases}$$

Proposition 10.1.0 (with other previous results) shows that v satisfies a weak Mean Value Property, and hence Lemma 10.1.4 shows that $v = 0$ in $\mathbb{D}$. □

It follows that there is a result like Theorem 10.1.6 valid in any disk. This is because it is clear that if u is harmonic and $v(z) = u(az+b)$ then v is harmonic as well. In fact something much stronger is true:

Proposition 10.1.7. *Suppose that D_1 and D_3 are open subsets of the plane. If $f : D_1 \to D_2$ is holomorphic and $u : D_2 \to \mathbb{C}$ is harmonic then $u \circ f$ is harmonic in D_1.*

- **Exercise 10.4.** Prove Proposition 10.1.7.

This is one of those exercises where you should actually give two solutions, because they are both important for different reasons. First, you should show directly that $\Delta(u \circ f) = 0$, using the chain rule and the Cauchy-Riemann equations. That is not very interesting, but it is a good idea to develop proficiency with that sort of calculation, for times when no more interesting solution is available.

You should also give a more interesting solution: First, explain why you may assume that u is real-valued. Now, if D_2 were simply connected then u would be the real part of a holomorphic function, and this would show that $u \circ f$ was also the real part of a holomorphic function, and hence that $u \circ f$ was harmonic. Explain why this argument gives a solution to the problem even though D_2 may *not* be simply connected.

Theorem 10.1.6.1. *Suppose that $u \in C(\overline{D}(a, R))$ satisfies a weak Mean Value Property in $D(a, R)$: For every $z \in D$ there exists $\rho = \rho(z) > 0$ such that $\overline{D}(z, \rho(z)) \subset D(a, R)$ and*

$$u(z) = \frac{1}{2\pi} \int_0^{2\pi} u(z + re^{it})\, dt \quad (0 < r < \rho(z)).$$

Then u is harmonic in $D(a, R)$.

And it follows that a weak Mean Value Property itself characterizes harmonic function in any open set, without any extra assumptions on continuity at the boundary:

10.1. Poisson Integrals and the Dirichlet Problem

Theorem 10.1.8. *Suppose that D is an open subset of the plane, $u \in C(D)$, and u satisfies a weak Mean Value Property (as in Theorem 10.1.6) in D. Then u is harmonic in D.*

Proof. The previous theorem shows that u is harmonic in $D(z,r)$ whenever $\overline{D}(z,r) \subset D$. □

So for example the hypothesis that $u \in C(\overline{\mathbb{D}})$ in Theorem 10.1.6 may be weakened to just $u \in C(\mathbb{D})$. All this is the beginning of a small industry: The game is to see exactly what versions of the hypotheses in Theorem 10.1.6 are needed. The reader may easily verify that the proof of Theorem 10.1.6 actually proves this:

Theorem 10.1.6.2. *Suppose that $u \in C(\overline{\mathbb{D}})$ satisfies a very weak version of the Mean Value Property in $\mathbb{D}$: For every $z \in \mathbb{D}$ there exists $r \in (0, 1 - |z|)$ such that*
$$u(z) = \frac{1}{2\pi} \int_0^{2\pi} u(z + re^{it})\, dt.$$
Then u is harmonic in $\mathbb{D}$.

Our "weak Mean Value Property" might be called a "small-radius Mean Value Property", while the "very weak Mean Value Property" used in Theorem 10.1.6.2 might be termed a "one-radius Mean Value Property". Note that Theorem 10.1.6.2 depends on the fact that u is continuous in the closed disk; it turns out that there exists $u \in C(\mathbb{D})$ which satisfies a one-radius Mean Value Property but which is not harmonic.

It often happens that one is in a situation where it is easy to verify that a small-radius Mean Value Property holds while it is not so clear that an unrestricted Mean Value Property holds, so knowing that a small-radius Mean Value Property characterizes harmonic functions is important (we will see an example later in this chapter, in the proof of the Schwarz Reflection Principle for harmonic functions). I am not aware of any reason that a one-radius Mean Value Property is really important.

One might wonder why a small-radius Mean Value Property is needed in Theorem 10.1.8: Since the proof proceeds by showing that u is harmonic in disks contained in D and a one-radius Mean Value Property is sufficient in disks, why is a one-radius Mean Value Property not sufficient in Theorem 10.1.8? The answer is that if we assume that u satisfies a small-radius Mean Value Property in D it follows that u satisfies a small-radius Mean Value Property in any open subset of D; on the other hand assuming that u satisfies a one-radius Mean Value Property in D does *not* imply that u satisfies a one-radius Mean Value Property in open subsets of D.

- **Exercise 10.5.** Suppose that D is an open subset of the plane and $u : D \to \mathbb{C}$.

    (i) Show that if u satisfies a small-radius Mean Value Property in D it follows that u satisfies a small-radius Mean Value Property in any open subset of D.

    (ii) Explain why the same argument does not work for a one-radius Mean Value Property. (You do not have to give an actual counterexample, just an explanation of why a one-radius Mean Value Property in D does not automatically imply a one-radius Mean Value Property for the restriction of u to an open subset of D.)

Summarizing: The Poisson integral gives the unique solution to the Dirichlet problem in the unit disk (with continuous boundary data). Harmonic functions satisfy the Mean Value Property, and under various sorts of technical hypotheses a function satisfying various versions of the Mean Value Property must be harmonic.

10.2. Poisson Integrals and $\text{Aut}(\mathbb{D})$

There are some interesting connections between the Poisson integral and the automorphisms of the unit disk. For example, one can use $\text{Aut}(\mathbb{D})$ to prove that a harmonic function in $\mathbb{D}$ which is continuous up to the boundary must be given by a Poisson integral:

Suppose that $u \in C(\overline{\mathbb{D}})$ is harmonic in $\mathbb{D}$. Fix $a \in \mathbb{D}$ and define $v = u \circ \phi_a$. Then v is also harmonic in $\mathbb{D}$ by Proposition 10.1.7 (note that the proof of Proposition 10.1.7 does not use Theorem 10.1.5). Thus Proposition 10.1.0 shows that

$$v(0) = \frac{1}{2\pi} \int_0^{2\pi} v(re^{it})\, dt$$

for $0 < r < 1$. If we note that $v(re^{it}) \to v(e^{it})$ uniformly as $r \to 1$ we see that

$$v(0) = \frac{1}{2\pi} \int_0^{2\pi} v(e^{it})\, dt,$$

which is to say that

$$u(a) = \frac{1}{2\pi} \int_0^{2\pi} u(\phi_a(e^{it}))\, dt.$$

In particular u is determined by its boundary values; if u_1 and u_2 are two functions satisfying the same hypotheses as u which agree on the boundary of the disk then $u_1 = u_2$. If we take $u_1 = u$ and $u_2 = P[u|_{\partial \mathbb{D}}]$ we see that $u = P[u|_{\partial \mathbb{D}}]$ as required.

10.3. Poisson Integrals and Cauchy Integrals

The Poisson integral commutes with the action of $\mathrm{Aut}(\mathbb{D})$:

Theorem 10.2.0. *If $f \in C(\partial \mathbb{D})$ and $\psi \in \mathrm{Aut}(\mathbb{D})$ then*
$$P[f \circ \psi] = P[f] \circ \psi.$$

Proof. The results in the previous section show that $P[f] \circ \psi$ is (or rather extends to) a continuous function in the closed disk which is harmonic in the open disk and has $f \circ \psi$ for boundary values. Results in the previous section also show that $P[f \circ \psi]$ is the only such function. $\square$

The results in the previous section can be useful. You should try to prove Theorem 10.2.0 "directly", taking $\psi = \phi_a$ for simplicity, by writing
$$P[f \circ \phi_a](re^{i\theta}) = \frac{1}{2\pi} \int_0^{2\pi} f(\phi_a(e^{it})) \frac{1 - r^2}{1 - 2r\cos(\theta - t) + r^2} \, dt$$
and then using the change of variables $\phi_a(e^{it}) = e^{is}$ to convert the integral to the integral defining $P[f](\phi_a(re^{i\theta}))$. (When you are trying to figure out what the dt is in terms of ds it may be useful to note that $\phi_a(e^{is}) = e^{it}$.)

10.3. Poisson Integrals and Cauchy Integrals

We should compare the Poisson integral with the Cauchy integral, to get a few things straight about how they both work. If $f \in C(\mathbb{D})$ then the *Cauchy integral* of f is the function $C[f] : \mathbb{D} \to \mathbb{C}$ defined by
$$C[f](z) = \frac{1}{2\pi i} \int_{\partial \mathbb{D}} \frac{f(w)}{w - z} \, dw.$$

We know that if f is holomorphic in a neighborhood of the closed unit disk then $f = C[f]$ in the open unit disk (or more carefully, $f = C[f|_{\partial \mathbb{D}}]$ in the open disk.) We note first that the hypotheses on f can be weakened somewhat:

Theorem 10.3.0. *Suppose that $f \in C(\overline{\mathbb{D}})$ is holomorphic in $\mathbb{D}$. Then*
$$f|_{\mathbb{D}} = C[f|_{\partial \mathbb{D}}].$$

Proof. For $0 < r < 1$ define $f_r(z) = f(rz)$. Then each f_r is holomorphic in a neighborhood of $\overline{\mathbb{D}}$, so we know that
$$f_r(z) = \frac{1}{2\pi i} \int_{\partial \mathbb{D}} \frac{f_r(w)}{w - z} \, dw$$
for $z \in \mathbb{D}$. Now the functions f_r tend to f uniformly on the closed disk as $r \to 1$, and it follows that
$$f(z) = \frac{1}{2\pi i} \int_{\partial \mathbb{D}} \frac{f(w)}{w - z} \, dw. \qquad \square$$

- **Exercise 10.6.** Show that $f_r \to f$ uniformly in $\overline{\mathbb{D}}$ in the proof of the theorem above.

You should note that Theorem 10.3.0 regarding holomorphic functions and Cauchy integrals is precisely analogous to Theorem 10.1.5 regarding harmonic functions and Poisson integrals, the theorem that gave uniqueness for the Dirichlet problem in $\mathbb{D}$. If we wanted something analogous to Theorem 10.1.3 (existence for the Dirichlet problem) we could say this:

Theorem 10.3.1. *If $f \in C(\partial \mathbb{D})$ then $C[f] \in H(\mathbb{D})$.*

Proof. This follows from Morera's Theorem as usual. □

So it appears that the Cauchy integral is to holomorphic functions exactly as the Poisson integral is to harmonic functions. It is actually easy to give a *wrong* proof that they are exactly the same thing:

Non-Theorem 10.3.2. *If $f \in C(\partial \mathbb{D})$ then*

$$C[f] = P[f].$$

Note. Non-Theorem 10.3.2 is *false*.

Incorrect Proof. We know that $C[f]$ is holomorphic, and that holomorphic functions are harmonic. So $C[f]$ is harmonic. We also know that $P[f]$ is the only harmonic function in the disk with boundary values given by f; hence $C[f]$ must equal $P[f]$. □

- **Exercise 10.7.** Explain what is wrong with the incorrect proof above.

Note. You need to explain *what* is wrong about the proof, not just show that the *result* is false and deduce that there is *something* wrong with the proof. The proof uses a false assumption about the Cauchy integral; this assumption is not explicitly stated in the proof, just hinted at.

10.4. Series Representations for Harmonic Functions in the Disk

In this section we will show that harmonic functions in the unit disk have a representation as an infinite series analogous to the power-series representation of a holomorphic function. This will be obtained from the Poisson integral in much the same way as the power series for a holomorphic function followed from the Cauchy Integral Formula. Then we will compare the

10.4. Series Representations for Harmonic Functions in the Disk

two and determine which continuous functions on the circle are boundary values of holomorphic functions.

If $f \in C(\partial \mathbb{D})$ we define the *Fourier coefficients* of f by the formula

$$\hat{f}(n) = \frac{1}{2\pi} \int_0^{2\pi} f(e^{it}) e^{-int} \, dt \quad (n \in \mathbb{Z}).$$

Proposition 10.4.0. *Suppose that $f \in C(\partial \mathbb{D})$. For $n \in \mathbb{Z}$ set $c_n = \hat{f}(n)$. Then*

$$P[f](re^{it}) = \sum_{n=-\infty}^{\infty} c_n r^{|n|} e^{int} = \sum_{n=0}^{\infty} c_n z^n + \sum_{n=1}^{\infty} c_{-n} \overline{z}^n$$

and

$$C[f](re^{it}) = \sum_{n=0}^{\infty} c_n r^n e^{int} = \sum_{n=0}^{\infty} c_n z^n$$

for all $z = re^{it} \in \mathbb{D}$.

(It's $\overline{z}^n$, the n-th power of the complex conjugate of z, that appears in the statement of the proposition, not z^{-n}.)

Proof. For the first part we use the definition of the Poisson integral, noting that the integral of the sum is the sum of the integrals because the series converges uniformly on the interval $[0, 2\pi]$:

$$\begin{aligned}
P[f](re^{it}) &= \frac{1}{2\pi} \int_0^{2\pi} f(e^{is}) P_r(t-s) \, ds \\
&= \frac{1}{2\pi} \int_0^{2\pi} f(e^{is}) \sum_{n=-\infty}^{\infty} r^{|n|} e^{in(t-s)} \, ds \\
&= \sum_{n=-\infty}^{\infty} r^{|n|} e^{int} \frac{1}{2\pi} \int_0^{2\pi} f(e^{is}) e^{-ins} \, ds \\
&= \sum_{n=-\infty}^{\infty} r^{|n|} e^{int} \hat{f}(n).
\end{aligned}$$

The second part is more or less contained in Theorem 2.6. The current hypotheses are different and we are writing the integral representing the coefficients as a Fourier coefficient instead of as a "Cauchy integral", so we may as well just repeat the argument:

$$C[f](z) = \frac{1}{2\pi i} \int_{\partial \mathbb{D}} \frac{f(w)}{w-z} dw$$
$$= \frac{1}{2\pi i} \int_{\partial \mathbb{D}} \frac{f(w)}{1 - \frac{z}{w}} \frac{dw}{w}$$
$$= \frac{1}{2\pi} \int_0^{2\pi} f(e^{is}) \sum_{n=0}^{\infty} r^n e^{in(t-s)} ds$$
$$= \sum_{n=0}^{\infty} r^n e^{int} \frac{1}{2\pi} \int_0^{2\pi} f(e^{is}) e^{-ins} ds$$
$$= \sum_{n=0}^{\infty} r^n e^{int} \hat{f}(n). \qquad \square$$

Note that the infinite series representing $P[f]$ above is actually the sum of a power series and the conjugate of a power series; thus the Poisson integral gives an explicit representation of a harmonic function in the disk as the sum of a holomorphic function and the conjugate of a holomorphic function:

$$P[f](z) = \sum_{n=0}^{\infty} c_n z^n + \sum_{n=1}^{\infty} c_{-n} \bar{z}^n = h(z) + \overline{g(z)},$$

where

$$h(z) = \sum_{n=0}^{\infty} c_n z^n, \quad g(z) = \sum_{n=1}^{\infty} \overline{c_{-n}} z^n.$$

The theorem allows us to say exactly which continuous functions on the boundary of the unit disk are the boundary values of holomorphic functions:

Theorem 10.4.1. *Suppose that $f \in C(\partial \mathbb{D})$. The following are equivalent:*

(i) *There exists a function $F \in C(\overline{\mathbb{D}})$ which is holomorphic in $\mathbb{D}$ and such that $F|_{\partial \mathbb{D}} = f$.*

(ii) *$P[f]$ is holomorphic in $\mathbb{D}$.*

(iii) *$P[f] = C[f]$.*

(iv) *$\hat{f}(n) = 0$ for all $n < 0$.*

Proof. Most of this is clear from previous results: Theorem 10.4.0 shows that (iv) implies (iii), the fact that $C[f]$ is always holomorphic (Theorem 10.3.1) shows that (iii) implies (ii), and Theorem 10.1.3 shows that (ii) implies (i).

10.4. Series Representations for Harmonic Functions in the Disk

It remains to show that (i) implies (iv). Suppose that $F \in C(\overline{\mathbb{D}})$ is holomorphic in $\mathbb{D}$ and that $F|_{\partial \mathbb{D}} = f$. For $0 < r < 1$ let $\gamma_r = \partial D(0, r)$, and fix a negative integer n. Then z^{-n-1} is holomorphic in the unit disk, so Cauchy's Theorem shows that

$$0 = \frac{1}{2\pi i} \int_{\gamma_r} F(z) z^{-n-1} \, dz$$

$$= \frac{r^{|n|}}{2\pi} \int_0^{2\pi} F(re^{it}) e^{-int} \, dt.$$

The fact that F is continuous in the closed disk shows that

$$F(re^{it}) \to F(e^{it}) = f(e^{it})$$

uniformly as $r \to 1$, so we obtain

$$\hat{f}(n) = \frac{1}{2\pi} \int_0^{2\pi} F(e^{it}) e^{-int} \, dt = 0. \qquad \square$$

We should mention in passing that Theorem 10.4.0 also shows that the Poisson integral gives a solution to one of the basic problems in the theory of Fourier series:

Suppose that $f \in C(\partial \mathbb{D})$ and set $c_n = \hat{f}(n)$. Fourier stated without proof that

$$f(e^{it}) = \sum_{n=-\infty}^{\infty} c_n e^{int}.$$

We now know that the reason he was unable to prove this is that it is not true. (It is true under somewhat stronger hypotheses on f.) The problem is then to find a way to recover f from its Fourier coefficients, given that f is not literally the sum of its Fourier series; people invent things called "summability methods" for exactly this purpose. The theorem above shows that

$$f(e^{it}) = \lim_{r \to 1} \sum_{n=-\infty}^{\infty} r^{|n|} c_n e^{int}$$

uniformly; in the language used in the study of Fourier series this says that while the Fourier series of $f \in C(\partial \mathbb{D})$ need not converge to f it *is* nonetheless "Abel summable" to f. (Abel's Theorem, proved in Chapter 21, states that a convergent series is Abel summable; the converse is false.)

The fact that the Fourier series for $f \in C(\partial \mathbb{D})$ is Abel summable to f has many important consequences, a few of which are developed in the following exercises. Note that a *trigonometric polynomial* is a function of the form

$$P(e^{it}) = \sum_{n=-N}^{N} c_n e^{int}.$$

Exercises

10.8. The trigonometric polynomials are dense in $C(\partial\mathbb{D})$. That is, if $f \in C(\partial\mathbb{D})$ and $\epsilon > 0$ then there exists a trigonometric polynomial P such that
$$|P(e^{it}) - f(e^{it})| < \epsilon$$
for all $t \in \mathbb{R}$.

Hint: For $0 < r < 1$ let $f_r(e^{it}) = \sum_{n=-\infty}^{\infty} r^{|n|} c_n e^{int}$, where c_n is the n-th Fourier coefficient of f. Theorems 10.4.0 and 10.1.3 show that there exists $r \in (0,1)$ such that $|f_r(e^{it}) - f(e^{it})| < \epsilon/2$ for all t. Fix such an r. Now show that (c_n) is bounded, and deduce that there exists N such that $\sum_{|n|>N} |c_n| r^{|n|} < \epsilon/2$.

10.9. Suppose that $\alpha \in \mathbb{R}$ and α/π is irrational. Show that
$$\lim_{n\to\infty} \frac{1}{n} \sum_{j=0}^{n-1} P(e^{ij\alpha}) = \frac{1}{2\pi} \int_0^{2\pi} P(e^{it})\,dt$$
for every trigonometric polynomial P. (It is enough to prove this assuming that $P(e^{it}) = e^{int}$ for some integer n. Now there are two cases ...)

10.10. Suppose that $\alpha \in \mathbb{R}$ and α/π is irrational. Show that
$$\lim_{n\to\infty} \frac{1}{n} \sum_{j=0}^{n-1} f(e^{ij\alpha}) = \frac{1}{2\pi} \int_0^{2\pi} f(e^{it})\,dt$$
for every $f \in C(\partial\mathbb{D})$.

10.11. Suppose that $\alpha \in \mathbb{R}$ and α/π is irrational. Show that $\{e^{ij\alpha} : j = 0, 1, \ldots\}$ is dense in $\partial\mathbb{D}$. (This follows from the previous exercise.)

The next exercise is a more precise version of the previous one; it says that not only are the numbers $e^{ij\alpha}$ dense in the unit circle if α/π is irrational, they are "uniformly distributed".

10.12. Suppose that $\alpha \in \mathbb{R}$ and α/π is irrational. Fix a and b with $0 < a < b < 2\pi$, and let $I = \{e^{it} : a < t < b\}$. For each positive integer n let $\#(n)$ be the number of integers $j \in [0, n)$ such that $e^{ij\alpha} \in I$. Show that
$$\lim_{n\to\infty} \frac{\#(n)}{n} = \frac{b-a}{2\pi}.$$

10.5. Green's Functions and Conformal Mappings

Hint: Let $\epsilon > 0$. Let $\chi : \partial \mathbb{D} \to \mathbb{C}$ be the (discontinuous) function defined by

$$\chi(e^{it}) = \begin{cases} 1 & (e^{it} \in I), \\ 0 & (e^{it} \notin I). \end{cases}$$

Show that there exist $f, g \in C(\partial \mathbb{D})$ such that $0 \le g \le \chi \le f \le 1$ everywhere and

$$\frac{b-a}{2\pi} - \frac{\epsilon}{2} < \frac{1}{2\pi}\int_0^{2\pi} g(e^{it})\,dt < \frac{1}{2\pi}\int_0^{2\pi} f(e^{it})\,dt < \frac{b-a}{2\pi} + \frac{\epsilon}{2}.$$

(A picture of f and g should suffice.) Now apply the previous exercise.

10.13. Suppose that $\alpha \in \mathbb{R}$ and α/π is irrational. Suppose that $f \in H(\mathbb{D})$, and for $n = 1, 2, \ldots$ define $f_n \in H(\mathbb{D})$ by

$$f_n(z) = \frac{1}{n}\sum_{j=0}^{n-1} f(e^{ij\alpha}z).$$

Show that $f_n \to f(0)$ uniformly on compact subsets of $\mathbb{D}$. (The previous exercises show that $f_n \to f(0)$ pointwise; you could show directly that the convergence is uniform on compact sets, or you could use Exercise 9.4.)

10.14. Suppose everything is as in the previous exercise except that α/π is rational. Show that there exists $g \in H(\mathbb{D})$ such that $f_n \to g$ uniformly on compact subsets of $\mathbb{D}$. Supposing that $f(z) = \sum_{j=0}^{\infty} c_j z^j$ and $\alpha/\pi = p/q$ in lowerst terms, give the power series for g.

10.5. Green's Functions and Conformal Mappings

Suppose that D is a bounded simply connected region in the plane. Solving the Dirichlet problem for D is closely related to finding a conformal equivalence mapping D onto the unit disk. Calling the two problems equivalent would be an exaggeration, but not by much.

First, it seems clear that if we can solve the Dirichlet problem in the disk (which we can) then we can solve it in any simply connected open set, by the Riemann Mapping Theorem. Alas, after a moment's reflection one sees that this is not *quite* as clear as it seems at first: Say D is a bounded simply connected open set, $f \in C(\partial D)$, and $\phi : D \to \mathbb{D}$ is a conformal equivalence. An attempt to use ϕ to transfer the problem to $\mathbb{D}$ begins by saying that $f \circ \phi^{-1} \in C(\partial \mathbb{D})$, but that does not make any sense, unless $\phi^{-1} : \mathbb{D} \to D$ extends continuously to $\overline{\mathbb{D}}$.

This can be fixed. It looks like we want to assume at least that $\overline{D}$ is homeomorphic to $\overline{\mathbb{D}}$, and in fact an assumption somewhat weaker than that suffices. Or the assumption appears weaker, in any case:

Theorem 10.5.0. *Suppose that D is a bounded simply connected open subset of the plane and ∂D is a simple closed curve. Then the Dirichlet problem in D has a unique solution, which can be obtained by transferring the problem to $\mathbb{D}$ using the Riemann Mapping Theorem as above.*

Theorem 10.5.0 is immediate from Theorem 10.5.1 below, which we will not prove (we will prove a slightly weaker version of Theorem 10.5.1 later, in Theorem 14.0):

Theorem 10.5.1 (Carathéodory). *Suppose D is a bounded simply connected region in the plane and ∂D is a simple closed curve. Let $\phi : D \to \mathbb{D}$ be a conformal equivalence. Then ϕ extends to a homeomorphism of $\overline{D}$ and $\overline{\mathbb{D}}$.*

It is clear that Theorem 10.5.0 follows from Theorem 10.5.1: If $f \in C(\partial D)$ then $f \circ \phi^{-1} \in C(\partial \mathbb{D})$, so there is a unique function U continuous in $\overline{\mathbb{D}}$ and harmonic in $\mathbb{D}$ which agrees with $f \circ \phi^{-1}$ on $\partial \mathbb{D}$. It follows that $u = U \circ \phi$ is continuous in $\overline{D}$, harmonic in D, and equal to f on ∂D.

So it's "obvious" that the Riemann Mapping Theorem shows that the Dirichlet problem can be solved in any simply connected open set, and if we get the hypotheses right and use Carathéodory's Theorem to fill in the details, it's not just obvious, it's even *true*. The converse is less obvious: If D is a bounded simply connected region in which we can solve the Dirichlet problem then it follows that there is a conformal mapping from D onto the unit disk!

How can we get a conformal equivalence out of the Dirichlet problem? We work backwards to motivate the proof: Suppose that $\phi : D \to \mathbb{D}$ is a conformal equivalence. We want a harmonic function g such that (i) we can define g as the solution to a certain instance of the Dirichlet problem and (ii) we can construct ϕ from g. Various things satisfy (ii) but not (i). For example ϕ itself is a harmonic function, but setting $g = \phi$ will not help, because when we start working forwards we do not know how to say what Dirichlet problem g is a solution to, because we do not know the boundary values of ϕ. What *do* we know about ϕ? Presumably it will be the case that $|\phi| = 1$ on ∂D. So we could ... no, that doesn't work because $|\phi|$ is not harmonic. Is there something like $|\phi|$ that *is* harmonic? Yes: $\log(|\phi|)$ will be harmonic except at the one point where $\phi = 0$.

That turns out to work: If we *had* a conformal equivalence ϕ then $\log(|\phi|)$ would be harmonic in D except for a logarithmic singularity at one point and it would vanish on ∂D. Assuming that we can solve the Dirichlet problem we will use it to get a function g which is harmonic in D except for a logarithmic singularity at one point and which vanishes on the boundary of D; then we will use g to construct a conformal equivalence of D and $\mathbb{D}$.

10.5. Green's Functions and Conformal Mappings

That's the idea, in any case. And that is probably how one should think of the proof, because g is a well-known thing called the "Green's function", and the idea is just that the solution to the Dirichlet problem gives us a Green's function and the Green's function then gives us our conformal mapping. If we try to carry out the plan above we get harmonic functions with "multi-valued" harmonic conjugates; the multi-valuedness goes away when we exponentiate. The proof below is a somewhat cleaned up version with no multi-valued functions — the Green's function no longer appears quite so explicitly in the cleaned-up version.

We will need a slightly more sophisticated version of the Maximum Modulus Theorem than the ones we have proved so far. If f is a function defined in an open set D and $\zeta \in \partial D$ we define

$$\limsup_{z \to \zeta} |f(z)| = \lim_{r \to 0} (\sup\{\,|f(z)| : z \in D, |z - \zeta| < r\,\});$$

note that the limit exists (in $[0, \infty]$) since $\sup\{\,|f(z)| : z \in D, |z - \zeta| < r\,\}$ is an increasing function of r. Another way to define $\limsup_{z \to \zeta} |f(z)|$ is this: It is the largest possible value of $\lim_{n \to \infty} |f(z_n)|$, where we consider all sequences (z_n) in D which converge to ζ, and for which $\lim_{n \to \infty} |f(z_n)|$ exists in $[0, \infty]$.

Lemma 10.5.2 (Maximum Modulus Theorem again). *Suppose that D is a bounded open set in the plane and $f \in H(D)$. If*

$$\limsup_{z \to \zeta} |f(z)| \le M$$

for all $\zeta \in \partial D$ then

$$|f(z)| \le M$$

for all $z \in D$.

The proof is very simple from versions of the Maximum Modulus Theorem that we know:

Proof. We may assume that D is connected and that f is not constant.

Suppose the theorem is false:

$$m = \sup_{z \in D} |f(z)| > M.$$

(Note that m could be ∞ here.) Choose a sequence (z_n) in D with $|f(z_n)| \to m$. Now replacing this sequence by a subsequence if necessary we may assume that $z_n \to \zeta \in \overline{D}$. We cannot have $\zeta \in D$ because this would show that $|f|$ had a local maximum in D. So we have $\zeta \in \partial D$. But this shows that $\limsup_{z \to \zeta} |f(z)| \geq m > M$, contrary to hypothesis. □

We will also be using the fact that any real-valued harmonic function in a simply connected open set is the real part of some holomorphic function, which we proved in Lemma 10.0.1(iii). (Note that the proof of Lemma 10.0.1 did not use the Riemann Mapping Theorem, although the Riemann Mapping Theorem *was* used in proving the converse in Theorem 10.0.2.)

Theorem 10.5.3. *Suppose that D is a bounded nonempty simply connected open subset of the plane, and suppose that the Dirichlet problem can be solved in D. Then D is conformally equivalent to the unit disk.*

(Of course the Riemann Mapping Theorem shows that D is conformally equivalent to the unit disk under weaker hypotheses; the point is to illustrate the connection with the Dirichlet problem.)

Proof. Fix $z_0 \in D$. Define $f \in C(\partial D)$ by

$$f(z) = -\log(|z - z_0|) \quad (z \in \partial D).$$

Now, since the Dirchlet problem can be solved in D, there exists a function $u \in C(\overline{D})$ such that u is harmonic in D and $u|_{\partial D} = f$.

(It's not directly relevant here, but in case anyone ever asks you how to get a Green's function from the solution to the Dirichlet problem you should note that the function $g(z) = \log(|z - z_0|) + u(z)$ is a Green's function: This means that g is continuous on $\overline{D} \setminus \{z_0\}$, harmonic in $D \setminus \{z_0\}$, equal to 0 on ∂D, and that $g(z) - \log(|z - z_0|)$ has a removable singularity at z_0.)

Since D is simply connected, it follows that there exists a holomorphic function $h \in H(D)$ such that $u = \operatorname{Re}(h)$ and $h(z_0) = u(z_0)$. (See the comment preceding the proof.) Define $\psi \in H(D)$ by

$$\psi(z) = (z - z_0)e^{h(z)}.$$

Now for $z \in D$ we have

$$|\psi(z)| = |z - z_0|e^{\operatorname{Re}(h(z))} = |z - z_0|e^{u(z)}.$$

Since u is continuous on $\overline{D}$, it follows that $|\psi|$ extends to a continuous function in $\overline{D}$, and so for any $\zeta \in \partial D$ we have

$$\limsup_{z \to \zeta} |\psi(z)| = |\zeta - z_0|e^{u(\zeta)} = |\zeta - z_0|e^{-\log|\zeta - z_0|} = 1.$$

10.5. Green's Functions and Conformal Mappings

So the previous theorem shows that $|\psi| \le 1$ in D, which of course implies that actually $|\psi| < 1$ in D, by the Maximum Modulus Theorem or the Open Mapping Theorem.

So $\psi : D \to \mathbb{D}$ is holomorphic, ψ has exactly one zero in D (at z_0), and $\lim_{z \to \partial D} |\psi(z)| = 1$. (Note that although $|\psi|$ extends to a continuous function on $\overline{D}$ the function ψ itself does not in general extend continuously to the boundary.) It follows from Rouché's theorem that ψ is actually a conformal equivalence from D onto $\mathbb{D}$:

We need only show that for every $a \in \mathbb{D}$ the function $\psi - a$ has exactly one zero in D. Fix $a \in \mathbb{D}$. Fix a number $r \in (|a|, 1)$. Now

$$K = \{ z \in \overline{D} : |\psi(z)| \le r \}$$

is a compact subset of D: The fact that $|\psi|$ is continuous in $\overline{D}$ shows that K is a (relatively) closed subset of $\overline{D}$, which implies that K is compact since $\overline{D}$ is closed and bounded. And K must actually be a subset of D since $|\psi|$ tends to 1 at the boundary of D.

Choose a smooth cycle Γ in D with the following properties: $\mathrm{Ind}(\Gamma, z) = 0$ for all $z \in \mathbb{C} \setminus D$, $\mathrm{Ind}(\Gamma, z) = 0$ or 1 for all $z \in D \setminus \Gamma^*$, $\Gamma^* \subset D \setminus K$, and $\mathrm{Ind}(\Gamma, z) = 1$ for all $z \in K$.

Now the fact that $\Gamma^* \subset D \setminus K$ shows that

$$|\psi| \ge r > |a| = |\psi - (\psi - a)|$$

on Γ^*, and so Rouché's Theorem (Theorem 5.4) shows that ψ and $\psi - a$ have the same number of zeroes in $\{ z \in D \setminus \Gamma^* : \mathrm{Ind}(\Gamma, z) = 1 \}$. Since $\psi - a$ cannot have any zeroes outside this set (because $|\psi| \ge r > |a|$ outside K), we see that $\psi - a$ has exactly one zero in D. $\square$

We should note that the conformal map ψ constructed in the proof satisfies $\psi(z_0) = 0$ and $\psi'(z_0) > 0$ (because $\psi'(z_0) = e^{h(z_0)} = e^{u(z_0)}$). Corollary 9.4.3 shows that it is the only conformal equivalence from D to $\mathbb{D}$ with these two properties.

There are two holes in the proof of the theorem that need to be filled in. First, Theorem 5.4 talks about a closed curve, not a cycle. This is of no consequence, since (as you can easily verify) the proof of Theorem 5.4 gives the same result with a cycle Γ instead of a closed curve γ. The second problem is that we never said how we know we can choose a cycle Γ with all the properties ascribed to it in the proof above. We prove this in Lemma 10.5.5 below; first we need a bit of combinatorics:

If S is a finite set then $\#S$ is the cardinality of S, that is, the number of elements.

A (finite) *directed graph* is an ordered pair $G = (V, E)$ where V is a finite set and E is a set of ordered pairs (a, b) with $a, b \in V$ and $a \ne b$. (It does not really matter whether edges of the form (a, a) are allowed — we will not allow them because they do not come up below.) The elements of V are the *vertices* of G and the elements of E are the *edges*. If $(a, b) \in E$ we say that G has an edge from a to b. (The edges are "directed": An edge from a to b is not the same as an edge from b to a.) The *order* of a vertex is the number of incoming edges minus the number of outgoing edges: If $a \in V$ then

$$\operatorname{order}(a) = \#\{\, b \in V : (b, a) \in E \,\} - \#\{\, b \in V : (a, b) \in E \,\}.$$

A *circuit* in G is a finite sequence of distinct vertices $\langle a_1, \ldots, a_n \rangle$ such that there is an edge from a_j to a_{j+1} for $j = 1, \ldots, n-1$ and also an edge from a_n to a_1. We will say that two circuits are *disjoint* if they have no *edges* in common ("disjoint" circuits are allowed to share vertices).

For example, if $V = \{a, b, c\}$ and $E = \{(a, b), (b, a), (b, c), (c, b), (a, c)\}$ then the graph $G = (V, E)$ might be represented "graphically" (as it were) as in Figure 10.1a or Figure 10.1b:

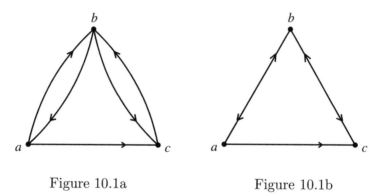

Figure 10.1a Figure 10.1b

(Figure 10.1a is probably a better way to represent the graph; we include Figure 10.1b because there is an example below where we are forced to use that style.)

In this graph we see that a has order $1 - 2 = -1$, b has order $2 - 2 = 0$, and c has order $2 - 1 = 1$. And we can see that $\langle a, b \rangle$, $\langle b, c \rangle$ and $\langle a, c, b \rangle$ are circuits ($\langle a, b \rangle$ and $\langle b, c \rangle$ are disjoint).

We need the following combinatorial fact:

Lemma 10.5.4. *A finite directed graph in which each vertex has order 0 may be written as a union of pairwise disjoint circuits.*

10.5. Green's Functions and Conformal Mappings

The alert reader may be wondering how a graph can be the union of anything, since a graph is an ordered pair, not a set. A more precise statement of the lemma follows.

Lemma 10.5.4.1. *Suppose $G = (V, E)$ is a finite directed graph in which each vertex has order 0. Then there exist finitely many circuits $C_1, \ldots, C_m$ such that E is the disjoint union of the edges occurring in $C_1, \ldots, C_m$.*

Here (as one would guess) the edges "occurring" in the circuit $\langle a_1, \ldots, a_n \rangle$ are the edges (a_j, a_{j+1}) for $j = 1, \ldots, n-1$ and the edge (a_n, a_1).

Proof. Note first that if $\langle a_1, \ldots, a_n \rangle$ is a circuit in such a graph then the graph we obtain by removing the edges (a_j, a_{j+1}) ($j = 1, \ldots, n-1$) and (a_n, a_1) from E still has the property that every vertex has order 0. Thus we need only show that G contains a circuit (or rather that if E is nonempty then G contains a nontrivial circuit); the lemma follows by induction on the number of edges.

If E is empty we are done; assume that E is nonempty and let (a_1, a_2) be an element of E. Now, the order of a_2 is 0 and (a_1, a_2) is an edge so there must be an edge of the form (a_2, b). If $b = a_1$ then $\langle a_1, a_2 \rangle$ is a circuit and we are done; if $b \neq a_1$ let $a_3 = b$.

We continue in this manner. Suppose that we have found distinct vertices $a_1, \ldots, a_n$ such that $(a_j, a_{j+1}) \in E$ for $j = 1, 2, \ldots, n-1$. Since a_n has degree 0, there must exist an edge of the form (a_n, b). If there exists $j < n$ such that $b = a_j$ then $\langle a_j, \ldots, a_n \rangle$ is a circuit and we are done; if not let $a_{n+1} = b$ and continue.

We must eventually find a circuit this way; otherwise we have constructed an infinite sequence of distinct vertices in a finite graph. □

We should perhaps point out that in the proof above we may not get back to a_1 when we find our circuit. For example, in Figure 10.2a below, if we choose a_j ($1 \leq j \leq 4$) as indicated then we find the circuit $\langle a_2, a_3, a_4 \rangle$. (Sure enough, when we remove the edges that occur in this circuit what remains is as in Figure 10.2b: Another graph where every vertex has order 0.)

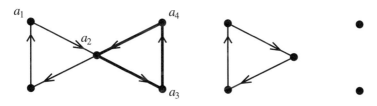

Figure 10.2a　　　　　　　　　　Figure 10.2b

Suppose that L is a finite set of (directed) line segments in the plane. There is a graph $G_L = (V_L, E_L)$ naturally associated with L, where the vertices are given by the endpoints,

$$V_L = \{\, a : [a, b] \in L \,\} \cup \{\, b : [a, b] \in L \,\},$$

and each element of L corresponds to an edge:

$$E_L = \{\, (a, b) : [a, b] \in L \,\}.$$

There is also a chain Γ_L naturally associated with L, defined by

$$\Gamma_L = \underset{[a,b]\in L}{\overset{\bullet}{\sum}} [a, b].$$

We should note that G_L and Γ_L are not the same thing. In particular, note that if $L_1 = \{[a, b], [b, a]\}$ and $L_2 = \emptyset$ then $\Gamma_{L_1} = \Gamma_{L_2}$ (because both are the zero chain) while $G_{L_1} \neq G_{L_2}$ (because G_{L_2} is the empty graph while G_{L_1} has two edges (a, b) and (b, a)). The fact that distinct graphs can correspond to the same chain in this way is important in the proof of the following lemma.

Generalizing the example in the previous paragraph, suppose that L_1 is a finite set of line segments and $L_2 = \{\, [a, b] \in L_1 : [b, a] \notin L_1 \,\}$. Then $\Gamma_{L_2} = \Gamma_{L_1}$, and while we may have $G_{L_2} \neq G_{L_1}$ it is nonetheless clear that every vertex in G_{L_2} has the same order in G_{L_2} as in G_{L_1} (because removing both (a, b) and (b, a) from the set of edges of a graph does not change the order of any vertex).

Recall that a *square of side length* δ is a set of the form $Q = [x, x + \delta] \times [y, y + \delta]$ for some $x, y \in \mathbb{R}$ and $\delta > 0$; the *boundary* of this square is the cycle

$$\partial Q = [z, z + \delta] \dot{+} [z + \delta, z + \delta + i\delta] \dot{+} [z + \delta + i\delta, z + i\delta] \dot{+} [z + i\delta, z]$$

(where $z = x + iy$.) The following lemma completes the proof of Theorem 10.5.3.

Lemma 10.5.5. *Suppose that D is an open subset of the plane and K is a compact subset of D. There exists a cycle Γ in D such that $\mathrm{Ind}(\Gamma, z) = 0$ for all $z \in \mathbb{C} \setminus D$, $\Gamma^* \subset D \setminus K$, $\mathrm{Ind}(\Gamma, z) = 0$ or 1 for all $z \in D \setminus \Gamma^*$, and $\mathrm{Ind}(\Gamma, z) = 1$ for all $z \in K$.*

Proof. Choose $\delta > 0$ such that if Q is any square of side length δ and $Q \cap K \neq \emptyset$ then $Q \subset D$. Let $\mathfrak{Q}$ be the collection of all squares of the form

$$Q = [j\delta, (j + 1)\delta] \times [k\delta, (k + 1)\delta] \quad (j, k \in \mathbb{Z})$$

10.5. Green's Functions and Conformal Mappings

such that $Q \cap K \ne \emptyset$, and define

$$\Gamma = \overset{\bullet}{\underset{Q \in \mathfrak{Q}}{\sum}} \partial Q.$$

This defines Γ as the sum of $4N$ line segments, 4 for each one of the N squares in $\mathfrak{Q}$. Let L_1 be the corresponding *set* of $4N$ line segments, and let $\mathcal{E} = \{\,[a,b] \in L_1 : [b,a] \notin L_1\,\}$. It follows that $\Gamma = \Gamma_{L_1} = \Gamma_{\mathcal{E}}$.

If $Q \in \mathfrak{Q}$ then each vertex in (the graph corresponding to) ∂Q has order 0; hence each vertex in G_{L_1} has order 0. As noted above this implies that each vertex in $G_{\mathcal{E}}$ has order 0. The previous lemma shows that $G_{\mathcal{E}}$ is a union of disjoint circuits; this says that the segments in $\mathcal{E}$ may be partitioned so as to form finitely many polygonal closed curves $\gamma_1, \ldots, \gamma_n$ such that γ_j and γ_k have no edges in common ($j \ne k$).

We now have various representations for Γ:

$$\Gamma = \overset{\bullet}{\underset{Q \in \mathfrak{Q}}{\sum}} \partial Q = \overset{\bullet}{\underset{E \in \mathcal{E}}{\sum}} E = \overset{\bullet}{\underset{1 \le j \le n}{\sum}} \gamma_j.$$

The reader may have noticed long ago that our definition of the trace of a cycle really does not make sense, or is at any rate not consistent with our definition of equality of two cycles: Our definitions allow $\Gamma_1 = \Gamma_2$ while $\Gamma_1^* \ne \Gamma_2^*$. This is an abuse of the equals sign: Any time $x = y$ and $f(x)$ is defined it should follow that $f(x) = f(y)$. In fact a cycle *per se* does not have a trace, it is a particular representation of a cycle as a sum of closed curves that has a trace.

So the assertion "$\Gamma^* \subset D \setminus K$" in the statement of the lemma must be interpreted as saying that Γ has a representation with a trace disjoint from K. It is clear that in general K will have nonempty intersection with the union of ∂Q^* for $Q \in \mathfrak{Q}$; we shall show however that

$$K \cap \bigcup_{j=1}^{n} \gamma_j^* = \emptyset.$$

It is clear that

$$\bigcup_{j=1}^{n} \gamma_j^* \subset \bigcup_{Q \in \mathfrak{Q}} \partial Q^*,$$

so we need only show that

$$\left(K \cap \bigcup_{Q \in \mathfrak{Q}} \partial Q^*\right) \cap \bigcup_{j=1}^{n} \gamma_j^* = \emptyset.$$

Suppose then that $Q \in \mathfrak{Q}$ and $z \in K \cap \partial Q^*$. It may be that z is one of the four vertices of Q. In this case our construction shows that there exist Q_1, Q_2, Q_3, $Q_4 \in \mathfrak{Q}$, with $Q = Q_1$, such that z is a vertex of each Q_j. Hence each edge at z occurs twice in the boundaries of the Q_j, with opposite orientation; hence each of these edges was removed in forming $\mathcal{E}$, and so $z \notin \bigcup_{j=1}^n \gamma_j^*$. Similarly if z is not a vertex of Q; in that case $z \in E^*$ where E is one of the edges of Q, and $\dot{-}E$ is an edge of another element of $\mathfrak{Q}$, so that E was removed and $z \notin \bigcup_{j=1}^n \gamma_j^*$.

The assertions about the index of Γ about various points follow similarly. If $z \in \mathbb{C} \setminus D$ then

$$\text{Ind}(\Gamma, z) = \sum_{Q \in \mathfrak{Q}} \text{Ind}(\partial Q, z) = \sum_{Q \in \mathfrak{Q}} 0 = 0,$$

as required.

Suppose that $z \in K$. There are three possibilities: (i) z lies in the interior of some $Q_1 \in \mathfrak{Q}$; (ii) z lies on the boundary of some $Q_1 \in \mathfrak{Q}$ but is not a vertex; (iii) z is a vertex of some $Q_1 \in \mathfrak{Q}$. In case (i) we have

$$\text{Ind}(\Gamma, z) = \text{Ind}(\partial Q_1, z) + \sum_{Q \in \mathfrak{Q}, Q \neq Q_1} \text{Ind}(\partial Q, z) = 1 + 0 = 1.$$

In case (ii) z lies on an edge common to two squares $Q_1, Q_2 \in \mathfrak{Q}$. We set $\gamma = \partial Q_1 \dot{+} \partial Q_2$ with the common edge removed, and we see that

$$\text{Ind}(\Gamma, z) = \text{Ind}(\gamma, z) + \sum_{Q \in \mathfrak{Q}, Q \neq Q_1, Q_2} \text{Ind}(\partial Q, z) = 1 + 0 = 1.$$

In case (iii) z is common to four squares $Q_1, Q_2, Q_3, Q_4 \in \mathfrak{Q}$ and

$$\text{Ind}(\Gamma, z) = \text{Ind}(\gamma, z) + \sideset{}{'}\sum_{Q \in \mathfrak{Q}, Q \neq Q_1, Q_2, Q_3, Q_4} \text{Ind}(\partial Q, z) = 1 + 0 = 1$$

where $\gamma = \partial Q_1 \dot{+} \partial Q_2 \dot{+} \partial Q_3 \dot{+} \partial Q_4$ with common edges removed.

And similarly $\text{Ind}(\Gamma, z) = 0$ or 1 for all $z \in D \setminus \bigcup_{j=1}^n \gamma_j^*$: It is clear that the index is 0 or 1 unless z lies on an edge of some $Q \in \mathfrak{Q}$, and the index is 0 or 1 in this case as well by continuity. □

Figures 10.3a and 10.3b illustrate the proof of the lemma; in the figures D is the lightly shaded annulus and K is the darker annulus.

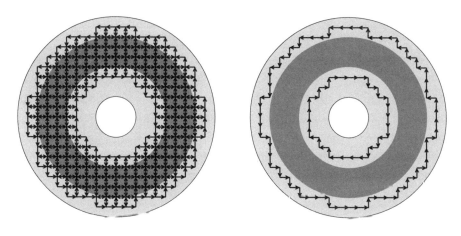

Figure 10.3a Figure 10.3b

I can't resist pointing out that the figures were "drawn" by a little computer program (a Python script generating PostScript code, for readers interested in such things) that actually implemented the proof of Lemma 10.5.5, with no cheating: First we set up a grid of squares small enough that any square intersecting K must lie entirely in D. Then we checked which squares intersected K, and drew each one in Figure 10.3a, one line at a time, attaching a little arrowhead to each line to indicate the direction. We kept track of what lines were drawn in Figure 10.3a; then for Figure 10.3b we redrew some of the lines, omitting the ones that occurred twice in Figure 10.3a, with opposite orientations. And sure enough the remaining lines in Figure 10.3b fit together into finitely many closed curves, with the right winding numbers about every point!

I'm always amazed when something like that actually works.

10.6. Intermission: Harmonic Functions and Brownian Motion

I decided to include a section on harmonic functions and Brownian motion even though we cannot possibly give actual proofs of the results — the subject is just too fascinating to skip. It is also very useful; although even giving precise definitions would take us too far afield, if one simply believes that there are precise definitions and proofs out there somewhere one can use the fuzzy definitions and proofs here to figure out a lot of stuff. Looking at things from the point of view of the current section gives the best intuition regarding various topics associated with harmonic functions.

In particular Brownian motion is really the best way to understand intuitively how the Dirichlet problem works in general open sets — even without precise definitions thinking about things as below allows one to see at a glance why various things work the way they do.

(You can find a traditional treatment of the Dirichlet problem in many places, for instance [**H**].)

Although we will be ignoring the formal measure-theoretic probability theory needed to make a discussion of Brownian motion rigorous, we should mention that the reader probably needs some familiarity with a more informal version of probability theory in order to make sense of the results in this section; we will be talking about things like "random variables", "distributions" and "expectations" without attempting to explain what these terms mean. Probably the most advanced notion that will arise is a simple version of the "conditional expectation", that is, the expected value of a random variable "given" an event.

Of course, since we do not actually *prove* anything in this section, the results discussed here will not be used in the rest of the book. To begin:

If $z \in \mathbb{C}$ the notation $B_t(z)$ will denote "Brownian motion started at z". This means that if we set $\gamma(t) = B_t(z)$ then $\gamma : [0, \infty) \to \mathbb{C}$ is (almost surely) a continuous curve with $\gamma(0) = z$ and such that γ is constantly changing direction "at random". (An event is said to occur "almost surely" if it happens with probability 1. If it sounds to you like this means the event is certain to occur, we assure you that it does not; the distinction between "almost surely" and "certainly" will become clear when you study formal measure-theory-based probability theory.)

Yes, that's much too fuzzy to be an actual definition. It is not too hard to give a precise definition of what Brownian motion is supposed to do, but it would take us a few pages, because the actual definition of phrases like "random variable" involves concepts from measure theory. Since there's no way we could hope to prove that there actually *is* something satisfying the precise definition, we may as well omit it; the fuzzy nondefinition above will suffice for our purposes. You should definitely try to understand what's going on here, because it really does give the best available intuition on a lot of things. But you should probably settle for an understanding something like the way a typical Calc 1 student understands calculus — sort of seeing how things work and how one can do various calculations, without worrying too much about low-level explanations of why the concepts actually make sense.

You can find a precise definition of Brownian motion and a proof that it exists in many places, including [**F**] in the section on "the Wiener Process" (that's the fancy name for Brownian motion; Wiener was the first to give a

10.6. Intermission: Harmonic Functions and Brownian Motion

mathematically rigorous proof that there is such a thing). It is well known that Brownian motion is important in physics (see for example the comments on the heat equation at the end of this section), but it may not be so clear why anyone would care about it from a purely mathematical point of view — the point to this section is to give one answer to that question.

(If you find this stuff interesting you can find other similar applications of Brownian motion to complex analysis, from a slightly informal point of view, in [**G**]. When you decide you actually want to get into the math there are lots of places to look, for example [**DU**], [**FB**], or what people working in the field call the mother of all books, [**DO**].)

So anyway, $B_t(z)$ is a random curve that starts at z and runs around the plane, continuously changing direction at random. It's not too hard to write a computer program that gives a moderately realistic idea of what a Brownian path looks like; see Figure 10.4.

Figure 10.4

While Figure 10.4 may have its own strange beauty it doesn't look much like a "curve", and in particular it's not easy to look at it and see where it starts and ends, etc. Hence when we need to illustrate something in this section we will settle for a much less realistic depiction of Brownian motion, something at worst like Figure 10.5.

Figure 10.5

It turns out that $B_t(z)$ is almost surely unbounded. (Think about it: If you want to remain in a bounded subset of the plane you probably don't want to make turns at random — if you run around at random you have to get infinitely lucky to avoid *ever* making a sequence of turns that takes you outside of the bounded set you're trying to stay inside.) Although $B_t(z)$ is almost surely unbounded it does not tend to infinity as $t \to \infty$; in fact the image is dense in the plane. (This changes in higher dimensions; Brownian motion in $\mathbb{R}^3$ does tend to infinity.)

Suppose that D is a bounded open set in the plane and $z \in D$. Since $B_t(z)$ is almost surely unbounded, it must leave D at some point. Since it starts at $z \in D$ and depends continuously on t, it follows that there is a first time τ such that $B_\tau(z) \in \partial D$; we will use the (non-standard) notation

$$\text{Exit}(D, z) = B_\tau(z)$$

to denote the first point of ∂D hit by $B_t(z)$. So for $z \in D$ we have

$$\text{Exit}(D, z) = B_\tau(z) \in \partial D,$$

where τ is such that

$$B_t(z) \in D \quad (0 \le t < \tau).$$

Now, if $z \in D$ is given then $B_t(z)$ is a random variable, whatever that means, so that τ and $\text{Exit}(D, z)$ are also random variables. Suppose that $f \in C(\partial D)$, and fix $z \in D$. Then $f(\text{Exit}(D, z))$ is a bounded complex-valued random variable, and as such it has an "expectation" or "expected value" $E(f(\text{Exit}(D, z)))$. We will see that setting $u(z) = E(f(\text{Exit}(D, z)))$ gives the solution to the Dirichlet problem, at least when the problem has a solution!

Suppose that $\overline{D}(z, r) \subset D$. Our Brownian particle starting at z must hit the boundary of the disk $D(z, r)$ on its way to the boundary of D; the place where it first hits the boundary of this disk is $\text{Exit}(D(z, r), z)$. Now, Brownian motion has no memory: Where it's about to be in the near future has something to do with where it is right now, but it has nothing at all to do with where it was in the past. This means that the following two ways of choosing a random point of ∂D are equivalent (technically, they give the same "distribution" on ∂D):

(i) Start at z and wander around D in a Brownian way until you hit ∂D.

(ii) Start at z and wander around until you hit $\partial D(z, r)$. Now "stop" and "restart" your Brownian motion, and wander until you hit ∂D.

Putting this a tiny bit more precisely, the two random variables $\text{Exit}(D, z)$ and $\text{Exit}(D, \text{Exit}(D(z, r), z))$ have the same "distribution".

Now if we start at z and wander until we hit $\partial D(z, r)$ what happens? Brownian motion in the plane treats all directions the same, so we are just as

10.6. Intermission: Harmonic Functions and Brownian Motion

likely to hit any point of $\partial D(z,r)$ as any other point. Well, the probability we hit any given point is of course 0; what we actually mean to say is that the probability that we hit $\partial D(z,r)$ in a given arc on the boundary depends only on the length of the arc: We have

$$\text{Exit}(D(z,r),z) = z + re^{i\theta}$$

for some random angle θ, and if $0 \le a < b \le 2\pi$ then the probability that $\theta \in [a,b]$ is just $(b-a)/(2\pi)$. (Officially: $\text{Exit}(D(z,r),z)$ is "uniformly distributed" on $\partial D(z,r)$. Which is clear by symmetry — why would one arc be more likely than another arc of the same size?) This says that (ii) above is also equivalent to

(iii) Choose a point $z+re^{i\theta} \in \partial D(z,r)$ "at random." Jump to that point and start Brownian motion, then wander until you hit ∂D.

The somewhat more precise version of this is the following: If Θ is a random variable uniformly distributed on $[0, 2\pi]$ and independent of our Brownian motion then the three ∂D-valued random variables $\text{Exit}(D,z)$, $\text{Exit}(D, \text{Exit}(D(z,r), z))$ and $\text{Exit}(D, z + re^{i\Theta})$ all have the same distribution.

Now for the Dirchlet problem: Suppose that $f \in C(\partial D)$ and define $u: D \to \mathbb{C}$ by

$$u(z) = E(f(\text{Exit}(D, z))).$$

The fact that $\text{Exit}(D, z)$ and $\text{Exit}(D, z + re^{i\Theta})$ have the same distribution shows that we also have

$$u(z) = E(f(\text{Exit}(D, z + re^{i\Theta}))).$$

But how does the point $\text{Exit}(D, z + re^{i\Theta})$ get chosen again? We start by choosing a random point of the boundary of the disk $D(z,r)$ and then start Brownian motion there, continuing until we hit the boundary of D. So we actually have

$$u(z) = E(f(\text{Exit}(D, z + re^{i\Theta})))$$
$$= \frac{1}{2\pi} \int_0^{2\pi} E(f(\text{Exit}(D, z + re^{is}))) \, ds$$
$$= \frac{1}{2\pi} \int_0^{2\pi} u(z + re^{is}) \, ds.$$

Wait, this looks familiar ... that's it, this is the Mean Value Property!

We've "shown" that if we define $u(z) = E(f(\text{Exit}(D, z)))$ then u has the Mean Value Property. One can also "show" that u is continuous, and it follows that u is harmonic in D. Let's record this for future reference:

Theorem 10.6.0. *If $D \subset \mathbb{C}$ is a bounded open set, $f \in C(\partial D)$ and*

$$u(z) = E(f(\text{Exit}(D, z))) \quad (z \in D)$$

then u is harmonic in D.

That's an actual true fact, by the way, not that we've actually *proved* it.

We will assume for the next few pages that D is a bounded open set in the plane, $f \in C(\partial D)$, and that $u : D \to \mathbb{C}$ has been defined by $u(z) = E(f(\text{Exit}(D, z)))$. We now turn to the question of whether

(∗) $$\lim_{z \to \zeta} u(z) = f(\zeta)$$

for every $\zeta \in \partial D$. The answer is yes or no, depending on D; if the answer is yes for every f and every ζ we say that D is a *Dirichlet domain* (or that D is "regular for the Dirichlet problem").

It seems intuitively clear that (∗) should hold for every f if and only if

(∗∗) The probability that $\text{Exit}(D, z)$ is close to $\zeta \in \partial D$ is close to 1 for every $z \in D$ which is close enough to ζ.

This is because f is continuous: If the probability that $\text{Exit}(D, z)$ is close to ζ is very large then the probability that $f(\text{Exit}(D, z))$ is close to $f(\zeta)$ will also be very large; since f is bounded, the low-probability possibility that $\text{Exit}(D, z)$ is far from ζ should not make much difference, and so $E(f(\text{Exit}(D, z)))$ should be close to $f(\zeta)$, which is exactly what (∗) says. (A more precise version of all this will appear below.)

This gives one illustration of our claim that Brownian motion gives the best intuition regarding the Dirichlet problem: The relevance of (∗∗) is clear, and even better, for many open sets it is clear whether or not (∗∗) holds! (One cannot say the same thing for the condition corresponding to (∗∗) with other methods of solving the Dirichlet problem.)

To give two examples: Suppose that D is bounded by finitely many (smooth) curves. Then (∗∗) is intuitively clear: If you start near a point of a curve and take a small step in a random direction there is a positive probability that you have already crossed the curve at a point near your starting point. If your first step is in the wrong direction then following that by *two* steps in the right direction suffices; in order to get far from your starting point without ever hitting that curve you need to get very unlucky, taking a large number of steps in the wrong direction. The same is true if we assume just that every component of the boundary of D contains more

10.6. Intermission: Harmonic Functions and Brownian Motion

than one point, so this condition should also imply that D is a Dirichlet domain.

On the other hand, consider a punctured disk $D = \mathbb{D} \setminus \{0\}$, more or less the simplest example of an open set where not every component of the boundary contains more than one point. It seems clear that if we set $\zeta = 0$ then (∗∗) is false: No matter how close to the origin we start, a single point is too small a target; we will almost surely hit the boundary of the unit disk (which is "far from" the origin) before we hit the origin. And indeed, a punctured disk is a simple example of an open set which is not a Dirichlet domain.

(The same sort of plausible argument will also make it easy to see why "every component of the boundary contains more than one point" is *not* sufficient to imply that D is a Dirichlet domain for $D \subset \mathbb{R}^n$, $n > 2$.)

We now turn to the proofs of the assertions in the last several paragraphs. Yes, the word "proof" is something of an exaggeration here, but you should note that the arguments below really are quite close to being actual proofs: If we did have precise definitions of the things we're talking about then the arguments below would become actual proofs, with essentially no modification (many things would need to be rephrased in more precise language, of course). We begin with a precise statement of (∗∗) and a proof that it characterizes Dirichlet domains; here and below the notation $\Pr(A)$ denotes the probability of the event A:

Lemma 10.6.1. *Suppose that D is a bounded open subset of $\mathbb{C}$. Then D is a Dirichlet domain if and only if for every $\zeta \in \partial D$ and every $\delta > 0$ we have*

$$(\ast\ast\ast) \qquad \lim_{D \ni z \to \zeta} \Pr(|\mathrm{Exit}(D,z) - \zeta| < \delta|) = 1.$$

The proof uses the notion of "conditional expectation": If X is a random variable and A is an event with positive probability then $E(X|A)$ is the expected value of X, given A. Thus for example if A^c is the complementary event "not A" and $p = \Pr(A)$ then

$$E(X) = pE(X|A) + (1-p)E(X|A^c).$$

Proof. Suppose first that (∗∗∗) holds. Assume that $f \in C(\partial D)$ and define $u(z) = E(f(\mathrm{Exit}(D,z)))$. Fix $\zeta \in \partial D$ and $\epsilon > 0$. Choose $\delta > 0$ so that

$$|f(\zeta') - f(\zeta)| < \epsilon/2$$

for all $\zeta' \in \partial D$ with $|\zeta' - \zeta| < \delta$. Suppose that $z \in D$, and let A be the event defined by the condition

$$|\mathrm{Exit}(D,z) - \zeta| < \delta.$$

Assume first that event A actually happened. It follows that $|f(\text{Exit}(D,z)) - f(\zeta)| < \epsilon/2$, since $|\text{Exit}(D,z) - \zeta| < \delta$. So the expected value of $|f(\text{Exit}(D,z)) - f(\zeta)|$, *given A*, is

$$E(|f(\text{Exit}(D,z)) - f(\zeta)| \, | A) \leq \epsilon/2.$$

Now choose M so that $|f| \leq M$ everywhere on ∂D. It follows that

$$|f(\text{Exit}(D,z)) - f(\zeta)| \leq 2M$$

regardless of whether A happened or not, so in particular

$$E(|f(\text{Exit}(D,z)) - f(\zeta)| \, | A^c) \leq 2M.$$

Let $p = \Pr(A)$. It follows that

$$\begin{aligned}
|u(z) - f(\zeta)| &= |E(f(\text{Exit}(D,z)) - f(\zeta))| \\
&\leq E(|f(\text{Exit}(D,z)) - f(\zeta)|) \\
&= pE(|f(\text{Exit}(D,z)) - f(\zeta)| \, | A) \\
&\quad + (1-p)E(|f(\text{Exit}(D,z)) - f(\zeta)| \, | A^c) \\
&< p\epsilon/2 + 2M(1-p) \leq \epsilon/2 + 2(1-p)M.
\end{aligned}$$

But our hypothesis (***) says that we can make p as close to 1 as we wish, merely by taking $z \in D$ close enough to ζ. In particular if z is close enough to ζ we will have $2(1-p)M < \epsilon/2$, and hence $|u(z) - f(\zeta)| < \epsilon$.

For the converse we assume that $\zeta \in \partial D$ and $\delta > 0$ are such that (***) does not hold: There exist $p < 1$ and a sequence $(z_n) \subset D$ with $z_n \to \zeta$, such that

$$\Pr(|\text{Exit}(D,z_n) - \zeta| < \delta) \leq p$$

for all n. Choose $f \in C(\partial D)$ such that $0 \leq f \leq 1$ everywhere on ∂D, $f(\zeta) = 1$, and $f(\zeta') = 0$ for every $\zeta' \in \partial D$ with $|\zeta' - \zeta| \geq \delta$. Fix n and let A be the event $|\text{Exit}(D,z_n) - \zeta| < \delta$. Now, if A does not happen then it follows that $f(\text{Exit}(D,z_n)) = 0$, so as before we see that

$$\begin{aligned}
u(z_n) &= \Pr(A)E(f(\text{Exit}(D,z_n))|A) + (1-\Pr(A))E(f(\text{Exit}(D,z_n))|A^c) \\
&= \Pr(A)E(f(\text{Exit}(D,z_n))|A) \leq \Pr(A) \leq p.
\end{aligned}$$

In particular, $u(z_n)$ does not tend to $f(\zeta) = 1$, so D is not a Dirichlet domain. □

If that seems complicated you should go back to our informal explanation of why (**) should characterize Dirichlet domains, and then try to see how the proof of the lemma is really just a more formal version of the same argument.

We can now show that an open set bounded by finitely many smooth curves is a Dirichlet domain:

10.6. Intermission: Harmonic Functions and Brownian Motion

Theorem 10.6.2. *If D is an open set in the plane bounded by finitely many continuously differentiable curves then D is a Dirichlet domain.*

Proof. We fix $\zeta \in \partial D$ and $\delta > 0$, and we show that $(***)$ holds. In the figures below we choose a coordinate system so that the tangent to the boundary at ζ is roughly horizontal; note that D is the gray area in the figures, and that the part of the boundary of D consisting of points at distance less than δ from ζ is drawn in black.

Suppose that $z \in D$ is close to ζ, and let Q be a square centered at z, such that the lower edge of Q lies in the complement of D and the part of the boundary of D within distance δ of ζ includes a curve in Q joining the left edge of Q to the right edge, as in Figure 10.6.

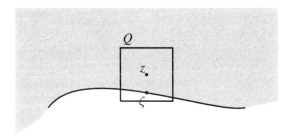

Figure 10.6

(The existence of a square Q with all the properties we require depends on the smoothness of the boundary of D.)

Now consider $B_t(z)$, Brownian motion started at z. It may happen that the first place $B_t(z)$ exits the square Q is on the lower edge of Q. If this happens then it follows that the first place $B_t(z)$ exits D is at a point of ∂D inside Q, and in particular at a point of ∂D within distance δ of ζ:

Figure 10.7

It is clear by symmetry that the probability that we exit Q along the lower edge is exactly $1/4$ (this is why we need z to be the center of Q). Since this event implies the event $|\text{Exit}(D,z) - \zeta| < \delta$, the latter event must have probability at least $1/4$ as well: $\Pr(|\text{Exit}(D,z) - \zeta| < \delta) \geq 1/4$.

We are trying to show that this probability tends to 1 as z tends to ζ, so showing it is at least $1/4$ is some progress. Now what happens if the point where B_t exits Q is *not* on the lower edge of Q? If that happens we can simply repeat the argument. Note that in fact the "bad" case is the case where the point where we exit Q is actually a point of D; assume that that happens:

Let $Q_1 = Q$. Let z_1 be the point where $B_t(z)$ exits Q_1, assume that $z_1 \in D$, and now let Q_2 be a square with center z_1, with lower edge lying outside D, and such that the part of the boundary of D within δ of ζ includes a curve joining the left edge of Q_2 to the right edge. If Brownian motion started at z_1 exits Q_2 at a point of the lower edge of Q_2 then it follows again that it exits D at a point within δ of ζ:

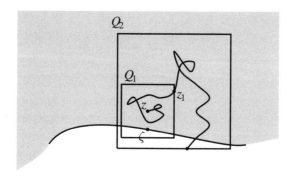

Figure 10.8

So we're happy if either of two things happens: Starting at z we exit Q_1 at a point outside D, *or* starting at z we exit Q_1 at a point $z_1 \in D$, and then starting at z_1 we exit Q_2 at a point outside D. As often happens, analyzing the probabilities is easier if we convert our "or" to "and" by taking complements: In order for things to go bad two things must happen: We must exit Q_1 at a point $z_1 \in D$, and then we must exit Q_2 at a point of D. These events are independent (because Brownian motion "has no memory"), so the probability that they both happen is the product of the probabilities, which is no larger than $(3/4)^2$. And so we have shown that if z is as in Figure 10.6 then in fact $\Pr(|\text{Exit}(D,z) - \zeta| < \delta) \geq 1 - \left(\frac{3}{4}\right)^2 = \frac{7}{16}$. This is an improvement over our previous estimate of $1/4$.

10.6. Intermission: Harmonic Functions and Brownian Motion

If z is as in Figure 10.9 we cannot repeat the argument a third time: If $z_2 \in D$ is the point where we exit Q_2 and we attempt to repeat the argument with Q_3 centered at z_2 we see that it is possible to exit Q_3 at a point of its lower edge without exiting D at a point within δ of ζ:

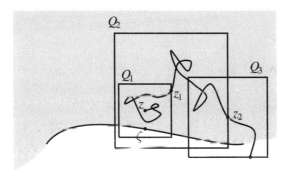

Figure 10.9

But, come to think of it, it is clear that the argument cannot work arbitrarily many times, because $\Pr(|\mathrm{Exit}(D, z) - \zeta| < \delta) < 1$ for any fixed z. The argument worked twice with z as in Figure 10.6. If you think about it you see that given a positive integer N, if z is close enough to ζ (no longer the z in Figure 10.6), then the argument works N times:

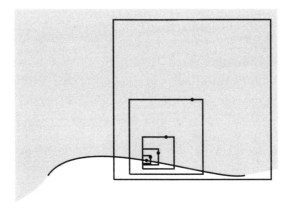

Figure 10.10

And so we see that for any N, if z is close enough to ζ then $\Pr(|\mathrm{Exit}(D, z) - \zeta| < \delta) \geq 1 - \left(\frac{3}{4}\right)^N$. Since $1 - (3/4)^N \to 1$ as $N \to \infty$, we are done. $\square$

Again, you are encouraged to compare the proof with our earlier informal explanation of why an open set bounded by smooth curves should satisfy (∗∗), noting in particular that the basic idea in the proof is simply "for things to go wrong you have to take a large number of steps in the wrong direction".

One can "prove" a lot of things about Brownian motion by similar crude arguments. In particular, one can give an analogous proof of the fact that if every component of the boundary of D contains more than one point then D is a Dirichlet domain. The basic idea is the same, but the obvious fact that if you start at the center of a square then the probability that you exit the square at a point of a given edge equals 1/4 is replaced by a somewhat more intricate lemma. Since this section is more than long enough already, we defer the proof to Chapter 22.

Next we show that a punctured disk is not a Dirichlet domain.

Theorem 10.6.3. *An isolated singularity for a bounded harmonic function is removable.*

We could prove the theorem using the "Poisson integral" for an annulus; the problem is that we do not know what the "Poisson integral" for an annulus *is!* But we can use the fact that $u(z) = E(f(\text{Exit}(D, z)))$ as a substitute for this Poisson integral that we don't know.

We begin with the waving of the hands: Fix $z \in \mathbb{C}$ with $0 < |z| < 1$, and then for $0 < r < |z|$ define $A_r = \{\, w \in \mathbb{C} : r < |w| < 1 \,\}$ and

$$p_r = \Pr(|\text{Exit}(A_r, z)| = r).$$

That is, p_r is the probability that Brownian motion started at z exits the annulus A_r at a point of the smaller of the two circles comprising ∂A_r. Then clearly

$$\lim_{r \to 0} p_r = 0.$$

That certainly seems clear in any case, and the theorem follows. We put half of the proof into a preliminary proposition:

Proposition 10.6.4. *Suppose that u is continuous in $\{\, z \in \mathbb{C} : 0 < |z| \leq 1 \,\}$ and harmonic and bounded in $\{\, z \in \mathbb{C} : 0 < |z| < 1 \,\}$. If $u(\zeta) = 0$ for all ζ with $|\zeta| = 1$ then $u(z) = 0$ for all z with $0 < |z| < 1$.*

Proof. Fix such a z and define A_r and p_r as above (for $0 < r < |z|$). It follows from Theorem 10.6.2 and Lemma 10.4.1.1 that

$$u(z) = E(u(\text{Exit}(A_r, z))).$$

Choose M so that $|u| \leq M$ everywhere, and let B be the event $|\text{Exit}(A_r, z)| = r$, so that $p_r = \Pr(B)$. Since the complementary event B^c implies that $u(\text{Exit}(A_r, z)) = 0$, we see that

$$u(z) = p_r E(u(\text{Exit}(A_r, z))|B) + (1 - p_r) E(u(\text{Exit}(A_r, z))|B^c)$$
$$= p_r E(u(\text{Exit}(A_r, z))|B).$$

Hence $|u(z)| \leq p_r M$ and so, since $p_r \to 0$ as $r \to 0$, it follows that $u(z) = 0$. □

Proof of Theorem 10.6.3. Suppose that u is harmonic in $D - \{z \subset \mathbb{C} : 0 < |z| < R\}$ for some $R > 1$ and that $|u| \leq M$ in D. Let $A = \{z \in \mathbb{C} : 0 < |z| < 1\}$, and define $v : A \to \mathbb{C}$ by

$$v(z) = u(z) - w(z) \quad (z \in A),$$

where $w \in C(\overline{\mathbb{D}})$ is harmonic in $\mathbb{D}$ and equal to u on $\partial \mathbb{D}$. Proposition 10.6.4 shows that $v = 0$ in A, which says that $u = w$ in A; hence u extends to a harmonic function in $\mathbb{D}$. □

Corollary 10.6.5. *The punctured disk $D = \{z \in \mathbb{C} : 0 < |z| < 1\}$ is not a Dirichlet domain.*

Proof. Suppose on the other hand that D is a Dirichlet domain. Then there exists a function $u \in C(\overline{D})$ such that u is harmonic in D, $u(z) = 0$ when $|z| = 1$, and $u(0) = 1$. But Proposition 10.6.4 shows that $u = 0$ in D, which shows that u is not continuous at the origin. □

We proved the last few results using the "obvious" fact that $p_r \to 0$ as $r \to 0$, where p_r is the probability that Brownian motion started at a fixed point z exits the annulus $\{w \in \mathbb{C} : r < |w| < 1\}$ at a point of the "inner" part of the boundary. We could actually give an argument that shows $p_r \to 0$ kind of like the "proof" we gave for the fact that Brownian motion started near the boundary of a set bounded by curves probably exits at a nearby point, by comparing p_r and $p_{r/2}$. But because we know exactly what the geometry looks like in the current situation it's easier to give a direct *proof* that $p_r \to 0$, using the connection between Brownian motion and the Dirichlet problem: This connection means that Brownian motion says something about harmonic functions, and it also means that harmonic functions say something about Brownian motion:

Lemma 10.6.6. *Suppose $0 < r < R$ and let A be the annulus $\{z : r < |z| < R\}$. Suppose $z \in A$ and let*

$$p = \Pr(|\text{Exit}(A, z)| = r).$$

Then
$$p = \frac{\log(R/|z|)}{\log(R/r)}.$$

(*In particular if z and R are fixed and $r \to 0$ then $p \to 0$.*)

Proof. Let $u(w) = \log(|w|)$. Then u is harmonic in A and continuous in $\overline{A}$, so
$$\log(|z|) = E(u(\text{Exit}(A,z))) = p\log(r) + (1-p)\log(R). \qquad \square$$

We "proved" Theorem 10.6.3 assuming without proof that $\lim_{r \to 0} p_r = 0$. Given the information we now have about the rate at which p_r tends to 0 the proof of Theorem 10.6.3 gives the following stronger result, in which u is not required to be bounded (although it's only allowed to blow up fairly slowly near the origin):

Theorem 10.6.3.1. *Suppose that u is harmonic in a punctured disk $\{z \in \mathbb{C} : 0 < |z| < R\}$ and*
$$\lim_{z \to 0} \frac{u(z)}{\log(|z|)} = 0.$$
Then u extends to a function harmonic in $D(0, R)$.

We leave the details of the proof to the interested reader; we note that the example $u(z) = \log(|z|)$ shows that the hypothesis of the theorem cannot be replaced by the barely weaker condition $|u(z)| \le c|\log(|z|)|$.

It is interesting to note that Lemma 10.6.6 can be used to show that $\{B_t(z) : t \ge 0\}$ is almost surely dense in the plane (that is, Brownian motion in the plane is *recurrent*.)

Proposition 10.6.7. *Suppose that $0 < r < |z|$. Then there almost surely exists $t > 0$ such that $|B_t(z)| = r$.*

Proof. If $R > |z|$ then Lemma 10.6.6 shows that the probability that there exists τ such that $|B_\tau(z)| = R$ while $|B_t(z)| > r$ for all $t \in [0, \tau)$ is
$$1 - \frac{\log(R/|z|)}{\log(R/r)} = \frac{\log(|z|/r)}{\log(R/r)}.$$

Now, if we *never* have $|B_t(z)| = r$ then since, as we are assuming without proof, $B_t(z)$ is almost surely unbounded, it follows that for *every* $R > |z|$ we reach a point where $|B_t(z)| = R$ before $|B_t(z)| = r$; the probability that this happens is no larger than
$$\lim_{R \to \infty} \frac{\log(|z|/r)}{\log(R/r)} = 0. \qquad \square$$

10.6. Intermission: Harmonic Functions and Brownian Motion

Corollary 10.6.8. *For every $r > 0$ and every $z \in \mathbb{C}$ there almost surely exists a sequence $t_n \to \infty$ such that $|B_{t_n}(z)| = r$.*

Proof. The proof of the proposition shows that for every $T > 0$ there almost surely exists $t > T$ with $|B_t(z)| = r$. □

Of course the corollary holds with any other disk $D(p, r)$ in place of $D(0, r)$. Now consider the countable collection of disks with rational radii, and such that the real and imaginary parts of the center are both rational. Using a fact from measure theory, namely that the intersection of a countable collection of events of probability 1 has probability 1, we obtain

Corollary 10.6.9. *$B_t(z)$ is almost surely dense in the plane.*

We begin the conclusion of this section with a few comments on what happens in $\mathbb{R}^d$ for $d > 2$. Many things work out more or less the same as in the plane. For example, the proof of Theorem 10.6.2 applies essentially without change (using cubes in place of squares) to show that if D is bounded by finitely many $(d-1)$-dimensional smooth surfaces then D is a Dirichlet domain. It is easy to see that the condition "every component of the boundary contains at least two points" does *not* imply that D is a Dirichlet domain:

Example 10.6.10. *Let B be the open unit ball of $\mathbb{R}^3$ and let $K = \{(0, 0, x_3) : 0 \leq x_3 \leq 1\}$. Then $D = B \setminus K$ is not a Dirichlet domain.*

Proof. Define $T : \mathbb{R}^3 \to \mathbb{R}^2$ by $T(x_1, x_2, x_3) = (x_1, x_2)$. It is easy to see from the (omitted) definitions that if $B_t(x)$ is three-dimensional Brownian motion then $T(B_t(x))$ is a two-dimensional Brownian motion started at $T(x)$.

Hence the fact that Brownian motion in the plane almost surely never hits the origin shows that Brownian motion in $\mathbb{R}^3$ almost surely never hits K; this shows that D is not a Dirichlet domain exactly as in the proof above that a punctured disk in the plane is not a Dirichlet domain. □

(An intuitive explanation for the difference between $\mathbb{R}^2$ and $\mathbb{R}^3$ here is that a connected subset of $\mathbb{R}^2$ containing more than one point must have "dimension" at least 1, in various senses which we are not going to go into; on the other hand a connected subset of $\mathbb{R}^3$ containing more than one point may have dimension less than 2.)

And it is also easy to see that if $d > 2$ then $|B_t(x)| \to \infty$ almost surely as $t \to \infty$ (which says that if $d > 2$ then Brownian motion in $\mathbb{R}^d$ is *transient* instead of recurrent):

Lemma 10.6.11. *Suppose that $d > 2$, $0 < r < R$, and let $A = \{ x \in \mathbb{R}^d : r < |x| < R \}$. Suppose $x \in A$, and let*

$$p = \Pr(|\text{Exit}(A, x)| = r).$$

Then
$$p = \frac{|x|^{2-d} - R^{2-d}}{r^{2-d} - R^{2-d}}.$$

Proof. Define $u : \mathbb{R}^d \setminus \{0\} \to \mathbb{R}$ by $u(w) = |w|^{2-d}$. A simple calculation (which you should actually *do* for practice) shows that u is harmonic; now repeat the proof of Lemma 10.6.6. □

Corollary 10.6.12. *Suppose $d > 2$, $x \in \mathbb{R}^d$ and $0 < r < |x|$. The probability that $|B_t(x)| > r$ for all $t \geq 0$ is greater than or equal to*

$$1 - \left(\frac{|x|}{r}\right)^{2-d}.$$

Proof. Choose $R_n > |x|$ with $R_n \to \infty$. The probability that $|B_t(x)| = R_n$ before $|B_t(x)| = r$ is

$$1 - \frac{|x|^{2-d} - R_n^{2-d}}{r^{2-d} - R_n^{2-d}}.$$

Since "$|B_t(x)| = R_n$ before $|B_t(x)| = r$" is a decreasing sequence of events, the probability that *all* of these events happen is the limit of the probabilities, which is

$$1 - \left(\frac{|x|}{r}\right)^{2-d}.$$

But if $|B_t(x)| = R_n$ before $|B_t(x)| = r$ for every n then it follows that $|B_t(x)| > r$ for all $t > 0$. □

Corollary 10.6.13. *Suppose $d > 2$ and $x \in \mathbb{R}^d$. Then almost surely $|B_t(x)| \to \infty$ as $t \to \infty$.*

Proof. Fix $r > 0$. Let p be the probability that there exists $T > 0$ such that $|B_t(x)| > r$ for all $t > T$. We need to show that $p = 1$. But now fix $R > \max(r, |x|)$. Since $B_t(x)$ is almost surely unbounded, there almost surely exists T_R such that $|B_{T_R}(x)| = R$. Now the previous corollary shows that the probability that $|B_t(x)| > r$ for all $t > T_R$ is at least

$$1 - \left(\frac{R}{r}\right)^{2-d}.$$

Hence
$$p \geq 1 - \left(\frac{R}{r}\right)^{2-d}.$$
Letting $R \to \infty$ shows that $p = 1$. □

We conclude with a comment on the connection between Brownian motion and physics. From a physical point of view it is not at all surprising that Brownian motion should have something to do with harmonic functions: The Laplace equation
$$\frac{\partial^2 u(x,y)}{\partial x^2} + \frac{\partial^2 u(x,y)}{\partial y^2} = 0$$
is what the so called "heat equation"
$$\frac{\partial^2 u(x,y,t)}{\partial x^2} + \frac{\partial^2 u(x,y,t)}{\partial y^2} = c\frac{\partial u(x,y,t)}{\partial t}$$
reduces to if we assume that
$$\frac{\partial u}{\partial t} = 0;$$
that is, harmonic functions are just equilibrium solutions to the heat equation. And it is not at all implausible that Brownian motion should have something to do with the heat equation — if you consider a bunch of infinitesimal atoms bumping into each other at random and define temperature in terms of the atoms' velocity then you can show that the path of an infinitesimal "particle of heat" is described by something like Brownian motion and you can also show that the temperature satisfies the heat equation. In fact Brownian motion is just the "diffusion" associated with the heat equation; other partial differential equations have other associated diffusions.

In Chapter 22 we give the proof that an open set in the plane such that every component of the boundary contains more than one point is a Dirichlet domain, and also a miscellaneous application to conformal mapping.

10.7. The Schwarz Reflection Principle and Harnack's Theorem

In this section we return to rigorous mathematics, with actual definitions and proofs; in particular we shall prove versions of the Schwarz Reflection Principle, Harnack's Inequality and Harnack's Theorem.

The Schwarz Reflection Principle is one of those things that come in many different flavors. It is often stated as a result about holomorphic functions, but the version below concerning holomorphic functions follows from a certain result about harmonic functions — it seems like a good thing to state the result about harmonic functions separately, since it works (with more or less the same proof) in $\mathbb{R}^n$, where there are no holomorphic functions.

Suppose that $D \subset \mathbb{C}$ is open. We will say that D is *symmetric about the real axis* if $\overline{z} \in D$ whenever $z \in D$. In this case we will write $D^+ = \{\, z \in D : \operatorname{Im}(z) > 0 \,\}$, $D^- = \{\, z \in D : \operatorname{Im}(z) < 0 \,\}$, and $D^0 = \{\, z \in D : \operatorname{Im}(z) = 0 \,\}$.

Theorem 10.7.0 (Schwarz Reflection for harmonic functions). *Suppose that D is an open subset of the plane which is symmetric about the real axis. Suppose that $u \in C(D^+ \cup D^0)$ is harmonic in D^+ and vanishes on D^0. Then u extends to a function v harmonic in D; the function v satisfies*

$$v(\overline{z}) = -v(z) \quad (z \in D).$$

Proof. It is clear how we need to define v if this is going to work: We set

$$v(z) = \begin{cases} u(z) & (z \in D^+), \\ 0 & (z \in D^0), \\ -u(\overline{z}) & (z \in D^-). \end{cases}$$

We need only show that v is harmonic in D.

It is clear that v is continuous in D, so we need only show that it satisfies the Mean Value Property in D; in fact we need only show that it satisfies a "small-radius" Mean Value Property. But this is clear: If $z \in D^+$ then we can choose $\rho(z) = \operatorname{Im}(z)$; now if $0 < r < \rho(z)$ it follows that

$$v(z) = u(z) = \frac{1}{2\pi} \int_0^{2\pi} u(z + re^{it})\, dt = \frac{1}{2\pi} \int_0^{2\pi} v(z + re^{it})\, dt$$

because $\overline{D}(z, r) \subset D^+$ and u is harmonic in D^+.

Similarly for $z \in D^-$. Now if $z \in D^0$ we can use any $\rho(z) > 0$ such that $\overline{D}(z, \rho(z)) \subset D$; then for $0 < r < \rho(z)$ we have $v(z + re^{-it}) = v(\overline{z + re^{it}}) = -v(z + re^{it})$. Since $a = b$ implies $a = b = (a+b)/2$, we see that

$$\frac{1}{2\pi} \int_0^{2\pi} v(z + re^{it})\, dt = \frac{1}{2\pi} \int_0^{2\pi} v(z + re^{-it})\, dt$$
$$= \frac{1}{2\pi} \int_0^{2\pi} \frac{v(z + re^{it}) + v(z + re^{-it})}{2}\, dt$$
$$= \frac{1}{2\pi} \int_0^{2\pi} 0\, dt = 0 = v(0). \qquad \square$$

The Schwarz Reflection Principle for holomorphic functions follows:

10.7. The Schwarz Reflection Principle and Harnack's Theorem

Theorem 10.7.1 (Schwarz Reflection for holomorphic functions).
Suppose that D is an open subset of the plane which is symmetric about the real axis. Suppose that $f \in H(D^+)$ satisfies

$$\lim_{n \to \infty} \mathrm{Im}\,(f(z_n)) = 0$$

for every sequence (z_n) in D^+ which converges to a point of D^0. Then f extends to a function $F \in H(D)$, which satisfies

$$F(\bar{z}) = \overline{F(z)} \quad (z \in D).$$

We should note that we are assuming that $\mathrm{Im}\,(f)$ extends continuously to $D^+ \cup D^0$, but we are not making any such assumption about $\mathrm{Re}\,(f)$. This is why the proof is a little more complicated than the proof of the previous theorem; for example we cannot simply define $F(z) = f(z)$ for $z \in D^0$ because f is not defined on D^0. (If we assume that f extends continuously to $D^+ \cup D^0$ and takes real values on D^0 then a somewhat simpler proof of the theorem is possible using Morera's Theorem.)

Proof. Let $f = u + iv$. The previous theorem shows that v extends to a function V harmonic in D, which satisfies

$$V(\bar{z}) = -V(z) \quad (z \in D).$$

For $t \in D^0$ we can choose a disk $D_t \subset D$ with center t. Now, V is harmonic in D_t and D_t is simply connected, so there is a function $f_t \in H(D_t)$ such that $\mathrm{Im}\,(f_t) = V$ in D_t. Any two such functions differ by a constant; since $\mathrm{Im}\,(f) = V$ in D_t^+, we may adjust f_t by adding a real constant so that $f = f_t$ in D_t^+.

Now, the fact that $V = 0$ on D^0 shows that f_t takes real values on D_t^0; thus each of the derivatives $f_t^{(k)}(t)$ is real, and so the power series representing f_t in D_t has real coefficients. This implies that

$$f_t(\bar{z}) = \overline{f_t(z)} \quad (z \in D_t),$$

and hence we may define $F : D \to \mathbb{C}$ by

$$F(z) = \begin{cases} f(z) & (z \in D^+), \\ f_t(z) & (z \in D_t), \\ \overline{f(\bar{z})} & (z \in D^-). \end{cases}$$

(We know that $f_t = f$ in $D^+ \cap D_t = D_t^+$, and it follows that $f_t(z) = \overline{f(\bar{z})}$ for $z \in D_t \cap D^- = D_t^-$; hence $f_t = f_s$ in $D_t \cap D_s$.) As in the proof of the

Cauchy Integral Formula, the reason we wrote the definition of F in such a curious way, with overlapping cases, is that the domain of the function defined in each case is an open set; thus to show F is holomorphic we need only show that $g \in H(D^-)$, where $g(z) = \overline{f(\bar z)}$ (since we already know that $f \in H(D^+)$ and $f_t \in H(D_t)$).

The simplest way to see that g is holomorphic in D^- is probably using power series. Suppose that $a \in D^-$. Then $\bar a \in D^+$, so, since $f \in H(D^+)$, there exist $r > 0$ and a sequence of complex numbers (c_n) such that

$$f(z) = \sum_{n=1}^{\infty} c_n (z - \bar a)^n \quad (|z - \bar a| < r).$$

It follows that

$$f(\bar z) = \sum_{n=1}^{\infty} c_n (\bar z - \bar a)^n \quad (|\bar z - \bar a| < r)$$

and hence that

$$g(z) = \overline{f(\bar z)} = \overline{\sum_{n=1}^{\infty} c_n (\bar z - \bar a)^n} = \sum_{n=0}^{\infty} \overline{c_n} (z - a)^n \quad (|z - a| < r),$$

which shows that $g \in H(D(a, r))$; since this works for any $a \in D^-$ we have $g \in H(D^-)$. □

The theorem above is an existence statement. The following might be regarded as the corresponding uniqueness statement; it's curious that it is an immediate corollary. Note that the corollary is also known as the Schwarz Reflection Principle:

Corollary 10.7.2. *Suppose that D is a connected open subset of the plane, D is symmetric about the real axis, $f \in H(D)$, and $f(z) \in \mathbb{R}$ for all $z \in D^0$. Then*

$$f(\bar z) = \overline{f(z)} \quad (z \in D).$$

Proof. The previous theorem shows that there exists $F \in H(D)$ such that

$$F(\bar z) = \overline{F(z)} \quad (z \in D)$$

and $F|_{D^+} = f|_{D^+}$. Since D is connected and D^+ has plenty of limit points in D, it follows that $F = f$. □

Thanks to Alan Noell for pointing out that the Reflection Principle gives the elegant solution to Exercise 8.7:

10.7. The Schwarz Reflection Principle and Harnack's Theorem 219

- **Exercise 8.7.** Suppose that $\psi \in \text{Aut}(\Pi^+)$. Show that there exist a, b, c, $d \in \mathbb{R}$, with $ad - bc = 1$, such that

$$\psi(z) = \frac{az+b}{cz+d}$$

for all $z \in \Pi^+$. Show that a, b, c, d are not unique but are almost unique (state and prove a precise statement regarding how close to unique they are).

First, if $\Phi : \mathbb{D} \to \Pi^+$ is the Cayley transform then $\phi = \Phi^{-1} \circ \psi \circ \Phi \in \text{Aut}(\mathbb{D})$, so in particular $\phi \in \mathcal{M}$, and hence $\psi = \Phi \circ \phi \circ \Phi^{-1} \in \mathcal{M}$; thus there exist a, b, c, $d \in \mathbb{C}$ with $\psi(z) = (az+b)/(cz+d)$, and we need only show that the coefficients may be taken to be real.

Since Φ is actually a homeomorphism from the closed disk to the closure of the upper half-plane in $\mathbb{C}_\infty$, taking the boundary of the disk to the (extended) boundary of the upper half-plane, and since the elements of $\text{Aut}(\mathbb{D})$ extend to homeomorphisms of the disk taking the boundary to the boundary, it follows that ψ extends to a homeomorphism of the closure (in $\mathbb{C}_\infty$) of the upper half-plane to itself, mapping the extended boundary to itself. That is, ψ maps $\mathbb{R} \cup \{\infty\}$ to itself; thus there exists $x_0 \in \mathbb{R}$ such that $\psi(\mathbb{R} \setminus \{x_0\}) \subset \mathbb{R}$. (If $\psi(\infty) = \infty$, you can take x_0 to be any element of $\mathbb{R}$.)

Let $D = \mathbb{C} \setminus \{x_0\}$. Then $\psi \in H(D)$, and Corollary 10.7.2 shows that

$$\overline{\psi(\overline{z})} = \psi(z)$$

for all $z \in D$. That is,

$$\frac{\overline{a}z + \overline{b}}{\overline{c}z + \overline{d}} = \frac{az+b}{cz+d}$$

for all $z \in D$. But if two matrices define the same linear-fractional transformation they must be scalar multiples of each other; there exists $\lambda \in \mathbb{C}$ such that $\overline{a} = \lambda a$, $\overline{b} = \lambda b$, $\overline{c} = \lambda c$, and $\overline{d} = \lambda d$.

Now if $\lambda = 1$ this shows that a, b, c and d are real. Suppose that $\lambda \neq 1$. Then if we set $\omega = (1-\lambda)/(2i)$ we have

$$\text{Im}(a) = \frac{a - \overline{a}}{2i} = \frac{1-\lambda}{2i} a = \omega a,$$

and similarly $\text{Im}(b) = \omega b$, $\text{Im}(c) = \omega c$, and $\text{Im}(d) = \omega d$. Since $\omega \neq 0$, this shows that

$$\psi(z) = \frac{az+b}{cz+d} = \frac{\text{Im}(a)z + \text{Im}(b)}{\text{Im}(c)z + \text{Im}(d)},$$

and again we have shown that ψ is given by a real matrix.

That was the tricky part. Now we change notation if necessary, and say that we have found a, b, c, $d \in \mathbb{R}$ such that $\psi(z) = (az+b)/(cz+d)$. The fact that $\operatorname{Im}(\psi(i)) > 0$ shows that $ad - bc > 0$ (do the math!), and hence we may divide the coefficients by a positive number to obtain $ad - bc = 1$. (We need to know that $ad - bc > 0$ here; if $ad - bc < 0$ we would need to divide the coefficients by an imaginary quantity to get the determinant to equal 1, and they would not be real any more.)

This finishes the existence part. For the uniqueness we note that A, B, C, and D define the same linear-fractional transformation as a, b, c, and d if and only if $A = \lambda a$, $B = \lambda b$, $C = \lambda c$ and $D = \lambda d$ for some nonzero $\lambda \in \mathbb{C}$. If $AD - BC = ad - bc$ we must have $\lambda^2 = 1$. Both solutions to $\lambda^2 = 1$ happen to be real numbers, so we get real A, B, C and D from either $\lambda = 1$ or $\lambda = -1$: a, b, c and d are unique up to multiplication by -1.

So $\operatorname{Aut}(\Pi^+)$ is isomorphic to $SL_2(\mathbb{R})/G$, where $SL_2(\mathbb{R})$ is the group of all real 2×2 matrices with determinant 1 and G is the subgroup $\{I, -I\}$.

Enough about the Schwarz Reflection Principle for now; on to Harnack's Inequality and Harnack's Theorem. Harnack's Inequality is kind of interesting; it says that if D is a connected open set and a, $b \in D$ then there exists a constant c such that

$$u(b) \leq cu(a)$$

for every positive function u harmonic in D. In fact such an inequality holds uniformly on compact sets, although the constant will blow up if we hold a fixed and let b approach the boundary. Harnack's Theorem follows: An increasing sequence of harmonic functions either tends to infinity or converges to a harmonic function (in either case the convergence is uniform on compact sets).

We begin with Harnack's Inequality for the unit disk:

Theorem 10.7.3 (Harnack's Inequality). *Suppose that $u \geq 0$ is harmonic in $\mathbb{D}$. Then*

$$\frac{1-r}{1+r} u(0) \leq u(re^{it}) \leq \frac{1+r}{1-r} u(0)$$

for all $re^{it} \in \mathbb{D}$.

Proof. First, we may assume that u is actually continuous on $\overline{\mathbb{D}}$: Suppose we know the result in that case, and suppose that $u \geq 0$ is harmonic in $\mathbb{D}$. For $0 < \rho < 1$ let $u_\rho(z) = u(\rho z)$; then $u_\rho \in C(\overline{\mathbb{D}})$ is harmonic and so

$$\frac{1-r}{1+r} u_\rho(0) \leq u_\rho(re^{it}) \leq \frac{1+r}{1-r} u_\rho(0),$$

10.7. The Schwarz Reflection Principle and Harnack's Theorem

which implies that
$$\frac{1-r}{1+r}u(0) \le u(re^{it}) \le \frac{1+r}{1-r}u(0)$$
(let $\rho \to 1$).

But if u is continuous in $\overline{\mathbb{D}}$ then we know that
$$u(re^{it}) = \frac{1}{2\pi}\int_0^{2\pi} u(e^{is})P_r(s-t)\,ds.$$

Now
$$P_r(s) = \frac{1-r^2}{1-2r\cos(s)+r^2} \le \frac{1-r^2}{1-2r+r^2} = \frac{1+r}{1-r}$$
and
$$P_r(s) = \frac{1-r^2}{1-2r\cos(s)+r^2} \ge \frac{1-r^2}{1+2r+r^2} = \frac{1-r}{1+r},$$
so, since $u \ge 0$, it follows that
$$u(re^{it}) = \frac{1}{2\pi}\int_0^{2\pi} u(e^{is})P_r(s-t)\,ds \le \frac{1}{2\pi}\int_0^{2\pi} u(e^{is})\frac{1+r}{1-r}\,ds = \frac{1+r}{1-r}u(0)$$
and
$$u(re^{it}) = \frac{1}{2\pi}\int_0^{2\pi} u(e^{is})P_r(s-t)\,ds \ge \frac{1}{2\pi}\int_0^{2\pi} u(e^{is})\frac{1-r}{1+r}\,ds = \frac{1-r}{1+r}u(0).$$
$\square$

Of course a similar inequality holds with any disk in place of $\mathbb{D}$:

Corollary 10.7.4. *If $u \ge 0$ is harmonic in $D(a, R)$ then*
$$\frac{R-r}{R+r}u(a) \le u(a+re^{it}) \le \frac{R+r}{R-r}u(a)$$
for all $a + re^{it} \in D(a, R)$.

Proof. Let $v(z) = u(a + Rz)$; then v is harmonic in $\mathbb{D}$. Note that $u(a + re^{it}) = v((r/R)e^{it})$ and apply the theorem. $\square$

It follows that a similar inequality holds for positive harmonic functions in any connected open set:

Theorem 10.7.5 (Harnack's Inequality). *Suppose that $D \in \mathbb{C}$ is a connected open set and $K \subset D$ is compact. There exists a constant c such that*
$$u(z) \le cu(w)$$
for all $z, w \in K$ and any function $u \ge 0$ harmonic in D.

Proof. Fix $a \in D$ and let A be the set of all $z \in D$ with the property that there exists a constant c (depending on z) such that

$$u(z) \leq cu(a)$$

for all $u \geq 0$ harmonic in D.

The set A is nonempty because $a \in A$. And Corollary 10.7.4 shows that A is open: Suppose that $z \in A$; choose c so that $u(z) \leq cu(a)$ for all harmonic $u \geq 0$, and choose $R > 0$ so that $D(z, R) \subset D$. If $0 < r < R/2$ the corollary shows that

$$u(z + re^{it}) \leq \frac{R+r}{R-r}u(z) \leq \frac{R+R/2}{R-R/2}u(z) = 3u(z) \leq 3cu(a),$$

and hence $D(z, R/2) \subset A$.

Note that we have shown not only that if $z \in A$ then some neighborhood of z lies in A, in fact we have shown that if $z \in A$ and $D(z, R) \subset D$ then there exists a constant c such that $u(w) \leq cu(a)$ for all w in $D(z, R/2)$.

It follows that the complement of A is also open: Suppose that $z \in D \setminus A$, and suppose that $D(z, R) \subset D$. Then we must have $D(z, R/3) \subset D \setminus A$: Suppose to the contrary that $w \in D(z, R/3) \cap A$. The triangle inequality shows that $D(w, 2R/3) \subset D$, so that $D(w, R/3) \subset A$ by the comments in the previous paragraph. But $z \in D(w, R/3)$, so $z \in A$, which is a contradiction.

Now, since D is connected, it follows that $A = D$: For every $z \in D$ there exists c such that $u(z) \leq cu(a)$ for all $u \geq 0$ harmonic in D, and in fact every point has a neighborhood in which the inequality holds uniformly (that is, with a fixed value of c). It follows that if $K \subset D$ is compact then there exists $c_1(K)$ such that $u(z) \leq c_1(K)u(a)$ for all $z \in K$.

The same argument shows that if $K \subset D$ is compact then there exists $c_2(K)$ such that $u(z) \geq c_2(K)u(a)$ for all $z \in K$. Now if $z, w \in K$ it follows that

$$u(z) \leq c_1(K)u(a) \leq \frac{c_1(K)}{c_2(K)}u(w) = cu(w). \qquad \square$$

The result known as Harnack's Theorem is often stated in two parts — the first part is important but kind of obvious, the second part is the interesting bit:

Theorem 10.7.6 (Harnack's Theorem). *Suppose that $D \subset \mathbb{C}$ is open.*

(i) The set of harmonic functions in D is a closed subset of $C(D)$. (In other words, if $u_1, u_2, \ldots$ are harmonic in D and $u_n \to u$ uniformly on compact subsets of D then u is harmonic in D.)

(ii) *Suppose that D is connected. If $u_1 \le u_2 \le \cdots$ and each u_n is harmonic in D then either $u_n \to \infty$ uniformly on compact subsets of D or $u_n \to u$ uniformly on compact subsets of D, where u is harmonic.*

Proof. The first part is clear: If u_n is harmonic and $u_n \to u$ uniformly on compact sets then u must be continuous and must also satisfy the Mean Value Property. Indeed, suppose that $\overline{D(z,r)} \subset D$. Then $u_n \to u$ uniformly on $\overline{D(z,r)}$, so that

$$u(z) = \lim_{n\to\infty} u_n(z) = \lim_{n\to\infty} \frac{1}{2\pi}\int_0^{2\pi} u_n(z+re^{it})\,dt = \frac{1}{2\pi}\int_0^{2\pi} u(z+re^{it})\,dt.$$

The second part is an easy consequence of the preceding theorem. Suppose first that $u_n \ge 0$. Let $u(z) = \lim_{n\to\infty} u_n(z)$. If there exists $z \in D$ with $u(z) = \infty$ then Harnack's inequality shows that $u(z) = \infty$ for all $z \in D$, and that the convergence is uniform on compact sets.

Suppose on the other hand that $u(z) < \infty$ for all $z \in D$, and suppose that $K \subset D$ is compact. Fix $a \in K$, and choose c so that

$$v(z) \le cv(a) \quad (z \in K)$$

whenever v is a non-negative harmonic function in D.

Now for $m < n$ we have

$$u_n(z) - u_m(z) \le c(u_n(a) - u_m(a))$$

for all $z \in K$; hence the fact that $(u_n(a))$ converges shows that the sequence (u_n) is uniformly Cauchy and hence uniformly convergent on K.

That finishes the proof in the case $u_n \ge 0$. In general, let $v_n = u_n - u_1$. Then (v_n) is an increasing sequence of non-negative harmonic functions, so we have shown that either $v_n \to \infty$ uniformly on compact sets, in which case $u_n \to \infty$ as well, or v_n converges to a harmonic function v uniformly on compact sets, in which case $u_n \to v + u_1$. $\square$

Essentially the same argument suffices to prove the following version of the second part of the theorem, with slightly different hypotheses:

Theorem 10.7.7 (Harnack's Theorem). *Suppose that $D \subset \mathbb{C}$ is a connected open set, $z_0 \in D$, and (u_n) is a sequence of real-valued harmonic functions in D.*

(i) *If $u_n \ge 0$ and $u_n(z_0) \to +\infty$ then $u_n \to +\infty$ uniformly on compact subsets of D.*

(ii) *If u is harmonic in D, $u_n \le u$ for all n and $u_n(z_0) \to u(z_0)$ then $u_n \to u$ uniformly on compact subsets of D.*

Exercises

10.15. For $\alpha \in \mathbb{R}$ let $L_\alpha = \{re^{i\alpha} : r \geq 0\}$. Suppose that $0 < \alpha < 2\pi$. Show that if α/π is rational then there exists a nontrivial function u harmonic in $\mathbb{C}$ which vanishes on L_0 and L_α, while if α/π is irrational then any such harmonic function must vanish identically.

Hint: To start, show that u vanishes on $L_{2\alpha}$. See Exercise 10.11 in Section 10.4.

10.16. Use Harnack's Theorem to give yet another proof of Hurwitz's Theorem: If D is connected, $f_n \in H(D)$ has no zero and $f_n \to f$ uniformly on compact sets then either f has no zero or f vanishes identically.

See Exercises 5.6 and 9.6.

10.17. Suppose f is entire, $f(x)$ is real for all $x \in \mathbb{R}$ and $f(iy)$ is purely imaginary for all $y \in \mathbb{R}$. Show that $f(-z) = -f(z)$.

Hint: Apply one of the results above to f and to g, where $g(z) = if(iz)$.

Chapter 11

Simply Connected Open Sets

The sole theorem in this chapter is in large part a summary of previous results:

Theorem 11.0. *Suppose that $D \subset \mathbb{C}$ is a connected open set. The following are equivalent*:

(i) *Either $D = \mathbb{C}$ or D is conformally equivalent to $\mathbb{D}$.*

(ii) *D is homeomorphic to $\mathbb{D}$.*

(iii) *D is simply connected.*

(iv) *$\operatorname{Ind}(\gamma, z) = 0$ for every smooth closed curve γ in D and every $z \in \mathbb{C} \setminus D$.*

(v) *$\mathbb{C}_\infty \setminus D$ is connected.*

(vi) *If $f \in H(D)$ then there exists a sequence of polynomials (P_n) such that $P_n \to f$ uniformly on compact subsets of D.*

(vii) *$\int_\gamma f(z)\,dz = 0$ for every $f \in H(D)$ and every smooth closed curve γ in D.*

(viii) *If $f \in H(D)$ then there exists $F \in H(D)$ with $F' = f$.*

(ix) *If $u : D \to \mathbb{R}$ is harmonic then $u = \operatorname{Re}(f)$ for some $f \in H(D)$.*

(x) *If $f \in H(D)$ has no zero in D then there exists $L \in H(D)$ with $f = e^L$.*

(xi) *If $f \in H(D)$ has no zero in D then there exists $g \in H(D)$ with $g^2 = f$.*

We should point out that $\mathbb{C} \setminus D$ need not be connected just because D is simply connected; if $D = \{z \in \mathbb{C} : 0 < \operatorname{Im}(z) < 1\}$ then D is simply connected but $\mathbb{C} \setminus D$ is not connected.

The theorem is interesting because it gives various connections between things that *a priori* do not appear to be related. For example, the fact that a simply connected open set in the plane must be homeomorphic to the unit disk is a not-quite-trivial purely topological fact which we have proved using complex analysis. The fact that a topological property of D should be equivalent to a topological property of $\mathbb{C}_\infty \setminus D$ is also very surprising.

Proof. It is clear that (i) implies (ii) and that (ii) implies (iii).

We saw long ago that (iii) implies (iv). Well, actually it was Exercise 4.4, so we give a sketch of how it goes: If D is simply connected and γ is a closed curve in D then there is a family of closed curves $(\gamma_t)_{0 \le t \le 1}$ such that $\gamma_1 = \gamma$, γ_0 is constant, and γ_t depends continuously on t. It follows that (for a given $z \in \mathbb{C} \setminus D$) $\phi(t) = \operatorname{Ind}(\gamma_t, z)$ is a continuous integer-valued function on $[0, 1]$; hence ϕ is constant and so $\phi(1) = \phi(0) = 0$.

The fact that (iv) implies (v) is one we really have not talked about at all. Suppose (iv): The index of any closed curve γ in D about any point $z \in \mathbb{C} \setminus D$ is 0. This implies that $\mathbb{C}_\infty \setminus D$ is connected:

Suppose to the contrary that $\mathbb{C}_\infty \setminus D$ is not connected. Then it is the union of two disjoint relatively closed nonempty subsets. But, since $\mathbb{C}_\infty \setminus D$ is compact, any relatively closed subset of $\mathbb{C}_\infty \setminus D$ is also compact; thus we have two nonempty compact sets $K_1, K_2 \subset \mathbb{C}_\infty$ such that $K_1 \cap K_2 = \emptyset$ and $K_1 \cup K_2 = \mathbb{C}_\infty \setminus D$. Since $D \subset \mathbb{C}$, we have $\infty \in \mathbb{C}_\infty \setminus D$, and so without loss of generality we have $\infty \in K_1$.

Now if we set
$$\Omega = \mathbb{C}_\infty \setminus K_1 = D \cup K_2$$
then Ω is an open subset of $\mathbb{C}$ and K_2 is a compact subset of Ω; thus Lemma 10.5.5 shows that there exists a cycle Γ in Ω such that
$$\Gamma^* \subset \Omega \setminus K_2 = D$$
and
$$\operatorname{Ind}(\Gamma, z) = 1 \quad (z \in K_2).$$
Since K_2 is nonempty, this shows that D does not satisfy (iv) (if $\operatorname{Ind}(\Gamma, z) = 1$ then there exists a closed curve γ in D such that $\operatorname{Ind}(\gamma, z) \ne 0$).

The fact that (v) implies (vi) is a special case of Runge's Theorem (Corollary 12.9 in the next chapter.)

The fact that (vi) implies (vii) is new but pretty easy: Suppose that every holomorphic function in D can be approximated uniformly on compact subsets of D by polynomials, and suppose that $f \in H(D)$ and γ is a smooth closed curve in D. Choose a sequence of polynomials (P_n) such that $P_n \to f$ uniformly on compact subsets of D.

Now $P_n \in H(\mathbb{C})$ and it is certainly true that $\text{Ind}(\gamma, z) = 0$ for all $z \in \mathbb{C} \setminus \mathbb{C}$, so Cauchy's Theorem shows that $\int_\gamma P_n(z)\, dz = 0$ for all n. But γ^* is compact, so $P_n \to f$ uniformly on γ^*; hence $\int_\gamma f(z)\, dz = 0$.

The fact that (vii) implies (viii) is something we have discussed at some length (Theorem 4.0).

We showed that (viii) implies (ix) in Lemma 10.0.1(iii). (There we assumed that D was simply connected, but the only property of simply connected open sets we used in the proof of the lemma was that every holomorphic function has an antiderivative: If u is a real-valued harmonic function then $\partial u/\partial z \in H(D)$, and if $f \in H(D)$ with $f' = \partial u/\partial z$ then it follows that $u = \text{Re}\,(f + c)$ for some constant c.)

Now suppose (ix), and suppose as well that $f \in H(D)$ has no zero in D. Let $u(z) = \log(|f(z)|)$. We know that u is harmonic in D (because if $D(z, r) \subset D$ then $f|_{D(z,r)}$ has a holomorphic logarithm and u is the real part of this logarithm, so u is harmonic in $D(z, r)$). So by hypothesis there exists $g \in H(D)$ with $\log(|f|) = \text{Re}\,(g)$ in D. It follows that

$$\left|\frac{e^g}{f}\right| = \frac{e^{\text{Re}\,(g)}}{|f|} = \frac{e^{\log(|f|)}}{|f|} = 1,$$

so that e^g/f is constant, and now if we choose a constant c so that $e^{g+c}/f = 1$ at one point it follows that $e^{g+c}/f = 1$ everywhere in D, giving (x).

It is clear that (x) implies (xi): If $e^L = f$ then $g^2 = f$ for $g = e^{L/2}$.

Finally, the fact that (xi) implies (i) is just the Riemann Mapping Theorem (the reason we stated Theorem 9.4.2 the way we did was so we could show that (xi) implies (i) here). □

We have not finished the proof of the theorem because we have not shown that (v) implies (vi); this we will do in the next chapter. But you should note that we *have* shown that all the conditions other than (vi) are equivalent, because it is clear that (iv) implies (vii) (this is exactly the statement of Cauchy's Theorem) and it is easy to see that (v) implies (iv) using Proposition 4.7.

The ambitious reader might regard the theorem as a large collection of exercises, each of the form "Give a direct proof that (j) implies (k)" for various values of j and k.

Chapter 12

Runge's Theorem and the Mittag-Leffler Theorem

If D is a disk and $f \in H(D)$ then the partial sums of the power series for f give a sequence of polynomials (P_n) such that $P_n \to f$ uniformly on compact subsets of D. Recalling that convergence in $H(D)$ is by definition uniform convergence on compact subsets of D, this says exactly that if D is a disk then the polynomials are dense in $H(D)$.

But the polynomials are not dense in $H(D)$ for every D. For example, let $D = \mathbb{C} \setminus \{0\}$, $K = \{\, z \in \mathbb{C} : |z| = 1 \,\}$ and define $f \in H(D)$ by $f(z) = 1/z$. Then there does not exist a sequence of polynomials converging to f uniformly on K, in fact the proof that (vi) implies (vii) in Theorem 11.0 shows the following:

- **Exercise 12.1.** If P is a polynomial then there exists z such that $|z| = 1$ and $|P(z) - \frac{1}{z}| \geq 1$.

We shall see in this chapter that the rational functions *are* dense in $H(D)$ for any open set D; in fact Runge's Theorem gives an optimal degree of control over the location of the poles. In particular it turns out that if D is simply connected then the rational functions with poles at ∞, that is to say the polynomials, are dense in $H(D)$; this shows that (v) implies (vi) in Theorem 11.0, finishing the proof of that theorem.

Runge's Theorem says that if D is an open set and $A \subset \mathbb{C}_\infty \setminus D$ intersects each component of $\mathbb{C}_\infty \setminus D$ then the rational functions with poles in A are dense in $H(D)$. The plan seems pretty simple: Suppose that K is a compact subset of D, and choose a cycle Γ that surrounds K in D in the sense of Lemma 10.5.5. Suppose that $f \in H(D)$. Then for any $z \in K$ we have

$$f(z) = \frac{1}{2\pi i} \int_\Gamma \frac{f(w)}{w-z}\,dw,$$

by the Cauchy Integral Formula. Now if we approximate the integral by Riemann sums we see that we have approximated $f(z)$ by a linear combination of sums of the form

$$R(z) = \frac{1}{2\pi i} \sum_{j=1}^N \frac{f(w_j)}{w_j - z}(w_j - w_{j-1}),$$

one such sum for each of the curves comprising Γ. Voila, we've found a rational function that is close to f on K. The Cauchy Integral Formula is amazing, we're done.

Except of course that the rational function above has its poles on Γ^*, so in particular it has poles in D, and we want rational functions with poles outside D. We do a thing called "pole pushing": It turns out that if R is a rational function with no poles on K then we can approximate R on K by another rational function with poles at arbitrary points near the poles of R. In fact we can push the poles of our rational functions anywhere we want within connected subsets of $\mathbb{C}_\infty \setminus K$; we will show how to choose K so that we can push poles along components of $\mathbb{C}_\infty \setminus K$ until they hit A.

We begin with a bit of trivia regarding rational functions, to clarify the meaning of the phrase "rational function with poles in A". Recall that if $\lim_{z\to\infty} R(z) = \infty$ then R is said to have a pole at infinity:

Lemma 12.0. *Suppose that $A \subset \mathbb{C}$ and R is a rational function such that all the finite poles of R are contained in A. Then there exists a polynomial P and complex numbers $(c_{a,j})_{j=1}^{N_a}$ for each $a \in A$ such that*

$$R(z) = P(z) + \sum_{a \in A} R_a(z) = P(z) + \sum_{a \in A} \sum_{j=1}^{N_a} c_{a,j}(z-a)^{-j}.$$

(Although A may be infinite this is really a finite sum; we have $c_{a,j} = 0$ unless $a \in A$ actually is a pole of R.)

Note that P is nonconstant if and only if R has a pole at infinity.

Proof. Let $R_a(z) = \sum_{j=1}^{N_a} c_{a,j}(z-a)^{-j}$ be the principal part of R at $a \in A$ (that is, the sum of the "negative terms" in the Laurent series about a) and set $P = R - \sum_{a \in A} R_a$. Then P is a rational function with poles only at points of A (and possibly at ∞) but the principal part of P at every point of A vanishes, so all the isolated singularities of P in $\mathbb{C}$ are removable. That is, P is a rational function with no poles in $\mathbb{C}$, so the Fundamental Theorem of Algebra (Corollary 3.7) shows that P is a polynomial. □

Note that the lemma shows that a rational function has a "partial-fractions decomposition", as is often assumed without proof in an elementary calculus class.

Of course, in Calc 101 the rational function typically has real coefficients and we want a partial-fractions decomposition that does not involve complex numbers. (In fact some of the applications could be simpler if terms like $1/(z+i)$ were allowed, except that we would need to talk about the subtleties of the complex logarithm. In any case, such terms are usually not wanted.) But it is not hard to get from Lemma 12.0 to the sort of decomposition that is given in the typical calculus class:

Note first that if R has real coefficients then any nonreal terms in Lemma 12.0 occur in conjugate pairs: If a term $c/(z-a)^j$ appears, where a and c are not both real, then the term $\bar{c}/(z-\bar{a})^j$ also appears. Now if $a = \alpha + i\beta$ then

$$\frac{c}{(z-a)^j} + \frac{\bar{c}}{(z-\bar{a})^j} = \frac{p(z)}{((z-\alpha)^2 + \beta^2)^j},$$

where p is a polynomial of degree j and $p(\bar{z}) = \overline{p(z)}$; this shows that p has real coefficients. Combining various pairs of terms this way in Lemma 12.0 gives the partial-fractions decomposition approved by calculus teachers.

We need some notation for certain spaces of rational functions. If $A \subset \mathbb{C}_\infty$ let $\mathcal{R}_A$ denote the set of rational functions R such that every pole of R (including the pole at infinity if there is one) is an element of A. Thus the previous lemma says that

$$\mathcal{R}_A = \sum_{a \in A} \mathcal{R}_{\{a\}}.$$

(If $(V_a)_{a \in A}$ is a family of vector spaces then $\sum_{a \in A} V_a$ denotes the linear span of the V_a, or in other words the set of all finite linear combinations of elements of $\bigcup_{a \in A} V_a$.)

We begin pushing poles. The next lemma says we can push them around *locally*, that is in a disk disjoint from K:

Lemma 12.1. *Suppose that $K \subset \mathbb{C}$ is compact and*

$$D(\alpha, r) \subset \mathbb{C} \setminus K.$$

Any function $f \in \mathcal{R}_{D(\alpha,r)}$ can be uniformly approximated on K by functions in $\mathcal{R}_{\{\alpha\}}$.

Proof. If f and g are continuous functions on K which can be uniformly approximated on K by elements of $\mathcal{R}_{\{\alpha\}}$ then the same is true of $f + g$, cf ($c \in \mathbb{C}$), and fg. Hence Lemma 12.0 shows that we may assume that $f(z) = 1/(z - \beta)$ for some $\beta \in D(\alpha, r)$.

Now if $z \in K$ then $|z - \alpha| \geq r > |\beta - \alpha|$, so

$$f(z) = \frac{1}{(z - \alpha) - (\beta - \alpha)}$$

$$= \frac{1}{z - \alpha} \frac{1}{1 - \dfrac{\beta - \alpha}{z - \alpha}}$$

$$= \frac{1}{z - \alpha} \sum_{n=0}^{\infty} \left(\frac{\beta - \alpha}{z - \alpha}\right)^n.$$

The sum converges uniformly on K, since

$$\left|\frac{\beta - \alpha}{z - \alpha}\right| \leq \frac{|\beta - \alpha|}{r} < 1$$

for all $z \in K$; thus the partial sums

$$R_N(z) = \frac{1}{z - \alpha} \sum_{n=0}^{N} \left(\frac{\beta - \alpha}{z - \alpha}\right)^n$$

give a sequence of elements of $\mathcal{R}_{\{\alpha\}}$ converging uniformly on K to f. $\square$

We used the following fact:

- **Exercise 12.2.** Suppose that $K \subset \mathbb{C}$ is compact. Suppose that (R_n) and and (S_n) are sequences of elements of $C(K)$ which converge to uniformly on K to f and g, respectively. Show that $R_n S_n$ converges to fg uniformly on K.

You should note that a correct solution to the preceding exercise must use the compactness of K:

- **Exercise 12.3.** Give an example of (R_n), $(S_n) \subset C(\mathbb{R})$ such that $R_n \to f$ and $S_n \to g$ uniformly on $\mathbb{R}$ but $R_n S_n$ does not converge to fg uniformly on $\mathbb{R}$.

The next lemma is the statement corresponding to Lemma 12.1 in the case $\alpha = \infty$. (There are two parts, where only one was needed in the

12. Runge's Theorem and the Mittag-Leffler Theorem

previous lemma. This is because of an asymmetry in the way things are set up: If $\beta \in D(\alpha, r)$ then it follows that $\alpha \in D(\beta, r)$, but $|\beta| > r$ (that is, β "close" to ∞) does not imply $\infty \in D(\beta, \rho)$, regardless of how we choose $\rho > 0$.)

Lemma 12.2. *Suppose that $K \subset \mathbb{C}$ is compact and*

$$A \subset \mathbb{C} \setminus K,$$

where

$$A = \{ z \in \mathbb{C} : |z| > r \}$$

for some $r > 0$. Then

(i) *Any $f \in \mathcal{R}_A$ can be uniformly approximated on K by polynomials.*

(ii) *If $|\beta| > 2r$ then any polynomial can be uniformly approximated on K by functions in $\mathcal{R}_{\{\beta\}}$.*

Proof. For the first part, as in the preceding proof we may assume that $f(z) = 1/(z - \beta)$ for some $\beta \in A$. Now if $z \in K$ then $|z| \le r < |\beta|$, so

$$f(z) = \frac{1}{z-\beta} = \frac{-1}{\beta} \frac{1}{1 - \frac{z}{\beta}} = \frac{-1}{\beta} \sum_{n=0}^{\infty} \left(\frac{z}{\beta}\right)^n;$$

the partial sums are polynomials converging uniformly to f on K because $|z|/|\beta| \le r/|\beta| < 1$ on K.

For the second part we need only show that if $|\beta| > 2r$ then the function z can be uniformly approximated on K by functions in $\mathcal{R}_{\{\beta\}}$. But for $z \in K$ we have

$$\beta - \beta \sum_{n=0}^{\infty} \left(\frac{z}{z-\beta}\right)^n = \beta - \beta \frac{1}{1 - \frac{z}{z-\beta}} = z,$$

and this sum converges uniformly on K since

$$\left|\frac{z}{z-\beta}\right| \le \frac{r}{|\beta| - r} < 1$$

for $z \in K$. $\square$

And now we can push poles in connected open sets disjoint from K:

Lemma 12.3 (Pole-Pushing Lemma). *Suppose that $K \subset \mathbb{C}$ is compact and V is a connected open subset of $\mathbb{C}_\infty \setminus K$. If $\alpha \in V$ then any element of $\mathcal{R}_V$ can be uniformly approximated on K by elements of $\mathcal{R}_{\{\alpha\}}$.*

Proof. Let W be the set of all $\beta \in V$ such that every element of $\mathcal{R}_{\{\beta\}}$ can be approximated uniformly on K by elements of $\mathcal{R}_{\{\alpha\}}$. The set W is nonempty since $\alpha \in W$.

The previous two lemmas show that W is open. In fact Lemma 12.1 shows that if $\beta \in W \cap \mathbb{C}$ and $D(\beta, r) \subset V$ then $D(\beta, r) \subset W$. (Suppose that $\omega \in D(\beta, r)$, $f \in \mathcal{R}_{\{\omega\}}$ and $\epsilon > 0$. Lemma 12.1 shows that there exists $g \in \mathcal{R}_{\{\beta\}}$ with
$$|f(z) - g(z)| < \epsilon/2 \quad (z \in K),$$
and now the fact that $\beta \in W$ shows that there exists $R \in \mathcal{R}_{\{\alpha\}}$ with
$$|g(z) - R(z)| < \epsilon/2 \quad (z \in K);$$
hence
$$|f(z) - R(z)| < \epsilon \quad (z \in K),$$
showing that $\omega \in W$.)

Similarly, Lemma 12.2(i) shows that if $\infty \in W$ and $\{z \in \mathbb{C} : |z| > r\} \subset V$ then $\{z \in \mathbb{C} : |z| > r\} \subset W$; thus W is open.

This is one of those situations where the argument showing that W is open also shows that $V \setminus W$ is open. Almost; in fact the same argument shows that if $\beta \in V \setminus W$ and $\beta \neq \infty$ then $V \setminus W$ contains a neighborhood of β, but it doesn't quite work that way for $\beta = \infty$, which is why we needed to add the second part to Lemma 12.2:

Suppose first that $\beta \in (V \setminus W) \cap \mathbb{C}$, and choose $r > 0$ such that $D(\beta, r) \subset V$. It follows that $D(\beta, r/2) \subset V \setminus W$: If $\omega \in D(\beta, r/2) \cap W$ then $D(\omega, r/2) \subset V$ and so the argument above shows that $D(\omega, r/2) \subset W$, which is a contradiction since $\beta \in D(\omega, r/2)$.

Suppose now that $\infty \in V \setminus W$. Choose $r > 0$ so that $\{z \in \mathbb{C} : |z| > r\} \subset V$. It follows from Lemma 12.2(ii) that $\{z \in \mathbb{C} : |z| > 2r\} \subset V \setminus W$: Suppose to the contrary that $|\omega| > 2r$ and $\omega \in W$; any element of $\mathcal{R}_{\{\alpha\}}$ can be approximated on K by elements of $\mathcal{R}_{\{\omega\}}$. Lemma 12.2(ii) shows that any element of $\mathcal{R}_{\{\omega\}}$ can be approximated on K by polynomials, and it follows that $\infty \in W$, a contradiction.

So W is a nonempty open subset of V and $V \setminus W$ is also open; since V is connected, this shows that $W = V$: If $\beta \in V$ then any element of $\mathcal{R}_{\{\beta\}}$ can be uniformly approximated on K by elements of $\mathcal{R}_{\{\alpha\}}$. As before Lemma 12.0 now shows that any element of $\mathcal{R}_V$ can be uniformly approximated on K by elements of $\mathcal{R}_\alpha$. □

We should point out that the fact that V is connected in the preceding lemma is essential; if α and β lie in different components of $\mathbb{C}_\infty \setminus K$ then elements of $\mathcal{R}_{\{\alpha\}}$ cannot be uniformly approximated on K by elements of $\mathcal{R}_{\{\beta\}}$ (Theorem 12.4 below).

Note first that the components of a locally connected space are open. (Suppose X is locally connected and C is a component of X. If $x \in C$ then x has a connected neighborhood V; now it follows that $V \subset C$.) In particular the components of an open subset of $\mathbb{C}_\infty$ are open.

We also need to note that if R is a rational function and $\phi \in \mathfrak{M}$ then $R \circ \phi$ is a rational function and $\alpha \in \mathbb{C}_\infty$ is a pole of R if and only if $\phi^{-1}(\alpha)$ is a pole of $R \circ \phi$. In particular for any $\alpha \in \mathbb{C}_\infty$ and $\phi \in \mathfrak{M}$ we have

$$\{\, R \circ \phi : R \in \mathcal{R}_{\{\alpha\}} \,\} = \mathcal{R}_{\{\phi^{-1}(\alpha)\}}.$$

Theorem 12.4. *Suppose that $K \subset \mathbb{C}$ is compact and $\alpha, \beta \in \mathbb{C}_\infty \setminus K$. Every function in $\mathcal{R}_{\{\alpha\}}$ can be uniformly approximated on K by elements of $\mathcal{R}_{\{\beta\}}$ if and only if α and β lie in the same component of $\mathbb{C}_\infty \setminus K$.*

Proof. Suppose first that α and β lie in the same component of $\mathbb{C}_\infty \setminus K$. Since this component is an open set, it follows from Lemma 12.3 that every function in $\mathcal{R}_{\{\alpha\}}$ can be unifomly approximated on K by elements of $\mathcal{R}_{\{\beta\}}$.

Now suppose that α and β lie in different components of $\mathbb{C}_\infty \setminus K$; we need to show that some elements of $\mathcal{R}_{\{\alpha\}}$ cannot be uniformly approximated on K by elements of $\mathcal{R}_{\{\beta\}}$.

Composing everything with an appropriate linear-fractional transformation we may assume that $\beta = \infty$. Let V be the component of $\mathbb{C}_\infty \setminus K$ containing α and let W be the union of the other components. Then V and W are open, $\alpha \in V$, $\infty \in W$, $V \cap W = \emptyset$, $V \cup W = \mathbb{C}_\infty \setminus K$, and we need to show that some element of $\mathcal{R}_{\{\alpha\}}$ cannot be uniformly approximated on K by polynomials. (Note that K is still a subset of the plane after our change of variables, because $\infty \in W$.)

In fact if $f \in \mathcal{R}_{\{\alpha\}}$ and f actually does have a pole at α then f cannot be approximated uniformly on K by polynomials; we will demonstrate this for $f(z) = 1/(z - \alpha)$.

Note that $V \cup K = \mathbb{C}_\infty \setminus W$ is a bounded closed subset of the plane (because $\infty \in W$ and W is open). Hence V is a bounded open subset of the plane and $\overline{V} \subset V \cup K$, which implies that $\partial V \subset K$.

Suppose that (P_n) is a sequence of polynomials and $P_n(z) \to 1/(z-\alpha)$ uniformly on K. It follows that $(z-\alpha)P_n(z) \to 1$ uniformly on K, and hence on ∂V. But now the Maximum Modulus Theorem (applied to the function $1 - (z-\alpha)P_n(z) \in H(V)$) shows that in fact $(z-\alpha)P_n(z) \to 1$ uniformly on V. In particular, setting $z = \alpha$ we see that $0 \to 1$ as $n \to \infty$. □

In any case, Lemma 12.3 shows that we can push poles anywhere we want inside connected subsets of the complement of K. The next lemma shows that if D is open then there is an exhaustion of D by compact sets K with the property that any connected component of the complement of K intersects the complement of D.

Lemma 12.5. *Suppose that D is an open subset of the plane. There exist compact sets K_j ($j = 1, 2, \dots$) such that $K_j \subset K_{j+1}^\circ$,*

$$D = \bigcup_{j=1}^{\infty} K_j,$$

and such that every component of $\mathbb{C}_\infty \setminus K_j$ contains a component of $\mathbb{C}_\infty \setminus D$.

Proof. The reader may have noticed that this is the same as Lemma 9.2.0 except for the last bit about components of complements. As previously we may assume that $D \ne \mathbb{C}$. The same sets K_j work; we use slightly different notation in defining them. We let

$$K_j = D \setminus \Omega_j,$$

where

$$\Omega_j = \{\, z \in \mathbb{C}_\infty : |z| > j \,\} \cup \bigcup_{p \in \mathbb{C} \setminus D} D(p, 1/j).$$

We need only show that every component of $\mathbb{C}_\infty \setminus K_j$ contains a component of $\mathbb{C}_\infty \setminus D$; the rest was done previously. Suppose that A is a component of $\mathbb{C}_\infty \setminus K_j$.

Now $A \ne \emptyset$; choose a point $z \in A$. The fact that $z \in \mathbb{C}_\infty \setminus K_j = \Omega_j$ shows that there exists a connected set $C \subset \mathbb{C}_\infty \setminus K_j$ such that $z \in C$ and $C \cap (\mathbb{C}_\infty \setminus D) \ne \emptyset$; in fact we can either use $C = \{\, w \in \mathbb{C}_\infty : |w| > j \,\}$ or $C = D(p, 1/j)$ for some $p \in \mathbb{C} \setminus D$. Let $\alpha \in C \cap (\mathbb{C}_\infty \setminus D)$, and let B be the component of $\mathbb{C}_\infty \setminus D$ containing α.

Since C is a connected subset of $\mathbb{C}_\infty \setminus K_j$ and $z \in C \cap A$, we must have $C \subset A$, and hence $\alpha \in A \cap B$. Since B is a connected subset of $\mathbb{C}_\infty \setminus D \subset \mathbb{C}_\infty \setminus K_j$, it must be contained in some component of $\mathbb{C}_\infty \setminus K_j$, and so, since B intersects A, we must have $B \subset A$. □

We are getting pretty close to Runge's Theorem — all that remains is a technical detail about approximating Cauchy integrals by Riemann sums (or rather Riemann-Stieltjes sums):

12. Runge's Theorem and the Mittag-Leffler Theorem

Lemma 12.6. *Suppose that γ is a smooth curve in the plane and $f \in C(\gamma^*)$. Define $F : \mathbb{C} \setminus \gamma^* \to \mathbb{C}$ by*

$$F(z) = \int_\gamma \frac{f(w)}{w - z}\, dw.$$

Suppose that $K \subset \mathbb{C} \setminus \gamma^*$ is compact and $\epsilon > 0$. Then there exist $w_0, \ldots, w_n \in \gamma^*$ such that

$$\left| F(z) - \sum_{j=1}^n \frac{f(w_j)}{w_j - z}(w_j - w_{j-1}) \right| < \epsilon \quad (z \in K).$$

Proof. Suppose that $\gamma : [a, b] \to \mathbb{C}$. Define $\Phi : [a, b] \times K \to \mathbb{C}$ by

$$\Phi(t, z) = \frac{f(\gamma(t))}{\gamma(t) - z}.$$

Now Φ is continuous and $[a, b] \times K$ is compact, so Φ is actually uniformly continuous on $[a, b] \times K$. Since γ' is bounded, it follows that there exists $\delta > 0$ such that if $s, t \in [a, b]$ and $|s - t| < \delta$ then

$$|(\Phi(s, z) - \Phi(t, z))\gamma'(s)| < \frac{\epsilon}{b - a} \quad (z \in K).$$

Choose finitely many points $a = t_0 < \cdots < t_n = b$ such that $|t_j - t_{j-1}| < \delta$ for $j = 1, \ldots, n$, and let $w_j = \gamma(t_j)$. Now for $j = 1, \ldots, n$ we have

$$\left| \int_{t_{j-1}}^{t_j} \Phi(t, z)\gamma'(t)\, dt - \Phi(t_j, z)(w_j - w_{j-1}) \right|$$

$$= \left| \int_{t_{j-1}}^{t_j} (\Phi(t, z) - \Phi(t_j, z))\gamma'(t)\, dt \right|$$

$$\leq \int_{t_{j-1}}^{t_j} |(\Phi(t, z) - \Phi(t_j, z))\gamma'(t)|\, dt$$

$$\leq \frac{\epsilon(t_j - t_{j-1})}{b - a};$$

hence for every $z \in K$ we have

$$\left| F(z) - \sum_{j=1}^n \frac{f(w_j)}{w_j - z}(w_j - w_{j-1}) \right|$$

$$= \left| \int_a^b \Phi(t, z)\gamma'(t)\, dt - \sum_{j=1}^n \Phi(t_j, z)(w_j - w_{j-1}) \right|$$

$$\leq \sum_{j=1}^n \frac{\epsilon(t_j - t_{j-1})}{b - a}$$

$$= \epsilon. \qquad \square$$

And now we can prove Runge's Theorem. We give two versions, one concerning approximation on one compact set and one concerning convergence on all compact subsets of an open set:

Theorem 12.7 (Runge's Theorem for one compact set). *Suppose that $K \subset \mathbb{C}$ is compact and $A \subset \mathbb{C}_\infty \setminus K$ intersects every component of $\mathbb{C}_\infty \setminus K$. If f is holomorphic in a neighborhood of K and $\epsilon > 0$ then there exists $R \in \mathcal{R}_A$ such that*

$$|f(z) - R(z)| < \epsilon \quad (z \in K).$$

Note that we are not saying that the rational function in the conclusion is holomorphic in the un-named set where f happens to be holomorphic; it has poles in A and so it is holomorphic in some neighborhood of K, but possibly a different one.

Proof. Suppose that D is an open set with $K \subset D$ and such that f is holomorphic in D. Choose a cycle Γ in D as in Lemma 10.5.5: $\Gamma^* \subset D \setminus K$, $\text{Ind}(\Gamma, z) = 0$ for all $z \in \mathbb{C} \setminus D$, and $\text{Ind}(\Gamma, z) = 1$ for all $z \in K$. The Cauchy Integral Formula shows that

$$f(z) = \frac{1}{2\pi i} \int_\Gamma \frac{f(w)}{w - z} \, dw \quad (z \in K),$$

and now Lemma 12.6 shows that there exists $R^1 \in \mathcal{R}_{\Gamma^*}$ such that

$$|f(z) - R^1(z)| < \epsilon/2 \quad (z \in K).$$

Now there exist finitely many points $w_1, \ldots, w_n \in \Gamma^*$ and functions $R_j^1 \in \mathcal{R}_{\{w_j\}}$ such that

$$R^1 = \sum_{j=1}^n R_j^1.$$

Since every component of $\mathbb{C}_\infty \setminus K$ intersects A, we can choose $\alpha_j \in A$ ($1 \le j \le n$) such that w_j and α_j lie in the same component of $\mathbb{C}_\infty \setminus K$. It follows from Theorem 12.4 that there exist $R_j \in \mathcal{R}_{\{\alpha_j\}}$ such that

$$|R_j^1(z) - R_j(z)| < \frac{\epsilon}{2n} \quad (z \in K).$$

Now let $R = \sum_{j=1}^n R_j$; then $R \in \mathcal{R}_A$ and the triangle inequality shows that

$$|f(z) - R(z)| < \epsilon \quad (z \in K). \qquad \square$$

And now Runge's Theorem:

Theorem 12.8 (Runge's Theorem). *Suppose that $D \subset \mathbb{C}$ is open and $A \subset \mathbb{C}_\infty \setminus D$ intersects every component of $\mathbb{C}_\infty \setminus D$. If $f \in H(D)$ then there exists a sequence of rational functions $(R_j) \subset \mathcal{R}_A$ such that $R_j \to f$ uniformly on compact subsets of D.*

Proof. Choose a sequence of compact sets (K_j) as in Lemma 12.5. Since every component of $\mathbb{C}_\infty \setminus K_j$ contains a component of $\mathbb{C}_\infty \setminus D$, it follows that every component of $\mathbb{C}_\infty \setminus K_j$ intersects A; now Theorem 12.7 shows that for each j there exists $R_j \in \mathcal{R}_A$ such that
$$|f(z) - R_j(z)| < 1/j \quad (z \in K_j).$$

If $K \subset D$ is compact then there exists N such that $K \subset K_N$, and hence $K \subset K_j$ for $j \geq N$. Thus we have
$$|f(z) - R_j(z)| < 1/j \quad (z \in K, j > N),$$
which shows that $R_j \to f$ uniformly on K. $\square$

This finishes the proof of Theorem 11.0:

Corollary 12.9. *If $D \subset \mathbb{C}$ is an open set such that $\mathbb{C}_\infty \setminus D$ is connected and $f \in H(D)$ then there exists a sequence of polynomials P_j with $P_j \to f$ uniformly on compact subsets of D.*

Proof. Take $A = \{\infty\}$ in Theorem 12.8. $\square$

Note that the following version of the corollary is actually a weaker statement, since D is not required to be connected in Corollary 12.9. (The fact that the sets K and D in Runge's Theorem need not be connected is very important for some applications, as we shall see.)

Corollary 12.10. *If $D \subset \mathbb{C}$ is a simply connected open set and $f \in H(D)$ then there exists a sequence of polynomials P_j with $P_j \to f$ uniformly on compact subsets of D.*

Note. One can use Runge's Theorem to do many things. It would be a mistake to forget about Theorem 12.7, thinking of it as just a preliminary step in the proof of Theorem 12.8, because often the one-compact-set version Theorem 12.7 is actually what you want to use.

We give a curious application. The question is this: Does there exist a sequence of polynomials P_n such that
$$\lim_{n \to \infty} P_n(z) = \begin{cases} 1 & (\operatorname{Re}(z) > 0), \\ 0 & (\operatorname{Re}(z) = 0), \\ -1 & (\operatorname{Re}(z) <)0? \end{cases}$$

This question has a curious property: If a student who had never seen applications of Runge's Theorem thought about it for a few seconds and guessed the right answer that would make it seem likely to me that the student was missing something fundamental; a student who understands the material we have presented so far "should" get this question wrong! It seems at first that the answer is clearly no, since a limit of a sequence of polynomials should be holomorphic, and that limit function is not even continuous, much less holomorphic. But the answer is actually yes:

For $n = 1, \dots$ define a compact set $K_n = K_n^+ \cup K_n^0 \cup K_n^-$, where K_n^+ is the closed rectangle with vertices $-n + i/n$, $n + i/n$, $n + in$, $-n + in$, K_n^0 is the segment $[-n, n] \subset \mathbb{R}$, and K_n^- is the closed rectangle with vertices $-n - in$, $n - in$, $n - i/n$, $-n - i/n$.

Now define $f_n : K_n \to \mathbb{C}$ by

$$f_n(z) = \begin{cases} 1 & (z \in K_n^+), \\ 0 & (x \in K_n^0), \\ -1 & (z \in K_n^-). \end{cases}$$

Since $\mathbb{C}_\infty \setminus K_n$ is connected and f_n is holomorphic in a neighborhood of K_n, Theorem 12.7 shows that there exists a polynomial P_n such that

$$|f(z) - P_n(z)| < 1/n \quad (z \in K_n),$$

which says that $|1 - P_n| < 1/n$ on K_n^+, $|P_n| < 1/n$ on K_n^0, and $|-1 - P_n| < 1/n$ on K_n^-. But if $\operatorname{Re}(z) > 0$ then there exists N such that $z \in K_n^+$ for all $n > N$; thus $P_n(z) \to 1$. Similarly if $\operatorname{Re}(z) = 0$ then $z \in K_n^0$ for large n and so $P_n(z) \to 0$, and if $\operatorname{Re}(z) < 0$ then $P_n(z) \to -1$.

- **Exercise 12.4.** Show that there exists a sequence of polynomials (P_n) such that $P_n(0) = 1$ for all n, while $P_n(z) \to 0$ as $n \to \infty$ if $z \in \mathbb{C}$ and $z \neq 0$.

The tricky part of the previous exercise is coming up with a sequence of compact sets K_n such that $\mathbb{C} \setminus K_n$ is connected, $K_n \subset K_{n+1}$, and $\bigcup_{n=1}^\infty K_n = \mathbb{C} \setminus \{0\}$.[1] A slightly more elaborate argument using Cauchy's Estimates suffices for the next exercise:

- **Exercise 12.5.** Show that there exists a sequence of polynomials (P_n) such that $P_n'(0) = 1$ for all n, although $P_n'(z) \to 0$ as $n \to \infty$ for all $z \in \mathbb{C} \setminus \{0\}$ and $P_n(z) \to 0$ for all $z \in \mathbb{C}$.

---

[1] Some readers may find the word "Pacman" to be a useful hint here ...

12. Runge's Theorem and the Mittag-Leffler Theorem

- **Exercise 12.6.** Fix a number $M < \infty$. Show that there does not exist a sequence of polynomials (P_n) such that
$$\lim_{n\to\infty} P_n(z) = \begin{cases} 0 & (z \neq 0), \\ 1 & (z = 0), \end{cases}$$
and such that
$$|P_n(z)| \leq M \qquad (|z| \leq 1,\ n = 1, 2, \dots).$$

The Mittag-Leffler Theorem is a more serious application of Runge's Theorem. Runge's Theorem is an approximation theorem, saying that certain holomorphic functions can be approximated in a certain sense by certain rational functions. It is interesting that this can be used to solve what might be called an interpolation problem: Given an open set D and a set $E \subset D$ with no limit point in D, find a meromorphic function in D which has poles precisely at the points of E, and such that the residue at each point of E is whatever we want it to be.

In fact the Mittag-Leffler Theorem does more: Not only can we prescribe the residues of a meromorphic function on an arbitrary set (with no limit point in the open set in question), we can actually prescribe the entire principal part:

Theorem 12.11 (Mittag-Leffler Theorem). *Suppose that $D \subset \mathbb{C}$ is open and $E \subset D$ has no limit point in D. Suppose that for each $a \in E$ we are given a positive integer N_a and a finite sequence of complex numbers $(c_{a,j})_{j=1}^{N_a}$. Then there exists a function f meromorphic in D such that f has no poles except at points of E and such that for each $a \in E$, f has a pole at a with principal part*
$$\sum_{j=1}^{N_a} \frac{c_{a,j}}{(z-a)^j}.$$

The obvious thing to try is to let
$$f(z) = \sum_{a \in E} \sum_{j=1}^{N_a} \frac{c_{a,j}}{(z-a)^j}.$$

The problem with that is that there is no reason to think the sum converges. We fix this by using Runge's Theorem to choose rational functions R_a so that
$$f(z) = \sum_{a \in E} \left(\sum_{j=1}^{N_a} \frac{c_{a,j}}{(z-a)^j} - R_a(z) \right)$$
does converge (more or less; the notation is going to look different because the terms get grouped differently in the actual proof):

Proof. Let (K_n) be a sequence of compact subsets of D such that $K_n \subset K_{n+1}^\circ$, $D = \bigcup_{n=1}^\infty K_n$, and every component of $\mathbb{C}_\infty \setminus K_n$ contains a component of $\mathbb{C}_\infty \setminus D$.

Let $E_1 = E \cap K_1$, and for $n \geq 2$ let $E_n = (E \cap K_n) \setminus K_{n-1}$; thus the sets E_n partition E. Since $E_n \subset K_n$ and E has no limit point in D, it follows that E_n is finite, so there is no problem in setting

$$f_n(z) = \sum_{a \in E_n} \sum_{j=1}^{N_a} \frac{c_{a,j}}{(z-a)^j}$$

(as always if $E_n = \emptyset$ we take this to mean that $f_n = 0$).

Choose a set $A \subset \mathbb{C}_\infty \setminus D$ which intersects every component of $\mathbb{C}_\infty \setminus D$. (For example, $A = \mathbb{C}_\infty \setminus D$ would work.) Let $R_1 = 0$. For $n > 1$ define R_n as follows:

Note that f_n is actually holomorphic in a neighborhood of K_{n-1}, since its poles lie in $\mathbb{C} \setminus K_{n-1}$. As in the proof of Theorem 12.8 we see that A intersects every component of the complement of $\mathbb{C}_\infty \setminus K_{n-1}$, and so Theorem 12.7 shows that there exists $R_n \in \mathcal{R}_A$ with

$$|f_n(z) - R_n(z)| < 2^{-n} \quad (z \in K_{n-1}).$$

Now define

$$f(z) = \sum_{n=1}^\infty (f_n(z) - R_n(z)),$$

for all z for which the series converges.

The series converges uniformly on compact subsets of $D \setminus E$. Indeed, suppose that $K \subset D \setminus E$ is compact; then there exists N so that $E \subset K_N$. Now for $z \in K$ we have

$$f(z) = \sum_{n=1}^{N+1} (f_n(z) - R_n(z)) + \sum_{n=N+2}^\infty (f_n(z) - R_n(z)).$$

Since $K \subset D \setminus E$, all the terms in the first sum are holomorphic in a neighborhood of K, and for $n > N+1$ we have $|f_n - R_n| < 2^{-n}$ on K, so the second sum converges uniformly on K. Thus the sum converges uniformly on K, showing that $f \in H(D \setminus E)$.

And the same decomposition shows that f has poles with the desired principal parts at every point of E: The first term is a rational function with the right principal part at every point of E_{N-1}, and the second term is holomorphic on K_N°, so the whole thing has the right principal part at every point of E_{N-1} (hence at every point of E). $\square$

Exercise

The Cauchy Integral Formula (Theorem 4.9) was a key step in the proof of Runge's Theorem. We saw in Chapters 2 and 4 that the Cauchy Integral Formula and Cauchy's Theorem are more or less equivalent. It is interesting to note that Cauchy's Theorem (Theorem 4.10) is in fact an immediate corollary of Runge's Theorem!

12.7. Assume Runge's Theorem and prove Theorem 4.10.

Hint: Suppose f is as in Theorem 4.10. Runge's Theorem shows (how?) that you may assume that f is a rational function. Now use Lemma 12.0, the hypothesis about winding numbers, and one of the versions of Cauchy's Theorem from Chapter 2 (I'm not going to tell you which one) to finish the proof.

Chapter 13

The Weierstrass Factorization Theorem

Suppose that $D \subset \mathbb{C}$ is a connected open set and $f \in H(D)$ is nonconstant. We know that the zero set $Z(f)$ cannot have any limit point in D. In this chapter we will prove a converse: If $E \subset D$ has no limit point in D then there exists $f \in H(D)$ such that $Z(f) = E$. This answers what seems like a very natural question; it also has various applications (for example we will use it to show that we can assign arbitrary values to a holomorphic function on a set with no limit point, and also to show that every meromorphic function is the quotient of two holomorphic functions).

We begin with the case $D = \mathbb{C}$. Suppose that (z_j) is a sequence of complex numbers and $z_j \to \infty$. The obvious way to construct an entire function with zeroes at the z_j and nowhere else is to say

$$f(z) = \prod_{j=1}^{\infty}(z - z_j).$$

The problem is that this product does not converge.

So we try to fix that by using different factors. If we could find entire functions f_j such that $f_j(z_j) = 0$ but such that $f_j(z)$ was very close to 1 for $|z| < |z_j|$ then we could try

$$f(z) = \prod_{j=1}^{\infty} f_j(z);$$

that might have a chance of converging since the factors tend to 1. It is convenient to normalize things differently: We will find entire functions f_j

such that $f_j(1) = 0$ and $f_j(z)$ is close to 1 for $|z| < 1$, and then set

$$f(z) = \prod_{j=1}^{\infty} f_j\left(\frac{z}{z_j}\right).$$

(This is more convenient because when we are trying to construct the f_j we do not need to worry about what z_j is.)

The functions that we are going to use here are known as the Weierstrass *elementary factors*. They are actually very natural things to try, if you look at it right: First, we want an entire function which has a simple zero at $z = 1$ and no other zeroes. Since the plane is simply connected any nonvanishing entire function has an entire logarithm, so our elementary factor must be of the form

$$E(z) = (1-z)e^{g(z)}$$

for some entire function g. Now we want to have $E(z)$ close to 1 for $|z| < 1$; this is the same as saying that $e^{g(z)}$ should be close to $1/(1-z)$. In other words, $g(z)$ should be close to $-\log(1-z)$.

Of course $\log(1-z)$ is not entire; we need an entire function which is close to $-\log(1-z)$ for $z \in \mathbb{D}$. But polynomials are entire, and it is easy to find polynomials that converge to $-\log(1-z)$ in $\mathbb{D}$; for example the partial sums of the power series should do that!

This raises the question of what that power series is. Suppose that we have a function $-\log(1-z) \in H(\mathbb{D})$ which has been chosen so that $-\log(1-0) = 0$, and consider its power series:

$$-\log(1-z) = \sum_{n=0}^{\infty} c_n z^n \quad (z \in \mathbb{D}).$$

If you take the derivative you get

$$\frac{1}{1-z} = \sum_{n=0}^{\infty} (n+1)c_{n+1} z^n;$$

comparing this with

$$\frac{1}{1-z} = \sum_{n=0}^{\infty} z^n$$

shows that $c_n = 1/n$ for $n = 1, 2, \ldots$. We know that $c_0 = -\log(1-0) = 0$, and so we know the c_n:

$$-\log(1-z) = \sum_{n=1}^{\infty} \frac{z^n}{n} \quad (z \in \mathbb{D}).$$

13. The Weierstrass Factorization Theorem

And this tells us how to define those elementary factors: We set $p_0(z) = 0$,
$$p_n(z) = \sum_{j=1}^n \frac{z^j}{j} \quad (n = 1, 2, \ldots),$$
and then we define the Weierstrass *elementary factors* by
$$E_n(z) = (1-z)e^{p_n(z)} \quad (n = 0, 1, \ldots).$$

It is clear that E_n is entire, that E_n has a simple zero at $z = 1$ and no other zeroes, and the considerations above show that in fact $E_n \to 1$ uniformly on compact subsets of $\mathbb{D}$. This would be sufficient for a very crude version of the Weierstrass Factorization Theorem; to get the more precise version below we need an estimate on just how fast $E_n \to 1$ in $\mathbb{D}$, which is provided by the following lemma.

Lemma 13.0. *For $z \in \mathbb{D}$ and $n = 0, 1, \ldots$ we have*
$$|1 - E_n(z)| \le |z^{n+1}|.$$

The proof is very simple but extremely clever. Weierstrass was a smart guy.

Proof. The main thing that makes the proof work is the fact that the coefficients in the power series (centered at the origin) for $1 - E_n$ are all non-negative. Since $1 - E_n$ vanishes at the origin, this is equivalent to saying that the coefficients in the power series for the derivative $(1 - E_n)'$ are all non-negative, and this is clear, since if you do the math you see that
$$(1 - E_n)'(z) = z^n e^{p_n(z)},$$
and the functions exp and p_n have non-negative power-series coefficients.

The above expression for $(1 - E_n)'$ also shows that $1 - E_n$ has a zero of order at least $n+1$ at the origin, so that $(1 - E_n(z))/z^{n+1}$ is holomorphic in $\mathbb{D}$ (or rather can be defined at the origin so as to give a function holomorphic in $\mathbb{D}$) and it also shows that the power series for $(1 - E_n(z))/z^{n+1}$ has all non-negative coefficients. It follows that for $z \in \mathbb{D}$ ($z \ne 0$) we have
$$\left|\frac{1 - E_n(z)}{z^{n+1}}\right| \le \frac{1 - E_n(1)}{1^{n+1}} = 1. \qquad \square$$

The Weierstrass theorem for entire functions is now quite simple (because we already did the prerequisites regarding infinite products in Chapter 6). First we show that an arbitrary subset of the plane with no limit point in the plane is the zero set of some entire function. Since this is clear for finite sets, we simplify the notation by restricting to infinite sets in the theorem:

Theorem 13.1. *Suppose that $(z_j)_{j=1}^{\infty}$ is a sequence of nonzero complex numbers which tend to infinity as $j \to \infty$; suppose N is a non-negative integer. If (n_j) is a sequence of non-negative integers such that*

$$\sum_{j=1}^{\infty} \left(\frac{r}{|z_j|}\right)^{n_j+1} < \infty$$

for all $r > 0$ then the product

$$f(z) = z^N \prod_{j=1}^{\infty} E_{n_j}\left(\frac{z}{z_j}\right)$$

converges uniformly on compact subsets of the plane to an entire function with a zero of order N at the origin, a zero at each z_k and no other zeroes. (The order of the zero of f at z_k is equal to the number of times the value z_k appears in the sequence (z_j).)

You should note that for any sequence (z_j) with $|z_j| \to \infty$ it is possible to choose a sequence n_j such that $\sum_{j=1}^{\infty} (r/|z_j|)^{n_j+1} < \infty$ for all $r > 0$; for example $n_j = j$ works, since for a given $r > 0$ we must have $r/|z_j| < 1/2$ for all but finitely many values of j. In typical applications one attempts to choose the n_j as small as possible.

Proof. By Lemma 6.1.4 we need only show that the sum

$$\sum_{j=1}^{\infty} \left|1 - E_{n_j}\left(\frac{z}{z_j}\right)\right|$$

converges uniformly on compact subsets of the plane. But this is clear: Suppose that $K \subset \mathbb{C}$ is compact, and choose $r > 0$ such that $|z| \leq r$ for all $z \in K$. Choose J so that $|z_j| > r$ for all $j > J$. Then for $z \in K$ and $j > J$ we have

$$\left|1 - E_{n_j}\left(\frac{z}{z_j}\right)\right| \leq \left|\frac{z}{z_j}\right|^{n_j+1} \leq \left(\frac{r}{|z_j|}\right)^{n_j+1}$$

by Lemma 13.0, so the fact that

$$\sum_{j=1}^{\infty} \left(\frac{r}{|z_j|}\right)^{n_j+1} < \infty$$

shows that the sum converges uniformly on K.

(Well, in any case Lemma 6.1.4 now shows that our product converges to an entire function with the right zero set, although the lemma does not say anything explicitly about the *order* of the zeroes. But the fact that the order works out right follows for free:

Suppose that $\alpha \in \mathbb{C}$, $\alpha \neq 0$, and let A be the (finite) set of j such that $z_j = \alpha$. Then $f = gh$, where

$$g(z) = z^N \prod_{j \notin A} E_{n_j}\left(\frac{z}{z_j}\right)$$

and

$$h(z) = \prod_{j \in A} E_{n_j}\left(\frac{z}{z_j}\right).$$

Lemma 6.1.4 shows that $g(\alpha) \neq 0$, and it is clear that h has a zero of the appropriate order at α.) □

That's more a "Weierstrass Multiplication Theorem" than a Weierstrass Factorization Theorem. The Weierstrass Factorization Theorem for entire functions is an immediate consequence:

Theorem 13.2 (Weierstrass Factorization Theorem for entire functions). *Suppose that $f \in H(\mathbb{C})$ is nonconstant. Suppose that f has a zero of order N at the origin $(0 \leq N < \infty)$. Suppose that f has M $(0 \leq M \leq \infty)$ zeroes away from the origin and suppose that $(z_j)_{j=1}^{M}$ is a sequence containing the nonzero zeroes of f, each listed according to its multiplicity. If $(n_j)_{j=1}^{M}$ is a sequence of positive integers such that*

$$\sum_{j=1}^{M}\left(\frac{r}{|z_j|}\right)^{n_j} < \infty$$

for all $r > 0$ then there exists an entire function g such that

$$f(z) = e^{g(z)} z^N \prod_{j=1}^{M} E_{n_j}\left(\frac{z}{z_j}\right)$$

for all $z \in \mathbb{C}$; the product converges uniformly on compact subsets of the plane.

We should make a few comments on the notation and terminology. Note first that the product of no factors at all equals 1, just as an empty sum equals 0:

$$\prod_{j=1}^{M} E_{n_j}\left(\frac{z}{z_j}\right) = 1$$

when $M = 0$. You should also note that if f has only finitely many zeroes then the result is trivial; the interesting case is when $M = \infty$. (Indeed, the fact that f may have only finitely many zeroes is sometimes overlooked: The statement of the Weierstrass Factorization Theorem in at least three well-known texts on complex analysis implies that every entire function has infinitely many zeroes! We leave it to you to show that this is not so ...)

Proof. We have shown that if we set

$$h(z) = z^N \prod_{j=1}^{M} E_{n_j}\left(\frac{z}{z_j}\right)$$

then h is an entire function with the same zeroes (with the same multiplicities) as f. (This is clear if $M < \infty$, and the previous theorem applies if $M = \infty$.) Thus f/h is an entire function with no zeroes; since the plane is simply connected, it follows that f/h has a holomorphic logarithm g. □

There is also a version of the theorem for functions holomorphic in an arbitrary open set:

Theorem 13.3 (Weierstrass Factorization Theorem in an open set). *Suppose that $D \subset \mathbb{C}$ is open, $A \subset D$ has no limit point in D, and m_α is a positive integer for each $\alpha \in A$. There exists $f \in H(D)$ such that f has a zero of order m_α at each $\alpha \in A$ and no other zeroes.*

This is one of those curious things where a slightly stronger statement is easier to prove (or rather it is easier to give a "natural" proof of the stronger statement). Recall that if $D \subset \mathbb{C}_\infty$ is open then $H(D)$ is the space of all holomorphic maps from D to $\mathbb{C}$.

Theorem 13.3.1 (Weierstrass Factorization Theorem in an open set). *Suppose that $D \subset \mathbb{C}_\infty$ is open, $D \neq \mathbb{C}_\infty$, $A \subset D$ has no limit point in D, and m_α is a positive integer for each $\alpha \in A$. There exists $f \in H(D)$ such that f has a zero of order m_α at each $\alpha \in A$ and no other zeroes.*

Note that we need to exclude the case $D = \mathbb{C}_\infty$, since $H(\mathbb{C}_\infty)$ contains only the constants, by Liouville's Theorem.

Proof. This is clear for finite A, so we assume that A is infinite. After a linear-fractional change of variables we may assume that $\infty \in D$ and $\infty \notin A$. Since $D \neq \mathbb{C}_\infty$, it follows that $\mathbb{C} \setminus D = \mathbb{C}_\infty \setminus D$ is a nonempty compact set.

Let (z_j) be a sequence consisting of the elements of A, in which each $\alpha \in A$ appears exactly m_α times. Let w_j be a point of $\mathbb{C} \setminus D$ at minimal distance from z_j. It follows that

$$f(z) = \prod_{j=1}^{\infty} E_j \left(\frac{z_j - w_j}{z - w_j} \right)$$

does what we want (note that $(z_j - w_j)/(\infty - w_j) = 0$, since w_j is finite and $z_j - w_j$ is finite and nonzero).

Note first that each factor $f_j(z) = E_j\left((z_j - w_j)/(z - w_j)\right)$ does in fact define a function holomorphic in D: It is certainly holomorphic in $D \cap \mathbb{C}$, and

$$f_j \left(\frac{1}{z} \right) = E_j \left(\frac{z_j - w_j}{\frac{1}{z} - w_j} \right) = E_j \left(\frac{(z_j - w_j)z}{1 - w_j z} \right)$$

is holomorphic near 0, which shows by definition that f_j is holomorphic near ∞. As before we need only show that the sum

$$\sum_{j=1}^{\infty} \left| 1 - E_j \left(\frac{z_j - w_j}{z - w_j} \right) \right|$$

converges uniformly on compact subsets of D (here of course a compact subset of D may well contain ∞).

Since D is open, $\infty \in D$, $\infty \notin A$, and A has no limit points in D it follows that there exists an $r > 0$ such that $A \subset D(0,r)$ and also $\mathbb{C}_\infty \setminus D \subset D(0,r)$. This implies that $|z_j - w_j| \to 0$ as $j \to \infty$: Since we have $|z_j - w_j| \le 2r$ for all j, if $|z_j - w_j|$ does not tend to zero we can find a sequence (j_k), a point $z \in \overline{D}(0,r)$, and a number $\delta > 0$ such that $z_{j_k} \to z$ and $|z_{j_k} - w_{j_k}| \to \delta$. Since $D(z_j, |z_j - w_j|) \subset D$ for all j, it follows that $D(z, \delta) \subset D$ and in particular that $z \in D$, which contradicts the fact that A has no limit point in D.

Suppose now that $K \subset D$ is compact. There exists $\delta > 0$ such that $|z - w| > \delta$ for all $z \in K$ and all $w \in \mathbb{C}_\infty \setminus D$, and there exists N such that $|z_j - w_j| < \delta/2$ for all $j > N$. So for $j > N$ we have

$$\left| 1 - E_j \left(\frac{z_j - w_j}{z - w_j} \right) \right| \le \left| \frac{z_j - w_j}{z - w_j} \right|^{j+1} \le \left(\frac{1}{2} \right)^{j+1}$$

for all $z \in K$; this shows that the sum converges uniformly on K, as required. $\square$

There are several interesting applications of these theorems.

Theorem 13.4. *If D is an open subset of the plane and f is meromorphic in D then there exist functions $g, h \in H(D)$ such that $f = g/h$.*

Proof. We may assume that $f \notin H(D)$. Let A be the set of poles of f, and suppose that f has a pole of order m_α at each $\alpha \in A$. The previous theorem shows that there exists $h \in H(D)$ such that h has a zero of order m_α at each $\alpha \in A$; it follows that $g = hf \in H(D)$. □

Readers who know some algebra should note that if D is a connected open subset of the plane then $H(D)$ is an integral domain (there are no zero divisors in $H(D)$ because ...), and Theorem 13.4 shows that the field of quotients of $H(D)$ is (isomorphic to) the space of all meromorphic functions in D.

We can use the Weierstrass Factorization Theorem together with the Mittag-Leffler Theorem to show that there exists $f \in H(D)$ with arbitrary values on an arbitrary subset of D with no limit point in D; in fact we can assign finitely many derivatives:

Theorem 13.5 (Holomorphic interpolation). *Suppose that $D \subset \mathbb{C}$ is open and $A \subset D$ has no limit point in D. Suppose that for each $\alpha \in A$ we are given a non-negative integer n_α and a sequence of complex numbers $(c_{\alpha,j})_{j=0}^{n_\alpha}$. There exists a function $f \in H(D)$ such that*

$$f^{(j)}(\alpha) = c_{j,\alpha} \quad (\alpha \in A,\ 0 \leq j \leq n_\alpha).$$

Proof. We begin by choosing $g \in H(D)$ such that g has a zero of order exactly $n_\alpha + 1$ at α, for each $\alpha \in A$. For each $\alpha \in A$ it follows that g has a power series expansion of the form

$$g(z) = \sum_{j=n_\alpha+1}^{\infty} b_{j,\alpha}(z-\alpha)^j$$

valid in some neighborhood of α, with $b_{n_\alpha+1,\alpha} \neq 0$.

Suppose now that h is meromorphic in D, with a pole of order no larger than $n_\alpha+1$ at each $\alpha \in A$ and no other poles; it follows that $f = gh \in H(D)$. The Mittag-Leffler Theorem (Theorem 12.11) shows that we may assign the principal part of h at $\alpha \in A$ arbitrarily — we will see that if we select h to have the right principal parts at its poles then f and its derivatives have the values we want.

Suppose that $\alpha \in A$ and the Laurent expansion of h about α is

$$h(z) = \sum_{j=-(n_\alpha+1)}^{\infty} d_{j,\alpha}(z-\alpha)^j.$$

13. The Weierstrass Factorization Theorem

Since both series are absolutely convergent near α, we can multiply the two series and combine terms, obtaining

$$f(z) = g(z)h(z) = \sum_{j=0}^{\infty} k_{j,\alpha}(z-\alpha)^j,$$

where

$$k_{j,\alpha} = \sum_{n+m=j} b_{n,\alpha} d_{m,\alpha} = \sum_{n=n_\alpha+1}^{j+n_\alpha+1} b_{n,\alpha} d_{j-n,\alpha}.$$

If we can arrange things so that $k_{j,\alpha} = c_{j,\alpha}/j!$ for $0 \le j \le n_\alpha$ then we will have $f^{(j)}(\alpha) = c_{j,\alpha}$ for the same values of j, as desired.

So: We are given numbers $c_{j,\alpha}$ ($0 \le j \le n_\alpha$) and $b_{j,\alpha}$ ($j \ge n_\alpha + 1$), we know that $b_{n_\alpha+1,\alpha} \ne 0$, and we need to show that there exist numbers $d_{j,\alpha}$ ($-(n_\alpha + 1) \le j \le -1$) such that

$$\sum_{n=n_\alpha+1}^{j+n_\alpha+1} b_{n,\alpha} d_{j-n,\alpha} = \frac{c_{j,\alpha}}{j!} \quad (0 \le j \le n_\alpha).$$

But this system of equations obviously has a solution, being upper triangular with nonzero coefficients on the diagonal. $\square$

Probably we should give an example to clarify the last claim in that proof. Fix an $\alpha \in A$, suppose $n_\alpha = 2$, and suppress the α in the notation. Assuming that $\alpha = 0$ to further simplify the notation, we have

$$g(z) = b_3 z^3 + b_4 z^4 + \cdots$$

with $b_3 \ne 0$, we are given c_0, c_1, and c_2, and we want to find d_{-3}, d_{-2}, and d_{-1} such that if

$$h(z) = \frac{d_{-3}}{z^3} + \frac{d_{-2}}{z^2} + \frac{d_{-1}}{z} + d_0 + d_1 z + \cdots$$

then

$$g(z)h(z) = c_0 + c_1 z + \frac{c_2}{2} z^2 + \cdots.$$

This leads to the following three equations:

$$b_3 d_{-3} = c_0,$$
$$b_3 d_{-2} + b_4 d_{-3} = c_1,$$
$$b_3 d_{-1} + b_4 d_{-2} + b_5 d_{-1} = c_2/2,$$

and it is clear that these equations have a solution, since $b_3 \neq 0$: We can solve the first equation for d_{-3}, then solve the second equation for d_{-2} and finally solve the third for d_{-1}.

So far we have seen two very interesting applications of the Weierstrass Factorization Theorem. We should consider what the theorem says about the factorization we already know for the function $\sin(\pi z)$. Recall that in Chapter 6 we showed that

$$\sin(\pi z) = \pi z \prod_{n=1}^{\infty} \left(1 - \frac{z^2}{n^2}\right).$$

The first thing we did was show that the product actually converges to an entire function. Other attempts didn't quite work; it seemed natural to try to show that

$$\frac{\sin(\pi z)}{\pi z} = \prod_{n \in \mathbb{Z},\, n \neq 0} \left(1 - \frac{z}{n}\right),$$

but this product doesn't quite converge unless we group the terms just right, because the sum $\sum 1/n$ diverges; the fact that the sum $\sum 1/n^2$ converges is what shows that the product in Euler's formula converges. We can understand what happened here in terms of Theorem 13.2 as follows: If $z_1, z_2, \ldots$ are the nonzero zeroes of the function $f(z) = \sin(\pi z)$, ordered in the natural way (that is, $z_1 = 1$, $z_2 = -1$, $z_3 = 2$, $z_4 = -2, \ldots$) then the sum

$$\sum \left(\frac{r}{|z_j|}\right)^2$$

converges for every $r > 0$, so the theorem shows that

$$\sin(\pi z) = z e^{g(z)} \prod_{j=1}^{\infty} E_1\left(\frac{z}{z_j}\right) = z e^{g(z)} \prod_{j=1}^{\infty} \left(1 - \frac{z}{z_j}\right) e^{z/z_j}$$

$$= z e^{g(z)} \prod_{j=1}^{\infty} \left(1 - \frac{z}{j}\right) e^{z/j} \prod_{j=-\infty}^{-1} \left(1 - \frac{z}{j}\right) e^{z/j}$$

for some entire function g. So that's why the product of $(1 - z/j)$ doesn't converge; the factors $e^{z/j}$ are missing. Now note that the product above converges absolutely, so we can rearrange the factors any way we like; if we group the factors in pairs in the obvious way we note that

$$\left(1-\frac{z}{j}\right)e^{z/j}\left(1-\frac{z}{-j}\right)e^{z/(-j)} = \left(1-\frac{z^2}{j^2}\right)$$

and so we obtain

$$\sin(\pi z) = ze^{g(z)}\prod_{j=1}^{\infty}\left(1-\frac{z^2}{j^2}\right),$$

as before.

All of which says that the Weierstrass Factorization Theorem is (of course) consistent with our previous results on factoring the sine function, but it really doesn't help all that much. One of the proofs in Chapter 6 essentially begins with the factorization above, and the hard part was determining that the function g is actually constant.

There exists a more sophisticated factorization theorem due to Hadamard. Roughly, if one knows something about the rate of growth of an entire function this says something about the rate of growth of the number of zeroes the function has in $D(0,r)$; this in turn says something about how small we can take the n_j in Theorem 13.2 and something about the rate of growth of the function g. In particular, if $f(z) = \sin(\pi z)$ then the Hadamard Factorization Theorem says that we have a factorization as above where g is a polynomial of degree 1 or less; getting from here to the fact that g must be constant is trivial (for example using the fact that e^g must be an even function does it). So the Hadamard Factorization Theorem really does contain Euler's infinite product for $\sin(\pi z)$, while the Weierstrass Factorization Theorem merely gives an indication that there may be such a factorization.

The detailed study of entire functions (in particular entire functions of finite "order") is one of those staggeringly wonderful aspects of complex analysis that we're not going to get into. You should visit the library some day, find the collected works of Polya, and note the ethereal glow emanating from the volume titled "Entire Functions".

More comments for readers who know some algebra: If D is an open set in the plane then $H(D)$ is a ring; it is easy to see that it is an integral domain if and only if D is connected. The Weierstrass Factorization Theorem is useful in studying the algebraic properties of this ring: See for example Chapter 15 in [**R**] for a proof that every finitely generated ideal in $H(D)$ is principal. (We are not going to get into this, but having said that every finitely generated ideal is principal we should note that not every ideal in $H(D)$ is principal: For example if I is the set of all $f \in H(\mathbb{C})$ such that $f(n) = 0$ for all but finitely many integers n then I is an ideal in $H(\mathbb{C})$ which is not finitely generated, and in particular not principal.)

Exercises

13.1. Suppose that D is a bounded and connected open subset of the plane. Show that there exists $f \in H(D)$ which cannot be extended to a function holomorphic in a strictly larger connected open set.

(The exercise says that f has ∂D as its "natural boundary".)

Hint: Show that there exists a set $E \subset D$ such that E has no limit point in D but every point of ∂D is a limit point of E.

13.2. Suppose that D is an open set in the plane and E and F are disjoint subsets of D, neither of which has a limit point in D. Show that there exists a function f meromorphic in D so that f has a simple pole at every point of E, a simple zero at every point of F, and no other poles or zeroes.

13.3. Suppose that D is a simply connected open subset of the plane and $f \in H(D)$ is not constant. Show that there exists $g \in H(D)$ such that $f = g^2$ if and only if every zero of f has even order.

Chapter 14

Carathéodory's Theorem

In a previous chapter we stated a theorem of Carathéodory without proof:

Theorem 10.5.1 (Carathéodory). *Suppose D is a bounded simply connected region in the plane and ∂D is a simple closed curve. Let $\phi : D \to \mathbb{D}$ be a conformal equivalence. Then ϕ extends to a homeomorphism of $\overline{D}$ and $\overline{\mathbb{D}}$.*

We are going to prove a somewhat weaker version of this theorem here; the weaker version will suffice for the application we have in mind (constructing a "modular function" for use in proving the Picard theorems).

Suppose that D is an open subset of the plane and $\zeta \in \partial D$. We will say that ζ is a *simple* boundary point if it has the following property: If (z_n) is any sequence of points in D with $z_n \to \zeta$ then there exists a continuous map $\gamma : [0,1) \to D$ and a sequence (t_n) in $[0,1)$ such that $t_n \leq t_{n+1}$, $t_n \to 1$, $\gamma(t_n) = z_n$, and $\lim_{t \to 1} \gamma(t) = \zeta$.

In other words, ζ is a simple boundary point if every sequence in D which tends to ζ lies on a curve in D which tends to ζ.

At first a person might think that every boundary point of any open set was simple. That would be because a person was only thinking about open sets bounded by simple closed curves: If $D = \mathbb{D} \setminus [0,1]$ then $1 \in \partial D$ gives a simple example of a nonsimple boundary point.

We will prove the following version of Carathéodory's Theorem:

Theorem 14.0 (Carathéodory). *Suppose D is a bounded simply connected region in the plane and every boundary point of D is simple. Let*

$\phi : D \to \mathbb{D}$ be a conformal equivalence. Then ϕ extends to a homeomorphism of $\overline{D}$ and $\overline{\mathbb{D}}$.

It follows that if D is a bounded open set and every boundary point is simple then the boundary of D is in fact a simple closed curve, because it is the image of the unit circle under (the extension of) ϕ^{-1}. The converse holds as well, although we will not prove it here: If the boundary of D is a simple closed curve then every boundary point of D is simple. (If we did prove this fact then our proof of Theorem 14.0 would give a proof of Theorem 10.5.1.)

(A person should not jump to conclusions: Supposing that D and ϕ are as in Theorem 14.0 it *does* follow that the boundary of D is equal to $\phi^{-1}(\partial \mathbb{D})$ as claimed in the previous paragraph; this is clear because we know that ϕ is a bijection from $\overline{D}$ onto $\overline{\mathbb{D}}$ and we also know that $\phi(D) = \mathbb{D}$. It also follows from the general fact that if $E_1, E_2 \subset \mathbb{R}^n$ and $f : E_1 \to E_2$ is a homeomorphism then $x \in E_1$ is in the interior of E_1 if and only if $f(x)$ lies in the interior of E_2, but this is a nontrivial theorem (the "Invariance of Domain" theorem) which luckily we don't need here.)

Theorem 14.0 will suffice for our applications, because when we have an explicit region bounded by a simple closed curve it will be clear that all the boundary points are simple. (In fact it is amusing to note that Theorem 14.0 does suffice for *every* explicit region to which we might wish to apply Theorem 10.5.1: This is because if ∂D is a simple closed curve then every boundary point of D *is* simple. So, although we are not going to prove that it's so, it will never happen that you have a region D to which you wish you could apply Theorem 10.5.1 because Theorem 14.0 does not work.)

We should at least give some criterion which allows us to actually prove that the typical boundary point is as simple as it looks:

- **Exercise 14.1.** Suppose that $f : [0, 1] \to (0, 1)$ is continuous, and let

$$D = \{\, x + iy : x \in (0, 1),\, f(x) < y < 1 \,\}.$$

Show that if $0 < x < 1$ then $x + if(x)$ is a simple boundary point of D.

Hint: One thing you could do is this: Given a sequence (z_n) in D converging to $\zeta = x + if(x)$, "connect" $z_n = x_n + iy_n$ to $z_{n+1} = x_{n+1} + iy_{n+1}$ by a polygonal path of the form

$$[z_n, x_n + i\delta_n, x_{n+1} + i\delta_n, z_{n+1}].$$

You need to show you can choose δ_n so that these polygons lie in D and also fit together to form a curve that actually tends to ζ; the continuity of f is going to be important here.

14. Carathéodory's Theorem

Exercise 14.1 suffices, after a suitable change of variables, to show that every boundary point of D is simple, for the sort of open sets D bounded by simple closed curves that are likely to arise in practice. We should note however that it is not so simple to prove that every boundary point of *every* open set bounded by a simple closed curve is simple. Consider the region in Figure 14.1:

Figure 14.1

It is clear that this region is bounded by a simple closed curve, and it is also clear that every boundary point is simple. But if ζ is the interesting boundary point, i.e., the center of the spiral, then the preceding exercise does not suffice to show that ζ is simple. Formulating a *proof* that every boundary point of every open set bounded by a simple closed curve is simple is not so easy.

We need a few preliminary results about bounded holomorphic functions in the unit disk.

Lemma 14.1. *Suppose that $f \in H(\mathbb{D})$ and $|f(z)| \leq 1$ for all $z \in \mathbb{D}$. Suppose that N is a positive integer, $a \in \mathbb{R}$, and let $S = \{\, re^{it} : 0 < r < 1,\, a < t < a + 2\pi/(2N)\,\}$. Suppose that $\gamma : [0,1] \to \mathbb{D}$ is continuous, $\gamma(0) = r_0 e^{ai}$ and $\gamma(1) = r_1 e^{(a + 2\pi/(2N))i}$ for some $r_j > 0$, and $\gamma(t) \in S$ for $0 < t < 1$. If $|f(\gamma(t))| < \epsilon < 1$ for all $t \in [0,1]$ then $|f(0)| < \epsilon^{1/(2N)}$.*

Proof. We may assume that $a = 0$. Define
$$g(z) = f(z)\overline{f(\bar{z})},$$
and then set
$$h(z) = \prod_{j=0}^{N-1} g(e^{2\pi i j/N} z).$$

It follows that $|g| < \epsilon$ at every point of the set $\gamma^* \cup \widetilde{\gamma^*}$, so that $|h| < \epsilon$ at every point of the set
$$\Gamma = \bigcup_{j=0}^{N-1} e^{-2\pi i j/N}(\gamma^* \cup \widetilde{\gamma^*})$$

(here we are using the standard notation $cA = \{\, cz : z \in A \,\}$ and the not-so-standard notation $\widetilde{A} = \{\, \overline{z} : z \in A \,\}$).

But there is a bounded open set $\Omega \subset \mathbb{D}$ such that $0 \in \Omega$ and such that $\partial\Omega \subset \Gamma$; hence the Maximum Modulus Theorem shows that $|h(0)| \leq \epsilon$, which implies that $|f(0)| < \epsilon^{1/(2N)}$. $\square$

The geometry of the proof is illustrated in Figure 14.2:

Figure 14.2

The existence of a bounded open set $\Omega \subset \mathbb{D}$ such that $0 \in \Omega$ and such that $\partial\Omega \subset \Gamma$ in the proof above is clear from Figure 14.2. It is generally a good thing to be suspicious of proofs by picture (for example, it is easy to imagine a picture making the highly nontrivial result that every boundary point of a region bounded by a simple closed curve is simple seem quite obvious). An outline of an actual *proof* of the existence of such a region follows:

Let Ω be the component of $\mathbb{C} \setminus \Gamma$ containing 0. We need to show that Ω is bounded, which is to say that Ω is not the unbounded component of $\mathbb{C} \setminus \Gamma$. But this is clear: There is a closed curve C with $C^* = \Gamma$. (We may assume that $\gamma(0) > 0$ as in the figure. First define $C : [-1, 1] \to \mathbb{C}$ by the requirements $C(t) = \gamma(t)$ for $t \in [0, 1]$ and $C(-t) = \overline{\gamma(t)}$, then extend C to a mapping $C : [-N, N] \to \mathbb{C}$ by the requirement $C(t+2) = e^{2\pi i j/N} C(t)$; we leave it to you to show that the hypotheses on γ imply that C is a continuous closed curve.) It is easy to see that $\operatorname{Ind}(C, 0) = 1$ using a collection of branches of the logarithm, one in each sector $e^{2\pi ji/(2N)}S$, as in the proof of Lemma 4.6. Hence 0 cannot lie in the unbounded component of $\mathbb{C} \setminus \Gamma$.

Lemma 14.2. *Suppose that $f \in H(\mathbb{D})$ is bounded. Suppose that $0 < \theta < \pi$ and let S be the region*
$$S = \{re^{it} : 1/2 < r < 1, 0 < t < \theta\}.$$
Suppose that $\gamma_n : [0,1] \to \mathbb{D}$ is continuous, $\gamma_n((0,1)) \subset S$, $\gamma_n(0) > 0$ and $\gamma_n(1) = r_n e^{i\theta}$ ($r_n > 0$) for $n = 1, 2, \ldots$. If $L \in \mathbb{C}$ and $f \circ \gamma_n \to L$ uniformly on $[0,1]$ as $n \to \infty$ then in fact $f = L$ everywhere in $\mathbb{D}$.

Proof. Subtracting a constant and dividing by another constant we may assume that $L = 0$ and that $|f| < 1$ in $\mathbb{D}$. Choose a positive integer N with $2\pi/(2N) < \theta$, and define $g, h \in H(\mathbb{D})$ as in the proof of Lemma 14.1.

Suppose $\epsilon > 0$, and choose n so that $|f| < \epsilon$ at every point of γ_n^*. The proof of Lemma 14.1 (with γ_n in place of γ) shows that $|h| \leq \epsilon$ at every point of $D(0, 1/2)$. Since this holds for all $\epsilon > 0$, it follows that h must vanish identically in $D(0, 1/2)$. Hence f is identically zero (if not then f would have only finitely many zeroes in $D(0, 1/2)$ and this would imply that h had only finitely many zeroes there). □

We should note that the lemma is not true of unbounded holomorphic functions in the disk; in fact Runge's Theorem allows us to construct a counterexample: Choose two sequences (r_n) and (r_n') such that $0 < r_n < r_n' < r_{n+1} < 1$ for all n and such that $r_n \to 1$. Let $D_n = \overline{D(0, r_n)}$, $A_n = \{r_n' e^{it} : 0 \leq t \leq \pi\}$, and set $K_n = D_n \cup A_n$. Let $P_1 = 2$, and define polynomials P_n for $n > 1$ as follows: Supposing that P_j has been defined for $1 \leq j < n$, set $f_n = P_1 + \cdots + P_{n-1}$. The function which equals 0 on D_n and $-f_n$ on A_n is holomorphic in a neighborhood of K_n and K_n does not separate the plane, so Runge's Theorem shows that there exists a polynomial P_n such that $|P_n| < 2^{-n}$ on D_n and $|P_n + f_n| < 2^{-n}$ on A_n.

Now if $K \subset \mathbb{D}$ is compact there exists N such that $K \subset D_n$ for all $n > N$; thus for all but finitely many values of n we have $|P_n| < 2^{-n}$ everywhere on K. So the sum $P_1 + P_2 + \cdots$ converges uniformly on compact subsets of $\mathbb{D}$ to a function $f \in H(\mathbb{D})$. If $z \in A_n$ then
$$f(z) = f_n(z) + P_n(z) + \sum_{j=n+1}^{\infty} P_j(z);$$
since $A_n \subset D_j$ for $j > n$, this shows that
$$|f(z)| \leq |f_n(z) + P_n(z)| + \sum_{j=n+1}^{\infty} |P_j(z)| \leq 2^{-n} + \sum_{j=n+1}^{\infty} 2^{-j} = 2^{-n+1} \quad (z \in A_n),$$

so $f \circ \gamma_n$ tends to zero uniformly, if γ_n is a curve with $\gamma_n^* = A_n$. On the other hand the fact that $P_1(0) = 2$ and $|P_n(0)| < 2^{-n}$ for $n \geq 1$ shows that $|f(0)| \geq 1$.

Runge's Theorem really is very useful for constructing examples of holomorphic functions.

We will also use a result known as Lindelöf's Theorem, to the effect that if a bounded holomorphic function in $\mathbb{D}$ has a limit along some curve ending at a point of $\partial \mathbb{D}$ then the function must have a radial limit with the same value at that point.

For $-1 < r < 1$ we define $\psi_r \in \mathrm{Aut}(\mathbb{D})$ by

$$\psi_r(z) = \frac{z + r}{1 + rz}.$$

To show that $\psi_r \in \mathrm{Aut}(\mathbb{D})$ one could note that $\psi_r(z) = -\phi_{-r}(z)$. Note that $\psi_r(-1) = -1$, $\psi_r(1) = 1$ and $\psi_r(0) = r$. The main reason we care about the ψ_r is that $\psi_r(z) \to 1$ as $r \to 1$ for all $z \in \overline{\mathbb{D}} \setminus \{-1\}$:

Lemma 14.3. *If K is a compact subset of $\overline{\mathbb{D}} \setminus \{-1\}$ then $\psi_r \to 1$ uniformly on K as $r \to 1^-$.*

Proof. Choose $\lambda \in (-1, 0)$ such that $\mathrm{Re}\,(z) \geq \lambda$ for all $z \in K$. Now if $z \in K$ and $0 \leq r < 1$ we have $\mathrm{Re}\,(1 + rz) \geq 1 + r\lambda \geq 1 + \lambda > 0$, hence $|1 + rz| \geq 1 + \lambda$; this shows that

$$|1 - \psi_r(z)| = \left| \frac{(1-z)(1-r)}{1+rz} \right| \leq \frac{|1-z|(1-r)}{1+\lambda} \leq \frac{2(1-r)}{1+\lambda}. \qquad \square$$

Lemma 14.4 (Lindelöf's Theorem). *Suppose that $\gamma : [0, 1] \to \overline{\mathbb{D}}$ is continuous, $\gamma(t) \in \mathbb{D}$ for $0 \leq t < 1$ and $\gamma(1) = 1$. Suppose that $f \in H(\mathbb{D})$ is bounded. If $f(\gamma(t)) \to L$ as $t \to 1$ then $\lim_{r \to 1^-} f(r) = L$.*

Proof. As before we may assume $L = 0$ and $|f| < 1$. Let $K = \{z \in \overline{\mathbb{D}} : \mathrm{Re}\,(z) \geq 0\}$, and fix $\epsilon \in (0, 1)$; now choose $\delta > 0$ such that $|f(\gamma(t))| < \epsilon$ for all $t \in [0, 1)$ with $|1 - \gamma(t)| < \delta$. The preceding lemma shows that there exists $R \in (0, 1)$ such that $|1 - \psi_r(z)| < \delta$ for all $z \in K$ whenever $R < r < 1$. We will show that $R < r < 1$ implies $|f(r)| < \epsilon^{1/6}$; since $\epsilon > 0$ was arbitrary, this shows that $\lim_{r \to 1^-} f(r) = 0$, as required. Fix $r \in (R, 1)$.

Let $C_0 = [-i, i] \subset \overline{\mathbb{D}}$, and let $C_r = \psi_r(C_0)$. Now there exists $t_r \in (0, 1)$ such that $\gamma(t_r) \in C_r$ but $\gamma(t) \notin C_r$ for $t_r < t < 1$. Let $\gamma_r = \gamma|_{[t_r, 1)}$. (Just to clarify things, note that $\gamma_r(t_r) \in C_r$ and that $\gamma_r^* \subset \psi_r(K)$.)

14. Carathéodory's Theorem

Define $\Gamma_r = \psi_r^{-1} \circ \gamma_r$ and $p_r = \psi_r^{-1}(\gamma_r(t_r)) = \Gamma_r(t_r)$. The fact that $\gamma_r(t_r) \in C_r = \psi_r([-i,i])$ shows that $p_r \in [-i,i]$. It may happen that $p_r = 0$; if so we are done, since then $f(r) = f(\gamma_r(t_r))$, so that $|f(r)| < \epsilon < \epsilon^{1/6}$. So we assume that $p_r \neq 0$; then $p_r = iy$ for some $y \neq 0$; assume without loss of generality that $y > 0$.

Now let $F = f \circ \psi_r$. We have $|F| < \epsilon$ on Γ_r^*. Since Γ_r is a curve in the right half-plane with one endpoint at 1 and the other endpoint at iy, it almost seems that we can apply Lemma 14.1 to conclude that $|F(0)| \leq \epsilon^{1/4}$. That's not quite right, because one endpoint of Γ_r does not lie in $\mathbb{D}$. However we can apply Lemma 14.1 with $S = \{\rho e^{it} : 0 < \rho < 1, \pi/6 < t < \pi/2\}$ to show that $|F(0)| < \epsilon^{1/6}$ (erase part of Γ_r, so that what remains is a curve lying in S connecting one edge of S to the other, as in the lemma). But $f(r) = F(0)$, so we have shown that $|f(r)| < \epsilon^{1/6}$ as promised. $\square$

Again we give an illustration:

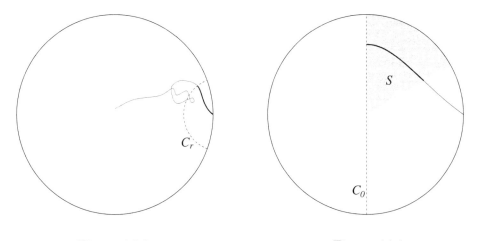

Figure 14.3   Figure 14.4

Figure 14.3 is a picture of γ, with γ_r drawn in bold; Figure 14.4 is a picture of Γ_r, with the part of Γ_r to which we applied the previous lemma drawn in bold.

The next exercise shows that the assumption that f is bounded is needed in Lindelöf's Theorem:

- **Exercise 14.2.** Define $f \in H(\mathbb{D})$ by

$$f(z) = e^{i/(z-1)^2}$$

and define $\gamma : [0,1] \to \mathbb{C}$ by $\gamma(t) = 1 + (1-t)e^{3\pi i/4}$. Draw a picture of γ, and show that $\gamma([0,1)) \subset \mathbb{D}$. Show that $\lim_{t \to 1^-} f(\gamma(t))$ exists but $\lim_{r \to 1^-} f(r)$

does not exist. Hence f must be unbounded in $\mathbb{D}$, by Lindelöf's Theorem; demonstrate explicitly that f is unbounded in $\mathbb{D}$ by defining $\gamma_1(t) =$??? and showing that $\lim_{t\to 1^-} |f(\gamma_1(t))| = \infty$.

The next proposition contains most of the proof of Theorem 14.0:

Proposition 14.5. *Suppose D is a bounded simply connected open set in the plane, and let $\phi : D \to \mathbb{D}$ be a conformal equivalence.*

(i) If ζ is a simple boundary point of D then there exists $L \in \partial \mathbb{D}$ such that $\phi(z_n) \to L$ whenever (z_n) is a sequence in D tending to ζ. (In other words, ϕ can be extended continuously to the set $D \cup \{\zeta\}$, and the extended function satisfies $\phi(\zeta) \in \partial\mathbb{D}$.)

(ii) Suppose that ζ_1 and ζ_2 are two simple boundary points; choose corresponding complex numbers L_1 and L_2 as in (i). If $\zeta_1 \neq \zeta_2$ then $L_1 \neq L_2$.

Proof. First we note that if (z_n) is any sequence in D tending to a point of the boundary then $|\phi(z_n)| \to 1$: If $r < 1$ then $K = \phi^{-1}(\overline{D}(0,r))$ is a compact subset of D and hence there exists N such that $z_n \notin K$ for all $n > N$; thus $|\phi(z_n)| > r$ for every $n > N$.

Now we prove (i). Suppose that ζ is a simple boundary point of D and that (z_n) is a sequence of points of D such that $z_n \to \zeta$. Now the compactness of the closed disk shows that there exists $L \in \overline{\mathbb{D}}$ and a subsequence (z_{n_j}) such that $\phi(z_{n_j}) \to L$ (hence $|L| = 1$ by the comments in the previous paragraph). To show that $\phi(z_n) \to L$ we need only show that every other convergent subsequence of $(\phi(z_n))$ also tends to L.

Suppose not: $(\phi(z_n))$ has two subsequences tending to different limits. Adjusting the notation a bit we may assume that $\phi(z_{2n}) \to L_1$ and $\phi(z_{2n+1}) \to L_2$, where $L_j \in \partial\mathbb{D}$ and $L_1 \neq L_2$. Now, since ζ is a simple boundary point, there exists a continuous curve $\gamma : [0,1) \to D$ and a sequence (t_n) in $[0,1)$ such that $t_n \leq t_{n+1}$, $t_n \to 1$, $\gamma(t_n) = z_n$, and $\lim_{t\to 1} \gamma(t) = \zeta$.

Now let $\Gamma = \phi \circ \gamma$. Then $\Gamma : [0,1) \to \mathbb{D}$, and an argument as above shows that $|\Gamma(t)| \to 1$ as $t \to 1$. But the curve Γ does not actually tend to a single point of $\partial\mathbb{D}$, since $\Gamma(t_{2n}) \to L_1$ while $\Gamma(t_{2n+1}) \to L_2$. Thus Γ must actually "cross" some sector S infinitely many times as it tends to $\partial\mathbb{D}$ (look at Figure 14.5: Γ must either cross S_1 infinitely many times, as shown, or cross S_2 infinitely many times). So, at least for large n, there exist numbers a_n and b_n such that $t_{2n} < a_n < b_n < t_{2n+1}$, $\gamma(a_n)$ lies on one edge of S, $\gamma(b_n)$ lies on the other edge of S, and $\gamma((a_n, b_n))$ lies in the interior of S. Let $\gamma_n = \gamma|_{[a_n, b_n]}$.

Let $f = \phi^{-1}$. Then $f \in H(\mathbb{D})$ and f is bounded. We have $f(\Gamma(t)) \to \zeta$ as $t \to 1$, since $f(\Gamma(t)) = \gamma(t)$; thus $f \circ \gamma_n \to \zeta$ uniformly, and so Lemma

14. Carathéodory's Theorem

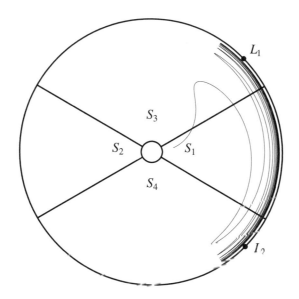

Figure 14.5

14.2 shows that f is constant. This is a contradiction, since f is actually a bijection from $\mathbb{D}$ onto D.

Now for part (ii), suppose that ζ_1 and ζ_2 are simple boundary points of D with $\zeta_1 \ne \zeta_2$. The first part shows that ϕ may be extended to a function continuous on $D \cup \{\zeta_1, \zeta_2\}$ and that $|L_j| = 1$ if $L_j = \phi(\zeta_j)$ for $j = 1, 2$. We need to show that $L_1 \ne L_2$.

For $j = 1, 2$ choose continuous curves $\gamma_j : [0, 1) \to D$ such that $\gamma_j(t) \to \zeta_j$ as $t \to 1$, and let $\Gamma_j = \phi \circ \gamma_j$. Let $f = \phi^{-1}$ as before. Now $f \in H(\mathbb{D})$ is bounded, Γ_j is a curve in $\mathbb{D}$ tending to L_j, and $f(\Gamma_j(t)) \to \zeta_j$ as $t \to 1$. Lindelöf's Theorem (Lemma 14.4) shows that $f(rL_j) \to \zeta_j$ as r increases to 1 and so, since $\zeta_1 \ne \zeta_2$, this shows that $L_1 \ne L_2$. □

Again, we give a few comments on the proof of the first part of the proposition for the benefit of readers who are (appropriately) nervous about relying on the picture (or who are just curious how the proof might be done without the figure):

Since $|\gamma(t)| \to 1$, we may assume that $|\gamma(t)| \ge \delta > 0$ for all $t \in [0, 1]$. Suppose that L_1, L_2, and S_j, $1 \le j \le 4$ are as in Figure 14.5 (we are willing to use a picture for *these* definitions since it's clear it would be entirely trivial to give picture-independent versions). Since $z_{2n} = \gamma(t_{2n}) \to L_1$, we may assume, passing to a subsequence, that $z_{2n} \in S_3$ for all n; similarly we may assume that $z_{2n+1} \in S_4$ for all n.

Now for each n let a_n be the largest element of $[t_{2n}, t_{2n+1}]$ such that $\gamma(a_n) \in \overline{S_3}$ and let b_n be the smallest element of $[t_{2n}, t_{2n+1}]$ such that $\gamma(b_n) \in \overline{S_4}$. We leave it to you to show that there do exist such a_n and b_n, and that then we have $t_{2n} < a_n < b_n < t_{2n+1}$. Then $\gamma((a_n, b_n)) \subset S_1^\circ \cup S_2^\circ$. But S_j° is open and $\gamma((a_n, b_n))$ is connected; hence for every n we must have either $\gamma((a_n, b_n)) \subset S_1^\circ$ or $\gamma((a_n, b_n)) \subset S_2^\circ$. So one case must happen for infinitely many n; passing to a subsequence we may assume there exists j such that $\gamma((a_n, b_n)) \subset S_j^\circ$ for all n. □

Now we should note that if ϕ is a continuous function in a bounded open set D and ϕ can be extended continuously to every boundary point then ϕ can be extended continuously to the entire boundary:

Lemma 14.6. *Suppose that D is a bounded open set in the plane, $\phi : D \to \mathbb{C}$ is continuous, and for every $\zeta \in \partial D$ there exists a function ϕ_ζ continuous on $D \cup \{\zeta\}$ which agrees with ϕ on D. Then ϕ extends to a function continuous on $\overline{D}$.*

Proof. Extend ϕ to a function $\psi : \overline{D} \to \mathbb{C}$ by

$$\psi(\zeta) = \phi_\zeta(\zeta) \quad (\zeta \in \partial D).$$

Suppose that $\zeta \in \partial D$; we need to show that ψ is continuous at ζ. Let $\epsilon > 0$. Since ϕ_ζ is continuous on $D \cup \{\zeta\}$, there exists $\delta > 0$ such that $|\psi(z) - \psi(\zeta)| < \epsilon$ for all $z \in D$ with $|z - \zeta| < \delta$. Suppose that $z \in \partial D$ and $|z - \zeta| < \delta$. Since $\psi(z) = \lim_{D \ni w \to z} \psi(w)$ it follows that $|\psi(z) - \psi(\zeta)| \leq \epsilon$. □

And now we can prove Carathéodory's Theorem:

Proof (Theorem 14.0). Proposition 14.5(i) shows that ϕ has a continuous extension to $D \cup \{\zeta\}$, for every $\zeta \in \partial D$. It follows from Lemma 14.6 that ϕ extends continuously to the closure of D; we will use the same letter to denote the extended function $\phi : \overline{D} \to \overline{\mathbb{D}}$.

Now $\phi(\overline{D})$ must be a compact subset of $\overline{\mathbb{D}}$ containing $\mathbb{D}$, so $\phi(\overline{D}) = \overline{\mathbb{D}}$. Proposition 14.3(ii) shows that ϕ is one-to-one on $\overline{D}$, so the inverse function $\phi^{-1} : \overline{\mathbb{D}} \to \overline{D}$ exists; now the inverse of ϕ is automatically continuous on $\overline{\mathbb{D}}$, by compactness (see Appendix 4). □

Chapter 15

More on Aut($\mathbb{D}$)

15.0. Classification of Elements of $\mathrm{Aut}(\mathbb{D})$

We define $\overline{\Pi_\infty^+}$ to be the closure of the upper half-plane in the extended plane

$$\overline{\Pi_\infty^+} = \{\, z \in \mathbb{C} : \mathrm{Im}(z) \geq 0 \,\} \cup \{\infty\}$$

and we define $\partial_\infty \Pi^+$ to be the extended boundary of $\overline{\Pi_\infty^+}$:

$$\partial_\infty \Pi^+ = \mathbb{R}_\infty = \mathbb{R} \cup \{\infty\}.$$

The Cayley transform $\Phi : \mathbb{D} \to \Pi^+$ extends to a homeomorphism of $\overline{\mathbb{D}}$ and $\overline{\Pi_\infty^+}$. The Cayley transform also induces an isomorphism between $\mathrm{Aut}(\mathbb{D})$ and $\mathrm{Aut}(\Pi^+)$: If $\phi \in \mathrm{Aut}(\mathbb{D})$ then the "corresponding" element of $\mathrm{Aut}(\Pi^+)$ is

$$\psi = \Phi \circ \phi \circ \Phi^{-1},$$

and we will often come close to speaking as though ϕ and ψ were the same thing (for example, if $\phi \in \mathrm{Aut}(\mathbb{D})$ and we say something like "In the upper half-plane ϕ becomes a dilation" that means that the corresponding $\psi \in \mathrm{Aut}(\Pi^+)$ is a dilation).

The nontrivial elements of $\mathrm{Aut}(\mathbb{D})$ can be classified as "parabolic", "hyperbolic", or "elliptic", according to the number and location of their fixed points. (Note that the number of fixed points of $\phi \in \mathrm{Aut}(\mathbb{D})$ in $\mathbb{D}$ is the same as the number of fixed points of the corresponding element of $\mathrm{Aut}(\Pi^+)$ in Π^+, and similarly for the number of fixed points on the (extended) boundary.) We begin with this classification; then we will see that each type of automorphism is conjugate to an automorphism with a very simple description, either in the disk or in the upper half-plane, and finally we shall use this description of the elements of $\mathrm{Aut}(\mathbb{D})$ to determine for which $\phi \in \mathrm{Aut}(\mathbb{D})$ there exists a (non-constant) bounded function $f \in H(\mathbb{D})$ such that $f \circ \phi = f$. These bounded holomorphic functions will be used in the proof of the Big Picard Theorem.

Theorem 15.0.0. *If $\phi \in \mathrm{Aut}(\mathbb{D})$ is not the identity then exactly one of the following holds*:

(i) *ϕ has exactly one fixed point in $\mathbb{D}$ and none on $\partial \mathbb{D}$;*

(ii) *ϕ has no fixed points in $\mathbb{D}$ and exactly one on $\partial \mathbb{D}$;*

(iii) *ϕ has no fixed points in $\mathbb{D}$ and exactly two on $\partial \mathbb{D}$.*

In case (i) we say that ϕ is *elliptic*, in case (ii) it is *parabolic*, and in case (iii) it is *hyperbolic*.

Proof. We prove the corresponding statement for $\phi \in \mathrm{Aut}(\Pi^+)$. Recall that any $\phi \in \mathrm{Aut}(\Pi^+)$ has the form

$$\phi(z) = \frac{az+b}{cz+d}$$

for some a, b, c, $d \in \mathbb{R}$ with $ad - bc = 1$ (cf. Exercise 8.6 and the solution given in Section 10.7).

It may happen that $c = 0$. In this case we have

$$\phi(z) = Az + B$$

for some $A, B \in \mathbb{R}$ with $(A, B) \neq (1, 0)$. This shows that $\phi(\infty) = \infty$, and if $A = 1$ then ∞ is the only fixed point of ϕ, while if $A \neq 1$ then there is exactly one fixed point in $\mathbb{C}$, and that fixed point is a real number. So if $c = 0$ then there is no fixed point in Π^+ and there are either one or two fixed points on $\partial_\infty \Pi^+$.

Suppose now that $c \neq 0$. Then $\phi(\infty) = a/c \in \mathbb{C}$, so ∞ is not a fixed point. For $z \in \mathbb{C}$ the condition that $\phi(z) = z$ is equivalent to a quadratic equation with real coefficients (the fact that $c \neq 0$ shows that the degree is exactly 2) and so there we have either one real solution, two real solutions, or two complex solutions; in the last case the solutions are complex conjugates of each other so exactly one lies in Π^+. □

Recall that if G is a group and $a, b \in G$ then b is said to be *conjugate to a* if $b = c^{-1}ac$ for some $c \in G$. Note that if G is a group of mappings from a set X to itself (with composition as the group operation) then this is the same as saying that b is equivalent to a after a "change of variables" given by c; for example $p \in X$ is a fixed point of a if and only if $c^{-1}(p)$ is a fixed point of b.

15.0. Classification of Elements of Aut($\mathbb{D}$)

Theorem 15.0.1. *Suppose that $\phi \in \mathrm{Aut}(\mathbb{D})$ is not the identity.*

(i) *If ϕ is elliptic then ϕ is conjugate to a rotation: There exists a complex number $\alpha \neq 1$ with $|\alpha| = 1$ such that ϕ is conjugate to ψ, where*

$$\psi(z) = \alpha z \quad (z \in \mathbb{D}).$$

(ii) *If ϕ is parabolic then ϕ is conjugate to a translation in Π^+: There exists $B \in \mathbb{R}$ with $B \neq 0$ such that the element of $\mathrm{Aut}(\Pi^+)$ corresponding to ϕ is conjugate to ψ, where*

$$\psi(z) = z + B \quad (z \in \Pi^+).$$

(iii) *If ϕ is hyperbolic then ϕ is conjugate to a dilation in Π^+: There exists $\delta > 0$ with $\delta \neq 1$ such that the element of $\mathrm{Aut}(\Pi^+)$ corresponding to ϕ is conjugate to ψ, where*

$$\psi(z) = \delta z \quad (z \in \Pi^+).$$

In other words, "up to a change of variables any nontrivial automorphism is either a rotation in the disk or a translation or dilation in the upper half-plane"; very convenient.

Proof. Suppose first that ϕ is elliptic. After a change of variables we may assume that $\phi(0) = 0$, and now Theorem 8.4.1 shows that ϕ is a rotation.

Suppose next that ϕ is either parabolic or hyperbolic. We switch to the upper half-plane: We may assume that $\phi \in \mathrm{Aut}(\Pi^+)$, ϕ has no fixed point in Π^+ and either one or two fixed points on $\partial_\infty \Pi^+$. In either case we may make a change of variables (using an element of $\mathrm{Aut}(\Pi^+)$) so that $\phi(\infty) = \infty$. As in the previous proof it follows that $\phi(z) = Az + B$ for some $A, B \in \mathbb{R}$ with $(A, B) \neq (1, 0)$.

If $A = 1$ then ϕ is a translation and hence ∞ is the only fixed point of ϕ; assume then that $A \neq 1$. Let $c = B/(1 - A)$, set $\psi(z) = z + c$, and define $\tilde{\phi} = \psi^{-1} \circ \phi \circ \psi$. You can check that in fact $\tilde{\phi}(z) = Az$ (and hence $A > 0$ and $A \neq 1$, since $\tilde{\phi} \in \mathrm{Aut}(\Pi^+)$ is not the identity); hence 0 and ∞ are the only fixed points of ψ. $\square$

This seems very interesting in itself, but the main reason we're doing it here is for the following result, which is a major step in a proof of the Big Picard Theorem:

Theorem 15.0.2. *Suppose that $\phi \in \text{Aut}(\mathbb{D})$.*

(i) If ϕ is either elliptic of finite order, parabolic, or hyperbolic then there exists a bounded nonconstant $f \in H(\mathbb{D})$ such that

$$f \circ \phi = f,$$

and in fact such that $f(z) = f(w)$ if and only if $w = \phi^n(z)$ for some integer n.

(ii) If ϕ is elliptic of infinite order and $f \in H(\mathbb{D})$ with $f \circ \phi = f$ then f is constant.

Proof. Suppose first that ϕ is elliptic. The previous theorem shows that we may assume that ϕ is a rotation: $\phi(z) = \alpha z$ where $|\alpha| = 1$. If ϕ has finite order N then $\alpha^N = 1$, $\alpha^k \neq 1$ for $k = 1, 2, \ldots, N-1$, and it follows that $f(z) = z^N$ works. On the other hand if ϕ has infinite order then the powers of α are dense in the unit circle, so that if $f \in H(\mathbb{D})$ and $f \circ \phi = f$ then f must be constant. (See Exercise 10.11 in Section 10.4.)

Now suppose that ϕ is parabolic; by the previous theorem we may assume that $\phi \in \text{Aut}(\Pi^+)$ has the form

$$\phi(z) = z + B$$

for some real $B \neq 0$. Let
$$f(z) = e^{2\pi i z/|B|}.$$

Then f is bounded in Π^+, since for $x + iy \in \Pi^+$ we have

$$|f(x+iy)| = e^{\text{Re}\,(2\pi i(x+iy)/|B|)} = e^{-2\pi y/|B|} < 1.$$

And $f(z) = f(w)$ if and only if $2\pi i w/|B| = 2\pi i z/|B| + 2\pi i k$ for some $k \in \mathbb{Z}$, which is the same as saying $w = z + nB$ (that is, $w = \phi^n(z)$) with $n = \pm k$.

Finally, if ϕ is hyperbolic we may assume that $\phi \in \text{Aut}(\Pi^+)$ has the form

$$\phi(z) = \delta z$$

for some positive $\delta \neq 1$. Define a branch of the logarithm in Π^+ by

$$\log(re^{it}) = \ln(r) + it \quad (r > 0, \, 0 < t < \pi)$$

and now set
$$f(z) = z^{ic} = e^{ic\log(z)}$$

for some real $c \neq 0$ to be determined.

Note that
$$|f(re^{it})| = e^{\operatorname{Re}(ic(\ln(r)+it))} = e^{-ct} \quad (r>0, 0<t<\pi),$$
so f is bounded in Π^+ for any $c \in \mathbb{R}$. For $z, w \in \Pi^+$ we have $f(z) = f(w)$ if and only if $ic\log(z) - ic\log(w) = 2\pi i k$ for some $k \in \mathbb{Z}$; this is equivalent to saying that $z = r_1 e^{it}$ and $w = r_2 e^{it}$ with $c\ln(r_1/r_2) = 2\pi k$, which is the same as
$$r_1 = e^{2\pi k/c} r_2$$
or equivalently
$$z = e^{2\pi k/c} w.$$
Now let $c = 2\pi/\ln(\delta)$; then we have $f(w) = f(z)$ if and only if $z = \delta^k w$ (which is to say $z = \phi^k(w)$) for some $k \in \mathbb{Z}$. □

15.1. Functions Invariant under Group Elements

We proved Theorem 15.0.2 by simply giving an example of an invariant function (after using a change of variables to put our automorphism into a standard form). That worked because the examples were fairly obvious; in this section we will digress and say a bit about more general ways of obtaining functions invariant under the action of elements of a group.

Just to emphasize that the sort of techniques we are considering in the section work in much greater generality we temporarily change the setting: Suppose that X is a set and V is a vector space of (complex-valued) functions defined on X. Suppose that G is a group of bijections of X: This means that each element of G is a one-to-one map from X onto X, and that the group operation in G is given by composition.

Suppose further that V is G-invariant: This means that $f \circ \phi \in V$ for every $f \in V$ and $\phi \in G$.

To begin, suppose that $\phi \in G$ has finite order n. Here is the "right" way to construct an $f \in V$ such that $f \circ \phi = f$: Choose any function $g \in V$ and let
$$f = \sum_{j=0}^{n-1} g \circ \phi^j = g + g \circ \phi + \cdots + g \circ \phi^{n-1}.$$
The fact that ϕ^n is the identity shows that $f \circ \phi = f$:
$$f \circ \phi = g \circ \phi + \cdots + g \circ \phi^{n-1} + g \circ \phi^n = g + g \circ \phi + \cdots + g \circ \phi^{n-1} = f.$$
So: If we define
$$Tg = \sum_{j=0}^{n-1} g \circ \phi^j$$

then every function f of the form $f = Tg$ for some $g \in V$ satisfies $f \circ \phi = f$. This is the "right" construction of functions that are invariant under ϕ because it is easy to see that every function invariant under ϕ arises this way: If $f \circ \phi = f$ then $f = Tg$ for $g = f/n$.

You shouldn't jump to conclusions: The argument in the previous paragraph does not prove that there *is* a nonconstant $f \in V$ with $f \circ \phi = f$; it may well happen that the only f invariant under ϕ are constant. But if there *are* any nonconstant ϕ-invariant functions in V then that construction is one way to get them.

What does this construction look like for $G = \text{Aut}(\mathbb{D})$? Suppose that $\phi \in \text{Aut}(\mathbb{D})$ has finite order n. Then results in the previous section show that ϕ must be elliptic, and hence that we may assume that ϕ is a rotation: $\phi(z) = \alpha z$, where $\alpha \in \mathbb{C}$ is such that $|\alpha| = 1$, $\alpha^n = 1$ and $\alpha^m \neq 1$ for $0 < m < n$.

Now one calculates that

$$\sum_{j=0}^{n-1} \alpha^{jm} = \begin{cases} n & (n \mid m), \\ 0 & (n \nmid m). \end{cases}$$

So if $g(z) = z^k$ we have $Tg = 0$ unless k is a multiple of n, in which case $Tg = ng$. It follows that if $g \in H(\mathbb{D})$ has a power series $g(z) = \sum_{k=0}^{\infty} c_k z^k$ then

$$Tg(z) = n \sum_{j=0}^{\infty} c_{jn} z^{jn}.$$

In particular $f(z) = z^n$ really is a "natural" example of an f with $f \circ \phi = f$, and it's really the simplest possible example of a nonconstant f with this property.

Switching back to the general situation where G is a group of bijections on the set X and V is a G-invariant vector space of functions on X: How would we construct an example of $f \in V$ with $f \circ \phi = \phi$ if $\phi \in G$ does not have finite order? A natural thing to try would be to choose $g \in V$ and then set

$$f = Tg = \sum_{j=-\infty}^{\infty} g \circ \phi^j.$$

This is of course much more problematic than the previous case, because the sum need not converge; this is more a "plan" than an actual "construction". But if little details like convergence work out then it is clear we should have $f \circ \phi = f$.

To give a specific example, let us consider a typical parabolic element $\phi \in \text{Aut}(\Pi^+)$, given by $\phi(z) = z + 1$. Then the general construction above

15.1. Functions Invariant under Group Elements

becomes
$$Tg(z) = \sum_{n=-\infty}^{\infty} g(z+n).$$

We need to impose some condition on g to make this converge; if we assume, for example, that there exists a continuous function $c : (0, \infty) \to (0, \infty)$ such that
$$|g(x+iy)| \leq \frac{c(y)}{1+x^2}$$
then the sum will converge to a function $f = Tg$ which will indeed satisfy
$$f(z+1) = f(z).$$

And (pretending that we don't already know the answer) it is not hard to see that this construction can be used to give a *nonconstant* bounded function f with $f(z+1) = f(z)$; for example if $g(z) = 1/(z+i)^2$ then it is not hard to show that f is bounded in Π^+ but has a pole at $-i$, and hence is nonconstant.

It is also possible to show that the infinite sum defining $f = Tg$ necessarily leads to a function of the form
$$f(z) = \sum_{n+-\infty}^{\infty} c_n e^{2\pi i n z},$$
by an argument analogous to our calculation with power series for an elliptic transformation above. Alas this would require an inappropriate amount of Fourier analysis, so we will not say any more about it here. (If you are familiar with the so-called "Poisson summation formula" you should see that it is relevant here.)

It is amusing to note that one can actually find the exact value of that infinite series in the special case $g(z) = 1/(z+i)^2$. Dropping the i to simplify the formula:

- **Exercise 15.1.** Using Theorem 6.1.2, determine the value of $\sum_{n=-\infty}^{\infty} \frac{1}{(z+n)^2}$ for $z \notin \mathbb{Z}$.

We can also show that any $g \in H(\Pi^+)$ with period 1 must have a series representation as above, using a little "complex" trick:

- **Exercise 15.2.** Suppose that $g \in H(\Pi^+)$ and $g(z+1) = g(z)$ for all $z \in \Pi^+$. Then g has an expansion
$$g(z) = \sum_{n=-\infty}^{\infty} c_n e^{2\pi i n z} \quad (z \in \Pi^+).$$

Hint: Show that there exists a function $G \in H(\mathbb{D}')$, where $\mathbb{D}' = \mathbb{D}\setminus\{0\}$ is the punctured disk, such that $g(z) = G(e^{2\pi i z})$. If you see why there should be such a G but you don't know quite how to state the proof precisely, you should try again after reading the next chapter on analytic continuation. Now the Laurent series for G gives the result.

Similarly for a hyperbolic automorphism:

- **Exercise 15.3.** Suppose that $\delta > 0$, $\delta \neq 1$, and $g \in H(\Pi^+)$ satisfies $g(z) = g(\delta z)$ for all z. Let $c = 2\pi/\ln(\delta)$, and show that g has an expansion of the form
$$g(z) = \sum_{n=-\infty}^{\infty} a_n z^{inc}.$$

Hint: The argument is similar to the previous exercise, using a certain mapping from Π^+ onto an annulus instead of the mapping from Π^+ onto the punctured disk. See the proof of Theorem 15.0.2.

Finally, let us consider the case of an elliptic automorphism of infinite order: Suppose that $\phi \in \text{Aut}(\mathbb{D})$ is of the form $\phi(z) = \alpha z$, where $\alpha^n \neq 1$ for any positive integer n. Then there is no nonconstant holomorphic function in the disk which is ϕ-invariant. What goes wrong with the sort of construction we've used here?

Suppose that $g \in H(\mathbb{D})$. We could try
$$f = Tg = \sum_{j=-\infty}^{\infty} g \circ \phi^j$$

as before, but it is clear why that doesn't work: The sum doesn't converge. It's not even close to converging; for example we get
$$f(0) = \sum_{j=-\infty}^{\infty} g(0).$$

Here is something else we could try (it works in other situations):
$$f = Tg = \lim_{n \to \infty} \frac{1}{n} \sum_{j=0}^{n-1} g \circ \phi^j.$$

In general, if that limit exists it should define a ϕ-invariant function; this is because one would typically have
$$\left(\frac{1}{n}\sum_{j=0}^{n-1} g \circ \phi^j\right) \circ \phi - \frac{1}{n}\sum_{j=0}^{n-1} g \circ \phi^j = \frac{1}{n}(g \circ \phi^n - g) \to 0.$$

15.1. Functions Invariant under Group Elements

We leave it as an exercise to figure out what goes wrong here when $\phi(z) = \alpha z$ as above: Does the limit fail to exist, does the limit exist but define a function which is not ϕ-invariant or what?

- **Exercise 15.4.** What goes wrong here? (This is the same as another exercise in a different chapter, in different notation.)

Chapter 16

Analytic Continuation

16.0. Introduction

Generally speaking the phrase "analytic continuation" refers to the process of extending a holomorphic function defined in an open set to some larger domain. The original and final domains are typically taken to be connected (because, for example, the "problem" of extending a function $f \in H(D(0,1))$ to a function holomorphic in $D(0,1) \cup D(42,1)$ is stupid: You choose any $g \in H(D(42,1))$ you wish and define F to equal f on $D(0,1)$ and g on $D(42,1)$.)

The domains of the original and extended function are typically taken to be connected, but the domain of the extended function is not always a subset of the plane (and to be precise, the "extension" is not always literally an extension; the original domain is not always literally a subset of the domain of the extended function. Why the "extension" nonetheless deserves to be called an "extension" will become clear below; it is analogous to the fact that (with the standard definitions) the real numbers are not literally a subset of the complex numbers, but everybody always talks as though they were.)

There are various ways one might show that a holomorphic function can be extended to a larger domain. The Schwarz Reflection Principle (Theorem 10.7.1) is an important example: If an open set is symmetric about the real axis then every function holomorphic in the "upper half" of the set with imaginary part tending to 0 on the real axis extends to the entire set. (We should note again that there is no hypothesis about the real part being continuous up to the boundary in Theorem 10.7.1; that is part of the conclusion instead.)

Sometimes one can show that a function satisfies a "functional equation" and then use that to do the extending. A simple example of this is given by the definition of the so-called "Gamma function": For $\operatorname{Re}(z) > 0$ one defines

$$\Gamma(z) = \int_0^\infty e^{-t} t^{z-1} \, dt.$$

This integral does not converge for $\operatorname{Re}(z) \leq 0$, but one can nonetheless define $\Gamma(z)$ for other values of z as follows: One begins by proving the functional equation

$$\Gamma(z+1) = z\Gamma(z)$$

for $\operatorname{Re}(z) > 0$; this is just an integration by parts, with a little care regarding the boundary terms. Now one defines $\Gamma(z)$ for $-1 < \operatorname{Re}(z) \leq 0$ ($z \neq 0$) using the formula $\Gamma(z) = \Gamma(z+1)/z$; then the same formula extends the definition to $-2 < \operatorname{Re}(z) \leq -1$ ($z \neq -1$), and so on, and one obtains a function Γ holomorphic in $\mathbb{C} \setminus \{0, -1, -2, \dots\}$ satisfying the same functional equation $\Gamma(z+1) = z\Gamma(z)$. It is easy to see that $\Gamma(1) = 1$; it follows from this and the functional equation that Γ actually has a pole at each nonpositive integer. It also follows that $n! = \Gamma(n+1)$ for $n = 0, 1, \dots$. (See Chapter 24 for much more on the Gamma function.)

People often speak of defining some special function by some integral or sum, then extending the definition to a larger domain "by analytic continuation". When one hears people speaking this way one can get the idea that there is some machine that does the extending automatically, when in fact typically the extension is done using a functional equation or something similar; there is nothing automatic about coming up with that functional equation.

It may happen that $f \in H(D)$ cannot be extended to any strictly larger connected open set; in this case ∂D is said to be the *natural boundary* for f. Exercise 13.1 says that for any connected open set D there exists $f \in H(D)$ having ∂D as its natural boundary.

There are many different ways one could do that exercise; one way is to construct a function which is discontinuous at every point of the boundary. We should mention that a holomorphic function which cannot be extended to a larger connected set need not be discontinuous; in fact it can be infinitely smooth up to the boundary.

For example, define $f \in H(\mathbb{D})$ by

$$f(z) = \sum_{n=1}^\infty 2^{-n^2} z^{2^n}.$$

We leave it to the reader to verify that this power series has radius of convergence 1, so that the series defines a function holomorphic in $\mathbb{D}$

which cannot be extended to be holomorphic in $D(0,r)$ for any $r > 1$. (See Chapter 1 for the "right" characterization of the radius of convergence.) We also leave it to the reader to verify that every derivative $f^{(k)}$ extends to a continuous function on $\overline{\mathbb{D}}$.

The part that is not just a routine calculation (unless you have some experience with "lacunary" power series) is this: Because of the special form of the power series for f, the fact that f cannot be extended to any larger disk centered at the origin actually implies that f cannot be extended to any larger connected open set! Note first that if n is a positive integer then there exists a polynomial P_n such that

$$f(e^{2\pi i/2^n} z) = P_n(z) + f(z).$$

Now suppose that f extends to a connected open set strictly containing $\mathbb{D}$. It follows that f extends to a neighborhood of some arc on the boundary of the unit circle. The magic formula above shows that if we rotate this arc through an angle $2\pi/2^n$ then f also extends to a neighborhood of the rotated arc. Now if we choose n large enough that $2\pi/2^n$ is less than the length of this arc and rotate repeatedly we cover the entire unit circle: It follows that f extends to a neighborhood of the unit circle, and hence to a disk $D(0, r)$ with $r > 1$, contradicting the fact that the radius of convergence is 1.

Back to our story: Given a function element (f, D) (see the next section for the definition) there is always a largest extension, that is, an extension which in some sense contains all the extensions of (f, D). We have seen a few examples where (f, D) is its own largest extension, but the situation is not always that simple; in fact the largest extension often cannot be realized as a holomorphic function in an open subset of the plane. It will take a few pages to say what the largest extension actually *is* in such a case; in this section we simply give an example showing why the largest extension, whatever it is, cannot be an ordinary holomorphic function in the plane:

Suppose that $f \in H(D(1,1))$ is defined to be the principal-value logarithm restricted to $D(1,1)$. There are many different ways to extend f to a larger open set; in fact there are many different *maximal* extensions (maximal among extensions to open sets in the plane, that is). One example is the principal-value logarithm itself: We set $D_1 = \mathbb{C} \setminus (-\infty, 0]$ and we define $f_1 \in H(D_1)$ by

$$f_1(re^{it}) = \ln(r) + it \quad (r > 0, \ -\pi < t < \pi.)$$

This is a maximal extension: If $D_1 \subset D \subset \mathbb{C}$, D is open, $g \in H(D)$ and $g|_{D_1} = f_1$ then $D = D_1$ and $g = f_1$.

But although f_1 is a maximal extension of f it is not a largest extension; for example if D_2 is the plane with the negative imaginary axis removed and we define $f_2 \in H(D_2)$ by

$$f_2(re^{it}) = \ln(r) + it \quad (r > 0, \; -\pi/2 < t < 3\pi/2)$$

then f_2 is an extension of f but f_1 is not an extension of f_2, so f_1 is not a largest extension. So if we want a holomorphic function in an open set to have a largest extension (we do and it does) we need to talk about some sort of extension other than extensions to larger subsets of the plane.

We will temporarily abandon the notion of extending functions to larger open sets and instead discuss the notion of the "continuation" of a function along a curve. The set of all continuations along curves (modulo a certain equivalence relation) will turn out to give the largest extension. (We will then see that under certain circumstances these continuations along curves can give an extension to a larger open set in the plane.)

16.1. Continuation along Curves

We are about to encounter a jump discontinuity in the level of abstraction; instead of talking about functions we will be talking about sets of functions and families of sets of classes of functions ... The reader who feels overwhelmed at some point in this chapter should be assured that the material here is really not all that difficult; we have already covered much harder material! The main difficulty is in keeping all the definitions straight — most of the ideas in most of the proofs were already encountered in the proof of Lemma 4.6 (the reader might want to review that proof as preparation for the current material, in fact).

But there will be an unavoidably large number of somewhat abstract definitions. We begin. With a definition, as it happens:

A *function element* is an ordered pair (f, D), where D is a nonempty connected open subset of the plane and $f \in H(D)$.

This seems a little curious, because if we think of a function as a set of ordered pairs then a function determines its domain: If f is a function with domain D then D is equal to the set of x such that $(x, y) \in f$ for some y. Hence there is no reason why a function element needs to be an ordered pair; simply saying a function element was a holomorphic function in a connected open subset of the plane would serve the same purpose. The notation and terminology goes back to an earlier notion of "function". Having said that, we should say that it is precisely in studying holomorphic functions where the earlier notion makes a lot of sense: If we define $f \in H(D(0,1))$ by $f(z) = \sin(z)$ and we define $g \in H(D(2,1))$ by $g(z) = \sin(z)$ then f and g are officially different functions, but they are both restrictions of one entire function to two different disks. From an older point of view f and g would not be different functions at all. For that matter they would probably look like the same function to most calculus students at present, and this is not because calculus students are stupid; if a function is a formula or a "rule" then f and g *are* the same function. Looking at things this way makes less sense when we start talking about nonanalytic functions, and eventually the modern general notion of a function as a set of ordered pairs developed.

Anyway, a function element is an ordered pair (f, D) where D is a nonempty connected open subset of the plane and $f \in H(D)$. Fix a point $a \in \mathbb{C}$ for a second, and consider the class of all function elements (f, D) with $a \in D$. We define an equivalence relation $\equiv_a$ on this set of function elements by saying that

$$(f, D) \equiv_a (g, E)$$

if $f = g$ in some neighborhood of a. An equivalence class with respect to $\equiv_a$ is a *germ of a holomorphic function at* a; we may call it just a *germ at* a, since (with a few exceptions) these are the only germs we will be talking about. If (f, D) is a function element with $a \in D$ then the $\equiv_a$ equivalence class containing (f, D) is the *germ of f at a*, and will be denoted $[f]_a$. The set of all germs at a will be denoted $\mathcal{O}_a$. If $D \subset \mathbb{C}$ is a connected open set we define

$$\mathcal{O}(D) = \bigcup_{a \in D} \mathcal{O}_a;$$

we will refer to $\mathcal{O}(D)$ as the sheaf of germs of holomorphic functions in D. (We may write $\mathcal{O} = \mathcal{O}(\mathbb{C})$.) Note that we will not be saying exactly what a "sheaf" is in general; we use the term here only because $\mathcal{O}(D)$ *is* in fact known as the sheaf of germs of holomorphic functions in D, so it would be silly to concoct another name for it.

Note that this is a disjoint union. If $a, b \in \mathbb{C}$ with $a \neq b$ then $\mathcal{O}_a \cap \mathcal{O}_b = \emptyset$, because a germ at a cannot also be a germ at b: If $g \in \mathcal{O}_a$ then g determines a by the formula

$$\{a\} = \bigcap_{(f,D) \in g} D,$$

and hence $[f_1]_a = [f_2]_b$ implies that $a = b$.

The fact that $\mathcal{O}_a \cap \mathcal{O}_b = \emptyset$ for $a \neq b$ means that we can define a map $p : \mathcal{O}(D) \to D$ by

$$p(g) = a \quad (g \in \mathcal{O}_a).$$

(That's p for "projection"; one thinks of the germ g as lying "above" the point $p(g) \in \mathbb{C}$.)

If $g \in \mathcal{O}_a$ then the value of g at a is defined by $g(a) = f(a)$, if (f, D) is any element of g; similarly the derivatives of g at a are defined by $g^{(k)}(a) = f^{(k)}(a)$. These are both well defined, since if $(f_1, D_1), (f_2, D_2) \in g$ then $f_1 = f_2$ in a neighborhood of a.

Suppose (f_1, D_1) and (f_2, D_2) are function elements with $a \in D_1 \cap D_2$. You should verify that all the following are equivalent:

(i) $(f_1, D_1) \equiv_a (f_2, D_2)$,

(ii) $(f_1, D_1) \in [f_2]_a$,

(iii) $(f_2, D_2) \in [f_1]_a$,

(iv) $[f_1]_a = [f_2]_a$.

Also:

- **Exercise 16.1.** Suppose that $g, h \in \mathcal{O}_a$. Show that $g = h$ if and only if
$$g^{(k)}(a) = h^{(k)}(a) \quad (k = 0, 1, \dots).$$

(This is primarily an exercise in boiling down the definitions, plus a tiny bit of complex analysis, so you should be explicit about reducing the statement to what it "really means".)

A *path from a to b in D* is a continuous map $\gamma : [0, 1] \to D$ such that $\gamma(0) = a$ and $\gamma(1) = b$; a path from a to b in $\mathbb{C}$ will be called just a path from a to b.

Suppose now that (f, D) is a function element, $a \in D$ and γ is a path from a to b. A *continuation of (f, D) along γ* is a family of function elements $(f_t, D_t)_{0 \le t \le 1}$ with the following properties:

(i) $(f_0, D_0) = (f, D)$,

(ii) $\gamma(t) \in D_t$ for all $t \in [0, 1]$,

(iii) for every $t \in [0, 1]$ there exists $\delta > 0$ such that if $s \in [0, 1]$ and $|s - t| < \delta$ then $[f_s]_{\gamma(s)} = [f_t]_{\gamma(s)}$.

If there exists a continuation of (f, D) along γ we will say that (f, D) can be continued along γ. (Or that (f, D) admits continuation along γ.)

Suppose that $(f_t, D_t)_{0 \le t \le 1}$ is a continuation of (f, D) along γ. The condition that $[f_s]_{\gamma(s)} = [f_t]_{\gamma(s)}$ for s near t says that if s is near t then $f_s = f_t$ near $\gamma(s)$. This may make analytic continuation along curves seem a little boring; we continue the function by extending it to exactly the same function. Indeed there is really not much going on here if we look at things locally, but things can be interesting globally. For example:

Let $D = D(1, 1)$, and let $f \in H(D)$ denote the principal-value logarithm restricted to D:
$$f(re^{it}) = \ln(r) + it \quad (re^{it} \in D(1, 1), -\pi < t < \pi).$$

16.1. Continuation along Curves

Now define a path γ from 1 to 1 by $\gamma(t) = e^{2\pi it}$. The function element (f, D) can be continued along γ, thus: Let $D_t = D(\gamma(t), 1)$, and let $f_t \in H(D_t)$ be the branch of the logarithm defined by

$$f_t(re^{i\theta}) = \ln(r) + i\theta \quad (re^{i\theta} \in D_t,\ 2\pi t - \pi < \theta < 2\pi t + \pi).$$

Now if $|s - t|$ is small then $f_s = f_t$ near $\gamma(s)$, but nonetheless we have $f_{2\pi} = f_0 + 2\pi i$.

The first thing we prove about the analytic continuation of a function element along a curve is that it is unique if it exists. Well, that cannot be literally true: Suppose that $(f_t, D_t)_{0 \le t \le 1}$ is a continuation of (f, D) along the curve γ. If we choose connected open sets D'_t such that $\gamma(t) \in D'_t \subset D_t$ for $0 \le t \le 1$ then $(f_t|_{D'_t}, D'_t)_{0 \le t \le 1}$ is another continuation of (f, D), which is not literally the same as the first one. But the continuation is essentially unique, in that any two continuations have the same germs:

Lemma 16.1.0. *Suppose that (f, D) is a function element, $a \in D$, and γ is a path from a to b. If $(f_t, D_t)_{0 \le t \le 1}$ and $(g_t, E_t)_{0 \le t \le 1}$ are continuations of (f, D) along γ then*

$$[f_1]_b = [g_1]_b$$

(that is, $f_1 = g_1$ in some neighborhood of b).

Proof. The idea is that we know that the two continuations are the same at $t = 0$ and it is easy to see that if the two continuations are the same at a given value of t then they are the same for slightly larger t; continuing this we see that they are the same everywhere. As always the way we make a proof out of that idea is via a connectedness argument:

Let

$$A = \{\, t \in [0, 1] : [f_t]_{\gamma(t)} = [g_t]_{\gamma(t)} \,\}.$$

We need to show that $A = [0, 1]$; since $[0, 1]$ is connected, it suffices to show that A is nonempty and (relatively) open and closed in $[0, 1]$.

It is clear that $0 \in A$, since in fact $f_0 = f = g_0$.

And it is easy to show that A is open: Suppose that $t \in A$. Let V be the component of $D_t \cap E_t$ containing $\gamma(t)$. By definition of analytic continuation along a curve and the continuity of γ there exists $\delta > 0$ such that if $s \in [0, 1] \cap (t - \delta, t + \delta)$ then $\gamma(s) \in V$, $[f_t]_{\gamma(s)} = [f_s]_{\gamma(s)}$ and $[g_t]_{\gamma(s)} = [g_s]_{\gamma(s)}$. But the fact that $t \in A$ shows that $[f_t]_{\gamma(t)} = [g_t]_{\gamma(t)}$, which is to say that $f_t = g_t$ in a neighborhood of $\gamma(t)$. Since V is connected, it follows that $f_t = g_t$ in all of V, and so if $s \in [0, 1] \cap (t - \delta, t + \delta)$ we have $[f_t]_{\gamma(s)} = [g_t]_{\gamma(s)}$; combining this with the previous equalities shows that $[f_s]_{\gamma(s)} = [g_s]_{\gamma(s)}$, so $s \in A$.

Now more or less the same argument shows that the complement of A is also open: Suppose that $t \in [0,1] \setminus A$. As before, let V be the component of $D_t \cap E_t$ containing $\gamma(t)$, and choose $\delta > 0$ such that if $s \in [0,1] \cap (t-\delta, t+\delta)$ then $\gamma(s) \in V$, $[f_t]_{\gamma(s)} = [f_s]_{\gamma(s)}$ and $[g_t]_{\gamma(s)} = [g_s]_{\gamma(s)}$. Now the fact that t is not in A shows that $[f_t]_{\gamma(t)} \neq [g_t]_{\gamma(t)}$, which implies that f_t and g_t cannot agree on any nonempty open subset of V. Supposing that $s \in [0,1] \cap (t-\delta, t+\delta)$, it follows that $[f_t]_{\gamma(s)} \neq [g_t]_{\gamma(s)}$ and hence that $[f_s]_{\gamma(s)} \neq [g_s]_{\gamma(s)}$, so $s \notin A$. □

It is of some interest to figure out how much of this analytic continuation stuff depends on complex analysis, and how much is just topology. In that regard we should note the following: We could define a "continuous function element" to be the same thing as a function element, except that the function was merely required to be continuous instead of holomorphic; then we could define "continuous continuation along a curve" in a manner analogous to the way we defined analytic continuation. It is clear that the preceding lemma is false for "continuous continuation"; this raises the question of where the proof fails. Curiously, the fact that A is open would go through in exactly the same way, but we would be unable to show that A was closed if we were talking about continuous continuation. (We will attempt to clarify what is really going on here in Section 16.4.)

So analytic continuation along a curve is unique, in the strongest possible sense, given that we defined the continuation in terms of function elements. (We could just as well have defined the continuation of a germ along a curve to be a family of germs, in which case the continuation would be literally unique.) The next lemma shows that continuation of a function element along a curve also depends continuously on the curve, in a very strong sense.

Lemma 16.1.1. *Suppose that (f, D) is a function element, $a \in D$, γ is a path from a to b, and $(f_t, D_t)_{0 \leq t \leq 1}$ is a continuation of (f, D) along γ. There exists $\delta > 0$ with the following property: If σ is another curve from a to b with $|\sigma(t) - \gamma(t)| < \delta$ for all t then (f, D) can be continued along σ as well; furthermore, if $(g_t, E_t)_{0 \leq t \leq 1}$ is a continuation of (f, D) along σ then $[g_1]_b = [f_1]_b$.*

We should make some comments first about why this is obvious, then why the obvious argument doesn't quite work. We make a temporary definition: If $\delta > 0$ and $(f_t, D_t)_{0 \leq t \leq 1}$ is a continuation of (f, D) along γ, let us say that it is a δ-*continuation* if $D(\gamma(t), \delta) \subset D_t$ for all $t \in [0,1]$.

The lemma is actually sort of obvious if you look at it right; if σ is close enough to γ then the very same function elements that give a continuation of (f, D) along γ should also give a continuation of (f, D) along σ. More

16.1. Continuation along Curves

or less: To make this literally correct we need to know that the continuation we started with is a δ-continuation for some $\delta > 0$, and not every continuation satisfies this property. (It is clear that there are continuations which are not δ-continuations for any $\delta > 0$, simply because as noted above if $(f_t, D_t)_{0 \leq t \leq 1}$ is a continuation of (f, D) along γ and we choose connected open sets D'_t such that $\gamma(t) \in D'_t \subset D_t$ for $0 \leq t \leq 1$ then $(f_t|_{D'_t}, D'_t)_{0 \leq t \leq 1}$ is another continuation of (f, D); we can certainly choose the D'_t so that the new continuation is not a δ-continuation for any $\delta > 0$.)

The proof of the lemma fixes the problem by showing that if a function element has a continuation along a curve then it also has a δ continuation along the same curve, for some $\delta > 0$.

Proof of Lemma 16.1.1. For every $t \in [0,1]$ there exists $\rho_t > 0$ such that if $s \in [0,1] \cap (t - \rho_t, t + \rho_t)$ then $\gamma(s) \in D_t$ and $[f_s]_{\gamma(s)} = [f_t]_{\gamma(s)}$. The interval $[0,1]$ is covered by finitely many intervals of the form $(t - \rho_t, t + \rho_t)$; we may discard (one at a time) any which lie in the union of the others, etc.; we finally obtain a finite sequence $(t_j)_{j=0}^n$ with $0 = t_0$, $t_j \leq t_{j+1}$, $t_n = 1$, such that if $I_j = [t_{j-1}, t_j]$ ($1 \leq j \leq n$) then there exists $s_j \in I_j$ such that $\gamma(I_j) \subset D_{s_j}$ for $1 \leq j \leq n$, and such that we also have $[f_t]_{\gamma(t)} = [f_{s_j}]_{\gamma(t)}$ for $t \in I_j$. Now, since $\gamma(I_j)$ is a compact subset of D_{s_j} and we are considering only finitely many values of j, we see that there exists $\delta > 0$ with this property:
$$D(\gamma(t), \delta) \subset D_{s_j} \quad (t_{j-1} \leq t \leq t_j, \ 1 \leq j \leq n).$$

Suppose that σ is another curve from a to b with $|\sigma(t) - \gamma(t)| < \delta$ for all $t \in [0,1]$. Let $E_t = D_{s_j}$ and $g_t = f_{t_j}$ for $t_{j-1} \leq t < t_j$, $1 \leq j \leq n$ (and set $E_1 = D_{s_n}$, $g_1 = f_{t_n}$). It is easy to see that $(g_t, E_t)_{0 \leq t \leq 1}$ is an analytic continuation of (f, D) along σ. $\square$

- **Exercise 16.2.** Finish the proof of Lemma 16.1.1: Show carefully that $(g_t, E_t)_{0 \leq t \leq 1}$ is an analytic continuation of (f, D) along σ.

An important consequence of the lemma involves analytic continuation along two homotopic paths. We need a flavor of homotopy that we have not previously defined:

Suppose that γ_0 and γ_1 are two paths from a to b. We say that γ_0 and γ_1 are *path-homotopic* if there is a continuous family of paths from a to b connecting γ_0 and γ_1. That is, γ_0 is path-homotopic to γ_1 if there exists a continuous function $\Gamma : [0,1] \times [0,1] \to \mathbb{C}$ such that $\Gamma(s, 0) = \gamma_0(s)$ for all s, $\Gamma(s, 1) = \gamma_1(s)$ for all s, $\Gamma(0, t) = a$ for all t and $\Gamma(1, t) = b$ for all t; we will often write $\gamma_t(s) = \Gamma(s, t)$ and refer to the family $(\gamma_t)_{0 \leq t \leq 1}$ as a path homotopy of γ_0 and γ_1. If γ_0 and γ_1 are paths in an open set D then we say that they are path-homotopic in D if there exists $\Gamma : [0,1] \times [0,1] \to D$

satisfying the above conditions. (Note that if we say "homotopic" without the "path-" we are referring to the previously defined notion of homotopy for closed curves.)

This notion of homotopy has a lot to do with the previous one. If γ_0 and γ_1 are two paths from a to b then $\gamma_0 - \gamma_1$ will denote the closed curve which starts at a, follows γ_0 from a to b and then follows γ_1 from b back to a; formally,
$$\gamma_0 - \gamma_1 : [0,1] \to \mathbb{C}$$
is defined by
$$(\gamma_0 - \gamma_1)(t) = \begin{cases} \gamma_0(2t) & (0 \leq t \leq 1/2), \\ \gamma_1(2 - 2t) & (1/2 \leq t \leq 1). \end{cases}$$

- **Exercise 16.3.** Suppose that γ_0 and γ_1 are paths from a to b in D. Show that γ_0 and γ_1 are path-homotopic in D if and only if the closed curve $\gamma_0 - \gamma_1$ is null-homotopic in D (that is, homotopic in D to a constant).

Hint: One approach would be to begin by showing that the closed curve $\gamma : [0,1] \to D$ is null-homotopic in D if and only if there exists a continuous map $\Gamma : \overline{\mathbb{D}} \to D$ such that $\gamma(t) = \Gamma(e^{2\pi i t})$ for all $t \in [0,1]$.

It follows from the exercise that the open set $D \subset \mathbb{C}$ is simply connected if and only if for every $a, b \in D$ there exists a path from a to b in D and any two such paths are path-homotopic in D.

The reason we care about path homotopy is the following theorem. You should note that Theorem 16.1.2 is sometimes called the Monodromy Theorem. (We will reserve that term for a corollary in Section 16.3.)

Theorem 16.1.2. *Suppose that D and G are open subsets of $\mathbb{C}$, $D \subset G$, $a \in D$, $b \in G$, and γ_0 and γ_1 are two paths in G from a to b which are path-homotopic in G. Suppose that $(\gamma_s)_{0 \leq s \leq 1}$ is a path homotopy of γ_0 and γ_1. Suppose that (f, D) is a function element which can be continued along each path γ_s, $0 \leq s \leq 1$. If $(f_t, D_t)_{0 \leq t \leq 1}$ is a continuation of (f, D) along γ_0 and $(g_t, E_t)_{0 \leq t \leq 1}$ is a continuation of (f, D) along γ_1 then*
$$[f_1]_b = [g_1]_b.$$

Proof. This is immediate from Lemma 16.1.1. In detail:

For $0 \leq s \leq 1$ let $(f_t^s, D_t^s)_{0 \leq t \leq 1}$ be a continuation of (f, D) along γ_s. Let $A = \{ s \in [0,1] : [f_1^s]_b = [f_1^0]_b \}$. The lemma shows that A is (relatively) open. The lemma also shows that $[0,1] \setminus A$ is relatively open: If $s \in [0,1] \setminus A$ then the lemma shows that there exists $\delta > 0$ such that $[f_1^{s'}]_b = [f_1^s]_b$ for

$|s - s'| < \delta$, implying that $s' \in [0,1] \setminus A$. Since $0 \in A$ and $[0,1]$ is connected, it follows that $1 \in A$. $\square$

The following proposition will be very useful when we start trying to prove that function elements can be continued along curves:

Proposition 16.1.3. *Suppose that (f, D) is a function element, $a \in D$, and γ is a path from a to $b \in \mathbb{C}$. If (f, D) cannot be continued along γ then there exists $t_0 \in (0, 1]$ such that (f, D) can be continued along $\gamma|_{[0,t_0)}$ but (f, D) cannot be continued along $\gamma|_{[0,t_0]}$.*

(The notion of "continuation along γ" was defined only for paths γ with parameter interval $[0, 1]$. We will let the reader guess what it means to continue (f, D) along $\gamma|_{[0,t_0)}$ or $\gamma|_{[0,t_0]}$.)

Proof of Proposition 16.1.3. We define t_0 in the obvious way:

$$t_0 = \sup\{\, t \in (0, 1] : (f, D) \text{ can be continued along } \gamma|_{[0,t]} \,\}.$$

We need to show that (f, D) cannot be continued along $\gamma|_{[0,t_0]}$. But this is clear, just like all this stuff is clear once you know what to try to prove:

First, note that if $t_0 = 1$ then $\gamma|_{[0,t_0]} = \gamma$, so the hypothesis implies that (f, D) cannot be continued along $\gamma|_{[0,t_0]}$. Suppose on the other hand that $t_0 < 1$. Then if we could continue (f, D) along $\gamma|_{[0,t_0]}$ it would follow immediately from the definition of analytic continuation that we could continue (f, D) along $\gamma|_{[0,t_0+\delta]}$ for some $\delta > 0$, which contradicts the choice of t_0. $\square$

Very convenient: If we want to construct a continuation $(f_t, D_t)_{0 \le t \le 1}$ of (f, D) along γ we can assume that $0 < t_0 \le 1$ and that $(f_t, D_t)_{0 \le t < t_0}$ is a continuation of (f, D) along $\gamma|_{[0,t_0)}$, and we need only show how to choose (f_{t_0}, D_{t_0}) so that $(f_t, D_t)_{0 \le t \le t_0}$ is a continuation of (f, D) along $\gamma|_{[0,t_0]}$. Almost like a proof by induction.

16.2. The Complete Analytic Function

Suppose that (f, D) is a function element. The *complete analytic function* of (f, D) (or "obtained from" (f, D)) is the set of all germs which can be obtained by analytic continuation of (f, D) along some curve.

More explicitly: Suppose that (f, D) is a function element, $a \in D$, γ is a path from a to b, and (f, D) has a continuation $(f_t, D_t)_{0 \le t \le 1}$ along γ. Then the germ $[f_1]_b$ is an element of the complete analytic function of (f, D), and every element of the complete analytic function of (f, D) is obtained this way.

The complete analytic function of (f, D) is the largest extension of (f, D) alluded to at the end of Section 16.0. It may not be clear why it deserves to be called an extension of (f, D), since it is not even a function. But there *is* an actual function associated with it in a natural way: Suppose that S is the complete analytic function of (f, D) and define a function $F : S \to \mathbb{C}$ by

$$F(g) = g(p(g)) \quad (g \in S).$$

(Tracing through the relevant definitions for readers who may feel lost: If g is a germ, that is, $g \in \mathcal{O}$, then $g \in \mathcal{O}_a$ for some $a \in \mathbb{C}$; we previously said that $p(g) = a$ in this case, and that $g(a) = f(a)$ for any $(f, D) \in g$ (that is, for any (f, D) such that $a \in D$ and $[f]_a = g$).)

Now F is an actual function, and everything there is to know about F is actually contained in S; this is the best excuse we can give for calling S a function. The function F is not literally an extension of (f, D), but it is essentially an extension: There is a natural embedding of D in S, via the mapping $\iota : D \to S$ defined by $\iota(z) = [f]_z$. For $z \in D$ we have $F(\iota(z)) = f(z)$, so if we identify D with $\iota(D) \subset S$ then f is identified with $F|_{\iota(D)}$ and F becomes an extension of f.

So F may naturally be regarded as an extension of (f, D), and hence S can also be so regarded. It is clear that F "contains" all the extensions of (f, D) so it is the largest extension. It is not clear at this point why F deserves to be called "holomorphic". We will explain this at some length in Section 16.4 (where we assume the reader knows a little bit of point-set topology). Briefly: There is a natural "topology" on $\mathcal{O}$, making S an open subset of $\mathcal{O}$. There is also a natural "complex structure" on $\mathcal{O}$, which is to say that $\mathcal{O}$ is a "complex manifold" in a natural way, so that there is a natural definition of "holomorphic function in an open subset of $\mathcal{O}$". Once we state all the relevant definitions it *will* be clear that $F : S \to \mathbb{C}$ is holomorphic.

16.3. Unrestricted Continuation — the Monodromy Theorem

Those complete analytic function things are elegant and interesting, as well as important, but we will not be using them for anything. On the other hand the notion of unrestricted continuation is something we are going to use in various applications:

Suppose that D and G are connected open subsets of $\mathbb{C}$ with $D \subset G$. If (f, D) is a function element which admits continuation along *every* path in G which starts at a point of D then we say that (f, D) *admits unrestricted continuation* in G.

16.3. Unrestricted Continuation — the Monodromy Theorem

Unrestricted continuation is going to be a hypothesis in the Monodromy Theorem. First we should give a few examples of situations where we might know that a function element admits unrestricted continuation. Suppose for the next few paragraphs that D and G are connected open subsets of $\mathbb{C}$ with $D \subset G$ and that (f, D) is a function element.

Example 0. The most obvious way (f, D) can admit unrestricted continuation in G is if the function actually *extends* to a holomorphic function in G: If $F \in H(G)$ and $F|_D = f$ then (f, D) admits unrestricted continuation in G. (If γ is any path in G starting at a point of D then $(f_t, D_t)_{0 \le t \le 1}$ is a continuation of (f, D) along γ, if we set $f_t = F$ and $D_t = G$ for $0 < t < 1$.)

This may be regarded as a trivial example of unrestricted analytic continuation. Given a function element that happens to admit unrestricted continuation in an open set Theorem 16.3.1 and Corollary 16.3.5 will give criteria for determining that the continuation is trivial in this sense, a question of great interest in many applications. (See also Proposition 16.3.7 for a special case with weaker hypotheses.)

Example 1. The second example is something we've known about for a long time: Antiderivatives admit unrestricted continuation. To be a little more precise: If $g \in H(G)$ and $f' = g$ in D then (f, D) admits unrestricted continuation in G (and each function element in a continuation is a local antiderivative for g).

If we were only talking about smooth curves we could construct the continuation of f along a curve just by integrating g. We could obtain the general case by approximating continuous curves by smooth curves, but this would involve a few details; it seems easier just to construct the continuation directly, using Proposition 16.1.3:

Suppose that $a \in D$ and γ is a path in G from a to b. Proposition 16.1.3 and the remarks following the proof show that we may assume as well that $0 < t_0 \le 1$ and we have already constructed (f_t, D_t) for $0 \le t < t_0$ so as to give a continuation of (f, D) along $\gamma|_{[0, t_0)}$; it is clear that we can also assume that $D_t \subset G$ for all $t \in [0, t_0)$. Note that $f'_t = g$ in D_t: This follows from the uniqueness of analytic continuation along a curve, since (as is easily shown) $(f'_t, D_t)_{0 \le t < t_0}$ is a continuation of (f', D) along $\gamma|_{[0, t_0)}$ and $(g|_{D_t}, D_t)$ is another continuation.

Choose $r > 0$ with $D(\gamma(t_0), r) \subset G$, and set $D_{t_0} = D(\gamma(t_0), r)$. Choose $\delta > 0$ so that $\gamma(t) \in D_{t_0}$ whenever $t_0 - \delta < t < t_0$, and fix $t_1 \in (t_0 - \delta, t_0)$. Now, since D_{t_0} is simply connected, the function g has an antiderivative in D_{t_0}: Thus we may choose $f_{t_0} \in H(D_{t_0})$ such that $f'_{t_0} = g$ and also $f_{t_0}(\gamma(t_1)) = f_{t_1}(\gamma(t_1))$.

It follows that $f_{t_1} = f_{t_0}$ near $\gamma(t_1)$, since the two functions have the same derivative and they agree at $\gamma(t_1)$. Now the uniqueness of analytic continuation shows that in fact $f_t = f_{t_0}$ near $\gamma(t)$, for all $t \in (t_0 - \delta, t_0)$, since $(f_{t_0}, D_t)_{t_0 - \delta \le t \le t_0}$ is another continuation of (f_{t_1}, D_{t_1}) along $\gamma|_{(t_0-\delta, t_0)}$. That is, $[f_t]_{\gamma(t)} = [f_{t_0}]_{\gamma(t)}$ for $t_0 - \delta < t < t_0$; thus we have an analytic continuation of (f, D) along $\gamma|_{[0, t_0]}$, as required.

(This stuff is all very simple and mostly obvious, honest. If it doesn't appear that way to you I assure you again that the problem is just the notation making things look much deeper than they really are — I don't know what to do about that without sacrificing precision. It would be a very useful exercise for you to re-read the last few paragraphs and convince yourself that all they really say is this: "It's *obvious* that antiderivatives admit unrestricted continuation: If you've continued an antiderivative along an initial segment of a path you can continue it a little farther by using an antiderivative defined in a disk that agrees with the antiderivatives you've already chosen at nearby points".

Similar comments apply to many if not most of the other arguments in this section and the next. I leave the "translations" to you; learning to look at a sea of notation and "see" what's really going on is an important skill.)

Example 2. Similarly harmonic conjugates admit unrestricted continuation: If $u : G \to \mathbb{R}$ is harmonic and $\operatorname{Re}(f) = u$ in D then (f, D) admits unrestricted continuation in G. The proof is the same as for the previous example (it is probably clear that there are many similar examples where one can show that a function element admits unrestricted continuation in an open set by the same argument).

Example 3. The third example of how a function element could admit unrestricted continuation in an open set takes a little space to describe precisely. The one-line description would be "local inverses for holomorphic covering maps admit unrestricted continuation".

Suppose that Ω and G are connected open subsets of the plane and $\phi : \Omega \to G$ is a holomorphic function with $\phi(\Omega) = G$. Suppose that every $z \in G$ has a connected neighborhood D with the property that

$$\phi^{-1}(D) = \bigcup_{\alpha \in A} \tilde{D}_\alpha,$$

where each $\tilde{D}_\alpha$ is an open subset of Ω, the $\tilde{D}_\alpha$ are pairwise disjoint, and $\phi|_{\tilde{D}_\alpha}$ is a homeomorphism from $\tilde{D}_\alpha$ onto D, for every $\alpha \in A$. Under these circumstances we say that ϕ is a *covering map* of G (or that ϕ covers G, or that Ω covers G, or Ω is a covering space of D, etc.) and that D is *elementary* for ϕ. Note that if D is elementary for ϕ then the sets $\tilde{D}_\alpha$ are

16.3. Unrestricted Continuation — the Monodromy Theorem

just the components of $\phi^{-1}(D)$. (Note as well that what we have defined here is just a very special case of a much more general notion of covering space.)

If D is elementary for ϕ, $z \in D$, $\tilde{z} \in \Omega$ and $\phi(\tilde{z}) = z$ then there exists a holomorphic function $f : D \to \Omega$ such that $f(z) = \tilde{z}$ and such that $\phi(f(w)) = w$ for all $w \in D$; the function element (f, D) will be called a *local inverse* for ϕ.

(Since $\tilde{z} \in \phi^{-1}(\{z\})$, there exists an $\alpha \in A$ with $\tilde{z} \in D_\alpha$. The fact that $\phi|_{D_\alpha}$ is a homeomorphism of D_α onto D shows that there exists a continuous function $f : D \to D_\alpha$ with $\phi(f(w)) = w$ for all $w \in D$, and it follows that f is holomorphic, by Theorem 7.5.)

Theorem 16.3.0. *Suppose that Ω and G are connected open subsets of the plane and $\phi : \Omega \to G$ is a holomorphic covering map. Any local inverse for ϕ admits unrestricted continuation in G; in fact if (f, D) is a local inverse for ϕ and γ is a path in G starting at a point of D then there is a continuation $(f_t, D_t)_{0 \le t \le 1}$ of (f, D) along γ such that each (f_t, D_t) is a local inverse for ϕ.*

(We cannot quite conclude that *every* analytic continuation of (f, D) as above consists of local inverses, just because there is no guarantee that $D_t \subset G$. But it follows from the essential uniqueness of analytic continuation that if $(f_t, D_t)_{0 \le t \le 1}$ is an analytic continuation of a local inverse for ϕ along γ then for every $t \in [0, 1]$ we have $\phi \circ f_t(z) = z$ near $\gamma(t)$.)

Proof of Theorem 16.3.0. Suppose that (f, D) is a local inverse for ϕ and γ is a curve in G starting at a point of D. As in Proposition 16.1.3 we may suppose that $0 < t_0 \le 1$, that $(f_t, D_t)_{0 \le t < t_0}$ is a continuation of (f, D) along $\gamma|_{[0,t_0)}$, and that each (f_t, D_t) is a local inverse for ϕ; now we need to define (f_{t_0}, D_{t_0}). (We say "As in Proposition 16.1.3" instead of "By Proposition 16.1.3" because there is nothing in the proposition about function elements satisfying extra conditions like being local inverses for a covering map.)

To begin, define $\tilde{\gamma} : [0, t_0) \to \Omega$ by $\tilde{\gamma}(t) = f_t(\gamma(t))$. (The "induction hypothesis" that (f_t, D_t) is a local inverse for ϕ shows that $f_t(\gamma(t)) \in \Omega$ for $t \in [0, t_0)$.) The first thing we need to show is that $\tilde{\gamma}$ is continuous. But given $t \in [0, t_0)$ we know that there exists $\delta > 0$ such that if $s \in [0, t_0)$ and $|s - t| < \delta$ then $[f_s]_{\gamma(s)} = [f_t]_{\gamma(s)}$. In particular $\tilde{\gamma}(s) = f_s(\gamma(s)) = f_t(\gamma(s))$ for s near t, so $\tilde{\gamma}$ is continuous.

Now let D_{t_0} be any (connected) neighborhood of $\gamma(t_0)$ which is elementary for ϕ. Since γ is continuous, there exists $\delta > 0$ such that $\gamma(t) \in D_{t_0}$ whenever $t_0 - \delta < t < t_0$. Now, since $\tilde{\gamma}$ is continuous, it follows that $\tilde{\gamma}((t_0 - \delta, t_0))$ is a connected subset of $\phi^{-1}(D_{t_0})$, so it must be contained in a component of $\phi^{-1}(D_{t_0})$. By the definition of "elementary" this shows that there is a connected open set $\tilde{D} \subset \Omega$ such that $\tilde{\gamma}((t_0 - \delta, t_0)) \subset \tilde{D}$ and such that $\phi|_{\tilde{D}}$ is a homeomorphism from $\tilde{D}$ onto D_{t_0}. We define f_{t_0} to be the inverse of this homeomorphism: $f_{t_0} = (\phi|_{\tilde{D}})^{-1}$.

Now the fact that f_{t_0} is continuous and $\phi(f_{t_0}(z)) = z$ shows that f_{t_0} is holomorphic, and so we are done if we can show that $[f_t]_{\gamma(t)} = [f_{t_0}]_{\gamma(t)}$ for $t < t_0$ near t_0. In fact this holds for every $t \in (t_0 - \delta, t_0)$, by another connectedness argument: Fix $t \in (t_0 - \delta, t_0)$, and let E_t be the component of $D_t \cap D_{t_0}$ containing $\gamma(t)$. Now $f_t(E_t)$ must be a connected subset of $\phi^{-1}(D_{t_0})$ (the fact that (f_t, D_t) is a local inverse for ϕ shows that $f_t(E_t) \subset \phi^{-1}(D_{t_0})$), so $f_t(E_t)$ must lie in a component of $\phi^{-1}(D_{t_0})$; since $\tilde{\gamma}(t) \in \phi^{-1}(E_t) \cap \tilde{D}$, this component must be $\tilde{D}$. So $f_t(E_t) \subset \tilde{D}$. It follows that $f_t = f_{t_0}$ in E_t, since both f_t and f_{t_0} map E_t into $\tilde{D}$ and ϕ is one-to-one on $\tilde{D}$. □

We note that a branch of the logarithm admits unrestricted continuation in $\mathbb{C} \setminus \{0\}$. This can be seen in various ways; in fact this is an instance of each of Examples 1 through 3 above! The fact that antiderivatives admit unrestricted continuation shows that a branch of the logarithm admits unrestricted continuation, because it is just an antiderivative of the function $1/z$. Similarly the fact that $\log(|z|)$ is harmonic in $\mathbb{C} \setminus \{0\}$ and $\operatorname{Re}(\log(z)) = \log(|z|)$ shows that the unrestricted continuation of a branch of the logarithm is an instance of Example 2. Finally, the exponential function is a covering map from $\mathbb{C}$ onto $\mathbb{C} \setminus \{0\}$ (you should verify this carefully) and a branch of the logarithm is nothing but a local inverse for the exponential, so Theorem 16.3.0 also shows that a branch of the logarithm admits unrestricted continuation in $\mathbb{C} \setminus \{0\}$.

So a branch of the logarithm admits unrestricted continuation in the punctured plane, and this gives a specific instance of each of Examples 1, 2, and 3. Of course this is not an instance of Example 0: There is no logarithm function holomorphic in the entire punctured plane. We now turn our attention to the question of when a function element admitting unrestricted continuation in a set G *can* be extended to a function holomorphic in G.

We already know the answer for antiderivatives: At least for smooth curves we can continue an antiderivative along a curve by integration, and Theorem 4.0 shows that the function element (f, D) has a global antiderivative in G if and only if whenever we continue a local antiderivative along a smooth closed curve we get back to where we started.

Put a little more precisely: Suppose that G is a connected open set, $f \in H(G)$, D is a connected open subset of G, and $F \in H(D)$ satisfies $F' = f$ in D. It follows immediately from Theorem 4.0 that F can be extended to a function holomorphic in G if and only if the integral of f about any smooth closed curve is 0. This is the same as saying that if

$(F_t, D_t)_{0 \leq t \leq 1}$ is the continuation of (F, D) about a smooth closed curve γ then $F_1 = F$ near $\gamma(0)$.

The general case is exactly the same. We invent some terminology for the sake of making the statement of the theorem more comprehensible: Suppose that $(f_t, D_t)_{0 \leq t \leq 1}$ is a continuation of a function element (f, D) along a *closed* curve γ. We will say that the continuation of (f, D) along γ is *trivial* if $[f_1]_{\gamma(0)} = [f_0]_{\gamma(0)}$.

Theorem 16.3.1. *Suppose that (f, D) is a function element, G is a connected open set with $D \subset G$, and (f, D) admits unrestricted continuation in G. Fix a point $a \in D$. There exists a function $F \in H(G)$ with $F|_D = f$ if and only if the continuation of (f, D) along every closed curve in G starting at a is trivial.*

It is clear that Theorem 4.0 is more or less a special case of this result; in fact a suitable generalization of the proof of Theorem 4.0 suffices here:

Proof. One direction is very easy: If f extends to $F \in H(G)$ then the uniqueness of analytic continuation shows that the continuation of f along any closed curve in G is trivial.

For the other direction: Assume that the continuation of (f, D) about any closed curve is trivial. As in the proof of Theorem 4.0, it follows that the continuation along a path depends only on the endpoints: If γ_1 and γ_2 are paths from a to b in G, and $(f_t^j, D_t^j)_{0 \leq t \leq 1}$ is a continuation of (f, D) along γ_j, then $[f_1^1]_b = [f_1^2]_b$.

Hence we may define a function $F : G \to \mathbb{C}$ by saying that $F(z)$ should be $f_1(z)$, where $(f_t, D_t)_{0 \leq t \leq 1}$ is a continuation of (f, D) along some path from a to z. The fact that this continuation is independent of the path we choose shows that F is well defined; it also shows that F is holomorphic. Indeed, fix $z \in G$ and fix a path γ from a to z. Choose $\delta > 0$ such that $D(z, \delta) \subset D_1$. Now if $w \in D(z, \delta)$ we can define a curve γ_w from a to w by

$$\gamma_w = \gamma + [z, w].$$

Here $\gamma + [z, w]$ denotes the curve which follows γ from a to z and then follows $[z, w]$ from z to w; you should give a more formal definition, somewhat like the definition of $\gamma_1 - \gamma_2$ preceding Exercise 16.3 above.

But f_1 is holomorphic in D_1 and $[z, w] \subset D_1$, so we can define a continuation of (f, D) along γ_w by starting with the given continuation along γ and then using f_1 to continue the continuation along the segment $[z, w]$. This shows that

$$F(w) = f_1(w) \quad (w \in D(z, \delta)),$$

and so F is holomorphic in $D(z, \delta)$. □

We will be giving two applications of Theorem 16.3.1; in both of them we will need to know that if a closed curve γ is null-homotopic and (f, D) admits unrestricted continuation then the continuation along γ is trivial. (A closed curve is *null-homotopic* if it is homotopic to a constant, as in the definition of simple connectedness.) The proof is just like the proof of Theorem 16.1.2; we organize it a little differently. First:

Lemma 16.3.2. *Suppose that (f, D) is a function element and G is a connected open set in the plane with $D \subset G$. Suppose that $\gamma : [0, 1] \to G$ is a closed curve in G starting at a point of D, and suppose that (f, D) can be continued along γ. Then there exists $\delta > 0$ such that if $\tilde{\gamma} : [0, 1] \to G$ is a closed curve with $|\tilde{\gamma}(t) - \gamma(t)| < \delta$ for all $t \in [0, 1]$ then (f, D) can be continued along $\tilde{\gamma}$; in this case the continuation along $\tilde{\gamma}$ is trivial if and only if the continuation along γ is trivial.*

Proof. As in the proof of Lemma 16.1.1, the fact that (f, D) can be continued along γ shows that there exists a continuation $(f_t, D_t)_{0 \le t \le 1}$ such that there exists $\delta > 0$ with $D(\gamma(t), \delta) \subset D_t$ for all $t \in [0, 1]$, and now if $\tilde{\gamma}$ is a curve with $|\tilde{\gamma}(t) - \gamma(t)| < \delta$ for all t then it follows that $(f_t, D_t)_{0 \le t \le 1}$ is also a continuation of (f, D) along $\tilde{\gamma}$.

Suppose first that the continuation along γ is trivial. Then $f_1 = f_0$ near $\gamma(0)$, which implies that $f_1 = f_0$ in $D(\gamma(0), \delta)$. In particular $f_1 = f_0$ near $\tilde{\gamma}(0)$, so the continuation of (f, D) along $\tilde{\gamma}$ is trivial. (Here we used a bit of complex analysis, but we didn't have to: Instead of using the fact that $f_1 = f_0$ near $\gamma(0)$ implies that $f_1 = f_0$ in $D(\gamma(0), \delta)$ we could have just restricted our attention to the neighborhood of $\gamma(0)$ where we knew that $f_1 = f_0$.)

Suppose on the other hand that the continuation along γ is not trivial: It is not true that $f_1 = f_0$ near $\gamma(0)$. It follows that we cannot have $f_1 = f_0$ in any nonempty open subset of $D(\gamma(0), \delta)$; in particular we cannot have $f_1 = f_0$ in a neighborhood of $\tilde{\gamma}(0)$, so the continuation of (f, D) along $\tilde{\gamma}$ is not trivial. (Here the bit of complex analysis we used was essential.) □

We also need a topological fact:

Lemma 16.3.3. *Suppose that G is an open set, $\gamma : [0, 1] \to G$ is a closed curve, and γ is null-homotopic in G. Then there exists a continuous family $(\gamma_s)_{0 \le s \le 1}$ of closed curves in G such that $\gamma_0 = \gamma$ and γ_1 is constant, with the further property that $\gamma_s(0) = \gamma(0)$ for all $s \in [0, 1]$.*

Proof. As noted previously, the fact that γ is null-homotopic in G shows that there exists a continuous map

$$\Gamma : \overline{\mathbb{D}} \to G$$

such that $\gamma(t) = \Gamma(e^{2\pi i t})$. Define a family of circles $(c_s)_{0\leq s\leq 1}$ by

$$c_s(t) = s + (1-s)e^{2\pi i t} \quad (t \in [0,1]).$$

Note that (c_s) is a continuous family of closed curves in $\overline{\mathbb{D}}$ such that $c_s(0) = c_s(1) = 1$ for all s; it follows that $(\gamma_s)_{0\leq s\leq 1}$ has the desired properties, if we set $\gamma_s(t) = \Gamma(c_s(t))$. □

It follows that continuation along a null-homotopic closed curve is trivial (at least when we have unrestricted continuation):

Proposition 16.3.4. *Suppose that (f, D) is a function element, G is a connected open set with $D \subset G$, and (f, D) admits unrestricted continuation in G. If γ is a closed curve in G starting at a point of D and γ is null-homotopic in G then the continuation along γ is trivial.*

Proof. Choose a continuous family $(\gamma_s)_{0\leq s\leq 1}$ of closed curves in G such that $\gamma_0 = \gamma$, γ_1 is constant, and $\gamma_s(0) = \gamma(0)$ for all $s \in [0, 1]$. Let A be the set of $s \in [0, 1]$ such that the continuation of (f, D) along γ_s is trivial.

Then $1 \in A$, since the continuation of a function element along a *constant* curve is certainly trivial. One half of Lemma 16.3.2 shows that A is relatively open in $[0, 1]$ and the other half of Lemma 16.3.2 shows that the complement of A is relatively open; hence $0 \in A$. □

The result that we will be calling the Monodromy Theorem is an immediate consequence:

Corollary 16.3.5 (the Monodromy Theorem). *Suppose that (f, D) is a function element, G is a simply connected open subset of the plane with $D \subset G$, and (f, D) admits unrestricted continuation in G. Then there exists a function $F \in H(G)$ such that $f = F|_D$.*

Proof. This is immediate from Theorem 16.3.1 and Proposition 16.3.4. □

The fact that our notion of triviality is preserved under homotopy as in Proposition 16.3.4 gives a stronger version of Theorem 16.3.1: If (f, D) admits unrestricted continuation and the continuation is trivial along every curve in a set of generators for the "fundamental group" of G then (f, D) extends to a function holomorphic in G. We are not going to explain what that means, but we are going to prove a special case. Readers who know a little bit of algebraic topology (actually just the definition of the fundamental group) should be able to prove the general case with no trouble.

If $\gamma : [0,1] \to G$ is a closed curve with $\gamma(0) = a$ and n is a positive integer, we define $\gamma^n : [0,1] \to G$ by
$$\gamma^n(t) = \gamma(nt - k) \quad (t \in [k/n, (k+1)/n], \, k = 0, \ldots, n-1);$$
that is, γ^n traces the same curve as γ, repeated n times. We set γ^0 to be the constant map defined by $\gamma^0(t) = a$ for all $t \in [0,1]$, and we define γ^n for negative integers n by
$$\gamma^n(t) = \gamma^{-n}(1-t),$$
so that γ^{-n} is the same curve as γ^n, except traced in the opposite direction.

For the next result we set $\mathbb{D}' = \mathbb{D} \setminus \{0\}$. The following lemma says that the "fundamental group" of $\mathbb{D}'$ is generated by one element.

Lemma 16.3.6. *Fix $a \in \mathbb{D}'$ and define $\gamma : [0,1] \to \mathbb{D}'$ by*
$$\gamma(t) = ae^{2\pi it} \quad (t \in [0,1]).$$
If σ is any closed curve in $\mathbb{D}'$ then σ is homotopic to γ^n for some integer n.

Proof. First, we may assume that $a > 0$ to simplify the notation, and then we may assume that $|\sigma(t)| = a$ for all t, since any closed curve in $\mathbb{D}'$ is easily seen to be homotopic to a curve with this property.

Now Lemma 4.6 shows that there is a continuous map $\theta : [0,1] \to \mathbb{R}$ such that $\theta(0) = 0$ and
$$\sigma(t) = ae^{i\theta(t)} \quad (t \in [0,1]).$$
It follows that $e^{i\theta(1)} = 1$, so that $\theta(1) = 2\pi n$ for some integer n. Define $\theta_s(t) = s2\pi nt + (1-s)\theta(t)$ and set $\sigma_s(t) = ae^{i\theta_s(t)}$. The fact that $\theta_s(1) = 2\pi n$ for all s shows that each σ_s is a closed curve; thus the family $(\sigma_s)_{0 \le s \le 1}$ shows that $\sigma = \sigma_0$ is homotopic to $\sigma_1 = \gamma^n$. $\square$

The lemma is essentially contained in Lemma 4.6. (The fact that the proof of Lemma 4.6 seems a lot like the arguments in this chapter is not a coincidence; as noted earlier, Lemma 4.6 is a special case of Theorem 16.4.2 in the next section, which is a basic tool in dealing with things like the lemma above.)

It follows that when $G = \mathbb{D}'$ we can give a version of Theorem 16.3.1 that talks about just one closed curve:

Proposition 16.3.7. *Suppose that D is a connected open subset of $\mathbb{D}'$ and (f, D) is a function element that admits unrestricted continuation in $\mathbb{D}'$. Fix $a \in D$ and define $\gamma : [0,1] \to \mathbb{D}'$ by*
$$\gamma(t) = ae^{2\pi it} \quad (t \in [0,1]).$$
There exists a function $F \in H(\mathbb{D}')$ such that $F|_D = f$ if and only if the continuation of (f, D) along γ is trivial.

Proof. The fact that the continuation along γ is trivial if f extends to a function holomorphic in $\mathbb{D}'$ is trivial (and also contained in Theorem 16.3.1). Suppose now that the continuation of (f, D) along γ is trivial.

It follows that the continuation along γ^n is trivial for any integer n; now Lemma 16.3.6 and Theorem 16.1.2 show that the continuation along any closed curve in $\mathbb{D}'$ starting at a is trivial, and so f extends to a function holomorphic in $\mathbb{D}'$, by Theorem 16.3.1. □

We deduced Proposition 16.3.7 from Theorem 16.3.1 because Theorem 16.3.1 really seems to be the fundamental result that explains why things work the way they do. We should note that we could also have given an *ad hoc* proof using only the Monodromy Theorem:

- **Exercise 16.4.** Show that Proposition 16.3.7 follows from Corollary 16.3.5, without using Theorem 16.3.1.

  Hint: Assume that D is a disk. Define two open sectors S_1 and S_2 such that each S_j is simply connected and contains D, and such that $\mathbb{D}' = S_1 \cup S_2$. Show that there exists $F_j \in H(S_j)$ such that $F_j|_D = f$, and now figure out why the hypothesis implies that you can piece F_1 and F_2 together to give one function $F \in H(\mathbb{D}')$.

16.4. Another Point of View

We should say at least something about a higher-level approach to all this, because although it adds one more level of abstraction it actually makes many things much simpler. The results in this section will not be used anywhere else; one reason for this is that in this section we assume that the reader knows a little bit of basic point-set topology, although that is not a prerequisite for most of the book. We are going to omit a lot of the proofs in this section, but this time it is not because the definitions are missing or the proofs are too hard: We will include all the definitions and the reader should regard omitted proofs as exercises (the techniques used in the proofs in the previous section suffice).

We begin by defining the topology on $\mathcal{O}$. Suppose that $g \in \mathcal{O}$. Then g is a set of function elements. If $(f, D) \in g$ then we define $S_{(f,D)} = \{\,[f]_z : z \in D\,\}$; a local basis for our topology on $\mathcal{O}$ at g is given by $\{\,S_{(f,D)} : (f, D) \in g\,\}$.

We leave it to the reader to recall what condition a set of subsets of X must satisfy to form a local basis for a topology on X, and to verify that the set of neighborhoods we just defined does in fact satisfy that condition.

What does this definition mean? It means that the germs "close" to a given germ are the germs for "the same function" at nearby points. In particular you should note the following: If (f, D) is a function element then $p|_{S_{(f,D)}}$ is a homeomorphism of $S_{(f,D)}$ onto D. (Recall the definition: If $g \in \mathcal{O}_a$ then $p(g) = a$.)

Suppose that (f, D) is a function element, $a \in D$, γ is a path starting at a, and $(f_t, D_t)_{0 \le t \le 1}$ is a continuation of (f, D) along γ. Define

$$\tilde{\gamma}(t) = [f_t]_{\gamma(t)}.$$

Then it is clear that $\tilde{\gamma} : [0, 1] \to \mathcal{O}$ and that $p \circ \tilde{\gamma} = \gamma$ (to say that $p(\tilde{\gamma}(t)) = \gamma(t)$ just says that $\tilde{\gamma}(t)$ is a germ at $\gamma(t)$, and this is indeed clear). What may not be quite so obvious is the fact that $\tilde{\gamma}$ is continuous. But it's true; this follows from the definition of analytic continuation and the definition of the topology on $\mathcal{O}$, and this one we are *not* going to explain, because you need to work it out yourself to get all the definitions straight. And in fact the converse holds:

Proposition 16.4.0. *Suppose that (f, D) is a function element, $a \in D$, and γ is a path starting at a. The function element (f, D) can be continued along γ if and only if there exists a continuous $\tilde{\gamma} : [0, 1] \to \mathcal{O}$ such that $\tilde{\gamma}(0) = [f]_a$ and $p \circ \tilde{\gamma} = \gamma$.*

Proof. Exercise. □

Looking at things this way makes a lot of things simpler. For example here is a proof of the uniqueness of analytic continuation: Suppose that (f, D) is a function element, $a \in D$, γ is a path starting at a, and $\tilde{\gamma}_1 : [0, 1] \to \mathcal{O}$ and $\tilde{\gamma}_2 : [0, 1] \to \mathcal{O}$ are both continuous functions satisfying $\tilde{\gamma}_j(0) = [f]_a$ and $p \circ \tilde{\gamma}_j = \gamma_j$. We do the obvious thing, as in the proof of Lemma 16.1.0: We let

$$A = \{\, t \in [0, 1] : \tilde{\gamma}_1(t) = \tilde{\gamma}_2(t) \,\}$$

and we try to show that $A = [0, 1]$. It is clear that $0 \in A$, and the fact that A is relatively open is easy to see from the definitions: Suppose that $t \in A$: $\tilde{\gamma}_1(t) = \tilde{\gamma}_2(t)$. Now let S be a basic neighborhood of $\tilde{\gamma}_1(t)$, as in the definition of the topology on $\mathcal{O}$ above. Since $\tilde{\gamma}_j$ is continuous, there exists $\delta > 0$ such that $\tilde{\gamma}_j(s) \in S$ whenever $s \in [0, 1]$ and $|s - t| < \delta$. But now note that p is one-to-one on S (recall that $p|_S$ is actually a homeomorphism from S onto an open subset of $\mathbb{C}$). Hence if $|s - t| < \delta$ the fact that

$$p(\tilde{\gamma}_1(s)) = \gamma(s) = p(\tilde{\gamma}_2(s))$$

shows that $\tilde{\gamma}_1(s) = \tilde{\gamma}_2(s)$, so $s \in A$.

16.4. Another Point of View

This shows that A is open. The fact that $\tilde{\gamma}_j$ is continuous for $j = 1, 2$ shows that A is closed, and so $A = [0, 1]$. □.

This proof seems somewhat simpler than the proof we gave for Lemma 16.1.0, which is a good thing.

In fact, if your error detector is working properly you should find this proof suspiciously simple. We claimed that the fact that A is closed in the original proof of Lemma 16.1.0 actually depended on some bit of complex analysis, and the proof above looks like nothing but topology. Not that there is anything wrong with a proof by nothing but topology, but here it is clear that a proof by nothing but topology is *impossible,* because it would show that "continuous continuation" for germs of continuous functions was also unique, and that is obviously false. So we'd better find something in the proof that fails for continuous continuation of germs of continuous functions.

One might suspect that for "continuous continuation" of germs of continuous functions the function $\tilde{\gamma}_j$ was not continuous. But that's false; if you trace through the details you see that $\tilde{\gamma}_j$ is continuous in this case, for exactly the same reason as in the analytic case (which might be motivation to figure out what the reason *is* in the analytic case ...).

So what *is* the problem with the proof in the continuous case? What goes wrong is the punchline "The fact that $\tilde{\gamma}_j$ is continuous for $j = 1, 2$ shows that A is closed". How can that be wrong? We are talking about a subset of $[0, 1]$, so we can just think about sequences: Suppose that (t_n) is a sequence in A and $t_n \to t$; we need to show that $t \in A$. But that is clear: $\tilde{\gamma}_1(t_n) = \tilde{\gamma}_2(t_n)$, $\tilde{\gamma}_1(t_n) \to \tilde{\gamma}_1(t)$, $\tilde{\gamma}_2(t_n) \to \tilde{\gamma}_2(t)$, so $\tilde{\gamma}_1(t) = \tilde{\gamma}_2(t)$ and hence $t \in A$. Very simple.

Very simple but wrong. We cannot conclude that $\tilde{\gamma}_1(t) = \tilde{\gamma}_2(t)$ just because they are both limits of the same sequence of germs unless we know that the space of germs is a *Hausdorff* space! In a non-Hausdorff space one sequence can have several different limits.

(Probably when we are talking about topology we'd rather think of continuity in terms of inverse images of open sets. It is true that if f is a continuous function from any topological space to any other then the inverse image of a closed set is closed. Suppose now that X and Y are topological spaces, $f_j : X \to Y$ is continuous for $j = 1, 2$, and let $A = \{\, x \in X : f_1(x) = f_2(x) \,\}$. Define $F : X \to Y \times Y$ by $F(x) = (f_1(x), f_2(x))$. Then F is continuous. Now if Y is Hausdorff then the diagonal $D = \{\, (y_1, y_2) \in Y \times Y : y_1 = y_2 \,\}$ is closed, and hence $A = F^{-1}(D)$ is closed. But if Y is not Hausdorff then D need not be closed; in fact D is closed if and only if Y is Hausdorff.)

So we'd better prove that $\mathcal{O}$, the space of germs of analytic functions, *is* Hausdorff. Luckily this is easy:

Proposition 16.4.1. $\mathcal{O}$ *is Hausdorff.*

Proof. Suppose that g_1, $g_2 \in \mathcal{O}$ and $g_1 \neq g_2$. If $p(g_1) \neq p(g_2)$ then it is easy to find disjoint neighborhoods of g_1 and g_2 by a purely topological argument (in particular this case works for germs of continuous functions as well). We consider the case where $p(g_1) = p(g_2) = a$.

By shrinking the domain of an element of g_1 and an element of g_2 we can obtain elements with the same domain: We have $(f_1, D) \in g_1$ and $(f_2, D) \in g_2$, where f_1 does not agree with f_2 on any neighborhood of a. Since D is a connected open set, it follows that f_1 cannot agree with f_2 on any nonempty open subset of D. This says that

$$[f_1]_z \neq [f_2]_z \quad (z \in D);$$

it follows that (using the same notation as in the definition of the topology on $\mathcal{O}$ above)

$$S_{(f_1, D)} \cap S_{(f_2, D)} = \emptyset,$$

and so we have found disjoint neighborhoods of g_1 and g_2. □

That fills the gap in the proof above that analytic continuation along a curve is unique. This is good stuff: When we look at things this way arguments from the previous section involving the definition of analytic continuation become arguments about the simpler, or at least more familiar, notion of continuity of curves, while arguments as in the proof of Proposition 16.4.1 (there were several of those "hence they cannot agree in any open set" arguments in the previous section) become just the fact that $\mathcal{O}$ is Hausdorff.

- **Exercise 16.5.** Show explicitly that the sheaf of germs of continuous functions is not Hausdorff: Give an example of two germs g_1 and g_2 and show that if S_j is a neighborhood of g_j then $S_1 \cap S_2 \neq \emptyset$.

- **Exercise 16.6.** Show that the sheaf of germs of continuous functions *is* a T_1 space: If g_1 and g_2 are germs with $g_1 \neq g_2$ then there is an open set S with $g_1 \in S$ but $g_2 \notin S$.

So the space of germs of continuous functions is an actual "natural" example of a non-Hausdorff space, and a T_1 space to boot. (I for one find this interesting: When I learned topology, things like non-Hausdorff T_1 spaces seemed like very artificial things that were constructed just to give topologists something to do.)

16.4. Another Point of View

What about complete analytic functions? We know that $\mathcal{O}$ is locally path-connected, which is to say that any point has a path-connected neighborhood; this is clear because any point of $\mathcal{O}$ has a neighborhood homeomorphic to a connected open subset of $\mathbb{C}$. It follows that an open set is connected if and only if it is path-connected.

But the germs in the complete analytic function for (f, D) are exactly the germs you can get to from a germ in (f, D) by analytic continuation. Since an analytic continuation along a curve in $\mathbb{C}$ is "the same thing as" a curve in $\mathcal{O}$, we see that the complete analytic functions are exactly the same as the connected components of $\mathcal{O}$.

The most interesting and useful results in the previous section involved the notion of unrestricted continuation. Since continuing a function element along a curve γ is the same as "lifting" γ to $\tilde{\gamma}$ (as in Proposition 16.4.0), the question of when every curve can be "lifted" is of some interest. This leads to the notion of covering spaces (in a more general setting than in Example 3 in the previous section):

Suppose that X and Y are topological spaces and $\pi : X \to Y$ is continuous and surjective. We say that π is a *covering map* if every y in Y has a local basis of connected neighborhoods each of which is elementary for π; as before we say that the (connected!) open set N is elementary if the inverse image is a disjoint union of open sets

$$\pi^{-1}(N) = \bigcup_{\alpha \in A} \tilde{N}_\alpha,$$

where $\pi|_{\tilde{N}_\alpha}$ is a homeomorphism from $\tilde{N}_\alpha$ onto N for each $\alpha \in A$.

If you have the right picture of $p : \mathcal{O} \to \mathbb{C}$ and the right idea of what a covering map is then p may seem like a covering map. It isn't:

- **Exercise 16.7.** Show that $p : \mathcal{O} \to \mathbb{C}$ is not a covering map.

(In fact if p were a covering map it would follow from results below that every function element would admit unrestricted continuation; you should give a more direct explanation of why the definition of "covering map" is not satisfied.)

Of the many basic facts about covering maps and curves, etc., for now we mention only the most fundamental:

Theorem 16.4.2. *Suppose that $\pi : X \to Y$ is a covering map, $\gamma : [0, 1] \to Y$ is continuous, $x \in X$, and $\pi(x) = \gamma(0)$. Then there exists a unique continuous function $\tilde{\gamma} : [0, 1] \to X$ such that $\tilde{\gamma}(0) = x$ and $\pi \circ \tilde{\gamma} = \gamma$.*

Proof. One can extract a proof of this from the proof of Theorem 16.3.0 above. You should do that. □

In fact covering maps typically *are* where unrestricted continuation comes from. We saw several examples of function elements admitting unrestricted continuation in the previous section. Example 0 was sort of trivial; if one insisted one could see that it had something to do with the trivial covering map $\pi : G \to G$ defined by $\pi(z) = z$. It seems clear that Example 3 has something to do with covering maps; we leave it as an exercise to deduce Theorem 16.3.0 from Theorem 16.4.2.

More interesting is that the fact that antiderivatives admit unrestricted continuation (Example 1 in the previous section) can also be deduced from Theorem 16.4.2. Define a map $\mathcal{D} : \mathcal{O} \to \mathcal{O}$ by
$$\mathcal{D}([f]_a) = [f']_a.$$
It is easy to show that $\mathcal{D}$ is a covering map, and if you apply Theorem 16.4.2 to this covering map you obtain precisely the fact that antiderivatives admit unrestricted continuation. You should work out the details. (Suppose that $D \subset G$, $f \in H(D)$, $g \in H(G)$, and $f' = g$ in D. If γ is a path in G starting at a point $a \in D$ define $\gamma_1(t) = [g]_{\gamma(t)}$. Then γ_1 is a path in $\mathcal{O}$, $[f]_a \in \mathcal{O}$, and $\mathcal{D}([f]_a) = [g]_a = \gamma_1(0)$. It follows from Theorem 16.4.2 that there exists a path $\tilde{\gamma}$ in $\mathcal{O}$ such that $\tilde{\gamma}(0) = [f]_a$ and $\mathcal{D} \circ \tilde{\gamma} = \gamma_1$.)

This is a good example of the power of abstraction. You may have noticed that many of the proofs in this chapter are sort of all the same; if we had started at a slightly higher level of abstraction, as in the current section, we could have derived various results as special cases of more general facts about covering maps, instead of repeating the arguments.

One can also show that Example 2 in the previous section follows from facts about covering maps. Define a map r from $\mathcal{O}_G$ to the sheaf of germs of *harmonic* functions in G (we leave the definition to you) by
$$r([f]_a) = [\operatorname{Re}(f)]_a.$$
It is easy to see that r is well defined and that r is a covering map, and now Theorem 16.4.2 implies that harmonic conjugates admit unrestricted continuation.

In fact one can show that *any* time a function element admits unrestricted continuation in an open set there is a covering map lurking in the background:

Theorem 16.4.3. *Suppose that (f, D) is a function element, G is a connected open set in the plane, $D \subset G$, and (f, D) admits unrestricted continuation in G. Fix $a \in D$ and let S be the connected component of $\mathcal{O}(G)$ containing $[f]_a$. Then $p|_S$ is a covering map.*

16.4. Another Point of View

Our second-last topic concerns something that came up informally several times in the previous section:

Suppose that (f, D) is a function element, γ is a path starting at a point of D, and $(f_t, D_t)_{0 \le t \le 1}$ is a continuation of (f, D) along γ. Then various other things are automatically continuations of various other things along γ. For example it follows that $(f_t', D_t)_{0 \le t \le 1}$ is a continuation of (f', D) along γ, if ϕ is an entire function then $(\phi \circ f_t, D_t)_{0 \le t \le 1}$ is a continuation of $(\phi \circ f, D)$ along γ, etc. (Of course ϕ need not be entire, just holomorphic in an open set containing the range of each f_t.)

This raises the question of what sort of mapping Φ from function elements to function elements has the property that $\Phi((f_t, D_t))_{0 \le t \le 1}$ is automatically a continuation of $\Phi((f, D))$ along γ (two examples above correspond to $\Phi((f, D)) = (f', D)$ and $\Phi((f, D)) = (\phi \circ f, D)$). It is clear that the same argument that shows those two examples work will show that there are many more examples; can we give a general criterion?

If you think about it you see that one fairly general criterion is as follows. Suppose that Φ is a mapping from the class of all function elements to itself. We shall say that Φ is *local* if it satisfies two properties: First, $\Phi((f, D))$ always has domain D, that is, for any D and any $f \in H(D)$ there exists $g \in H(D)$ such that $\Phi((f, D)) = (g, D)$, and second, if (f_1, D_1) and (f_2, D_2) are two function elements such that $D_1 \subset D_2$ and $f_1 = f_2|_{D_1}$ then we have $\Phi((f_j, D_j)) = (g_j, D_j)$ where

$$g_1 = g_2|_{D_2}.$$

Easy Fact 16.4.4. *Suppose that Φ is a mapping from the class of all function elements to itself, and suppose that Φ is local, as above. If $(f_t, D_t)_{0 \le t \le 1}$ is a continuation of (f, D) along γ then $\Phi((f_t, D_t))_{0 \le t \le 1}$ is a continuation of $\Phi((f, D))$ along γ.*

Proof. Easy. □

Now, while 16.4.4 is easy it is not *obvious*. You should contrast this with the following obvious fact:

Obvious Fact 16.4.5. *Suppose that $F : \mathcal{O} \to \mathcal{O}$ is continuous and satisfies $p \circ F = p$. Suppose that $\gamma : [0, 1] \to \mathbb{C}$ is continuous, $\tilde{\gamma} : [0, 1] \to \mathcal{O}$ is continuous, and $p \circ \tilde{\gamma} = \gamma$. Then $F \circ \tilde{\gamma}$ is continuous and satisfies $p \circ F \circ \tilde{\gamma} = \gamma$.*

That really is obvious. You should convince yourself that Easy Fact 16.4.4 and Obvious Fact 16.4.5 really say the same thing, in different language. (*Hint:* If Φ is a mapping from function elements to function elements and Φ is local then we can define a map $\tilde{\Phi} : \mathcal{O} \to \mathcal{O}$ by

$$\tilde{\Phi}(g) = [\Phi((f, D))]_{p(g)} \quad ((f, D) \in g);$$

the fact that Φ is local shows that $\tilde{\Phi}$ is well defined.)

We have not yet explained why the total analytic function associated with a function element deserves to be called holomorphic. For this we need to introduce another concept, the notion of "one-dimensional complex manifold" (or "Riemann surface").

Informally, a one-dimensional complex manifold is a topological space such that in a neighborhood of any point there is a "complex coordinate system", making that neighborhood look like an open subset of $\mathbb{C}$. There need not be a single global complex coordinate system; as we move from point to point we may be forced to switch from one coordinate system to another. It is required that the function effecting the change of coordinates be holomorphic.

Formally, suppose that M is a topological space. A *chart* on M is an ordered pair (O, c), where O is an open subset of M and c is a homeomorphism from O onto a connected open subset of $\mathbb{C}$; O is the *domain* of (O, c). An *atlas* on M is a collection A of charts such that for every $x \in M$ there exists a chart (O, c) with $x \in O$, and such that the overlap maps are holomorphic; this last condition means that if (O_1, c_1) and (O_2, c_2) are two elements of A such that $O = O_1 \cap O_2 \ne \emptyset$ then the function $c_1 \circ (c_2^{-1})$ is holomorphic in $c_2(O)$. A *one-dimensional complex manifold*, also known as a *Riemann surface*, is an ordered pair (M, A) where M is a topological space and A is an atlas on M. (Of course the atlas is usually understood, in which case one typically refers to M itself as a complex manifold.)

Suppose that M is a one-dimensional complex manifold. Then $f : M \to \mathbb{C}$ is said to be holomorphic if $f \circ c^{-1}$ is holomophic in $c(O)$, for every chart (O, c).

More generally, if M and N are one-dimensional complex manifolds then $f : M \to N$ is holomorphic if for every chart (O_1, c_1) on M, every $x \in O_1$, and every chart (O_2, c_2) on N such that $f(x) \in O_2$, the function $c_2 \circ f \circ c_1^{-1}$ is holomorphic on the open subset of the plane where it is defined (namely, $c_1(O_1 \cap f^{-1}(O_2))$).

If V is a connected open subset of $\mathbb{C}$ then V is a one-dimensional complex manifold in a trivial way, via the single chart (V, ι), where $\iota(z) = z$. Any open subset of a one-dimensional complex manifold is also a one-dimensional complex manifold; we leave the formal definition to you.

The Riemann sphere $\mathbb{C}_\infty$ is a less trivial example of a one-dimensional complex manifold; you should verify that if $\iota(z) = z$ and $j(z) = 1/z$ (where

16.4. Another Point of View

$1/\infty = 0$ and $1/0 = \infty$) then $A = \{(\mathbb{C}, \iota), (\mathbb{C}_\infty \setminus \{0\}, j)\}$ is an atlas on $\mathbb{C}_\infty$ and that the definition of "holomorphic mapping from $\mathbb{C}_\infty$ to $\mathbb{C}_\infty$" given in Section 8.0 is consistent with the definition above.

Now suppose that (f, D) is a function element. Define $O = \{[f]_z : z \in D\}$ and define $c : O \to \mathbb{C}$ by $c([f]_z) = z$. It is trivial to verify that (O, c) is a chart on O, and that if we let A be the collection of all such charts (O, c) obtained from various function elements (f, D) then A is an atlas on O, making O into a one-dimensional complex manifold.

Note. "Trivial" does not necessarily imply "easy"! You should verify the "trivial" assertions in the previous paragraph carefully, precisely because it can be confusing sorting through all the various definitions to determine exactly what needs to be proved. The assertions are nonetheless trivial, in spite of being possibly confusing, because once you figure out what needs to be proved you're done — you find that the proofs are immediate from the definitions.

Now define $F : O \to \mathbb{C}$ by $F(g) = g(p(g))$, as in Section 16.2. It is trivial to verify that F is holomorphic. If (f, D) is a function element, $a \in D$ and S is the connected component of O containing $[f]_a$, then $f^* = F|_S$, the complete analytic function derived from (f, D), is holomorphic.

Chapter 17

Orientation

There are some details about orientation of triples of points on the boundary of the disk that we need to get straight.

Suppose that a, b and c are three distinct points on $\partial \mathbb{D}$. We know that there is a unique linear-fractional transformation ϕ such that $\phi(a) = 1$, $\phi(b) = i$, and $\phi(c) = -1$. The question is whether $\phi \in \text{Aut}(\mathbb{D})$. The fact that linear-fractional transformations take lines and circles to lines and circles shows that $\phi(\partial \mathbb{D}) = \partial \mathbb{D}$, and then since ϕ is a homeomorphism of $\mathbb{C}_\infty$ and the connected components of $\mathbb{C}_\infty \setminus \partial \mathbb{D}$ are $\mathbb{D}$ and $\mathbb{C}_\infty \setminus \overline{\mathbb{D}}$, it follows that $\phi(\mathbb{D})$ is either $\mathbb{D}$ or $\mathbb{C}_\infty \setminus \overline{\mathbb{D}}$, but either can happen. We need a definition of "positively oriented triple" so that $\phi \in \text{Aut}(\mathbb{D})$ if and only if the triple (a, b, c) is positively oriented.

If all we cared about was the unit disk we could just define "positively oriented" in terms of the result: Write down what ϕ is and say that (a, b, c) is positively oriented if and only if $\phi(0) \in \mathbb{D}$. The problem with that is that it only works for the disk; we need a definition that will make some sense in other open sets. One useful version of the definition is this: "The triple (a, b, c) is positively oriented if when you traverse the boundary starting at a, with the set to your left, you hit b before c." That's a useful way to look at it; the obvious problem with using it as the official definition is that we do not know officially what "the set lies to your left" means. But we can give an acceptable definition in terms of winding numbers. The actual math starts here:

First, a *parametrization* of $\partial \mathbb{D}$ is a continuous map $\gamma : [0, 1] \to \partial \mathbb{D}$ such that $\gamma(0) = \gamma(1)$, γ is one-to-one on $[0, 1)$, and $\gamma([0, 1]) = \partial \mathbb{D}$. Lemma 4.6 shows that if γ is a parametrization of $\partial \mathbb{D}$ then there exists a continuous function $\theta : [0,1] \to \mathbb{R}$ such that

$$\gamma(t) = e^{i\theta(t)};$$

the fact that γ is one-to-one on $[0,1)$ shows that θ must be monotone and that $\theta(1) = \theta(0) \pm 2\pi$; in fact $\mathrm{Ind}(\gamma, 0) = 1$ if $\theta(1) = \theta(0) + 2\pi$ and $\mathrm{Ind}(\gamma, 0) = -1$ if $\theta(1) = \theta(0) - 2\pi$.

Suppose now that (a, b, c) is a triple of distinct points of $\partial \mathbb{D}$. We say that (a, b, c) is *positively oriented* if there exists a smooth parametrization γ of $\partial \mathbb{D}$ with $\gamma(0) = a$, such that the $s, t \in [0, 1]$ with $\gamma(s) = b$ and $\gamma(t) = c$ satisfy $s < t$, and such that $\mathrm{Ind}(\gamma, 0) = 1$.

Suppose that (a, b, c) is a positively oriented triple in $\partial \mathbb{D}$, and let γ be as in the definition above. Choose a continuous function $\theta : [0, 1] \to \mathbb{R}$ such that $\gamma(t) = e^{i\theta(t)}$ as above. Now, the fact that $\mathrm{Ind}(\gamma, 0) = 1$ shows that θ is monotone increasing (as opposed to decreasing; we already know it is monotone). Hence there exist real numbers α, β and δ such that

$$\alpha < \beta < \delta < \alpha + 2\pi$$

with $a = e^{i\alpha}$, $b = e^{i\beta}$ and $c = e^{i\delta}$.

And so:

Proposition 17.0. *Suppose that (a_0, b_0, c_0) and (a_1, b_1, c_1) are two positively oriented triples of distinct points of $\partial \mathbb{D}$. There exists $\phi \in \mathrm{Aut}(\mathbb{D})$ such that $\phi(a_0) = a_1$, $\phi(b_0) = b_1$ and $\phi(c_0) = c_1$.*

Proof. For $j = 0, 1$ choose real numbers α_j, β_j, and δ_j such that

$$\alpha_j < \beta_j < \delta_j < \alpha_j + 2\pi$$

and such that $a_j = e^{i\alpha_j}$, $b_j = e^{i\beta_j}$ and $c_j = e^{i\delta_j}$. Now for $0 < t < 1$ define $\alpha_t = (1-t)\alpha_0 + t\alpha_1$, $\beta_t = (1-t)\beta_0 + t\beta_1$, and $\delta_t = (1-t)\delta_0 + t\delta_1$. It follows that

$$\alpha_t < \beta_t < \delta_t < \alpha_t + 2\pi$$

for all $t \in [0, 1]$, so that (a_t, b_t, c_t) is a triple of *distinct* points, if we set $a_t = e^{i\alpha_t}$, $b_t = e^{i\beta_t}$, and $c_t = e^{i\delta_t}$. Note that a_t, b_t and c_t depend continuously on $t \in [0, 1]$.

For $0 \le t \le 1$ let ϕ_t be the linear-fractional transformation such that $\phi_t(a_t) = a_1$, $\phi_t(b_t) = b_1$, and $\phi_t(c_t) = c_1$. We need to show that $\phi_0 \in \mathrm{Aut}(\mathbb{D})$, which is the same as saying that $\phi_0(0) \in \mathbb{D}$.

But it is clear that $\phi_1(0) \in \mathbb{D}$, because in fact $\phi_1(z) = z$ for all z. It is easy to see that $\phi_t(0)$ depends continuously on t, since one can write down an explicit formula for ϕ_t in terms of t and (a_1, b_1, c_1), and we have $|\phi_t(0)| \neq 1$ for all t. Hence $|\phi_0(0)| < 1$. □

The reason we gave the definition of "positively oriented triple" the way we did is because it is something we could hope to show is in some sense invariant under conformal maps. Proposition 17.2 below shows we succeeded. First a lemma about winding numbers being invariant:

Lemma 17.1. *Suppose that Ω is an open subset of the plane, γ is a smooth closed curve in Ω such that $\mathrm{Ind}(\gamma, z) = 0$ for all $z \in \mathbb{C} \setminus \Omega$, and $p \in \Omega \setminus \gamma^*$. Suppose that $f \in H(\Omega)$ is one-to-one and let $\Gamma = f \circ \gamma$. Then*

$$\mathrm{Ind}(\Gamma, f(p)) = \mathrm{Ind}(\gamma, p).$$

Proof. Suppose that γ has parameter interval $[0, 1]$. The lemma is immediate from the Residue Theorem:

$$\begin{aligned}
\mathrm{Ind}(\Gamma, f(p)) &= \frac{1}{2\pi i} \int_\Gamma \frac{dw}{w - f(p)} \\
&= \frac{1}{2\pi i} \int_0^1 \frac{\Gamma'(t)\, dt}{\Gamma(t) - f(p)} \\
&= \frac{1}{2\pi i} \int_0^1 \frac{f'(\gamma(t))\gamma'(t)\, dt}{f(\gamma(t)) - f(p)} \\
&= \frac{1}{2\pi i} \int_\gamma \frac{f'(w)\, dw}{f(w) - f(p)} \\
&= \mathrm{Res}\left(\frac{f'}{f - f(p)}, p\right) \mathrm{Ind}(\gamma, p) \\
&= \mathrm{Ind}(\gamma, p).
\end{aligned}$$

We leave a few details to the reader (such as showing that $f'/(f - f(p))$ has only one pole and calculating the residue at the pole). □

The lemma allows us to show that the notion of "positively oriented triple of distinct boundary points" is conformally invariant (under suitable hypotheses). First we need a definition.

Suppose that Ω is a bounded simply connected open set and every boundary point is simple. Carathéodory's Theorem (Theorem 14.0) shows that any conformal equivalence of Ω and $\mathbb{D}$ extends to a homeomorphism of $\overline{\Omega}$ and $\overline{\mathbb{D}}$ (mapping the interior to the interior and the boundary to the boundary), and hence that $\partial\Omega$ is a simple closed curve. Suppose that (a, b, c) is a triple of distinct points of $\partial\Omega$. We will say that (a, b, c) is positively oriented (with respect to Ω) if there exists a parametrization γ of $\partial\Omega$ which starts at a, hits b before c, and is such that $\mathrm{Ind}(\gamma, z) = 1$ for $z \in \Omega$. (Note that if this holds for one $z \in \Omega$ then it holds for all, by continuity.)

If ϕ is a conformal equivalence of such an Ω with the unit disk we will use the same letter ϕ to denote the extension to the closure that Carathéodory's Theorem tells us must exist.

Proposition 17.2. *Suppose that Ω is a bounded simply connected open set such that every boundary point of Ω is simple, and let ϕ be a conformal equivalence of Ω and $\mathbb{D}$. If (a, b, c) is a positively oriented triple of distinct points of $\partial \Omega$ then $(\phi(a), \phi(b), \phi(c))$ is a positively oriented triple of distinct points of $\partial \mathbb{D}$.*

Proof. Suppose that $\gamma : [0, 1] \to \mathbb{C}$ is a parametrization of $\partial \Omega$ that starts at a, reaches b before c, and has the property that $\mathrm{Ind}(\gamma, z) = 1$ for $z \in \Omega$. Let $\Gamma = \phi \circ \gamma$. Then Γ is a parametrization of $\partial \mathbb{D}$ that starts at $\phi(a)$ and reaches $\phi(b)$ before $\phi(c)$, so we are done if we can show that $\mathrm{Ind}(\Gamma, 0) = 1$. This is immediate from the lemma, plus a little limiting argument:

Choose $p \in \Omega$ with $\phi(p) = 0$. For $0 < r < 1$ define a curve $\Gamma_r : [0, 1] \to \mathbb{D}$ by
$$\Gamma_r(t) = r\Gamma(t),$$
and let $\gamma_r = \phi^{-1} \circ \Gamma_r$. The lemma shows that $\mathrm{Ind}(\gamma_r, p) = \mathrm{Ind}(\Gamma_r, 0)$. We know that the index of a curve about a point depends continuously on the curve (this follows from the second and third parts of Lemma 4.6), so we have
$$\mathrm{Ind}(\Gamma, 0) = \lim_{r \to 1} \mathrm{Ind}(\Gamma_r, 0) = \lim_{r \to 1} \mathrm{Ind}(\gamma_r, p) = \mathrm{Ind}(\gamma, p) = 1. \qquad \square$$

Finally:

Theorem 17.3. *Suppose that Ω is a bounded simply connected open set and every boundary point of Ω is simple. If (a, b, c) is a positively oriented triple of distinct points of $\partial \Omega$ and (A, B, C) is a positively oriented triple of distinct points of $\partial \mathbb{D}$ then there is a conformal equivalence ϕ of Ω and $\mathbb{D}$ such that $\phi(a) = A$, $\phi(b) = B$, and $\phi(c) = C$.*

Proof. Proposition 17.2 says that if ψ_1 is a conformal equivalence of Ω and $\mathbb{D}$ then $(\psi_1(a), \psi_1(b), \psi_1(c))$ is a positively oriented triple of distinct points of $\partial \mathbb{D}$. Proposition 17.0 shows that there exists $\psi_2 \in \mathrm{Aut}(\mathbb{D})$ such that $\psi_2(\psi_1(a)) = A$, $\psi_2(\psi_1(b)) = B$, and $\psi_2(\psi_1(c)) = C$. Let $\phi = \psi_2 \circ \psi_1$. $\qquad \square$

Theorem 17.3 will suffice for the application in a later chapter. But just for fun, let's see if we can make sense of the traditional "if you traverse the

17. Orientation

boundary in the positive direction then the set lies to your left". Of course the first question is what this means, exactly.

Suppose that γ is a differentiable curve and $p = \gamma(t_0)$. Suppose that $\gamma'(t_0) \neq 0$. Then if you "traverse" γ, the vector $\gamma'(t_0)$ is pointing "ahead" of you when you arrive at p. So it seems that $i\gamma'(t_0)$ should be a vector pointing "to your left" and $-i\gamma'(t_0)$ should be a vector pointing "to your right". We may as well take this as the definition: A point lying to the left of γ at p will be by definition a point of the form $p + i\alpha\gamma'(t)$ for small $\alpha > 0$, and similarly for "right".

With this definition of "to the left" and "to the right" the next proposition says that as you cross γ from right to left the index increases by 1.

Supposing that γ is smooth, we will say that $p = \gamma(t_0)$ is a *regular point* of γ if γ' is continuous at t_0 (here we assume, as we may, that t_0 is an interior point of the parameter interval of γ; the definition would be slightly different for an endpoint), $\gamma'(t_0) \neq 0$, and p is not a point of self-intersection of γ, which is to say that t_0 is the only t such that $\gamma(t) = p$.

Proposition 17.4. *Suppose that γ is a smooth closed curve and $p = \gamma(t_0)$ is a regular point of γ. For $\alpha \in \mathbb{R}$ define $a = a(\alpha) = p + i\alpha\gamma'(t_0)$ and $b = b(\alpha) = p - i\alpha\gamma'(t_0)$. If $\alpha > 0$ is small enough then both a and b are elements of $\mathbb{C} \setminus \gamma^*$ and*

$$\mathrm{Ind}(\gamma, a) = \mathrm{Ind}(\gamma, b) + 1.$$

The idea of the proof is very simple: We look carefully at the formula for $\mathrm{Ind}(\gamma, a) - \mathrm{Ind}(\gamma, b)$, and it turns out that we can show that it tends to 1 as $\alpha \to 0$. Here we will give a sloppy version of the proof, simply replacing this or that with something that is "close" to it, without worrying about the size of the error. We give a rigorous proof at the end of the chapter.

"Proof". The fact that $a, b \in \mathbb{C} \setminus \gamma^*$ if $\alpha > 0$ is small enough seems clear; we defer an actual proof of that until later. Since $\mathrm{Ind}(\gamma, z)$ depends continuously on $z \in \mathbb{C} \setminus \gamma^*$ and takes only integer values, it is enough to show that

$$\lim_{\alpha \to 0^+} (\mathrm{Ind}(\gamma, a) - \mathrm{Ind}(\gamma, b)) = 1.$$

To simplify the notation we assume that $\gamma : [-1, 1] \to \mathbb{C}$ and $t_0 = 0$. Since γ' is continuous at 0, we may choose $\delta > 0$ so that $\gamma'(t)$ is close to $\gamma'(0)$ for all $t \in [-\delta, \delta]$. Since

$$\frac{1}{\gamma(t) - a} - \frac{1}{\gamma(t) - b} \to 0$$

as $\alpha \to 0$, uniformly for $t \in [-1, 1] \setminus [-\delta, \delta]$, we need only show that

$$\lim_{\alpha \to 0^+} \frac{1}{2\pi i} \int_{-\delta}^{\delta} \left(\frac{1}{\gamma(t) - a} - \frac{1}{\gamma(t) - b} \right) \gamma'(t) \, dt = 1.$$

The serious hand-waving starts here: The fact that $\gamma'(t)$ is close to $\gamma'(0)$ shows that $\gamma(t)$ is close to $p + t\gamma'(0)$ for $t \in [-\delta, \delta]$, and hence the integral above is close to

$$\frac{1}{2\pi i} \int_{-\delta}^{\delta} \left(\frac{1}{p + t\gamma'(0) - (p + i\alpha\gamma'(0))} - \frac{1}{p + t\gamma'(0) - (p - i\alpha\gamma'(0))} \right) \gamma'(0) \, dt$$

$$= \frac{1}{2\pi i} \int_{-\delta}^{\delta} \frac{2i\alpha\gamma'(0)}{(t^2 + \alpha^2)\gamma'(0)^2} \gamma'(0) \, dt$$

$$= \frac{1}{\pi} \int_{-\delta}^{\delta} \frac{\alpha \, dt}{t^2 + \alpha^2}$$

$$= \frac{1}{\pi}(\arctan(\delta/\alpha) - \arctan(-\delta/\alpha)),$$

which tends to 1 as $\alpha \to 0^+$. □

We've committed a multitude of sins here, but this version of the proof does have the virtue that it's probably easier to understand than the rigorous proof below, where we fear the basic idea may be obscured by the details.

One can use the proposition, together with the fact that the index is constant on components of the complement, to determine the value of $\text{Ind}(\Gamma, z)$ for any given z, for just about any cycle Γ that you're likely to run into: Start at a point of the unbounded component of $\mathbb{C} \setminus \Gamma^*$, where you know that the index vanishes, and move towards z; each time you cross one of the curves making up Γ note whether you're crossing it from right to left or left to right, and record the change this makes in the index. A few applications:

Proposition 17.5. *Suppose γ is a smooth simple closed curve in $\mathbb{C}$. Then $\mathbb{C} \setminus \gamma^*$ is not connected.*

(Of course the Jordan Curve Theorem says that if γ is any simple closed curve then $\mathbb{C} \setminus \gamma^*$ has exactly two components. But the proof of the Jordan Curve Theorem is far from trivial.)

Proof. There exists t such that $\gamma'(t) \neq 0$. Hence $\gamma(t)$ is a regular point of γ, and now Proposition 17.4 shows that $\text{Ind}(\gamma, z)$ is not constant for $z \in \mathbb{C} \setminus \gamma^*$. □

A similar argument proves

Proposition 17.6. *Suppose that γ is a smooth simple closed curve such that $\mathbb{C} \setminus \gamma^*$ has exactly two components. Then either $\mathrm{Ind}(\gamma, z) = 1$ for all z in the bounded component of $\mathbb{C} \setminus \gamma^*$ or $\mathrm{Ind}(\gamma, z) = -1$ for all z in the bounded component of $\mathbb{C} \setminus \gamma^*$.*

If γ is as in the previous proposition we will say that γ is positively or negatively oriented depending on whether $\mathrm{Ind}(\gamma, z)$ is 1 or -1 for z in the bounded component of $\mathbb{C} \setminus \gamma^*$. And we have now shown that, at least for regions bounded by simple smooth closed curves, if you traverse the boundary in the positive direction then the region lies to your left!

We now turn to the actual proof of Proposition 17.4. Squeamish readers may wish to stop reading at this point, but readers who want to improve their epsilon delta skills may find a few useful tricks.

The following lemma can be useful in many places (readers who have studied measure theory will recognize it as a rudimentary version of the Dominated Convergence Theorem, in which regard recall Lemma 9.1.4 above):

Lemma 17.7. *Suppose that f, g, and f_n, $n = 1, 2, \ldots$, are piecewise-continuous functions from $\mathbb{R}$ to $\mathbb{C}$. Suppose that $g \geq 0$ and $\int_{-\infty}^{\infty} g(x)\, dx < \infty$. If $|f_n| \leq g$ for all n and $f_n \to f$ uniformly on compact subsets of $\mathbb{R}$ then*

$$\lim_{n \to \infty} \int_{-\infty}^{\infty} f_n(x)\, dx = \int_{-\infty}^{\infty} f(x)\, dx.$$

Readers wondering why such a lemma might be needed should consider the following example: Define

$$f_n(x) = \begin{cases} \frac{1}{n} & (0 < x < n), \\ 0 & \text{otherwise.} \end{cases}$$

Then $f_n \to 0$ uniformly on $\mathbb{R}$, although $\int_{-\infty}^{\infty} f_n$ does not tend to 0. Uniform convergence is no help when dealing with improper integrals.

Proof. Let $\epsilon > 0$. Since the integral of g over the entire line is finite, there exists $A > 0$ such that
$$\int_A^{\infty} g(x)\, dx < \epsilon$$
and
$$\int_{-\infty}^{-A} g(x)\, dx < \epsilon.$$

Now, since $f_n \to f$ uniformly on $[-A, A]$, there exists N such that
$$|f_n(x) - f(x)| < \epsilon/A \qquad (n > N, x \in [-A, A]).$$

Note that $|f(x)| \le g(x)$ for all x. It follows from the triangle inequality that for every $n > N$ we have

$$\left| \int_{-\infty}^{\infty} (f_n - f) \right| \le \int_{-\infty}^{-A} |f_n - f| + \int_{-A}^{A} |f_n - f| + \int_{A}^{\infty} |f_n - f|$$

$$\le \int_{-\infty}^{-A} 2g + \int_{-A}^{A} \epsilon/A + \int_{A}^{\infty} 2g$$

$$< 2\epsilon + 2\epsilon + 2\epsilon. \qquad \square$$

And now we dive in:

Proof (Proposition 17.4). As before we assume that $\gamma : [-1,1] \to \mathbb{C}$ and $t_0 = 0$. We choose $\delta > 0$ so that

$$|\gamma'(t) - \gamma'(0)| < \frac{1}{4}|\gamma'(0)|$$

for all $t \in [-\delta, \delta]$. It follows that

$$\left| \frac{\gamma(t) - \gamma(0)}{t} - \gamma'(0) \right| < \frac{1}{4}|\gamma'(0)|$$

whenever $0 < |t| < \delta$. We prove this assuming that $0 < t < \delta$; the notation would be different for $-\delta < t < 0$:

$$\left| \frac{\gamma(t) - \gamma(0)}{t} - \gamma'(0) \right| = \left| \frac{1}{t} \int_0^t (\gamma'(s) - \gamma'(0))\, ds \right|$$

$$\le \frac{1}{t} \int_0^t |\gamma'(s) - \gamma'(0)|\, ds$$

$$< \frac{1}{t} \int_0^t \frac{1}{4}|\gamma'(0)| = \frac{1}{4}|\gamma'(0)|.$$

Hence $0 < |t| < \delta$ implies that

$$\left| 1 - \frac{\gamma(t) - \gamma(0)}{t\gamma'(0)} \right| < \frac{1}{4}.$$

The triangle inequality implies $\left| 1 + \frac{\gamma(t)-\gamma(0)}{t\gamma'(0)} \right| < \frac{9}{4}$, and hence

$$\left| 1 - \left(\frac{\gamma(t) - \gamma(0)}{t\gamma'(0)}\right)^2 \right| = \left| \left(1 - \frac{\gamma(t) - \gamma(0)}{t\gamma'(0)}\right)\left(1 + \frac{\gamma(t) - \gamma(0)}{t\gamma'(0)}\right) \right| < \frac{9}{16},$$

finally showing that

$$\operatorname{Re}\left(\left(\frac{\gamma(t)-\gamma(0)}{t\gamma'(0)}\right)^2\right) > \frac{7}{16} \quad (0 < |t| < \delta).$$

Now, before we can start talking about $\operatorname{Ind}(\gamma, a)$ and $\operatorname{Ind}(\gamma, b)$ we do need to verify that $a, b \notin \gamma^*$ for small $\alpha > 0$. We need to show that if $\alpha > 0$ is small enough then

$$\gamma(t) \neq \gamma(0) \pm i\alpha\gamma'(0)$$

for all $t \in [-1, 1]$. If $t = 0$ this is clear since $\alpha \neq 0$ and $\gamma'(0) \neq 0$. If $0 < |t| \leq \delta$ then the inequality $\left|1 - \frac{\gamma(t)-\gamma(0)}{t\gamma'(0)}\right| \leq \frac{1}{4}$ shows that

$$\frac{\gamma(t)-\gamma(0)}{t\gamma'(0)} \neq \pm\frac{i\alpha}{t},$$

since $i\alpha/t$ is purely imaginary. Finally, $K = \gamma([-1,1] \setminus (-\delta, \delta))$ is compact, and the fact that $\gamma(0)$ is not a point of self-intersection of γ shows that $\gamma(0) \notin K$. Hence there exists $r > 0$ such that $K \cap D(\gamma(0), r) = \emptyset$, which shows that $\gamma(t) \neq a$ and $\gamma(t) \neq b$ for $\delta \leq |t| \leq 1$ if α is small enough.

As before, our conclusion follows if we can show that

$$\lim_{\alpha \to 0^+} \frac{1}{2\pi i} \int_{-\delta}^{\delta} \left(\frac{1}{\gamma(t)-a} - \frac{1}{\gamma(t)-b}\right) \gamma'(t)\, dt = 1.$$

Now

$$\frac{1}{2\pi i} \int_{-\delta}^{\delta} \left(\frac{1}{\gamma(t)-a} - \frac{1}{\gamma(t)-b}\right) \gamma'(t)\, dt$$

$$= \frac{1}{2\pi i} \int_{-\delta}^{\delta} \frac{2i\alpha\gamma'(0)}{(\gamma(t)-\gamma(0))^2 + \alpha^2\gamma'(0)^2} \gamma'(t)\, dt$$

$$= \frac{1}{\pi} \int_{-\delta}^{\delta} \frac{\gamma'(t)}{\gamma'(0)} \frac{\alpha}{\left(\frac{\gamma(t)-\gamma(0)}{t\gamma'(0)}\right)^2 t^2 + \alpha^2}\, dt.$$

We need to make a change of variables in order to use the preceding lemma to see what this converges to; we obtain

$$\frac{1}{2\pi i} \int_{-\delta}^{\delta} \left(\frac{1}{\gamma(t)-a} - \frac{1}{\gamma(t)-b}\right) \gamma'(t)\, dt$$

$$= \frac{1}{\pi} \int_{-\delta/\alpha}^{\delta/\alpha} \frac{\gamma'(\alpha t)}{\gamma'(0)} \frac{1}{\left(\frac{\gamma(\alpha t)-\gamma(0)}{\alpha t\gamma'(0)}\right)^2 t^2 + 1}\, dt.$$

Now let α tend to 0. It certainly *appears* as though

$$\frac{1}{\pi}\int_{-\delta/\alpha}^{\delta/\alpha} \frac{\gamma'(\alpha t)}{\gamma'(0)} \frac{1}{\left(\frac{\gamma(\alpha t)-\gamma(0)}{\alpha t \gamma'(0)}\right)^2 t^2 + 1}\, dt \to \frac{1}{\pi}\int_{-\infty}^{\infty} \frac{1}{t^2+1}\, dt = 1.$$

We can justify this using the preceding lemma. First, the fact that γ' is continuous at 0 shows that

$$\frac{\gamma'(\alpha t)}{\gamma'(0)} \to 1$$

uniformly on compact subsets of $\mathbb{R}$, while the fact that $\gamma'(0)$ exists (and is nonzero) shows that

$$\frac{1}{\left(\frac{\gamma(\alpha t)-\gamma(0)}{\alpha t \gamma'(0)}\right)^2 t^2 + 1} \to \frac{1}{t^2+1}$$

uniformly on compact sets.

So we are done if we can show that the integrand is dominated by an appropriate function g. Note that if $t \in [-\delta/\alpha, \delta/\alpha]$ then $\alpha t \in [-\delta, \delta]$. Hence we can apply the inequalities we derived at the start of the proof: The fact that $|\gamma'(t) - \gamma'(0)| < \frac{1}{4}|\gamma'(0)|$ for $t \in [-\delta, \delta]$ shows that $|\gamma'(\alpha t)| < \frac{5}{4}|\gamma'(0)|$ for $|t| < \delta/\alpha$, so that

$$\left|\frac{\gamma'(\alpha t)}{\gamma'(0)}\right| < \frac{5}{4}.$$

Similarly, we know that

$$\mathrm{Re}\left(\left(\frac{\gamma(\alpha t) - \gamma(0)}{\alpha t \gamma'(0)}\right)^2\right) > \frac{7}{16},$$

which shows that

$$\left|\frac{1}{\left(\frac{\gamma(\alpha t)-\gamma(0)}{\alpha t \gamma'(0)}\right)^2 t^2 + 1}\right| < \frac{1}{\frac{7}{16}t^2 + 1},$$

since t is real. So we can apply the lemma, with

$$g(t) = \frac{5}{4}\frac{1}{\frac{7}{16}t^2 + 1}. \qquad \square$$

Well, you were warned. Actually that wasn't so bad now, was it?

There are various other ways to prove things like Proposition 17.4. One reason we chose the method we did is that it is related to an interesting connection between Cauchy integrals and Poisson integrals:

17. Orientation

- **Exercise 17.1.** Suppose that $f \in C(\partial \mathbb{D})$, and for $|z| \neq 1$ define the Cauchy integral of f by
$$C[f](z) = \frac{1}{2\pi i} \int_{\partial \mathbb{D}} \frac{dw}{z - w}$$
(where of course we choose a positive orientation for $\partial \mathbb{D}$). Show that if $0 < r < 1$ and $e^{it} \in \partial \mathbb{D}$ then
$$C[f](re^{it}) - C[f]\left(\frac{1}{r}e^{it}\right) = P[f](re^{it}).$$
Here $P[f]$ is the Poisson integral of f, as in Section 10.1. Note that it follows that
$$C[f](re^{it}) - C[f]\left(\frac{1}{r}e^{it}\right) \to f(e^{it})$$
uniformly as $r \to 1^-$.

Chapter 18

The Modular Function

Actually "A Modular Function" would be a more accurate title for the chapter. Roughly speaking, a modular function is a meromorphic function in Π^+ which is invariant under some nontrivial subgroup of the modular group Γ defined below. We are going to construct one.

The *modular group* Γ is the group of all linear-fractional transformations ϕ which can be represented in the form

$$\phi(z) = \frac{az+b}{cz+d}$$

where a, b, c, and d are integers satisfying $ad - bc = 1$. (It is clear that Γ is closed under composition since the product of two matrices with integral entries and determinant one is another such matrix. Similarly the inverse of any element of Γ is another element of Γ: If ϕ is as above then

$$\phi^{-1}(z) = \frac{dz-b}{-cz+a}.$$

So Γ is a group.) Note that Γ is a subgroup of $\text{Aut}(\Pi^+)$.

In particular Γ is the subgroup of $\text{Aut}(\Pi^+)$ consisting of all transformations corresponding to matrices in the group $SL_2(\mathbb{Z})$. We will often ignore the distinction between the elements of $SL_2(\mathbb{Z})$ and Γ, stating something for one group and proving it in the other group without comment. This could possibly lead to problems since an element of Γ actually corresponds to *two* matrices in $SL_2(\mathbb{Z})$, but in fact everything we say about $M \in SL_2(\mathbb{Z})$ will apply equally well to $-M$.

We will be concerned with a subgroup of Γ sometimes denoted $\Gamma(2)$; this is the subgroup consisting of all ϕ as above, subject to the additional

condition that a and d are odd and b and c are even. Note that if $\begin{bmatrix} a & b \\ c & d \end{bmatrix} \in SL_2(\mathbb{Z})$ then the extra condition required for membership in $\Gamma(2)$ is equivalent to
$$\begin{bmatrix} a & b \\ c & d \end{bmatrix} \equiv \begin{bmatrix} 1 & 0 \\ 0 & 1 \end{bmatrix} \mod 2.$$

This shows that $\Gamma(2)$ is indeed a subgroup of Γ. (If you're not happy with that argument you can easily prove directly that $\Gamma(2)$ is a subgroup of Γ; in any case the condition about being congruent to the identity matrix mod 2 is a useful mnemonic for keeping the definition of $\Gamma(2)$ straight.)

The main goal of the chapter is to construct a certain function $\lambda \in H(\Pi^+)$ such that $\lambda(\Pi^+) = \mathbb{C} \setminus \{0, 1\}$, and such that $\lambda(z) = \lambda(w)$ if and only if $z = \phi(w)$ for some $\phi \in \Gamma(2)$. (In the terminology to be introduced in the next chapter it will turn out that λ is a holomorphic covering map from Π^+ onto $\mathbb{C} \setminus \{0, 1\}$, with $\mathrm{Aut}(\Pi^+, \lambda) = \Gamma(2)$.)

The first thing we will show is that $\Gamma(2)$ is generated by σ and τ, where
$$\sigma(z) = \frac{z}{-2z+1}, \quad \tau(z) = z + 2.$$

(Our σ is the inverse of the transformation often called σ in this context; although of course it makes no real difference, we will see below that a certain formula that we will be using a great deal has a more elegant appearance in our notation.) We will see a little later that $\Gamma(2)$ is in fact "freely generated" by σ and τ.

Note first that
$$\tau^n = \begin{bmatrix} 1 & 2n \\ 0 & 1 \end{bmatrix},$$
so that
$$\tau^n \begin{bmatrix} a & b \\ c & d \end{bmatrix} = \begin{bmatrix} a+2nc & b+2nd \\ c & d \end{bmatrix};$$
similarly
$$\sigma^n \begin{bmatrix} a & b \\ c & d \end{bmatrix} = \begin{bmatrix} a & b \\ c-2na & d-2nb \end{bmatrix}.$$

Lemma 18.0. *The group $\Gamma(2)$ is generated by σ and τ.*

Proof. It is clear that σ and τ are elements of $\Gamma(2)$; let G be the subgroup of $\Gamma(2)$ that they generate. We need to show that $\Gamma(2) = G$.

Define $\chi : \Gamma(2) \to \mathbb{Z}^+$ by
$$\chi\left(\begin{bmatrix} a & b \\ c & d \end{bmatrix}\right) = |a| + |c|.$$

Suppose that $\phi \in \Gamma(2)$; we need to show that $\phi \in G$. To this end, define

$$G \circ \phi = \{\psi \circ \phi : \psi \in G\}.$$

Now, since χ takes only positive integer values, there must exist $\phi_0 = \begin{bmatrix} a & b \\ c & d \end{bmatrix} \in G \circ \phi$ with $\chi(\phi_0)$ minimal.

Suppose that $c \neq 0$. If $|a| > |c|$ this leads to a contradiction: There exists $n \in \mathbb{Z}$ such that $-|c| \leq a + 2nc < |c|$; this implies that $-|c| < a + 2nc < |c|$ since a is odd and c is even, and hence that

$$\chi(\tau^n \phi_0) = |a + 2nc| + |c| < |c| + |c| < |a| + |c| = \chi(\phi_0),$$

contradicting our choice of ϕ_0. On the other hand, if $|a| \leq |c|$ then we must have $0 < |a| < |c|$ since a is odd, and this shows by a similar argument that there exists $n \in \mathbb{Z}$ such that $\chi(\sigma^n \phi_0) < \chi(\phi)$, again contradicting our choice of ϕ_0.

So we must have $c = 0$. Since $ad - bc = 1$, this implies that $a = d = \pm 1$, so that $\phi_0 = \tau^n$ for some $n \in \mathbb{Z}$. In particular $\phi_0 \in G$, and hence $\phi \in G$. □

It turns out that σ and τ are conjugate in Γ (although they are not conjugate in $\Gamma(2)$). Define $j \in \Gamma$ by

$$j(z) = \frac{-1}{z}.$$

Then you can easily check that

$$\sigma = j^{-1} \circ \tau \circ j.$$

(Noting that j is its own inverse, we will often write this relation as $\sigma = j \circ \tau \circ j$.) We will find this fact very useful; it will often happen that assertions that are obvious for τ lead to assertions about σ which are not quite so obvious. (This is the reason for our nontraditional choice of σ; the traditional σ is conjugate in Γ to τ^{-1} instead of to τ.)

We begin the study of the way σ and τ act on the upper half-plane with the next lemma. Define

$$T = \{z \in \Pi^+ : -1 \leq \operatorname{Re}(z) < 1\}$$

and

$$S = \left\{z \in \Pi^+ : \left|z + \frac{1}{2}\right| > \frac{1}{2}, \left|z - \frac{1}{2}\right| \geq \frac{1}{2}\right\}.$$

These two regions are illustrated in Figure 18.1 and Figure 18.2:

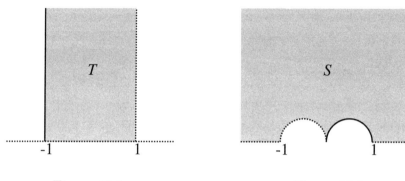

Figure 18.1   Figure 18.2

- **Exercise 18.1.** Show that $S = j(T)$. (You should do this without a lot of calculation, using the fact that j takes lines and circles to lines and circles and preserves angles.)

Lemma 18.1. (i) *The sets $\tau^n(T)$ $(n \in \mathbb{Z})$ are disjoint, and their union is Π^+.*

(ii) *The sets $\sigma^n(S)$ $(n \in \mathbb{Z})$ are disjoint, and their union is Π^+.*

(iii) *If $n \in \mathbb{Z}$ and $n \neq 0$ then $\tau^n(T) \subset S$.*

(iv) *If $n \in \mathbb{Z}$ and $n \neq 0$ then $\sigma^n(S) \subset T$.*

Part (i) says that T is a *fundamental domain* for the subgroup of $\Gamma(2)$ generated by τ.

Proof. Parts (i) and (iii) are obvious from Figures 18.1 and 18.2 (of course they are also very easy to prove algebraically). On the other hand, parts (ii) and (iv) may well not be obvious. But they follow immediately from parts (i) and (ii), since $\sigma = j \circ \tau \circ j$ and $S = j(T)$. □

Sure enough, the fact that σ is conjugate (in Γ) to τ is very useful. Just this once you should write out the details:

- **Exercise 18.2.** Explain carefully how parts (ii) and (iv) of the lemma follow from the other two parts. (Recall that j is its own inverse.)

It is quite obvious how the sets $\tau^n(T)$ tile the upper half-plane; Figure 18.3 shows what the sets $\sigma^n(S)$ look like:

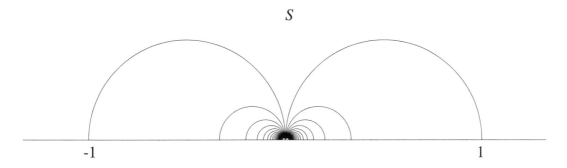

Figure 18.3

Since σ and τ are conjugate in Γ, it seems that pictures involving σ should look like pictures involving τ in some way. If the reader is bothered by the fact that the two seem to look very different we point out that this is caused by our point of view; if we look at the same families of sets in the unit disk we can see the symmetry more easily. Figure 18.4 is an illustration of how the sets $\sigma^n(S)$ appear in the unit disk, while Figure 18.5 shows the sets $\tau^n(T)$ in the unit disk:

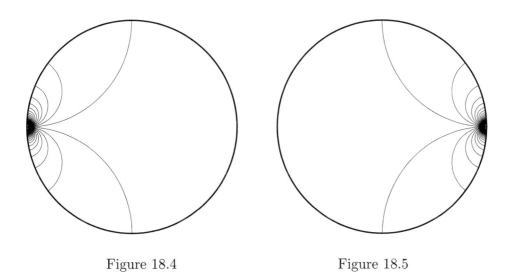

Figure 18.4   Figure 18.5

That's better. (The reason the symmetry is more apparent in the disk is that there j becomes $j(z) = -z$; this symmetry looks more symmetric to the eye than $j(z) = -1/z$.) We work in Π^+ instead of $\mathbb{D}$ here in spite of the fact that the pictures look nicer in the disk, largely because the formula for τ is so simple in Π^+.

Lemma 18.1 will be used in determining a fundamental domain for $\Gamma(2)$. The connection is given by the next lemma, which will also show that $\Gamma(2)$ is freely generated by σ and τ. (I stumbled upon the lemma while thinking about how one might prove that $\Gamma(2)$ is freely generated by σ and τ. Of course I realized immediately that the lemma must be well known; imagine my delight when I was told that it's a version of something commonly known as the "Ping Pong Lemma"!)

First we should probably explain what it *means* to say a group is freely generated by two elements:

Suppose that G is a group and G is generated by $a, b \in G$. A *nontrivial reduced word* in a and b is an expression of the form

$$g_1^{n_1} g_2^{n_2} \cdots g_k^{n_k},$$

where $k \geq 1$, each n_j is a nonzero integer, each g_j is either a or b, and the a's and b's alternate (that is, $g_{j+1} = b$ if and only if $g_j = a$). For example, a, $a^2 b^{-1} ab$ and bab are nontrivial reduced words in a and b. On the other hand, the identity element e, corresponding to $k = 0$, is the trivial word, and an expression like abb^2 is a word but not a reduced word; any word is equal to a reduced word, for example $abb^2 = ab^3$.

We say that G is *freely generated* by a and b if G is generated by a and b and no nontrivial reduced word in a and b is equal to the identity element of G. (In this case G is a *free group on two generators*.) Since G is a group, it follows that any two reduced words are different; two expressions formed from a and b are not the same unless they look the same after "combining factors" to obtain reduced words (of course $abb^2 = ab^3$ in any group, but if G is freely generated by a and b then we know that $a^2 baba \neq abab$ because both sides are reduced words and they don't look the same).

Lemma 18.2 (Ping Pong). *Suppose that G is a group of bijections of a set X (with composition as the group operation), and G is generated by a, $b \in G$. Suppose that $A, B \subset X$ satisfy the following:*

(i) *The sets $a^n(A)$ ($n \in \mathbb{Z}$) are disjoint, and their union is X.*

(ii) *The sets $b^n(B)$ ($n \in \mathbb{Z}$) are disjoint, and their union is X.*

(iii) *If $n \in \mathbb{Z}$ and $n \neq 0$ then $a^n(A) \subset B$.*

(iv) *If $n \in \mathbb{Z}$ and $n \neq 0$ then $b^n(B) \subset A$.*

Let $E = A \cap B$, and assume that $E \neq \emptyset$. If w is a nontrivial reduced word in a and b then

$$E \cap w(E) = \emptyset.$$

18. The Modular Function

Corollary 18.3. *The hypotheses of Lemma 18.2 imply that G is freely generated by a and b.*

Corollary 18.4. *The hypotheses of Lemma 18.2 imply that*
$$g_1(E) \cap g_2(E) = \emptyset$$
for all $g_1, g_2 \in G$ with $g_1 \neq g_2$.

- **Exercise 18.3.** Prove the two corollaries.

Proof of Lemma 18.2. We may assume that w begins with a power of a. If w also ends with a power of a, so that $w = a^{n_1} b^{n_2} \ldots a^{n_k}$, then repeated applications of (iii) and (iv) show that
$$w(A) = a^{n_1} b^{n_2} \ldots b^{n_{k-1}} a^{n_k}(A) \subset a^{n_1} b^{n_2} \ldots b^{n_{k-1}}(B) \subset \cdots \subset a^{n_1}(A);$$
now (i) shows that $A \cap w(A) = \emptyset$, which implies that $(A \cap B) \cap w(A \cap B) = \emptyset$. Similarly if w ends in a power of b, so $w = a^{n_1} b^{n_2} \ldots b^{n_k}$, then we see that $w(B) \subset a^{n_1}(A)$, so that $A \cap w(B) = \emptyset$, which again implies that $(A \cap B) \cap w(A \cap B) = \emptyset$. □

Our current goal is to construct a fundamental domain for $\Gamma(2)$. This means we want a set $Q \subset \Pi^+$ such that

(i) The sets $\phi(Q)$ ($\phi \in \Gamma(2)$) are disjoint.

(ii) $\Pi^+ = \bigcup_{\phi \in \Gamma(2)} \phi(Q)$.

It is clear from Lemma 18.1 and Lemma 18.2 that if we set $Q = T \cap S$ then (i) holds (this strikes me as very interesting, since Lemma 18.1 is "obvious" and Lemma 18.2 is "just algebra"). The lemmas do not imply (ii) directly; in particular the set E in Lemma 18.2 need not be a fundamental domain for G (see Appendix 6 for a counterexample if you're curious). However: Since $\Gamma(2)$ is a group, proving (ii) is the same as showing that if $z \in \Pi^+$ then there exists $\phi \in \Gamma(2)$ with $\phi(z) \in Q$. The lemmas *do* give a procedure for finding such a ϕ if one exists:

Suppose for example that $w = \tau \circ \sigma^{-1} \circ \tau^2(z) \in Q = T \cap S$. In particular $w \in T$, so it follows from Lemma 18.1 that $\sigma^{-1} \circ \tau^2(z) = \tau^{-1} w \in S$. And hence $\tau^2(z) = \sigma \circ \tau^{-1}(w) \in T$, so $z \notin T$, which implies that $z \in S$. Note how the points z, $\tau^2(z)$, $\sigma^{-1} \circ \tau^2(z)$ and $w = \tau \circ \sigma^{-1} \circ \tau^2(z)$ jump back and forth from S to T and vice versa, until they reach a point of the intersection Q. Since $\Gamma(2)$ is freely generated by σ and τ, this is really the only way to get from z to a point of Q!

Let us define functions
$$\text{ping} : \Pi^+ \to T, \quad \text{pong} : \Pi^+ \to S$$
by saying that $\text{ping}(z)$ is the unique element of T such that $\text{ping}(z) = \tau^n(z)$ for some $n \in \mathbb{Z}$ and $\text{pong}(z)$ is the unique element of S such that $\text{pong}(z) = \sigma^n(z)$ for some $n \in \mathbb{Z}$. Note that $\text{ping}(z) = \text{pong}(z) = z$ if and only if $z \in Q$. If we are given $z \in \Pi^+$ and we want to find $\phi \in \Gamma(2)$ such that $\phi(z) \in Q$ then a reasonable approach is to set $z_1 = \text{ping}(z)$, $z_2 = \text{pong}(z_1)$, etc., and hope that eventually we reach Q. This is a reasonable approach because if there is such a ϕ then this sequence must in fact terminate at a point of Q. Proving that it actually does so takes a little work.

The idea is that first, ping and pong both move points closer to i; this is obvious for ping, and it will follow immediately for pong if we can define a suitable notion of "close to i" which is invariant under j. Hence the ping-pong sequence defined above must be eventually constant, since if not we shall see that it would tend to $\partial_\infty \Pi^+$, contradicting the fact that it remains close to i.

First, define $\rho : \Pi^+ \to [0, \infty)$ by
$$\rho(z) = \left| \frac{z-i}{z+i} \right|.$$

Note that if $\Phi : \Pi^+ \to \mathbb{D}$ is the Cayley transform then $\rho(z) = |\Phi(z)|$; hence $0 \leq \rho(z) < 1$ for all $z \in \Pi^+$, and the set of all z with $\rho(z) \leq r$ is a compact subset of Π^+ for any $r \in (0,1)$.

Lemma 18.5. *For every $z \in \Pi^+$ we have*

(i) $\rho(j(z)) = \rho(z)$,

(ii) $\rho(\text{ping}(z)) \leq \rho(z)$,

(iii) $\rho(\text{pong}(z)) \leq \rho(z)$.

Proof. The first part is a simple calculation. For (ii), note that if $z = x+iy$ then
$$\rho(z)^2 = 1 - \frac{4y}{x^2 + (y+1)^2}.$$
We may assume that $z \notin T$, since if $z \in T$ then $\text{ping}(z) = z$. But now $z = x + iy$ and $\text{ping}(z) = x' + iy$ where $|x| \geq 1$ and $|x'| \leq 1$; (ii) follows.

And (iii) is immediate from (i) and (ii), since $\sigma = j \circ \tau \circ j$. $\square$

(Note that if $x = 1$ then $\rho(\text{ping}(z)) = \rho(z)$ although $z \notin T$.)

The second half of the proof outlined above begins with a lemma most conveniently stated in the disk:

18. The Modular Function

Lemma 18.6. *Suppose that $f_n : \mathbb{D} \to \mathbb{D}$ is holomorphic. If $z \in \mathbb{D}$ and $|f_n(z)| \to 1$ then $|f_n(w)| \to 1$ for every $w \in \mathbb{D}$.*

Proof. We could prove this in any number of ways. For example, it follows easily from Montel's Theorem:

Suppose the conclusion fails. Then there exists $w \in \mathbb{D}$ and a subsequence (f_{n_j}) such that $f_{n_j}(w) \to \alpha$ with $|\alpha| < 1$. Now Montel's Theorem shows that (f_{n_j}) has a further subsequence which converges uniformly on compact sets to $f \in H(\mathbb{D})$. It is clear that $|f| \leq 1$ in $\mathbb{D}$. But $|f(z)| = 1$, so the Maximum Modulus Theorem implies that f is constant, contradicting the fact that $|f(w)| < 1$. □

If $(z_n) \subset \Pi^+$ we will say $z_n \to \partial_\infty \Pi^+$ if for every compact $K \subset \Pi^+$ there exists N so that $z_n \in \Pi^+ \setminus K$ for all $n > N$. Let $\Phi : \Pi^+ \to \mathbb{D}$ be the Cayley transform, and note that $z_n \to \partial_\infty \Pi^+$ if and only if $|\Phi(z_n)| \to 1$. Hence the previous lemma implies the following:

Corollary 18.7. *Suppose that $f_n : \Pi^+ \to \Pi^+$ is holomorphic, $z \in \Pi^+$ and $f_n(z) \to \partial_\infty \Pi^+$. Then $f_n(w) \to \partial_\infty \Pi^+$ for all $w \in \Pi^+$.*

As before we define

$$\rho(z) = \left|\frac{z-i}{z+i}\right| = |\Phi(z)|$$

for $z \in \Pi^+$. If $\begin{bmatrix} a & b \\ c & d \end{bmatrix} \in SL_2(\mathbb{Z})$ then a simple calculation, noting that $ad - bc = 1$, shows that

$$\rho\left(\frac{ai+b}{ci+d}\right)^2 = \frac{a^2+b^2+c^2+d^2-2}{a^2+b^2+c^2+d^2+2}.$$

Corollary 18.8. *If (ϕ_n) is a sequence of distinct elements of Γ (that is, $\phi_n \neq \phi_m$ for $n \neq m$) then $\phi_n(z) \to \partial_\infty \Pi^+$ for all $z \in \Pi^+$.*

Proof. The formula above shows that $\rho(\phi_n(i)) \to 1$, since for a given $R < \infty$ there are only finitely many $(a,b,c,d) \in \mathbb{Z}^4$ with $a^2+b^2+c^2+d^2 < R$. This says that $|\Phi(\phi_n(i))| \to 1$, so that $\phi_n(i) \to \partial_\infty \Pi^+$. Hence $\phi_n(z) \to \partial_\infty \Pi^+$ for all $z \in \Pi^+$, by the preceding lemma. □

It's time to wrap all this into a theorem:

Theorem 18.9. *The group $\Gamma(2)$ is freely generated by τ and σ. Further, $\Gamma(2)$ is a normal subgroup of Γ, every nontrivial element of $\Gamma(2)$ is either parabolic or hyperbolic, and if $Q = S \cap T$ then*

(i) *the sets $\phi(Q)$ $(\phi \in \Gamma(2))$ are disjoint*

(ii) $\bigcup_{\phi \in \Gamma(2)} \phi(Q) = \Pi^+$.

Proof. As noted above, Lemmas 18.1 and 18.2 show that $\Gamma(2)$ is freely generated by σ and τ and that (i) holds. To prove (ii) we need to show that if $z \in \Pi^+$ then there exists $\phi \in \Gamma(2)$ such that $\phi(z) \in Q$ (this shows that $z \in \phi^{-1}(Q)$). Assume that $z \in \Pi^+ \setminus Q$.

Since $\Pi^+ = S \cup T$, we must have $z \in S \setminus T$ or $z \in T \setminus S$. Since the two cases are equivalent via j, we may assume that $z \in S \setminus T$. Now we implement the algorithm suggested above for finding ϕ: We let $z_1 = \text{ping}(z)$, $z_2 = \text{pong}(z_1)$, $z_3 = \text{ping}(z_2)$, etc. So in particular $z_1 \in T$, $z_2 \in S$, $z_3 \in T$, etc., and there exist $\phi_n \in \Gamma(2)$ with $z_n = \phi_n(z)$.

If there exists n with $z_n \in Q$ we are done. Suppose not. Then (since, as noted above, $\text{ping}(w) = w$ if and only if $w \in T$ and $\text{pong}(w) = w$ if and only if $w \in S$) we have $z_{n+1} \neq z_n$ for all $n \geq 2$.

Now, the definition of ping and pong shows that for every m, n with $m > n$ there exists w, a nontrivial word in σ and τ, such that $\phi_m = w \circ \phi_n$. Since $\Gamma(2)$ is freely generated by σ and τ, this shows that $\phi_m \neq \phi_n$ for $m > n$. Hence Lemma 18.8 shows that $z_n \to \partial_\infty \Pi^+$. But this is impossible, since Lemma 18.5 shows that $\rho(z_n) \leq \rho(z) < 1$ for all n, so that the z_n are all contained in some fixed compact subset of Π^+.

That finishes the proof of (ii). Of course one could show that every nontrivial element of $\Gamma(2)$ is hyperbolic or parabolic by direct calculation, but it is more interesting to note that this fact follows from (i) and (ii): Suppose that $\psi \in \Gamma(2)$, $z \in \Pi^+$, and $\psi(z) = z$. Choose $\phi \in \Gamma(2)$ so that $z \in \phi(Q)$. Then we also have $z = \psi(z) \in \psi \circ \phi(Q)$, so (i) implies that $\phi = \psi \circ \phi$; hence ψ is the identity.

Finally, we need to show that $\Gamma(2)$ is a normal subgroup of Γ. Suppose that $A = \begin{bmatrix} a & b \\ c & d \end{bmatrix} \in \Gamma(2)$. Then

$$A = I + 2M$$

for some matrix M with integer entries. Suppose that $B \in SL_2(\mathbb{Z})$. Then

$$BAB^{-1} = I + 2BMB^{-1};$$

since B^{-1} also has integer entries it follows that $BAB^{-1} \in \Gamma(2)$. $\square$

18. The Modular Function

Note that the proof does give an actual algorithm for finding $\phi \in \Gamma(2)$ with $\phi(z) \in Q$, which could be easily implemented on a computer: It is clear how to find $n \in \mathbb{Z}$ such that $\tau^n(z) \in T$; then to find $n \in \mathbb{Z}$ such that $\sigma^n(z) \in S$ it suffices to find n such that $\tau^n(j(z)) \in T$.

Readers who know a little algebra can give essentially the same proof that $\Gamma(2)$ is a normal subgroup of Γ in a more interesting form:

- **Exercise 18.4.** Let $\mathbb{Z}_2 = \{0, 1\}$ be the field with two elements. There is a natural ring homomorphism $\chi : \mathbb{Z} \to \mathbb{Z}_2$, given by $\chi(n) = 0$ if n is even, $\chi(n) = 1$ if n is odd. Show that χ induces a group homomorphism $\tilde{\chi} : SL_2(\mathbb{Z}) \to SL_2(\mathbb{Z}_2)$ in a natural way. Find the kernel of $\tilde{\chi}$, and conclude that $\Gamma(2)$ is a normal subgroup of Γ.

Figure 18.6 below illustrates a few of the sets $\phi(Q)$ ($\phi \in \Gamma(2)$) as they appear in the unit disk.

It may be interesting to note that it is possible to determine from Figure 18.6 exactly *which* $\phi \in \Gamma(2)$ maps Q onto one of the regions pictured. Of course if you happen to know the value of some point in the region you could use the ping-pong algorithm described after the proof of Lemma 18.2 above, but you can also read this information directly from the figure, as follows:

Consider Figure 18.7 below, where a few of the regions have been labelled. Suppose we want to determine which $\phi \in \Gamma(2)$ gives $Q_{10} = \phi(Q)$.

First note that (in Π^+) we have $i \in Q$. It follows that $2 + i \in \tau(Q)$ and $-2 + i \in \tau^{-1}(Q)$, and hence that $j(2+i) \in \sigma(Q)$ and $j(-2+i) \in \sigma^{-1}(Q)$. Applying the Cayley transform it follows that in Figure 18.7 we have $Q_0 = Q$, $Q_1 = \tau(Q)$, $Q_2 = \tau^{-1}(Q)$, $Q_3 = \sigma(Q)$ and $Q_4 = \sigma^{-1}(Q)$.

Now, each $\phi(Q)$ is a "circular polygon" with four circular arcs for edges. Say $\phi(Q)$ and $\psi(Q)$ are *adjacent* if they have have an edge in common. Now, since ϕ is continuous, it follows that the regions adjacent to $\phi(Q)$ are $\phi \circ \tau(Q)$, $\phi \circ \tau^{-1}(Q)$, $\phi \circ \sigma(Q)$ and $\phi \circ \sigma^{-1}(Q)$. And, since ϕ preserves orientation, it actually follows that the four regions $\phi \circ \tau(Q)$, $\phi \circ \tau^{-1}(Q)$, $\phi \circ \sigma(Q)$ and $\phi \circ \sigma^{-1}(Q)$ traverse the boundary of $\phi(Q)$ *in counterclockwise order*, with respect to a point of $\phi(Q)$.

Thus, for example, since the four regions Q_0, Q_5, Q_6 and Q_7 are adjacent to $Q_1 = \tau(Q)$, we know that those four regions must be the same as the four regions $\tau^2(Q)$, Q, $\tau \circ \sigma(Q)$ and $\tau \circ \sigma^{-1}(Q)$, in some order, and in fact we know that (Q_0, Q_5, Q_6, Q_7) must be a *cyclic* permutation of $(\tau^2(Q), Q, \tau \circ \sigma(Q), \tau \circ \sigma^{-1}(Q))$! And we already know that $Q_0 = Q$; it follows that $Q_5 = \tau \circ \sigma(Q)$, $Q_6 = \tau \circ \sigma^{-1}(Q)$ and $Q_7 = \tau^2(Q)$.

Now repeat the argument with the regions adjacent to $Q_5 = \tau \circ \sigma(Q)$. We see that (Q_1, Q_8, Q_9, Q_{10}) must be a cyclic permutation of $(\tau \circ \sigma \circ \tau(Q), \tau \circ$

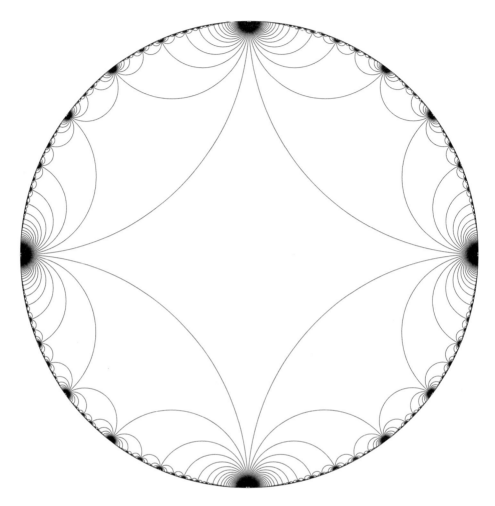

Figure 18.6.

$\sigma \circ \tau^{-1}(Q), \tau \circ \sigma^2(Q), \tau(Q))$. But we already know that $\tau(Q) = Q_1$, and it follows that $\tau \circ \sigma \circ \tau(Q) = Q_8$, $\tau \circ \sigma \circ \tau^{-1}(Q) = Q_9$ and $\tau \circ \sigma^2(Q) = Q_{10}$. In particular $Q_{10} = \tau \circ \sigma^2(Q)$, which is what we wanted to determine.

Lest it appear that we just got lucky here we should explain why this always works. If $\phi \in \Gamma(2)$ let the *length* of ϕ be $\ell(\phi)$, the sum of the absolute values of the exponents when ϕ is written as a reduced word in σ and τ, so for example $\ell(\tau\sigma^{-2}\tau^3) = 6$. Say $\ell(\phi) = 0$ if ϕ is the identity. Then it is clear that for any ϕ with $\ell(\phi) = n > 0$ exactly one of the four transformations $\phi \circ \tau$, $\phi \circ \tau^{-1}$, $\phi \circ \sigma$ and $\phi \circ \sigma^{-1}$ has length $n-1$ (and the other three have length $n+1$). So: We started by locating $\phi(Q)$ in Figure 18.7 for all ϕ with $\ell(\phi) \leq 1$; the procedure above then allows us to locate $\phi(Q)$ for all ϕ with $\ell(\phi) = 2$, since each such $\phi(Q)$ is adjacent to $\psi(Q)$ with $\ell(\psi) = 1$. Now we can locate the $\phi(Q)$ with $\ell(Q) = 3$, etc.

18. The Modular Function

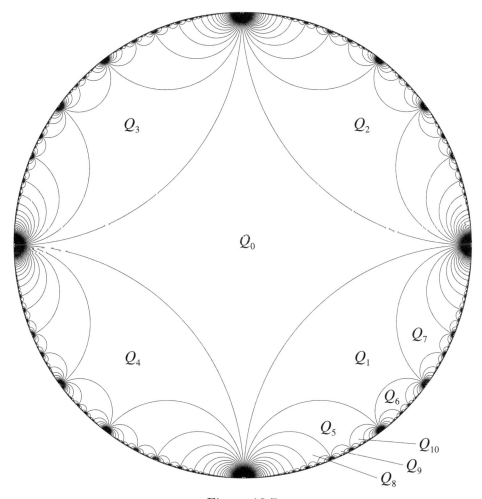

Figure 18.7.

This is an important algorithm, if only because it gives a reason to include those figures in the text. We now return to work, constructing our modular function. Recall that the notion of "covering map" was defined immediately before the statement of Theorem 16.3.0.

Theorem 18.10. *There exists a holomorphic covering map λ from Π^+ onto $\mathbb{C} \setminus \{0, 1\}$, with the following properties*:

(i) $\lambda \circ \phi = \lambda$ *for all* $\phi \in \Gamma(2)$.

(ii) λ *is one-to-one on* Q.

(iii) *If (z_j) is a sequence in Π^+ with $\operatorname{Im}(z_j) \to \infty$ then $\lambda(z_j) \to \infty$*.

We should note again that Q is a fundamental domain for $\Gamma(2)$; thus (for example) Theorem 18.9 above implies that $\lambda(Q) = \mathbb{C}\setminus\{0,1\}$, and that given $z, w \in \Pi^+$ we have $\lambda(z) = \lambda(w)$ if and only if $w = \phi(z)$ for some $\phi \in \Gamma(2)$.

In the proof we will be applying results that were proved for bounded open sets, such as Carathéodory's Theorem, to certain unbounded open sets in Π^+. If you're concerned about that you should note that we could use the Cayley transform to turn our open sets into subsets of the unit disk and then to transfer the conclusion back to Π^+.

Proof of Theorem 18.10. Let Q^+ be the interior of the right half of Q:

$$Q^+ = \left\{ z = x + iy : y > 0,\ 0 < x < 1,\ \left|z - \frac{1}{2}\right| > \frac{1}{2} \right\}.$$

The Riemann Mapping Theorem shows that there exists a conformal map h from Q^+ onto Π^+. Carathéodory's Theorem (Theorem 14.0) shows that h extends to a homeomorphism of the extended closure of Q^+, that is, $\overline{Q^+}\cup\{\infty\}$, onto $\overline{\Pi^+_\infty}$; we will denote this extension again by h. Since $(0, 1, \infty)$ is a positively oriented triple in the extended boundary of Q^+, Theorem 17.3 shows that we may choose h in such a way that $h(0) = 0$, $h(1) = 1$, and $h(\infty) = \infty$. (The triple $(0, 1, \infty)$ certainly *appears* to be a positively oriented triple in the extended boundary of Q^+; you can actually prove this fact using Exercise 4.20.)

In particular $h(iy) \in \mathbb{R}$ for all $y > 0$. The Schwarz Reflection Principle (Theorem 10.7.1) shows that there exists a function holomorphic in the interior of Q, which we will again denote by h, such that

$$h(x + iy) = \overline{h(-x + iy)}$$

whenever $x + iy \in Q^\circ$. It follows that h maps the interior of the left half of Q onto the lower half-plane, and also that h extends to a continuous map from $\overline{Q}\cup\{\infty\}$ onto $\mathbb{C}_\infty$. Now, since $h^{-1}(\{1\}) = \{1, -1\}$, $h^{-1}(\{0\}) = \{0\}$, $h^{-1}(\{\infty\}) = \{\infty\}$ and $(\overline{Q}\cup\{\infty\})\setminus\Pi^+ = \{-1, 0, 1, \infty\}$, this shows that $h(\overline{Q}\cap\Pi^+) = \mathbb{C}\setminus\{0,1\}$. But note that if $x + iy \in \partial Q \cap \Pi^+$ then $h(x+iy) = \overline{h(-x+iy)} = h(-x+iy)$ and one of the two points $x+iy$, $-x+iy$ lies in Q; hence we have

$$h(Q) = \mathbb{C}\setminus\{0,1\}.$$

Also note that h is one-to-one on Q: The construction of h shows that it is one-to-one on the closure of Q^+, and then the fact that $h(-x+iy) = \overline{h(x+iy)}$ shows that it is one-to-one on all of Q.

Note as well that if z_n is a sequence of points in Q and $\operatorname{Im}(z_n) \to \infty$ then $h(z_n) \to \infty$.

18. The Modular Function

Below we will need to know what $h(j(z))$ is for $z \in Q$; we may as well determine this now, while we're talking about h. (Note first that

$$j(Q) = j(T \cap S) = j(T) \cap j(S) = S \cap T = Q,$$

so that there is such a thing as $h \circ j$.) Define $\tilde{h} : Q^+ \to \mathbb{C}$ by

$$\tilde{h}(z) = \frac{1}{h(j(z))}.$$

Now, the construction shows that h is a conformal map from the interior of the left half of Q onto the lower half-plane. Since j defines a conformal map from Q^+ onto the interior of the left half of Q, it follows that $h \circ j$ defines a conformal map from Q^+ onto the lower half-plane, and hence that $\tilde{h}$ is a conformal map from Q^+ onto Π^+. The map $\tilde{h}$ extends continuously to the extended closure, and one easily verifies that $\tilde{h}(1) = 1 = h(1)$, $\tilde{h}(0) = 0 = h(0)$ and $\tilde{h}(\infty) = \infty = h(\infty)$. Hence $\tilde{h} = h$ in Q^+. By continuity and uniqueness of analytic continuation it follows that

$$h(j(z)) = \frac{1}{h(z)} \quad (z \in Q).$$

We now define $\lambda : \Pi^+ \to \mathbb{C}$ by

$$\lambda(z) = h(\phi^{-1}(z)) \quad (\phi \in \Gamma(2),\ z \in \phi(Q)).$$

Theorem 18.9 shows that λ is well defined, and much of what we need to prove about λ is immediate from Theorem 18.9 together with things we have proved about h: It is clear that $\lambda(\Pi^+) = \lambda(Q) = \mathbb{C} \setminus \{0, 1\}$, that $\lambda \circ \phi = \lambda$ for $\phi \in \Gamma(2)$, and that λ is one-to-one on Q. Condition (iii) is also quite simple:

Note that if $\mathrm{Im}(z) > 1$ then there exists $k \in \mathbb{Z}$ such that $\tau^k(z) \in Q$. If $\mathrm{Im}(z_n) \to \infty$ we may assume that $\mathrm{Im}(z_n) > 1$ for all n; now if we let $\tilde{z}_n = \tau^k z_n$ where $\tilde{z}_n \in Q$ then $\lambda(z_n) = \lambda(\tilde{z}_n) \to \infty$, since $\mathrm{Im}(\tilde{z}_n) \to \infty$.

It remains only to show that λ is a holomorphic covering map, and in particular that λ is holomorphic. We begin by showing that λ is holomorphic in $(Q \cup \tau^{-1}(Q))^\circ$: Suppose that $-1 - x + iy \in \tau^{-1}(Q)$ and $-1 - x > -3$. Since $\tau(-1 - x + iy) = 1 - x + iy \in Q$, we see that

$$\lambda(-1 - x + iy) = \lambda(1 - x + iy) = h(1 - x + iy) = \overline{h(-1 + x + iy)}.$$

So the Schwarz Reflection Principle (together with the uniqueness of analytic continuation) shows that λ is holomorphic in a neighborhood of any point z with $\mathrm{Im}(z) = -1$, and hence that λ is holomorphic in $(Q \cup \tau^{-1}(Q))^\circ$.

We could prove that λ is holomorphic in $(Q\cup\sigma^{-1}(Q))^\circ$ by a similar argument. This would require that we discuss the Schwarz Reflection Principle for reflections in circles instead of lines; in any case it is more interesting to note that the fact that λ is holomorphic in $(Q\cup\sigma^{-1}(Q))^\circ$ actually *follows* immediately from the fact that λ is holomorphic in $(Q\cup\tau^{-1}(Q))^\circ$, as it should:

Note first that the function $\lambda\circ j$ is invariant under $\Gamma(2)$. This is because $\Gamma(2)$ is a normal subgroup of Γ: If $\phi\in\Gamma(2)$ then there exists $\psi\in\Gamma(2)$ with $j\circ\phi=\psi\circ j$, and hence

$$\lambda\circ j\circ\phi=\lambda\circ\psi\circ j=\lambda\circ j.$$

But we already know that $\lambda\circ j=1/\lambda$ in Q; since both sides are invariant under $\Gamma(2)$, it follows that $\lambda\circ j=1/\lambda$ in all of Π^+. (One reason we took this route is to give an example showing why one might care about the fact that $\Gamma(2)$ is a normal subgroup of Γ; we will see later that this implies that $\lambda\circ g$ satisfies a similar identity for any $g\in\Gamma$.)

Now, since

$$j(Q\cup\tau^{-1}(Q))=j(Q)\cup j(\tau^{-1}(Q))=j(Q)\cup\sigma^{-1}(j(Q))=Q\cup\sigma^{-1}(Q)$$

and j is a homeomorphism, it follows that $j((Q\cup\tau^{-1}(Q))^\circ)=(Q\cup\sigma^{-1}(Q))^\circ$, and hence that $\lambda\circ j$ is holomorphic in $(Q\cup\sigma^{-1}(Q))^\circ$. But $\lambda\circ j=1/\lambda$, so λ itself is also holomorphic in $(Q\cup\sigma^{-1}(Q))^\circ$. Hence λ is holomorphic in $\Omega=(Q\cup\tau^{-1}(Q))^\circ\cup(Q\cup\sigma^{-1}(Q))^\circ$. Since λ is invariant under $\Gamma(2)$, this shows that λ is in fact holomorphic in $\bigcup_{\phi\in\Gamma(2)}\phi(\Omega)$.

Now, it is clear from Figures 18.1 and 18.2 that

$$Q\setminus T^\circ=\{-1+iy:y>0\},$$

and hence that $Q\setminus T^\circ\subset(Q\cup\tau^{-1}Q)^\circ$. (If you want to give an algebraic proof of this fact you might note that if k is an odd integer then $|-1-k/2|\geq 1/2$, and hence $|-1+iy-k/2|>1/2$ for $y>0$.) Applying j it follows that $Q\setminus S^\circ\subset(Q\cup\sigma^{-1}Q)^\circ$; since $Q^\circ=T^\circ\cap S^\circ$, this shows that

$$Q\setminus Q^\circ=(Q\setminus T^\circ)\cup(Q\setminus S^\circ)\subset(Q\cup\tau^{-1}(Q))^\circ\cup(Q\cup\sigma^{-1}(Q))^\circ=\Omega.$$

Since we certainly have $Q^\circ\subset\Omega$, it follows that $Q\subset\Omega$; hence $\Pi^+=\bigcup_{\phi\in\Gamma(2)}\phi(Q)\subset\bigcup_{\phi\in\Gamma(2)}\phi(\Omega)$, so that λ is holomorphic in Π^+.

And finally, it is not hard to show that λ is a covering map. Let $N=\lambda(Q^\circ)$. Then N is open and, since

$$\lambda^{-1}(N)=\bigcup_{\phi\in\Gamma(2)}\phi(Q^\circ),$$

$\lambda^{-1}(N)$ is the union of a collection of disjoint open sets O_ϕ such that the restriction of λ to any O_ϕ is a homeomorphism onto N; thus N is an elementary neighborhood of each of its points.

This takes care of most points of $\mathbb{C} \setminus \{0, 1\}$. For the other points we simply use a different fundamental domain; the most natural way to obtain this other fundamental domain in the present context is to use the image of Q under an element of Γ, as follows:

Suppose that $g \in \Gamma$, and let $Q' = g(Q)$. Suppose that $\phi_0 \in \Gamma(2)$ is not the identity. Since $\Gamma(2)$ is a normal subgroup of Γ, there exists $\psi \in \Gamma(2)$ with $\phi_0 \circ g = g \circ \psi$. Hence

$$\phi_0(Q') = \phi_0(g(Q)) = g(\psi(Q));$$

since $Q \cap \psi(Q) = \emptyset$ and g is a bijection, it follows that $Q' \cap \phi_0(Q') = \emptyset$. Hence the sets $\phi(Q')$ ($\phi \in \Gamma(2)$) are pairwise disjoint, and as above this shows that if $N' = \lambda((Q')^\circ)$ then N' is elementary for λ.

Now define $g_1 \in \Gamma$ by $g_1(z) = z - 1$; set $Q_1 = g(Q)$. The previous paragraph shows that every point of $\lambda(Q_1^\circ)$ has an elementary neighborhood. Let $g_2 = j \circ g_1$ and $Q_2 = g_2(Q)$; then every point of $\lambda(Q_2^\circ)$ has an elementary neighborhood.

So every point of $\lambda(Q^\circ \cup Q_1^\circ \cup Q_2^\circ)$ has an elementary neighborhood. But it is clear that $Q \setminus T^\circ \subset Q_1^\circ$. It follows that $Q \setminus S^\circ \subset Q_2^\circ$, and now as before the fact that $Q = Q^\circ \cup (Q \setminus T^\circ) \cup (Q \setminus S^\circ)$ shows that $Q \subset Q^\circ \cup Q_1^\circ \cup Q_2^\circ$. Hence every point of $\lambda(Q) = \mathbb{C} \setminus \{0, 1\}$ has an elementary neighborhood; λ is a covering map. $\square$

Exercises

18.5. Show that Lemma 18.6 follows from the Schwarz Lemma (Theorem 8.5.0). (First show that there exists $r < 1$ such that $d(z, w) \leq r$ for all $w \in K$ ...)

18.6. Show that Lemma 18.6 follows from Harnack's Inequality (Theorem 10.7.5). (Consider $u_n = 1 - \operatorname{Re} f_n$.)

18.7. Find a different proof of Lemma 18.6. (There are a few easy variations available, using results that are closely related to the results we used in the three proofs we have given so far. An essentially different proof would be better.)

Chapter 19

Preliminaries for the Picard Theorems

19.0. Holomorphic Covering Maps

The notion of "holomorphic covering map" was defined in Section 16.3, Example 3; covering maps in general were discussed very briefly in Section 16.4. We need a little more of the basic theory than we gave previously. (A standard reference for more than you want to know about general covering maps is [**MA**].)

For the rest of this section we will assume without further comment that X and Y are path-connected and locally path-connected Hausdorff spaces. (A topological space is *locally path-connected* if it has a basis consisting of path-connected sets.) If you don't know what a topological space is you can just assume that X and Y are connected open subsets of the plane; that case will suffice for the application to the Picard Theorems in the next chapter.

Suppose that $p : X \to Y$ is continuous and surjective. We say that p is a *covering map* if every y in Y has an "elementary" neighborhood; we say $N \subset Y$ is *elementary* (or "elementary for p") if N is a connected open set and the inverse image of N is a disjoint union of open sets

$$p^{-1}(N) = \bigcup_{\alpha \in A} \tilde{N}_\alpha,$$

where $p|_{\tilde{N}_\alpha}$ is a homeomorphism from $\tilde{N}_\alpha$ onto N for each $\alpha \in A$. You should note that the $\tilde{N}_\alpha$ are precisely the connected components of $p^{-1}(N)$. We will refer to the maps $(p|_{\tilde{N}_\alpha})^{-1} : N \to \tilde{N}_\alpha$ as *local inverses* for p.

We are about to prove various things about covering maps. You may want to keep in mind the example $\exp : \mathbb{C} \to \mathbb{C} \setminus \{0\}$; having an example in mind may clarify things.

The following lemma is immediate from the definition:

Lemma 19.0.0. *Suppose that $p : X \to Y$ is a covering map, (x_j) is a sequence in X, $y \in Y$, and $p(x_j) \to y$ in Y. Then there exist a sequence (x'_j) in X and $x \in X$ such that $p(x'_j) = p(x_j)$ for all j and such that $x'_j \to x$ in X.*

The proof of the lemma is an easy exercise. We will give only outlines of the proofs of the next two theorems; the details are essentially the same as in the proofs of various special cases we have seen previously (for example in Theorem 4.0, Lemma 4.6, and Theorems 16.1.2 and 16.3.0).

Theorem 19.0.1. *Suppose that $p : X \to Y$ is a covering map.*

(i) If $\gamma : [0,1] \to Y$ is continuous, $x_0 \in X$, and $p(x_0) = \gamma(0)$ then there exists a unique continuous function $\tilde{\gamma} : [0,1] \to X$ such that $\tilde{\gamma}(0) = x_0$ and $p \circ \tilde{\gamma} = \gamma$.

(ii) Suppose $\gamma_0, \gamma_1 : [0,1] \to Y$ are continuous, $x_0 \in X$, and $p(x_0) = \gamma_0(0) = \gamma_1(0)$. Let $\tilde{\gamma}_0$ and $\tilde{\gamma}_1$ be the continuous functions mapping $[0,1]$ to X such that $\tilde{\gamma}_j(0) = x_0$ and $p \circ \tilde{\gamma}_j = \gamma_j$ ($j = 0,1$) as in (i). If γ_0 and γ_1 are path-homotopic then

$$\tilde{\gamma}_0(1) = \tilde{\gamma}_1(1).$$

Note. If γ is a curve in Y then a curve $\tilde{\gamma}$ as in part (i) of the theorem is known as a *lifting* of γ to X.

Proof. To prove existence in part (i), let A be the set of $t_0 \in [0,1]$ such that there exists a continuous $\tilde{\gamma} : [0, t_0] \to X$ with $\tilde{\gamma}(0) = x_0$ and $p(\tilde{\gamma}(t)) = \gamma(t)$ for all $t \in [0, t_0]$. Show that A is both open and closed. To prove uniqueness, given two liftings of γ, show that the set of t_0 such that the two liftings agree on $[0, t_0]$ is open and closed. For part (ii), suppose that $(\gamma_t)_{0 \le t \le 1}$ is a path homotopy of γ_0 and γ_1. Let $\tilde{\gamma}_t$ be a lifting of γ_t with $\tilde{\gamma}_t(x_0) = x_0$, and show that the set of all t_0 such that $\tilde{\gamma}_t(1) = \tilde{\gamma}_0(1)$ for all $t \in [0, t_0]$ is both open and closed. $\square$

A topological space is *simply connected* if it is path-connected and any two paths with the same endpoints are path-homotopic (equivalently, if any closed curve is null-homotopic).

Theorem 19.0.2. *Suppose that $p : X \to Y$ is a covering map, D is a path-connected, locally path-connected and simply connected topological space, and*

19.0. Holomorphic Covering Maps

$f : D \to Y$ is continuous. Suppose that $a \in D$. Fix $x_0 \in X$ with $p(x_0) = f(a)$. There exists a unique continuous function $\tilde{f} : D \to X$ such that $\tilde{f}(a) = x_0$ and $p \circ \tilde{f} = f$.

(Note that Theorem 19.0.1(i) is a special case, with $D = [0, 1]$.)

Proof. Given $b \in D$, let $\gamma : [0, 1] \to D$ be continuous with $\gamma(0) = a$ and $\gamma(1) = b$. Let $\tilde{\gamma}$ be a lifting of $f \circ \gamma$ with $\tilde{\gamma}(0) = x_0$, and define $\tilde{f}(b) = \tilde{\gamma}(1)$. Since D is simply connected, part (ii) of the previous theorem shows that $\tilde{f}$ is well defined. □

Complete proofs can be found in many books on algebraic topology, for example [**MA**].

There is a simple corollary that we will use several times:

Corollary 19.0.2.1. *Suppose that $p : X \to Y$ is a covering map and X is simply connected. If $f : X \to X$ is continuous, $p \circ f = p$, and f has a fixed point in X then $f(x) = x$ for all $x \in X$.*

Proof. Choose $a \in X$ with $f(a) = a$. Define $g : X \to Y$ by $g = p$. Theorem 19.0.2 (with g in place of f, $D = X$ and $x_0 = a$) says that there is a unique $\tilde{g} : X \to X$ such that $\tilde{g}(a) = a$ and $p \circ \tilde{g} = g$. Since the two functions $\tilde{g} = f$ and $\tilde{g}(x) = x$ both satisfy these two conditions, they must be the same function. □

We get something interesting if we apply the previous theorem again with $f = p$:

Theorem 19.0.3. *Suppose that $p : X \to Y$ is a covering map and X is simply connected. There exists a group $\mathrm{Aut}(X, p)$ of homeomorphisms of X such that for $x_0, x_1 \in X$ we have $p(x_0) = p(x_1)$ if and only if $x_1 = g(x_0)$ for some $g \in \mathrm{Aut}(X, p)$.*

We will call $\mathrm{Aut}(X, p)$ the *covering group* of p; the elements of $\mathrm{Aut}(X, p)$ are sometimes called *deck transformations*.

Proof. Suppose that $x_0, x_1 \in X$ and $p(x_0) = p(x_1)$. Theorem 19.0.2 shows that there exists a continuous $g : X \to X$ such that $g(x_0) = x_1$ and $p \circ g = p$. (It follows that if $x \in X$ and $y = g(x)$ then $p(y) = p(x)$.)

Similarly there exists a continuous $g' : X \to X$ such that $g'(x_1) = x_0$ and $p \circ g' = p$. Then $g \circ g'(x_1) = x_1$ and so Corollary 19.0.2.1 shows that $g \circ g'$ is the identity. Similarly $g' \circ g$ is the identity, and so g is a homeomorphism of X onto X.

Let $\text{Aut}(X, p)$ be the set of all homeomorphisms g that arise as above. The previous paragraph shows that $\text{Aut}(X, p)$ contains the inverse of each of its elements, and as in the previous paragraph the uniqueness in Theorem 19.0.2 shows that $\text{Aut}(X, p)$ is closed under composition. □

See Appendix 6 for a simple example showing that the conclusion may fail if X is not assumed to be simply connected. (I found the existence of such an example surprising — it showed me that my mental picture of what a covering space looks like wasn't quite right.)

Corollary 19.0.3.1. *Suppose that $p : X \to Y$ is a covering map and X is simply connected. If $f : X \to X$ is continuous and $p \circ f = p$ then $f \in \text{Aut}(X, p)$.*

Proof. Fix $x_0 \in X$. Since $p(f(x_0)) = p(x_0)$, Theorem 19.0.3 shows that there exists $g \in \text{Aut}(X, p)$ such that $f(x_0) = g(x_0)$. Now Corollary 19.0.2.1 shows that $f \circ g^{-1} \in \text{Aut}(X, p)$. □

The next corollary is just a rephrasing of Corollary 19.0.2.1:

Corollary 19.0.3.2. *Suppose that $p : X \to Y$ is a covering map and X is simply connected. If $g \in \text{Aut}(X, p)$ has a fixed point in X then g is the identity.*

- **Exercise 19.1.** Give an example showing that Theorem 19.0.2 is false without the hypothesis that D is simply connected.

- **Exercise 19.2.** Show that Corollary 19.0.2.1 remains true even if X is not simply connected (recall our standing assumption that X is path-connected). Of course this version is no longer a corollary of Theorem 19.0.2.

It turns out that if X is simply connected then a covering map p with domain X is essentially determined by $\text{Aut}(X, p)$, in the following sense:

Theorem 19.0.4. *Suppose that X is simply connected and $p_0, p_1 : X \to Y$ are covering maps. If $\text{Aut}(X, p_0) = \text{Aut}(X, p_1)$ then there exists a homeomorphism χ of Y such that*

$$p_1 = \chi \circ p_0.$$

Proof. Informally, we simply let $\chi = p_1 \circ p_0^{-1}$; we note that p_0^{-1} is determined up to elements of $\text{Aut}(X, p_0)$ and p_1 is invariant under $\text{Aut}(X, p_1) = \text{Aut}(X, p_0)$, so χ is well defined.

A little more carefully: Define $\chi : Y \to Y$ by $\chi(y) = p_1(x)$, where $x \in X$ satisfies $p_0(x) = y$. This is well defined: If $p_0(x) = p_0(x') = y$ then $x' = g(x)$

19.0. Holomorphic Covering Maps

for some $g \in \operatorname{Aut}(X, p_0) = \operatorname{Aut}(X, p_1)$, so that $p_1(x') = p_1(x)$. The fact that p_0 has continuous local inverses shows that χ is locally continuous, hence continuous.

So there exists a continuous $\chi : Y \to Y$ such that $p_1 = \chi \circ p_0$. For the same reason there is a continuous $\chi' : Y \to Y$ such that $p_0 = \chi' \circ p_1$. It follows that $p_0 = \chi' \circ \chi \circ p_0$; since $p_0(X) = Y$, this shows that $\chi' \circ \chi$ is the identity. Similarly $\chi \circ \chi'$ is the identity, and so χ is a homeomorphism. □

For future reference we should note that essentially the same proof gives the following generalization:

Theorem 19.0.4.1. *Suppose that X is simply connected and $p_j : X \to Y_j$ ($j = 0, 1$) are covering maps. If $\operatorname{Aut}(X, p_0) = \operatorname{Aut}(X, p_1)$ then there exists a homeomorphism $\chi : Y_0 \to Y_1$ such that*

$$p_1 = \chi \circ p_0.$$

So much for covering maps. The previous few results easily imply corresponding results for holomorphic covering maps.

Theorem 19.0.5. *Suppose that Ω and V are connected open subsets of the plane and $p : \Omega \to V$ is a holomorphic covering map. Suppose that D is a simply connected open subset of the plane, and $f : D \to V$ is holomorphic. Suppose that $a \in D$. Fix $z \in \Omega$ with $p(z) = f(a)$. There exists a unique holomorphic function $\tilde{f} : D \to \Omega$ such that $\tilde{f}(a) = z$ and $p \circ \tilde{f} = f$.*

Proof. Theorem 19.0.2 shows that there exists a unique continuous $\tilde{f} : D \to \Omega$ such that $\tilde{f}(a) = z$ and $p \circ \tilde{f} = f$; we need only show that $\tilde{f}$ is holomorphic. But this is immediate from the definitions:

Fix $w \in D$. Let $W \subset V$ be an elementary neighborhood of $f(w)$. The definition of "holomorphic covering map" shows that there exists a holomorphic function $\phi : W \to \Omega$ such that $p(\phi(\alpha)) = \alpha$ for all $\alpha \in W$, and such that $\tilde{f}(w) \in \phi(W)$. It follows that

$$\tilde{f} = \phi \circ f$$

in $f^{-1}(W)$, which is an open set containing w; thus $\tilde{f}$ is holomorphic near w. □

Theorem 19.0.6. *Suppose that Ω and V are connected open subsets of the plane, Ω is simply connected, and $p : \Omega \to V$ is a holomorphic covering map. Then $\operatorname{Aut}(\Omega, p) \subset \operatorname{Aut}(\Omega)$.*

Proof. If $g \in \mathrm{Aut}(\Omega, p)$ then $p \circ g = p$; it follows by arguments as in the proof of the previous theorem that g is holomorphic. □

Theorem 19.0.7. *Suppose that Ω and V are connected open subsets of the plane, Ω is simply connected, and p_0, $p_1 : \Omega \to V$ are holomorphic covering maps. If $\mathrm{Aut}(\Omega, p_0) = \mathrm{Aut}(\Omega, p_1)$ then*

$$p_1 = \chi \circ p_0$$

for some $\chi \in \mathrm{Aut}(V)$.

Proof. Theorem 19.0.4 shows that there is a homeomorphism χ of V such that $p_1 = \chi \circ p_0$; as before we need only show that χ is holomorphic, and as before this is immediate from the fact that p_1 is holomorphic and p_0 has holomorphic local inverses. □

Of course you might wonder how it might happen that you had two covering maps with the same covering group. Here is one way:

Theorem 19.0.8. *Suppose that Ω and V are connected open subsets of the plane, Ω is simply connected, and $p : \Omega \to V$ is a holomorphic covering map. Suppose that G is a subgroup of $\mathrm{Aut}(\Omega)$ and that $\mathrm{Aut}(\Omega, p)$ is a normal subgroup of G. Suppose that $g \in G$. Then $p \circ g$ is a holomorphic covering map with $\mathrm{Aut}(\Omega, p \circ g) = \mathrm{Aut}(\Omega, p)$, and hence there exists $\chi \in \mathrm{Aut}(V)$ such that*

$$p \circ g = \chi \circ p.$$

Proof. Suppose $g \in G$. It is immediate from the definitions that $p \circ g$ is a holomorphic covering map, so by Theorem 19.0.7 we need only show that $\mathrm{Aut}(\Omega, p \circ g) = \mathrm{Aut}(\Omega, p)$. This follows from the fact that $\mathrm{Aut}(\Omega, p)$ is a normal subgroup of G:

Suppose first that $\phi \in \mathrm{Aut}(\Omega, p)$. Then $g \circ \phi \circ g^{-1} \in \mathrm{Aut}(\Omega, p)$, so $p \circ (g \circ \phi \circ g^{-1}) = p$. Hence $(p \circ g) \circ \phi = p \circ g$, so $\phi \in \mathrm{Aut}(\Omega, p \circ g)$.

Conversely, if $\phi \in \mathrm{Aut}(\Omega, p \circ g)$ then $p \circ (g \circ \phi \circ g^{-1}) = p$, which shows that $g \circ \phi \circ g^{-1} \in \mathrm{Aut}(\Omega, p)$ and hence $\phi \in \mathrm{Aut}(\Omega, p)$. □

We have already seen a special case of Theorem 19.0.8 in the proof of Theorem 18.10, when we showed that $\lambda \circ j = 1/\lambda$. We will see several applications of the result below.

We should include a result analogous to Theorem 19.0.7, with weaker hypotheses and a weaker conclusion:

Theorem 19.0.9. *Suppose that Ω and V are connected open subsets of the plane and $p : \Omega \to V$ is a holomorphic covering map. Suppose that $f \in H(\Omega)$. There exists $F \in H(V)$ with*
$$f = F \circ p$$
if and only if
$$f \circ \phi = f$$
for all $\phi \in \mathrm{Aut}(\Omega, p)$.

We leave the proof as an exercise.

Corollary 19.0.10. *Let $\mathbb{D}' = \mathbb{D} \setminus \{0\}$ and suppose that $f \in H(\Pi^+)$. There exists $F \in H(\mathbb{D}')$ with $f(z) = F(e^{2\pi i z})$ for all $z \in \Pi^+$ if and only if $f(z+1) = f(z)$ for all $z \in \Pi^+$.*

- **Exercise 19.3.** Suppose that $D \subset \mathbb{C}$ is simply connected, $D \neq \mathbb{C}$, and $p : D \to V$ is a holomorphic covering map. Suppose that $f : D \to V$ is holomorphic, $a \in D$, and $f(a) = p(a)$. Show that $|f'(a)| \leq |p'(a)|$.

Note/Hint. The conclusion is false for $D = \mathbb{C}$: Let $V = \mathbb{C} \setminus \{0\}$, $p(z) = e^z$, $f(z) = e^{2z}$, $a = 0$.

19.1. Examples of Holomorphic Covering Maps

We already know several examples of holomorphic covering maps, many of which will come up in the proof of the Big Picard Theorem.

The first one we have known for a long time: $\exp : \mathbb{C} \to \mathbb{C} \setminus \{0\}$ is a holomorphic covering map. We have also known for a long time that $\mathrm{Aut}(\mathbb{C}, \exp)$ is the group generated by the translation $z \mapsto z + 2\pi i$.

The results in Section 15.0 give various examples of holomorphic covering maps from the upper half-plane to various open sets. The next three theorems are rephrasings of those results, better suited to the present context. Recall that $\partial_\infty \Pi^+$ is the extended boundary of Π^+, that is, the boundary in $\mathbb{C}_\infty$, including ∞.

Theorem 19.1.0. *If $\phi \in \mathrm{Aut}(\Pi^+)$ is not the identity then exactly one of the following holds*:

(i) *ϕ has exactly one fixed point in Π^+ and none on $\partial_\infty \Pi^+$;*

(ii) *ϕ has no fixed points in Π^+ and exactly one on $\partial_\infty \Pi^+$;*

(iii) *ϕ has no fixed points in Π^+ and exactly two on $\partial_\infty \Pi^+$.*

In case (i) we say that ϕ is *elliptic*, in case (ii) it is *parabolic*, and in case (iii) it is *hyperbolic*.

Theorem 19.1.1. *Suppose that $\phi \in \mathrm{Aut}(\Pi^+)$ is not the identity.*

(i) If ϕ is elliptic then the element of $\mathrm{Aut}(\mathbb{D})$ corresponding to ϕ is conjugate to a rotation: If Φ is the Cayley transform then there exists a complex number $\alpha \neq 1$ with $|\alpha| = 1$ such that $\Phi \circ \phi \circ \Phi^{-1}$ is conjugate in $\mathrm{Aut}(\mathbb{D})$ to ψ, where

$$\psi(z) = \alpha z \quad (z \in \mathbb{D}).$$

(ii) If ϕ is parabolic then ϕ is conjugate to a translation: There exists $B \in \mathbb{R}$ with $B \neq 0$ such that ϕ is conjugate in $\mathrm{Aut}(\Pi^+)$ to ψ, where

$$\psi(z) = z + B \quad (z \in \Pi^+).$$

(iii) If ϕ is hyperbolic then ϕ is conjugate to a dilation: There exists $\delta > 0$ with $\delta \neq 1$ such that ϕ is conjugate in $\mathrm{Aut}(\Pi^+)$ to ψ, where

$$\psi(z) = \delta z \quad (z \in \Pi^+).$$

Theorem 19.1.2. *Suppose that $\phi \in \mathrm{Aut}(\Pi^+)$, and let G be the subgroup of $\mathrm{Aut}(\Pi^+)$ generated by ϕ. If ϕ is not elliptic then there exists a covering map h mapping Π^+ onto a bounded open set in the plane, such that $\mathrm{Aut}(\Pi^+, h) = G$. In more detail:*

(i) If ϕ is the identity then we can take h to be a covering map onto $\mathbb{D}$ (in fact h is a homeomorphism in this case).

(ii) If ϕ is elliptic of infinite order then there is no nonconstant $h \in H(\Pi^+)$ with $h \circ \phi = h$. Suppose that ϕ is elliptic of finite order, and let $a \in \Pi^+$ be the fixed point of ϕ. There exists a holomorphic map $h : \Pi^+ \to \mathbb{D}$ such that $h|_{\Pi^+ \setminus \{a\}}$ is a covering map onto $\mathbb{D}' = \mathbb{D} \setminus \{0\}$, with $\mathrm{Aut}(\Pi^+, h|_{\Pi^+ \setminus \{a\}}) = G$.

(iii) If ϕ is parabolic then we can take h to be a covering map onto $\mathbb{D}' = \mathbb{D} \setminus \{0\}$.

(iv) If ϕ is hyperbolic then we can take h to be a covering map onto a proper annulus $\{z \in \mathbb{C} : r < |z| < R\}$ for some r, R with $0 < r < R < \infty$.

Proof. If $\phi(z) = z$ we can define $h(z) = (z-i)/(z+i)$; we have known for a long time that this gives a homeomorphism onto $\mathbb{D}$, and a homeomorphism is certainly a covering map. For (iii) and (iv) one need only inspect the proof of Theorem 15.0.2 to verify that the maps constructed there are covering maps; similarly for part (ii). □

(Note that if ϕ is elliptic then there is no covering map $h : \Pi^+ \to D$ with $h \circ \phi = h$; this follows from Corollary 19.0.3.2 above.)

And of course the modular function λ constructed in Chapter 18 is a covering map from Π^+ onto $\mathbb{C} \setminus \{0, 1\}$; Theorem 18.9 says that $\operatorname{Aut}(\Pi^+, \lambda) = \Gamma(2)$. Again, Corollary 19.0.3.2 shows that $\Gamma(2)$ does not contain any elliptic transformations (this is really the same argument as was given in the proof of Theorem 18.9, in different language).

Recall that $\Gamma(2)$ is a *normal* subgroup of Γ. By Theorem 19.0.8 this implies the following result, which will be a key step in the proof of the Big Picard Theorem in the next chapter:

Theorem 19.1.3. *If $g \in \Gamma$ then there exists $\chi \in \operatorname{Aut}(\mathbb{C} \setminus \{0, 1\})$ such that*

$$\lambda \circ g = \chi \circ \lambda.$$

In the next section we will see that $\operatorname{Aut}(\mathbb{C} \setminus \{0, 1\})$ is naturally isomorphic to the group of all permutations of the set $\{0, 1, \infty\}$.

19.2. Two More Automorphism Groups

We begin with the automorphisms of the punctured disk $\mathbb{D}' = \mathbb{D} \setminus \{0\}$:

Theorem 19.2.0. *If $\phi \in \operatorname{Aut}(\mathbb{D}')$ then ϕ is a rotation: There exists $\alpha \in \mathbb{C}$ with $|\alpha| = 1$ such that*

$$\phi(z) = \alpha z \quad (z \in \mathbb{D}).$$

Proof. Since ϕ is bounded, it has a removable singularity at the origin; thus it extends to a function holomorphic in $\mathbb{D}$, which we will denote again by ϕ.

It follows from the continuity of ϕ and the Maximum Modulus Theorem that $|\phi(0)| < 1$. Now the fact that ϕ is one-to-one in $\mathbb{D}'$ (and $\phi(\mathbb{D}') = \mathbb{D}'$) shows that $\phi(0) = 0$: If $0 < |\phi(0)| < 1$ then there exists $a \in \mathbb{D}'$ with $\phi(0) = \phi(a)$, and now the Open Mapping Theorem shows that values near $\phi(0)$ must also be attained twice, at points near 0 and also at points near a; this contradicts the fact that ϕ is one-to-one in $\mathbb{D}'$.

Since $\phi(0) = 0$, it follows that ϕ is a bijection from $\mathbb{D}$ to $\mathbb{D}$. That is, $\phi \in \operatorname{Aut}(\mathbb{D})$; now Theorem 8.4.1 shows that ϕ is a rotation. □

We will use a similar argument to identify the automorphisms of $\mathbb{C} \setminus \{0, 1\}$: It turns out that they all extend to automorphisms of $\mathbb{C}_\infty$, that is, to linear-fractional transformations.

Note first that if f has an essential singularity at a then f cannot be one-to-one in any deleted neighborhood of a: The Open Mapping Theorem shows that $f(D(a+\delta,\delta/2))$ is a nonempty open set, so it follows from Weierstrass' Theorem (part (ii) of Proposition 3.14) that

$$f(D(a,\delta/2) \setminus \{a\}) \cap f(D(a+\delta,\delta/2)) \neq \emptyset.$$

Theorem 19.2.1. *If $\phi \in \mathrm{Aut}(\mathbb{C}\setminus\{0,1\})$ then ϕ is a linear-fractional transformation mapping the set $\{0,1,\infty\}$ onto itself.*

The converse is clear; now, since a linear-fractional transformation is determined by its values at three points, it follows that $\mathrm{Aut}(\mathbb{C} \setminus \{0,1\})$ is isomorphic to the group of permutations of $\{0,1,\infty\}$. One can easily find linear-fractional transformations which give arbitrary permutations of $\{0,1,\infty\}$: In fact

$$\mathrm{Aut}(\mathbb{C} \setminus \{0,1\}) = \{\phi_0, \phi_1, \phi_2, \phi_3, \phi_4, \phi_5\},$$

where $\phi_0(z) = z$, $\phi_1(z) = 1/z$, $\phi_2(z) = 1 - z$, $\phi_3(z) = (z-1)/z$, $\phi_4(z) = 1/(1-z)$ and $\phi_5(z) = z/(z-1)$.

Proof (Theorem 19.2.1). As noted above, the fact that ϕ is one-to-one in $\mathbb{C} \setminus \{0,1\}$ shows that it must have either a pole or removable singularity at 0, 1 and ∞. Thus ϕ extends to a holomorphic map from $\mathbb{C}_\infty$ to $\mathbb{C}_\infty$, which we will again denote by ϕ.

Now we must have $\phi(0) \in \{0,1,\infty\}$: If not then there exists $a \in \mathbb{C}\setminus\{0,1\}$ such that $\phi(0) = \phi(a)$, and as in the proof of the previous theorem the Open Mapping Theorem shows that ϕ is not one-to-one in $\mathbb{C} \setminus \{0,1\}$. The same argument shows that $\phi(1) \in \{0,1,\infty\}$ and $\phi(\infty) \in \{0,1,\infty\}$, and the same argument then shows that the three values $\phi(0)$, $\phi(1)$, and $\phi(\infty)$ must be distinct. Thus $\phi : \mathbb{C}_\infty \to \mathbb{C}_\infty$ is a bijection, and now Theorem 8.1.1 shows that ϕ is a linear-fractional transformation. $\square$

If we combine this with the results in the previous section we get a result about the boundary behavior of the modular function:

Theorem 19.2.2. *Suppose $g \in \Gamma$ and (w_j) is a sequence of points in Π^+ such that*

$$\mathrm{Im}\,(g(w_j)) \to \infty.$$

Then

$$\lambda(w_j) \to L$$

for some $L \in \{0,1,\infty\}$.

19.2. Two More Automorphism Groups

Proof. Theorem 18.10 shows that $\lambda(g(w_j)) \to \infty$. But Theorem 19.1.3 shows that there exists $\chi \in \operatorname{Aut}(\mathbb{C} \setminus \{0,1\})$ such that $\lambda \circ g = \chi \circ \lambda$; since χ^{-1} is a continuous map from $\mathbb{C}_\infty$ to $\mathbb{C}_\infty$, it follows from Theorem 19.2.1 that

$$\lambda(w_j) = \chi^{-1}(\lambda(g(w_j))) \to \chi^{-1}(\infty) \in \{0,1,\infty\}. \qquad \square$$

This will be used to handle an exceptional case in the proof of the Big Picard Theorem in the next chapter. We give the application here, in order to avoid technicalities in the presentation of that proof:

Theorem 19.2.3. *Suppose that $\phi \in \Gamma(2)$ is parabolic; let $h : \Pi^+ \to \mathbb{D}' = \mathbb{D} \setminus \{0\}$ be a covering map such that $\operatorname{Aut}(\Pi^+, h)$ is generated by ϕ, as in Theorem 19.1.2. If $(z_n) \subset \Pi^+$ and $h(z_n) \to 0$ then there exists $L \in \{0,1,\infty\}$ such that $\lambda(z_n) \to L$.*

Proof. Let $\alpha \in \partial_\infty \Pi^+$ be the unique fixed point of ϕ in $\overline{\Pi^+_\infty}$. Lemma 19.2.4 below shows that $\alpha \in \mathbb{Q} \cup \{\infty\}$, and that hence there exists $g \in \Gamma$ with $g(\infty) = \alpha$.

Now $h \circ g : \Pi^+ \to \mathbb{D}'$ is a covering map, and $\operatorname{Aut}(\Pi^+, h \circ g)$ is generated by $\psi = g^{-1} \circ \phi \circ g$. Since ∞ is the only fixed point of ψ in $\overline{\Pi^+_\infty}$, it follows (see the proof of Theorem 15.0.1) that ψ is a translation: $\psi(z) = z + a$ for some nonzero $a \in \mathbb{R}$.

Define
$$H(z) = e^{2\pi i z/|a|}.$$

Then $H : \Pi^+ \to \mathbb{D}'$ is a covering map, and $\operatorname{Aut}(\Pi^+, H)$ is also generated by ψ. So Theorem 19.0.7 shows that there exists $\chi \in \operatorname{Aut}(\mathbb{D}')$ with $h \circ g = \chi \circ H$. Now Theorem 19.2.0 shows that χ is a rotation; thus there exists β with $|\beta| = 1$ such that

$$h \circ g(z) = \beta e^{2\pi i z/|a|}.$$

Since $h \circ g(g^{-1}(z_n)) \to 0$, this implies that $\operatorname{Im}(g^{-1}(z_n)) \to \infty$, and now Theorem 19.2.2 shows that $\lambda(z_n) \to L$ for some $L \in \{0,1,\infty\}$. $\qquad \square$

We used the following lemma in the proof:

Lemma 19.2.4. (i) *Suppose that $\phi \in \Gamma$ is parabolic and let α be the unique fixed point of ϕ in $\overline{\Pi^+_\infty}$. Then $\alpha \in \mathbb{Q} \cup \{\infty\}$.*

(ii) *If $\alpha \in \mathbb{Q}$ then there exists $g \in \Gamma$ with $g(\infty) = \alpha$.*

Proof. For (i), suppose that $\phi(z)$ is represented by the matrix $\begin{bmatrix} a & b \\ c & d \end{bmatrix} \in SL_2(\mathbb{Z})$. If $c = 0$ then $\alpha = \infty$. Suppose $c \neq 0$. Then $\phi(z) = z$ is equivalent to a quadratic equation, the roots of which are given by

$$z = \frac{p \pm \sqrt{r}}{q}$$

for certain integers p, q, r. If $r < 0$ then ϕ has a fixed point in Π^+, while if $r > 0$ then ϕ has two real fixed points. So we must have $r = 0$, and hence $\alpha \in \mathbb{Q}$.

For (ii), note first that if $\alpha = 0$ then $g(\infty) = \alpha$ if $g(z) = -1/z$. Suppose $\alpha \neq 0$, and write $\alpha = p/q$ where $p, q \in \mathbb{Z}$ are relatively prime. There exist $n, m \in \mathbb{Z}$ with $np + mq = 1$, and now if $g(z) = (pz - m)/(qz + n)$ then $g \in \Gamma$ and $g(\infty) = \alpha$. $\square$

Note. If $\phi(z) = z/(2z + 1)$ then $\phi \in \Gamma(2)$ and $\alpha = 0$ is the only fixed point of ϕ, so ϕ is parabolic. On the other hand, if $g \in \Gamma(2)$ it is easy to see that $g(\infty) \neq 0$. So we cannot obtain $g \in \Gamma(2)$ in the proof of Theorem 19.2.3.

For the sake of completeness we include a proof of the lemma from elementary number theory used above, although most readers have probably seen it before:

Lemma 19.2.5. *Suppose that p and q are nonzero integers with greatest common divisor d. Then there exist $n, m \in \mathbb{Z}$ such that $np + mq = d$.*

Proof. Let $I = \{ np + mq : n, m \in \mathbb{Z} \}$, and let d' be the smallest positive element of I. It follows that $I = I'$, where $I' = \{ nd' : n \in \mathbb{Z} \}$: First, it is clear that $I' \subset I$, since I is closed under addition. Suppose then that $k \in I \setminus I'$. Then we have $nd' < k < (n+1)d'$ for some $n \in \mathbb{Z}$. The fact that I is closed under addition shows that $k - nd' \in I$. But $0 < k - nd' < d'$, contradicting our choice of d'. So $I = I'$.

The fact that $I \subset I'$ shows that d' divides both p and q, while the fact that $I' \subset I$ shows that any common divisor of p and q must also divide d'. Hence $d' = d$, and hence $d \in I$. $\square$

This finishes the material that will be used in the next chapter. In the next two sections we give some fairly simple and very interesting further results on holomorphic covering maps; you're encouraged to read them even if you're in a hurry to get to the Picard theorems, if only because it would be a shame to miss the proof of Theorem 19.4.4. In Chapter 25 we discuss a somewhat more technical topic regarding covering spaces.

19.3. Normalizers of Covering Groups

Suppose that Ω and V are connected open subsets of the plane, Ω is simply connected, and $p : \Omega \to V$ is a holomorphic covering map. Theorem 19.0.8 says that if G is a subgroup of $\operatorname{Aut}(\Omega)$ and $\operatorname{Aut}(\Omega, p)$ is a normal subgroup of G then for each $g \in G$ there exists a $\chi \in \operatorname{Aut}(V)$ such that

$$p \circ g = \chi \circ p.$$

This raises the question of whether *every* $\chi \in \operatorname{Aut}(V)$ can be obtained in this way (for some g in some G). The answer is yes; we will see a converse to Theorem 19.0.8 below. (We used Theorem 19.0.8 in Theorem 19.2.2 to prove something about a certain covering map in a case where we knew what $\operatorname{Aut}(V)$ was. We can use the converse to calculate automorphism groups if we know covering maps; this is useful because, as we have seen, you can never know too many automorphism groups.) Now, if we want to prove a converse to Theorem 19.0.8 it seems reasonable to try to take G as large as possible, so as to pick up all the elements of $\operatorname{Aut}(V)$. Luckily there is a natural candidate for "largest possible G" and it turns out to work:

Suppose that S is a subset of the group G. The *normalizer* of S (in G) is the set

$$N_S = \{\, g \in G : gSg^{-1} = S \,\}.$$

It is easy to verify that if S is a subgroup of G then S is a normal subgroup of N_S, and it is evident that N_S is the largest subgroup of G that contains S as a normal subgroup. So if Theorem 19.0.8 has a converse of the sort we are seeking, it must be the following:

Theorem 19.3.0. *Suppose that Ω and V are connected open subsets of the plane, Ω is simply connected, and $p : \Omega \to V$ is a holomorphic covering map. Let G be the normalizer of $\operatorname{Aut}(\Omega, p)$ in $\operatorname{Aut}(\Omega)$. Then for every $\chi \in \operatorname{Aut}(V)$ there exists $g \in G$ such that*

$$p \circ g = \chi \circ p.$$

Corollary 19.3.1. *Under the same hypotheses, the map*

$$g \mapsto p \circ g \circ p^{-1}$$

is a homomorphism from G onto $\operatorname{Aut}(V)$, with kernel $\operatorname{Aut}(\Omega, p)$; thus $\operatorname{Aut}(V)$ is canonically isomorphic to the quotient group $G/\operatorname{Aut}(\Omega, p)$.

Note. The notation $p \circ g \circ p^{-1}$ in the statement of the corollary is at best somewhat informal, since there is no such thing as p^{-1}; if $z \in V$ then the intended interpretation is $p \circ g \circ p^{-1}(z) = p(g(w))$, where $w \in \Omega$ satisfies $p(w) = z$.

Proof of the corollary. Note first that if $g \in G$ then $p \circ g \circ p^{-1} : V \to V$ is well defined: Suppose that $z \in V$, and suppose that w, $w' \in \Omega$ satisfy $p(w) = p(w') = z$. Then there exists $\phi \in \text{Aut}(\Omega, p)$ with $w' = \phi(w)$, and the fact that $\text{Aut}(\Omega, p)$ is normal in G shows that there exists $\phi' \in \text{Aut}(\Omega, p)$ with $g \circ \phi = \phi' \circ g$; hence

$$p(g(w')) = p(g(\phi(w))) = p(\phi'(g(w))) = p(g(w)).$$

It is clear that $(p \circ g \circ p^{-1}) \circ (p \circ g' \circ p^{-1}) = p \circ (g \circ g') \circ p^{-1}$. (At least the notation makes this *appear* to be clear. Since $p \circ g \circ p^{-1}$ is not literally the composition of three functions (as the notation suggests), it might be worth verifying this identity carefully.) Taking $g' = g^{-1}$ this shows that $p \circ g \circ p^{-1}$ is a bijection, so that $p \circ g \circ p^{-1} \in \text{Aut}(V)$ for all $g \in G$; so we have defined a homomorphism from G into $\text{Aut}(V)$. Theorem 19.3.0 says precisely that this homomorphism is surjective, so we are done if we can show that the kernel is $\text{Aut}(\Omega, p)$.

But this is clear: If $g \in G$ and $p \circ g \circ p^{-1}$ is the identity element of $\text{Aut}(V)$ then $p \circ g = p$, which says that $g \in \text{Aut}(\Omega, p)$ by definition. □

Proof of Theorem 19.3.0. Fix $a \in \Omega$. Suppose that $\chi \in \text{Aut}(V)$. Then $\chi \circ p : \Omega \to V$ is holomorphic; since Ω is simply connected, it follows from Theorem 19.0.5 that if we fix $b \in \Omega$ with $p(b) = \chi(p(a))$ then there exists a holomorphic $g : \Omega \to \Omega$ with

$$\chi \circ p = p \circ g$$

and $g(a) = b$. Now, since $\chi^{-1}(p(b)) = p(a)$, the same argument shows that there exists a holomorphic $g' : \Omega \to \Omega$ with $\chi^{-1} \circ p = p \circ g'$ and $g'(b) = a$. It follows that

$$p \circ (g \circ g') = (p \circ g) \circ g' = (\chi \circ p) \circ g' = \chi \circ (p \circ g') = \chi \circ (\chi^{-1} \circ p) = p$$

and $g \circ g'(a) = a$.

So Corollary 19.0.2.1 shows that $g \circ g'(w) = w$ for all $w \in \Omega$. Similarly $g' \circ g(w) = w$, so that g is invertible: $g \in \text{Aut}(\Omega)$.

It remains only to show that $g \in G$. Since G was defined to be the normalizer of $\text{Aut}(\Omega, p)$, this means that we need to show that $g \circ \phi \circ g^{-1} \in \text{Aut}(\Omega, p)$ for all $\phi \in \text{Aut}(\Omega, p)$. But this is clear, since

$$p \circ g \circ \phi \circ g^{-1} = \chi \circ p \circ \phi \circ g^{-1} = \chi \circ p \circ g^{-1} = p \circ g \circ g^{-1} = p. \quad □$$

Let's use this stuff to find the automorphism group of a punctured disk again, just to illustrate what's going on:

19.3. Normalizers of Covering Groups

Let $\mathbb{D}' = \mathbb{D} \setminus \{0\}$. Define $p : \Pi^+ \to \mathbb{D}'$ by

$$p(w) = e^{2\pi i w}.$$

We know that p is a covering map and that $\text{Aut}(\Pi^+, p) = \{\tau_n : n \in \mathbb{Z}\}$, where $\tau_r \in \text{Aut}(\Pi^+)$ is defined for $r \in \mathbb{R}$ by

$$\tau_r(w) = w + r.$$

In order to apply Corollary 19.3.1 to find $\text{Aut}(\mathbb{D}')$ we need to find the normalizer of $\text{Aut}(\Pi^+, p)$ in $\text{Aut}(\Pi^+)$. So assume that $\phi \in \text{Aut}(\Pi^+)$ is in the normalizer of $\text{Aut}(\Pi^+, p)$, and choose a, b, c, and $d \in \mathbb{R}$ with $ad - bc = 1$, such that $\phi(w) = (aw + b)/(cw + d)$.

It follows that for every $n \in \mathbb{Z}$ there exists $m \in \mathbb{Z}$ such that

$$\phi \circ \tau_n = \tau_m \circ \phi,$$

and also that for every $m \in \mathbb{Z}$ there exists $n \in \mathbb{Z}$ satisfying the same equation. Using the standard identification of $\text{Aut}(\Pi^+)$ with $SL_2(\mathbb{R})/\{I, -I\}$ the equation above becomes

$$\begin{bmatrix} a & b \\ c & d \end{bmatrix} \begin{bmatrix} 1 & n \\ 0 & 1 \end{bmatrix} = \pm \begin{bmatrix} 1 & m \\ 0 & 1 \end{bmatrix} \begin{bmatrix} a & b \\ c & d \end{bmatrix},$$

or

$$\begin{bmatrix} a & na+b \\ c & nc+d \end{bmatrix} = \pm \begin{bmatrix} a+mc & b+md \\ c & d \end{bmatrix}.$$

The equation cannot hold with a minus sign regardless of the values of n and m (it would follow that $c = 0$ and then that $a = 0$, contradicting the assumption that $ad - bc = 1$).

So for every $n \in \mathbb{Z}$ there exists $m \in \mathbb{Z}$ such that

$$mc = 0, \quad nc = 0, \quad na = md,$$

and for every $m \in \mathbb{Z}$ there exists $n \in \mathbb{Z}$ satisfying the same condition. Taking $n = 1$ shows that $c = 0$; now the fact that $ad - bc = 1$ shows that $ad = 1$. Again taking $n = 1$ shows that $a^2 = amd = m \in \mathbb{Z}$, and taking $m = 1$ shows that $d^2 \in \mathbb{Z}$.

So we must have $a = d = \pm 1$ and we may assume that $a = 1$; thus

$$\phi(w) = w + b.$$

So the normalizer of $\text{Aut}(\Pi^+, p) = \{\tau_n : n \in \mathbb{Z}\}$ is exactly $\{\tau_b : b \in \mathbb{R}\}$, and Corollary 19.3.1 shows that $\text{Aut}(\mathbb{D}') = \{p \circ \tau_b \circ p^{-1} : b \in \mathbb{R}\}$.

Since $p(w) = e^{2\pi i w}$, it follows that $p^{-1}(z) = \log(z)/(2\pi i)$, and so
$$p \circ \tau_b \circ p^{-1}(z) = e^{\log(z) + 2\pi i b} = e^{2\pi i b} z.$$
So sure enough, $\text{Aut}(\mathbb{D}')$ consists of rotations, just like it did in the previous section.

We can also see from the above why some $\phi \in \text{Aut}(\Pi^+)$ lead to automorphisms of $\mathbb{D}'$ and some do not. For example, $\phi = \tau_b$ works because it leads to the formula above; although $\log(z)$ is not well defined the quantity $e^{\log(z)+2\pi i b}$ is well defined, since if you add $2\pi i k$ to $\log(z)$ (with $k \in \mathbb{Z}$), you do not change the value of $e^{\log(z)+2\pi i b}$. On the other hand, if, say, $\phi(z) = -1/z$ then we would get
$$p \circ \phi \circ p^{-1}(z) = e^{4\pi^2/\log(z)},$$
which is *not* well defined; adding $2\pi i k$ to $\log(z)$ has a major effect on the value of $e^{4\pi^2/\log(z)}$.

- **Exercise 19.4.** Define $\phi \in \text{Aut}(\Pi^+)$ by $\phi(w) = 2w$. Define $p : \Pi^+ \to \mathbb{D}'$ by $p(w) = e^{2\pi i w}$ as above. Show that the expression $p \circ \phi \circ p^{-1}$ *does* give a well defined holomorphic map from $\mathbb{D}'$ to $\mathbb{D}'$, which nonetheless is not an automorphism of $\mathbb{D}'$. Explain. (Evidently ϕ is not in the normalizer of $\text{Aut}(\Pi^+, p)$, which is to say that $\phi \text{Aut}(\Pi^+, p) \phi^{-1} \neq \text{Aut}(\Pi^+, p)$. There is nonetheless an inclusion between these two sets; which one is a subset of the other?)

- **Exercise 19.5.** Find the automorphism group of the punctured plane $\mathbb{C}' = \mathbb{C} \setminus \{0\}$. Do this twice, once using arguments as in the proofs of Theorems 19.2.0 and 19.2.1 and once using Corollary 19.3.1 as above. (The automorphisms are not all rotations.)

- **Exercise 19.6.** Find the automorphism group of a proper annulus
$$A = \{z \in \mathbb{C} : r < |z| < R\}$$
for $0 < r < R < \infty$.

- **Exercise 19.7.** Find the normalizer of $\Gamma(2)$ in $\text{Aut}(\Pi^+)$.

Note. The answer turns out to be Γ. But I do not think that this is *quite* immediate from anything we have proved: As far as I can see, what we have proved shows that $\Gamma/\Gamma(2)$ is isomorphic to a subgroup of $\text{Aut}(\mathbb{C} \setminus \{0, 1\})$, but I do not think we have quite shown that it is isomorphic to $\text{Aut}(\mathbb{C} \setminus \{0, 1\})$ itself. It would be enough (why?) to show that the cardinality of $\Gamma/\Gamma(2)$ is at least six. Or you can munge about with matrices and find the normalizer directly.

19.4. Covering Groups and Conformal Equivalence

There are fascinating connections between covering groups and conformal equivalence, some of which we will explore here. As an application we will give a very simple proof of the fact that not all annuli are conformally equivalent.

We begin with the fact that a simply connected covering space for an open set is essentially unique. Here "essentially unique" means of course "unique up to conformal equivalence":

Theorem 19.4.0. *Suppose that $V \subset \mathbb{C}$ is an open set, and for $j = 0, 1$ suppose that Ω_j is a simply connected open subset of the plane and that $p_j : \Omega_j \to V$ is a holomorphic covering map. Then there exists a conformal equivalence $\chi : \Omega_1 \to \Omega_0$ such that $p_1 = p_0 \circ \chi$.*

Proof. Theorem 19.0.5 shows that there exist holomorphic maps $\chi_0 : \Omega_0 \to \Omega_1$ and $\chi_1 : \Omega_1 \to \Omega_0$ such that $p_1 = p_0 \circ \chi_1$ and $p_0 = p_1 \circ \chi_0$. Hence $p_1 = p_1 \circ \chi_0 \circ \chi_1$, so Theorem 19.0.3.1 shows that $\chi_0 \circ \chi_1 \in \text{Aut}(\Omega_1, p_1)$. In particular χ_1 is one-to-one and χ_0 is onto. Similarly χ_0 is one-to-one and χ_1 is onto. So χ_1 is a conformal equivalence. $\square$

Corollary 19.4.1. *If Ω is simply connected and $p_j : \Omega \to V$ is a holomorphic covering map for $j = 0, 1$ then $\text{Aut}(\Omega, p_0)$ and $\text{Aut}(\Omega, p_1)$ are conjugate in $\text{Aut}(\Omega)$. (That is, there exists $\chi \in \text{Aut}(\Omega)$ such that $\text{Aut}(\Omega, p_0) = \chi \text{Aut}(\Omega, p_1) \chi^{-1}$.)*

Proof. The theorem shows that there exists $\chi \in \text{Aut}(\Omega)$ such that $p_1 = p_0 \circ \chi$. Suppose that $\phi \in \text{Aut}(\Omega, p_1)$. Then

$$p_0 \circ (\chi \circ \phi \circ \chi^{-1}) = p_1 \circ \phi \circ \chi^{-1} = p_1 \circ \chi^{-1} = p_0,$$

so $\chi \circ \phi \circ \chi^{-1} \in \text{Aut}(\Omega, p_0)$. Similarly, if $\psi \in \text{Aut}(\Omega, p_0)$ then $\chi^{-1} \circ \psi \circ \chi \in \text{Aut}(\Omega, p_1)$. $\square$

Corollary 19.4.2. *Suppose Ω is simply connected and $p_j : \Omega \to V_j$ is a holomorphic covering map for $j = 0, 1$. If V_1 and V_0 are conformally equivalent then $\text{Aut}(\Omega, p_0)$ and $\text{Aut}(\Omega, p_1)$ are conjugate in $\text{Aut}(\Omega)$.*

- **Exercise 19.8.** Prove Corollary 19.4.2.

This gives the main result in this section:

Theorem 19.4.3. *Suppose Ω is simply connected and $p_j : \Omega \to V_j$ is a holomorphic covering map for $j = 0, 1$. Then V_1 and V_0 are conformally equivalent if and only if $\text{Aut}(\Omega, p_0)$ and $\text{Aut}(\Omega, p_1)$ are conjugate in $\text{Aut}(\Omega)$.*

Proof. One direction is contained in Corollary 19.4.2.

For the converse, suppose that $\chi \in \operatorname{Aut}(\Omega)$ and
$$\operatorname{Aut}(\Omega, p_1) = \chi \operatorname{Aut}(\Omega, p_2) \chi^{-1}.$$
Let $q_0 = p_0$ and $q_1 = p_1 \circ \chi$. Then q_j is a covering map from Ω onto V_j and one may easily check that $\operatorname{Aut}(\Omega, q_0) = \operatorname{Aut}(\Omega, q_1)$; hence Theorem 19.0.4.1 (noting as previously that local inverses for holomorphic covering maps are holomorphic) shows that V_0 and V_1 are conformally equivalent. □

For $r > 1$ define $A_r = \{z : 1 < |z| < r\}$. The next theorem gives a demonstration of the power of Theorem 19.4.3.

Theorem 19.4.4. *If $1 < r_1 < r_2$ then A_{r_1} and A_{r_2} are not conformally equivalent.*

Recall that two square matrices A and B are said to be *similar* if there exists a matrix C such that $A = CBC^{-1}$. Recall as well the trivial fact that similar matrices have the same eigenvalues. And recall from Section 10.7 that $\operatorname{Aut}(\Pi^+)$ is isomorphic to $SL_2(\mathbb{R})/G$, where $SL_2(\mathbb{R})$ is the group of all real 2×2 matrices with determinant 1 and G is the subgroup $\{I, -I\}$.

Proof. For $\delta > 0$ define $\phi_\delta \in \operatorname{Aut}(\Pi^+)$ by $\phi_\delta(z) = \delta z$; note that the matrix in $SL_2(\mathbb{R})$ corresponding to ϕ_δ (or rather one of the two matrices in $SL_2(\mathbb{R})$ corresponding to ϕ_δ) is
$$M_\delta = \begin{bmatrix} \delta^{1/2} & 0 \\ 0 & \delta^{-1/2} \end{bmatrix}.$$

Theorem 19.1.2 shows that for $j = 1, 2$ there is a holomorphic covering map $p_j : \Pi^+ \to A_{r_j}$ such that $\operatorname{Aut}(\Pi^+, p_j)$ is generated by ϕ_{δ_j}, where $\delta_j > 1$ and $\delta_1 \ne \delta_2$. We need only show that $\operatorname{Aut}(\Pi^+, p_1)$ is not conjugate to $\operatorname{Aut}(\Pi^+, p_2)$ in $\operatorname{Aut}(\Pi^+)$.

Suppose to the contrary that $\operatorname{Aut}(\Pi^+, p_1) = \chi \operatorname{Aut}(\Pi^+, p_2) \chi^{-1}$ for some $\chi \in \operatorname{Aut}(\Pi^+)$. Then $\chi \circ \phi_{\delta_2} \circ \chi^{-1}$ must be a generator of $\operatorname{Aut}(\Pi^+, \delta_1)$, so that $\chi \circ \phi_{\delta_2} \circ \chi^{-1} = \phi_{\delta_1}$ or $\chi \circ \phi_{\delta_2} \circ \chi^{-1} = \phi_{1/\delta_1}$. Hence M_{δ_2} must be similar to one of the four matrices $\pm M_{\delta_1}, \pm M_{1/\delta_1}$. But this is impossible because similar matrices have the same eigenvalues. □

This business of "abstraction" can be very useful — it makes your head hurt sometimes but it's worth it! Here we have a bit of essentially trivial abstract results about covering maps, plus the trivial fact that similar matrices have the same eigenvalues, giving the fact that two annulli are not conformally equivalent. Utterly keen.

Of course Theorem 19.4.4 can also be derived by other methods; see for example Theorem 14.22 in [**R**].

Exercises

19.9. Suppose that $r > 1$ and show that no two of the sets $\mathbb{D} \setminus \{0\}$, $\mathbb{C} \setminus \{0\}$ and A_r are conformally equivalent.

Note. Of course the fact that $\mathbb{C} \setminus \{0\}$ is not conformally equivalent to either of the other two sets is clear from Liouville's Theorem (with the fact that an isolated singularity of a bounded holomorphic function must be removable). To derive this using the methods of this section, recall that the exponential is a covering map from the plane onto the punctured plane.

For the next exercise, note that if $M \in SL_2(\mathbb{R})$ has eigenvalues λ_1, λ_2 then the fact that $\det(M) = 1$ shows that $\lambda_1 \lambda_2 = 1$, while the fact that M has real entries shows that $\lambda_1 + \lambda_2 \in \mathbb{R}$. Hence if λ_1 and λ_2 are not both real then they must be complex numbers of absolute value 1.

19.10. Suppose that $M \in SL_2(\mathbb{R})$ has eigenvalues λ_1, λ_2. Suppose that ϕ, the element of $\mathrm{Aut}(\Pi^+)$ corresponding to M, is not the identity. Show that ϕ is elliptic if and only if $|\lambda_j| = 1$ and $\lambda_j \ne \pm 1$, ϕ is parabolic if and only if $\lambda_1 = \lambda_2 = \pm 1$, and ϕ is hyperbolic if and only if $\lambda_j \in \mathbb{R}$ and $\lambda_j \ne \pm 1$.

The next two exercises deal with a basic aspect of covering-space theory that curiously did not arise in the chapter. Exercise 19.11 is for students who know a little algebraic topology; other readers should still do Exercise 19.12, assuming that X and Y are connected open subsets of the plane:

19.11. Suppose that X and Y are connected and locally path-connected Hausdorff spaces, X is simply connected, and $p : X \to Y$ is a covering map. Show that the fundamental group of Y is naturally isomorphic to $\mathrm{Aut}(X, p)$.

Hint: Fix a point $a \in Y$ and a point $b \in X$ with $p(b) = a$. Suppose that $\gamma_1, \gamma_2 : [0, 1] \to Y$ are two closed curves with $\gamma_1(0) = \gamma_2(0) = a$. Let $\tilde{\gamma}_j : [0, 1] \to X$ be the unique curve with $\tilde{\gamma}_j(0) = b$ and $p \circ \tilde{\gamma}_j = \gamma_j$. Show that γ_1 and γ_2 are homotopic if and only if $\tilde{\gamma}_1(1) = \tilde{\gamma}_2(1)$.

Note. If we assume now that both X and Y are simply connected it follows that $\mathrm{Aut}(X, p)$ is trivial and hence that p is one-to-one, so that X and Y are homeomorphic. One could also deduce this from Theorem 19.4.0 (or rather from a purely topological version of the theorem): Since X and Y are both simply connected covering spaces of Y, they must be equivalent (conformally equivalent or topologically equivalent, depending on the context). The next exercise shows that the same conclusion holds without assuming that X is simply connected: A covering space for a simply connected space Y must be trivial.

19.12. If Y is simply connected and $p : X \to Y$ is a covering map then p is a homeomorphism.

Hint: It is sufficient to show that p is one-to-one. Suppose $x, x' \in X$ and $p(x) = p(x')$. Let $\tilde{\gamma} : [0, 1] \to X$ be a curve with $\tilde{\gamma}(0) = x$ and $\tilde{\gamma}(1) = x'$, and let $\gamma = p \circ \tilde{\gamma}$ (so that γ is a closed curve in Y). It follows from Lemma 16.3.3 and Theorem 19.0.1 that $\tilde{\gamma}(1) = \tilde{\gamma}(0)$.

Chapter 20

The Picard Theorems

Notation will be as in the previous two chapters: Γ is the group of all $\phi \in \text{Aut}(\Pi^+)$ which can be written in the form $\phi(z) = (az+b)/(cz+d)$ with $a, b, c, d \in \mathbb{Z}$ and $ad - bc = 1$, $\Gamma(2)$ is the subgroup of Γ where a and d are odd and b and c are even, and $\lambda : \Pi^+ \to \mathbb{C} \setminus \{0, 1\}$ is a holomorphic covering map with $\text{Aut}(\Pi^+, \lambda) = \Gamma(2)$.

The fact that λ is a holomorphic covering map from Π^+ onto $\mathbb{C} \setminus \{0, 1\}$ makes the Little Picard Theorem very simple:

Theorem 20.0 (Little Picard Theorem). *If f is a nonconstant entire function then the range $f(\mathbb{C})$ is either $\mathbb{C}$ or $\mathbb{C} \setminus \{\alpha\}$ for some $\alpha \in \mathbb{C}$.*

In other words, if the range of an entire function omits two complex values then the function is constant.

Proof. Suppose that f is an entire function and f omits the two values α, $\beta \in \mathbb{C}$, with $\alpha \neq \beta$. We need to show that f is constant.

Considering $(f - \alpha)/(\beta - \alpha)$ in place of f, we may assume that the two values omitted are 0 and 1; thus

$$f(\mathbb{C}) \subset \mathbb{C} \setminus \{0, 1\}.$$

Now Theorem 19.0.5 shows that there exists $\tilde{f} : \mathbb{C} \to \Pi^+$ with $\lambda \circ \tilde{f} = f$. Liouville's Theorem shows that $1/(i + \tilde{f})$ is constant; hence $\tilde{f}$ is constant and so f is constant. $\square$

I think it was Littlewood who said that this would be the world's shortest PhD thesis. A one-line proof of a very deep theorem. (Of course the existence of a covering map from the upper half-plane onto $\mathbb{C}\setminus\{0,1\}$ was not entirely trivial.)

Now for the Big Picard Theorem. There are various statements that commonly go by this name:

Theorem 20.1 (Big Picard Theorem). *If f is an entire function and f is not a polynomial then, with at most one exception, f attains every complex value infinitely many times.*

That's the way the theorem is often stated; a more explicit version of the conclusion would be this: There exists $\alpha \in \mathbb{C}$ such that $f - \beta$ has infinitely many zeroes for every $\beta \in \mathbb{C}\setminus\{\alpha\}$.

This is an immediate consequence of the following stronger result, also known as the Big Picard Theorem:

Theorem 20.2 (Big Picard Theorem). *If f has an essential singularity at z_0 then, with at most one exception, f attains every complex value infinitely many times in every neighborhood of z_0.*

To derive Theorem 20.1 from Theorem 20.2 note that if f is an entire function and f is not a polynomial then the function $f(1/z)$ has an essential singularity at the origin (the power series for f becomes the Laurent series for $f(1/z)$).

As previously we set $\mathbb{D}' = \mathbb{D}\setminus\{0\}$. Theorem 20.2 is equivalent to the following

Theorem 20.3 (Big Picard Theorem). *If $f \in H(\mathbb{D}')$ and $f(\mathbb{D}') \subset \mathbb{C}\setminus\{0,1\}$ then f has a pole or a removable singularity at 0.*

It is clear that Theorem 20.3 is a special case of Theorem 20.2. The proof of the converse involves just a few changes of variables: Assume Theorem 20.3, and suppose that f has an essential singularity at z_0 and there are two values α and β, each of which is attained only finitely many times by f in some neighborhood of z_0. Considering the restriction of f to $D(z_0, r)$ for sufficiently small $r > 0$ we may assume that f never attains the values α and β. Now if $f_1 = (f-\alpha)/(\beta-\alpha)$ then f_1 never attains the values 0 or 1 in $D(0, r)$. So if $f_2(z) = f_1(z_0 + rz)$ then $f_2 \in H(\mathbb{D}')$ and $f_2(\mathbb{D}') \subset \mathbb{C}\setminus\{0,1\}$. Now Theorem 20.3 shows that f_2 has a pole or removable singularity at the origin, and hence the same is true of f at z_0.

We will prove Theorem 20.3 below, after one final rephrasing. Traditional proofs of the Big Picard Theorem are totally unlike the usual proof of the Little Picard Theorem, but we shall see that it is possible to give a proof of the Big Picard Theorem that is very much like a generalization of the proof of the Little Picard Theorem; where the first proof used nothing but the most basic properties of covering maps, the second proof uses the slightly more sophisticated results from Chapter 19.

The idea behind the proof may be clearer if we start with two results dealing with more familiar notions like harmonic conjugates and exponentials: Our proof of the Big Picard Theorem is to the traditional proof of the Little Picard Theorem exactly as the proof of Theorem B below is to the proof of Theorem A:

Theorem A. *If $u : \mathbb{C} \to \mathbb{R}$ is a bounded harmonic function then u is constant.*

Proof. Since $\mathbb{C}$ is simply connected, there exists a real-valued harmonic function v such that $u + iv$ is holomorphic. Let $h = e^{u+iv}$. Then h is an entire function, and h is bounded, since $|h| = e^u$. Thus h is constant, and hence $u = \log|h|$ is constant. $\square$

Theorem B. *If $f : \mathbb{D}' \to \mathbb{R}$ is harmonic and bounded then u extends to a function harmonic in $\mathbb{D}$.*

The problem with Theorem B is that we cannot say that u has a harmonic conjugate, since $\mathbb{D}'$ is not simply connected. (This is precisely analogous to the problem with Theorem 20.3: We cannot say that there exists $\tilde{f} : \mathbb{D}' \to \Pi^+$ with $\lambda \circ \tilde{f} = f$ because $\mathbb{D}'$ is not simply connected.)

One can give a slightly informal proof of Theorem B as follows: There does exist a "multi-valued" harmonic conjugate v (this is an informal way of saying that there is a holomorphic function $u + iv$ defined in some disk contained in $\mathbb{D}'$ which admits unrestricted continuation in $\mathbb{D}'$). Now it is not hard to see that in fact there exists a multi-valued harmonic conjugate v and a real number c such that any two branches of v differ by a multiple of c. If $c = 0$ we are done, so we assume that $c \neq 0$, and define $h = e^{2\pi(u+iv)/c}$. The fact that any two branches of v differ by a multiple of c shows that h is an actual single-valued holomorphic function. As in the proof of Theorem A we see that h is bounded; hence h has a removable singularity, and hence u has a removable singularity at the origin as well.

If the "informal" parts of that proof bother you don't worry, we will give a more formal version soon. The corresponding informal proof of Theorem 20.3 is this: Although we cannot assert that there exists $\tilde{f} : \mathbb{D}' \to \Pi^+$ with $\lambda \circ \tilde{f} = f$, there does exist a "multi-valued function" $\tilde{f}$ with this property (that is, an $\tilde{f}$ defined in some disk contained in $\mathbb{D}'$ which admits unrestricted continuation in $\mathbb{D}'$). It turns out that there exists $\phi \in \Gamma(2)$ such that any two branches of $\tilde{f}$ differ by a power of ϕ, and, since ϕ cannot be elliptic, we know that there exists a nonconstant bounded function $h \in H(\Pi^+)$ such that $h \circ \phi = h$. It follows that $h \circ \tilde{f}$ is single-valued, and hence has a removable singularity at the origin.

It is now quite plausible, and not hard to prove using the results in the previous chapter, that in fact f has a (possibly infinite) limit at the origin, so that in particular f does not have an essential singularity (for the details here see the formal proof below).

The actual proof is going to use a covering-map argument in place of the analytic continuation — I think this makes the details of the proof easier to follow, possibly at the price of making it harder to see where the proof comes from. So for the sake of illustrating how the covering-map argument corresponds to the analytic continuation argument, here's the "formal" proof of Theorem B:

Proof of Theorem B. Define a function $U : \Pi^+ \to \mathbb{R}$ by $U(z) = u(e^{2\pi i z})$. Then U is harmonic, being the composition of a harmonic function and a holomorphic function. Since Π^+ is simply connected, there exists $F \in H(\Pi^+)$ such that $U = \operatorname{Re}(F)$. Now, although $U(z+1) = U(z)$ we need not have $F(z+1) = F(z)$. But $\operatorname{Re}(F(z+1) - F(z)) = U(z+1) - U(z) = 0$, and so $F(z+1) - F(z)$ must be an imaginary constant: There exists $c \in \mathbb{R}$ such that $F(z+1) - F(z) = ic$ for all $z \in \Pi^+$. Choose a real number $\alpha \ne 0$ so that $c\alpha$ is a multiple of 2π (you can take $\alpha = 2\pi/c$ unless $c = 0$). Let

$$E(z) = e^{\alpha F(z)}.$$

It follows that $E(z+1) = E(z)$, and so there exists $f \in H(\mathbb{D}')$ with

$$E(z) = f(e^{2\pi i z}).$$

Now the fact that u is bounded shows that f is bounded, so that f has a removable singularity at the origin. The fact that u is bounded also shows that f is bounded away from 0, so in particular $f(0) \ne 0$. Hence u has a limit at 0, since $u = \log(|f|)/\alpha$. (Since $f(e^{2\pi i z}) = e^{\alpha F(z)}$ it follows that

$$\log(|f(e^{2\pi i z})|) = \alpha \operatorname{Re}(F(z)) = \alpha U(z) = \alpha u(e^{2\pi i z}).) \qquad \square$$

If we do the same thing with the Big Picard Theorem, replacing functions in $\mathbb{D}'$ by the corresponding periodic functions in the upper half-plane, we see that we need to prove this:

Theorem 20.4. *Suppose that $F : \Pi^+ \to \mathbb{C} \setminus \{0, 1\}$ is holomorphic and satisfies $F(z+1) = F(z)$ for all $z \in \Pi^+$. Then there exists $L \in \mathbb{C}_\infty$ such that $F(z_j) \to L$ for every sequence (z_j) in Π^+ with $\operatorname{Im}(z_j) \to \infty$.*

- **Exercise 20.1.** Show that Theorem 20.3 follows from Theorem 20.4.

Proof of Theorem 20.4. Since Π^+ is simply connected and $\lambda : \Pi^+ \to \mathbb{C} \setminus \{0, 1\}$ is a covering map, it follows from Theorem 19.0.5 that there exists a holomorphic function $\tilde{F} : \Pi^+ \to \Pi^+$ satisfying

$$F = \lambda \circ \tilde{F}.$$

Now $\lambda(\tilde{F}(i+1)) = F(i+1) = F(i) = \lambda(\tilde{F}(i))$, so there exists $\phi \in \Gamma(2) = \mathrm{Aut}(\Pi^+, \lambda)$ such that

$$\tilde{F}(i+1) = \phi(\tilde{F}(i)).$$

Define $\tau_1(z) = z + 1$. Since

$$\lambda \circ (\tilde{F} \circ \tau_1) = F \circ \tau_1 = F = \lambda \circ \tilde{F} = \lambda \circ (\phi \circ \tilde{F})$$

and the two functions $\tilde{F} \circ \tau_1$ and $\phi \circ \tilde{F}$ have the same value at the point $i \in \Pi^+$, the uniqueness assertion in Theorem 19.0.5 shows that

$$\tilde{F} \circ \tau_1 = \phi \circ \tilde{F}.$$

Now, since $\phi \in \Gamma(2)$, it cannot be elliptic, and so by Theorem 19.1.2 there exists a covering map h from Π^+ onto a bounded open set in the plane such that $\mathrm{Aut}(\Pi^+, h) = G$, the subgroup of $\mathrm{Aut}(\Pi^+)$ generated by ϕ. Let $H = h \circ \tilde{F}$.

The fact that h is bounded shows that H is bounded, and we also have

$$H \circ \tau_1 = h \circ \tilde{F} \circ \tau_1 = h \circ \phi \circ \tilde{F} = h \circ \tilde{F} = H.$$

So Corollary 19.0.10 shows that there exists a function $H^* \in H(\mathbb{D}')$ with $H(z) = H^*(e^{2\pi i z})$. The fact that H^* is bounded shows it has a removable singularity at the origin; let $L = H^*(0)$.

Suppose that (z_j) is a sequence in Π^+ with $\mathrm{Im}\,(z_j) \to \infty$, and let $w_j = \tilde{F}(z_j)$; since $\lambda(w_j) = F(z_j)$, we need to show that the sequence $(\lambda(w_j))$ has a limit in $\mathbb{C}_\infty$.

Now, $h(w_j) = h(\tilde{F}(z_j)) = H(z_j) = H^*(e^{2\pi i z_j}) \to L$, since $\mathrm{Im}\,(z_j) \to \infty$. Assume for a second that L is in the range of h; we will see below that this holds in most cases. If $L \in h(\Pi^+)$ then, since h is a covering map onto $h(\Pi^+)$, it follows from Lemma 19.0.0 that there exists a sequence (w_j') in Π^+ such that $h(w_j') = h(w_j)$ and $w_j' \to w \in \Pi^+$. Since $\mathrm{Aut}(\Pi^+, h) \subset \mathrm{Aut}(\Pi^+, \lambda)$, it follows that

$$\lambda(w_j) = \lambda(w_j') \to \lambda(w);$$

in particular the sequence $(\lambda(w_j))$ is convergent in $\mathbb{C}_\infty$, which is what we needed to prove. (If you trace through the definitions you see that we have shown that f has a removable singularity at the origin, with $f(0) = \lambda(w)$; in particular $f(0) \in \mathbb{C} \setminus \{0, 1\}$.)

So if $L \in h(\Pi^+)$ we are done. In most cases the Maximum Modulus Theorem shows that we do in fact have $L \in h(\Pi^+)$. There are various cases, depending on ϕ:

First, if ϕ is the identity then

$$H^*(\mathbb{D}') = H(\Pi^+) \subset h(\Pi^+) = \mathbb{D};$$

since $L = H^*(0)$, the Maximum Modulus Theorem shows that $|L| < 1$, and so $L \in h(\Pi^+)$. Similarly if ϕ is hyperbolic then h is a covering map onto an annulus defined by $r < |z| < R$ for some $0 < r < R < \infty$; now the Maximum Modulus Theorem, applied to H^* and also to $1/H^*$, shows that $r < |L| < R$ and so $L \in h(\Pi^+)$.

Suppose finally that ϕ is parabolic. Now $h(\Pi^+)$ is the punctured disk $\mathbb{D}'$, and the Maximum Modulus Theorem shows that $|L| < 1$ as before. It may happen that $0 < |L| < 1$; if so then $L \in h(\Pi^+)$ and we are finished.

There remains only the case where ϕ is parabolic and $L = 0$. (It's a good thing that this exceptional case exists; otherwise we would have a proof that functions cannot have poles!) We took care of this case in the previous chapter: Theorem 19.2.3 shows that the sequence $(\lambda(w_j))$ converges in $\mathbb{C}_\infty$. □

Most of the concepts that arose in this book are very old; however, the terminology and style used in even slightly older literature is often very different. If you know a little bit of French and you want to improve your mathematical-language skills you should look at a version of the original proof of the Big Picard Theorem [J] and convince yourself that it really is exactly the same as the proof above, expressed very differently. (I'll give you one hint: It took me a little while to figure out what a "uniform function" was in [J]. It turns out that the word "uniform" there is the opposite of "multiform" ...)

Exercises

20.2. Suppose that D is a connected open subset of the plane and $f \in H(D)$ has no zero in D. Show that there exists $g \in H(D)$ with $f = e^g$ if and only if

$$\frac{1}{2\pi i} \int_\gamma \frac{f'(z)}{f(z)} \, dz = 0$$

for every smooth closed curve $\gamma \subset D$.

20.3. Suppose that D is a connected open subset of the plane and $f \in H(D)$ has no zero in D. Show that there exists $g \in H(D)$ with $f = g^2$ if and only if

$$\frac{1}{2\pi i} \int_\gamma \frac{f'(z)}{f(z)} \, dz$$

is an even integer for every smooth closed curve $\gamma \subset D$. (For the harder direction: Suppose that $D(a, r) \subset D$. The function element $(\log(f), D(a, r))$ admits unrestricted continuation in D. Use an argument like the proof of Theorem B above.)

20.4. Show how Exercise 13.3 follows from Exercise 20.3.

20.5. Let $D = \mathbb{C} \setminus [0, 1]$ and define $f_1, f_2, f_3 \in H(D)$ by $f_1(z) = z/(z-1)$, $f_2(z) = z(z-1)$, $f_3(z) = z$. Show that f_1 has a holomorphic logarithm in D, f_2 does not a have a holomorphic logarithm but does have a holomoprhic square root in D, and f_3 does not have a holomorphic square root.

20.6. Suppose that f is a nonconstant entire function and f is *odd* (that is, $f(-z) = -f(z)$). Show that $f(\mathbb{C}) = \mathbb{C}$.

20.7. Give an example of a nonconstant even entire function f with $f(\mathbb{C}) \neq \mathbb{C}$.

Part 2

Further Results

Chapter 21

Abel's Theorem

Abel's Theorem gives a relation between convergence of a power series at a point of the boundary of the disk of convergence and the behavior of the function defined by the power series at nearby points in the interior of the disk. First an exercise:

- **Exercise 21.1.** Suppose that $a_j \in \mathbb{C}$ for $0 \leq j \leq N$ and define $s_n = \sum_{j=0}^{n} a_j$. Define $f(r) = \sum_{j=0}^{N} a_j r^j$, and show that

$$f(r) = s_N r^N + \sum_{j=0}^{N-1} s_j (r^j - r^{j+1}).$$

The exercise is an example of a technique known as "summation by parts", the analogue for sums of integration by parts. You should give a formal proof, perhaps by induction on N; we illustrate why the result is true with the case $N = 2$:

$$s_2 r^2 + \sum_{j=0}^{2-1} s_j(r^j - r^{j+1})$$
$$= (a_0 + a_1 + a_2)r^2 + (a_0)(1-r) + (a_0 + a_1)(r - r^2)$$
$$= a_0(r^2 + r - r^2 + 1 - r) + a_1(r^2 + r - r^2) + a_2 r^2$$
$$= a_0 + a_1 r + a_2 r^2.$$

As a corollary we have

- **Exercise 21.2.** Suppose that $a_j \in \mathbb{C}$ for $0 \le j \le N$. Define $s_n = \sum_{j=0}^n a_j$ and $f(r) = \sum_{j=0}^N a_j r^j$. Suppose that $|s_n| \le M$ for $0 \le n \le N$ and show that
$$|f(r)| \le M \quad (0 \le r \le 1).$$

(Make certain it is clear where the fact that $0 \le r \le 1$ is used!)

Theorem 21.0 (Abel's Theorem). *Suppose that $a_j \in \mathbb{C}$ for $j \ge 0$ and the sum $\sum_{j=0}^\infty a_j$ converges to s. Define $f(r) = \sum_{j=0}^\infty a_j r^j$ for $0 \le r \le 1$. Then*
$$\lim_{r \to 1^-} f(r) = s.$$

(Note that the convergence of $\sum_{j=0}^\infty a_j$ implies that the a_j are bounded, so that the sum defining $f(r)$ converges, in fact converges absolutely, for $0 \le r < 1$. The sum also converges for $r = 1$ by hypothesis, although the convergence when $r = 1$ need not be absolute. In fact the theorem would be much simpler if we assumed that $\sum_{j=0}^\infty a_j$ converged absolutely; the fact that it applies when this sum converges conditionally is sort of the whole point.)

Proof. Let $s_n = \sum_{j=0}^n a_j$ and define $f_j(r) = a_j r^j$. The hypothesis says that the sequence (s_n) is Cauchy; we will show that this implies that the partial sums of $\sum f_j$ are uniformly Cauchy on $[0, 1]$, so that $f = \sum f_j$ is continuous on $[0, 1]$ and in particular $\lim_{r \to 1} f(r) = f(1) = s$.

Let $\epsilon > 0$, and choose N so that $|s_n - s_m| < \epsilon$ for all $n, m > N$. Fix n, m with $N < n < m$, and define b_j for $0 \le j \le m - n - 1$ by $b_j = a_{j+n+1}$ (so that $(b_0, \ldots, b_{m-n-1}) = (a_{n+1}, \ldots, a_m)$).

Now if $0 \le k \le m - n - 1$ then
$$\sum_{j=0}^k b_j = s_{k+n+1} - s_n,$$

so our choice of N shows that
$$\left| \sum_{j=0}^k b_j \right| < \epsilon.$$

Hence the previous exercise shows that
$$\left| \sum_{j=0}^{m-n-1} b_j r^j \right| < \epsilon \quad (0 \le r \le 1),$$

so that for every $r \in [0,1]$ we have

$$\left|\sum_{j=n+1}^{m} a_j r^j\right| = r^{n+1}\left|\sum_{j=0}^{m-n-1} b_j r^j\right| < \epsilon.$$

Since this holds whenever $N < n < m$, we have shown that the partial sums of $\sum f_j$ are uniformly Cauchy on $[0,1]$, as required. □

You should note that the converse of Abel's Theorem is false; for example if $a_j = (-1)^j$ then $\lim_{r \to 1^-} f(r)$ exists (Exercise 21.3 below) although $\sum a_j$ diverges. This says that $\sum (-1)^j$ is "Abel summable"; various "summability methods" are often useful substitutes for convergence.

Although the converse of Abel's Theorem is false, it becomes true if we add various "tauberian" conditions. For example Tauber's Theorem says that the converse of Abel's Theorem holds under the additional assumption that $ja_j \to 0$, while Littlewood's Theorem replaces this with the condition $|ja_j| \leq K$. These are examples of "tauberian" theorems, which are very useful in analysis and in applications to other fields like number theory.

We will give a proof of Tauber's Theorem, because it is not very hard and illustrates a few generally useful techniques. Littlewood's theorem is much deeper; we will not attempt to give a proof here. We need a few inequalities:

Lemma 21.1. *Suppose that n is a positive integer and let $r = 1 - \frac{1}{n}$. Then*

$$\sum_{j=1}^{n} \frac{1}{j}(1 - r^j) \leq 1$$

and

$$\sum_{j=n}^{\infty} \frac{1}{j} r^j \leq 1.$$

Proof. It follows either from the Mean Value Theorem or the factorization $1 - r^j = (1-r)(1 + r + \cdots + r^{j-1})$ that $1 - r^j \leq j(1-r)$; hence

$$\sum_{j=1}^{n} \frac{1}{j}(1 - r^j) \leq \sum_{j=1}^{n}(1-r) = 1.$$

The second inequality is even easier:

$$\sum_{j=n}^{\infty} \frac{1}{j} r^j \leq \frac{1}{n}\sum_{j=n}^{\infty} r^j = \frac{r^n}{n(1-r)} = r^n \leq 1. \quad \square$$

Theorem 21.2 (Tauber's Theorem). *Suppose that (a_j) is a sequence of complex numbers such that $ja_j \to 0$ as $j \to \infty$. Let*

$$s_n = \sum_{j=0}^{n} a_j$$

and

$$f(r) = \sum_{j=0}^{\infty} a_j r^j \quad (0 \le r < 1)$$

(note that a_j is bounded, so the series defining f converges). If $\lim_{r \to 1} f(r)$ exists then $\lim_{n \to \infty} s_n$ exists and has the same value.

Note again that the hypothesis can be weakened to $|ja_j| \le K$, although that makes the proof much harder.

Proof. Suppose $\epsilon > 0$, and write

$$a_j = b_j + c_j,$$

where $b_j = 0$ except for finitely many values of j and $|jc_j| < \epsilon$ for all j. Write $s'_n = \sum_{j=0}^{n} b_j$, $s''_n = \sum_{j=0}^{n} c_j$, $g(r) = \sum_{j=0}^{\infty} b_j r^j$ and $h(r) = \sum_{j=0}^{\infty} c_j r^j$.

For $n = 1, 2, \ldots$ let $r_n = 1 - \frac{1}{n}$. Since $b_j = 0$ except for finitely many values of j, it is clear that

$$\lim_{n \to \infty} (s'_n - g(r_n)) = 0.$$

On the other hand, the lemma shows that

$$|s''_n - h(r_n)| \le 2\epsilon$$

for all n. Since

$$|s_n - f(r_n)| \le |s'_n - g(r_n)| + |s''_n - h(r_n)|,$$

it follows that

$$\limsup_{n \to \infty} |s_n - f(r_n)| \le 2\epsilon.$$

Since this holds for all $\epsilon > 0$, we have

$$\lim_{n \to \infty} (s_n - f(r_n)) = 0. \qquad \square$$

Note that the argument actually proves the following:

21. Abel's Theorem

Corollary 21.3. *Suppose that* $ja_j \to 0$. *Set* $s_n(e^{it}) = \sum_{j=0}^{n} a_j e^{ijt}$ *and* $f(z) = \sum_{j=0}^{\infty} a_j z^j$ ($|z| < 1$). *Then* $s_n(e^{it}) - f((1 - \frac{1}{n})e^{it}) \to 0$ *uniformly.*

I claimed that all this was supposed to be instructive. The moral is this:

Moral 21.4. *If you want to find an r such that $s_n - f(r)$ is small then $r = 1 - \frac{1}{n}$ is often a good choice.*

The proof was too easy to really demonstrate the truth of the moral. The real reason the proof works is that, as we recall from calculus,

$$\lim_{n \to \infty} \left(1 - \frac{1}{n}\right)^n = e^{-1}.$$

Roughly speaking this means that if $r = 1 - \frac{1}{n}$ then r^j is close enough to 1 for $0 \leq j \leq n$ and close enough to 0 for $n \leq j < \infty$ to make $f(r)$ close to s_n. Of course if we had taken $r = 1 - \frac{2}{n}$ or $r = 1 - \frac{1}{2n}$ or something similar then things would have worked out just as well, with different constants. But if we took r such that $1 - r$ was not comparable to $1/n$ then Lemma 21.1 would not work. For example:

Example 21.5. *If $n \geq 2$ and $r = 1 - 1/n^2$ then*

$$\sum_{j=n}^{\infty} \frac{1}{j} r^j \geq e^{-2} \log(n).$$

Proof. The Mean Value Theorem (applied to $\log(1 - t)$) shows that

$$r^{n^2} \geq e^{-2}.$$

Hence

$$\sum_{j=n}^{\infty} \frac{1}{j} r^j \geq \sum_{j=n}^{n^2} \frac{1}{j} r^j \geq e^{-2} \sum_{j=n}^{n^2} \frac{1}{j} \geq e^{-2} \int_{n}^{n^2} \frac{1}{t} \, dt = e^{-2} \log(n)$$

(cf. Section 6.2). □

So the second part of Lemma 21.1 would not work if $r = 1 - 1/n^2$. Similarly for a modification of the first part of the lemma:

Example 21.6. *There exists a $c > 0$ such that if $r = 1 - 1/n^{1/2}$ then*

$$\sum_{j=1}^{n} \frac{1}{j}(1 - r^j) \geq c \log(n).$$

Proof. Noting that $r^{n^{1/2}} \leq e^{-1}$ we see that

$$\sum_{j=1}^{n} \frac{1}{j}(1-r^j) \geq \sum_{n^{1/2} \leq j \leq n} \frac{1}{j}(1-r^j) \geq (1-e^{-1}) \sum_{n^{1/2} \leq j \leq n} \frac{1}{j} \geq c\log(n). \qquad \square$$

Exercises

21.3. Show that $\lim_{r \to 1^-} \sum_{j=0}^{\infty} (-1)^j r^j$ exists (you can do this by evaluating the sum in closed form).

The next exercise shows that Abel's Theorem holds for any boundary point, not just $\zeta = 1$:

21.4. Suppose that $f \in H(\mathbb{D})$ and the power series for f converges at the point $\zeta \in \partial \mathbb{D}$. Show that $\lim_{r \to 1^-} f(r\zeta)$ exists (and is equal to the sum of the power series at ζ). (This follows easily from the version of Abel's Theorem given above; you don't need to repeat the proof!)

21.5. Find the value of $\sum_{j=1}^{\infty} (-1)^j/j$ (you can do this using Abel's Theorem plus something that came up in Chapter 13).

Here is another simple converse to Abel's Theorem, with a different extra hypothesis:

21.6. Suppose that $f \in H(\mathbb{D})$, the derivative $f^{(k)}(0)$ is non-negative for every k, and $\lim_{r \to 1^-} f(r)$ exists. Show that the power series for f converges at the point $z = 1$. Then show that in fact the power series converges uniformly on $\overline{\mathbb{D}}$.

The next exercise gives a condition other than boundedness of ja_j under which $r = 1 - 1/n$ is the right choice:

21.7. Suppose a_j is bounded. Define $s_n = \sum_{j=0}^{n} a_j$ and $f(r) = \sum_{j=0}^{\infty} a_j r^{2^j}$ ($0 \leq r < 1$). Let $r_n = 1 - 2^{-n}$.

(i) There exists a constant c such that if $|a_j| \leq 1$ for all j then $|s_n - f(r_n)| \leq c$ for all n.

(ii) If $a_j \to 0$ as $j \to \infty$ then $s_n - f(r_n) \to 0$ as $n \to \infty$.

21. Abel's Theorem

Note. If $r_n = 1 - 2^{-n}$ in the previous exercise seems inconsistent with our motto that $r = 1 - 1/n$ is the way to go that's just an illusion, caused by the fact that a_j is the coefficient of r^{2^j} rather than of r^j: If we choose b_j so that

$$\sum_{j=0}^{\infty} a_j r^{2^j} = \sum_{j=0}^{\infty} b_j r^j$$

then

$$s_n = \sum_{j=0}^{2^n} b_j.$$

The following sharper versions of the two examples show that $1-r \approx 1/n$ really is necessary in Lemma 21.1:

21.8. Suppose that $r_n \in (0,1)$ satisfies $n(1 - r_n) \to \infty$ as $n \to \infty$. Then

$$\sum_{j=0}^{n} \frac{1}{j}(1 - r_n^j) \to \infty$$

as $n \to \infty$.

21.9. Suppose that $r_n \in (0,1)$ satisfies $n(1 - r_n) \to 0$ as $n \to \infty$. Then

$$\sum_{j=n}^{\infty} \frac{1}{j} r_n^j \to \infty$$

as $n \to \infty$.

21.10. Explain what Tauber's Theorem says about convergence of the Fourier series of $f \in C(\partial \mathbb{D})$. (See Section 10.4; note that you may need a slightly different version of Theorem 21.2.)

Chapter 22

More on Brownian Motion

In this chapter we give a proof of a fact stated in Section 10.6: If D is a bounded open subset of the plane and every component of the boundary contains more than one point then D is a Dirichlet domain. We also give a miscellaneous application of our results on Brownian motion to conformal mapping. (Of course all the disclaimers in Section 10.6 about how our "proofs" are not quite proofs since we haven't even given the *definition* of Brownian motion apply equally well to the material in this chapter.)

Suppose that $z \in \mathbb{C}$, $r > 0$, and $0 < \delta < 2\pi$. The phrase "an arc of opening δ on $\partial D(z,r)$" will refer to any set of the form

$$I = \{\, z + re^{it} : a < t < b\,\},$$

where $a, b \in \mathbb{R}$ and $b - a = \delta$.

Lemma 22.0. *Suppose that $0 < \rho < 1$ and $0 < \delta < \pi$. There exists $p > 0$ with the following property: If $z \in \mathbb{C}$, $r > 0$, I is an arc of opening δ on $\partial D(z,r)$ and $w \in D(z, \rho r)$ then*

$$\Pr(\mathrm{Exit}(D(z,r), w) \in I) \geq p.$$

Proof. A change of variables shows that we may assume that $D(z,r) = \mathbb{D}$. Choose $f \in C(\partial \mathbb{D})$ so that $0 \leq f \leq 1$ everywhere, $f = 0$ on $\partial \mathbb{D} \setminus I$, and $f = 1$ at at least one point of I. Let $u \in C(\overline{\mathbb{D}})$ be harmonic in $\mathbb{D}$ and equal to f on $\partial \mathbb{D}$. It follows by compactness that there exists $p > 0$ such that $u \geq p$ on

$\overline{D(0,\rho)}$. Suppose that $|w| \le \rho$, and let A be the event $\mathrm{Exit}(\mathbb{D}, w) \in I$. Then

$$\begin{aligned}
p \le u(w) &= E(f(\mathrm{Exit}(\mathbb{D}, w))) \\
&= \Pr(A) E(f(\mathrm{Exit}(\mathbb{D}, w)) | A) + (1 - \Pr(A)) E(f(\mathrm{Exit}(\mathbb{D}, w)) | A^c) \\
&= \Pr(A) E(f(\mathrm{Exit}(\mathbb{D}, w)) | A) \\
&\le \Pr(A). \quad \square
\end{aligned}$$

Now suppose that $\gamma : [0,1] \to \mathbb{C}$ is a continuous curve. We will say that "$B_t(z)$ traces γ within ϵ" if there exist $T > 0$ and an increasing continuous bijection $\phi : [0, T] \to [0, 1]$ such that

$$|B_t(z) - \gamma(\phi(t))| < \epsilon$$

for all $t \in [0, T]$.

Proposition 22.1. *Suppose that $\gamma : [0,1] \to \mathbb{C}$ is continuous, $\gamma(0) = z$ and $\epsilon > 0$. Then the probability that $B_t(z)$ traces γ within ϵ is positive.*

Taking $\epsilon > 0$ small enough that you cannot see the difference between two points separated by a distance of less than ϵ we see that for any curve γ there is a positive probability that the image of $B_t(z)$ for $0 \le t \le T$ (for some T) looks just like the image of γ. (In fact more is true: With probability 1 there exists $[a, b] \subset [0, 1]$ such that $B_t(z)$ for $a \le t \le b$ looks like a suitably translated, rotated and rescaled version of γ. If you look at B_t at the right resolution you will almost surely see the complete text of Hamlet.)

Proof. We can cover the image of γ by finitely many disks $D_j = D(a_j, \epsilon)$ ($1 \le j \le N$) with the following property: There exist $\delta > 0$ and $\rho \in (0, 1)$ such that for each j with $1 \le j \le N - 1$ there is an arc I_j in ∂D_j of opening δ, such that $I_j \subset D(a_{j+1}, \rho\epsilon)$. See Figure 22.1:

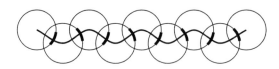

Figure 22.1

Choose $p > 0$ as in the preceding lemma, and set $z_0 = z$. Let $z_1 = \mathrm{Exit}(D_1, z_0)$. The lemma shows that the probability that $z_1 \in I_1$ is at least p. Now assuming that $z_1 \in I_1$ it follows that $|a_2 - z_1| < \rho\epsilon$, so if $z_2 = \mathrm{Exit}(D_2, z_1)$ then the lemma shows that the probability that $z_2 \in I_2$ is also at least p. Continuing in this way, we see that the probability that $z_j \in I_j$ for $1 \le j \le N - 1$, where $z_j = \mathrm{Exit}(D_j, z_{j-1})$, is at least p^{N-1}. $\quad \square$

22. More on Brownian Motion

Lemma 22.2. *There exists $p > 0$ with the following property: If $K \subset \mathbb{C}$ is connected, $z \in \mathbb{C}$, $r > 0$, and there exist $w_1 \in K$ with $|z - w_1| \leq r/10$ and $w_2 \in K$ with $|z - w_2| \geq r$ then the probability that $B_t(z)$ hits K before leaving $D(z, r)$ is at least p. In other words,*

$$\Pr(\mathrm{Exit}(D(z,r) \setminus K, z) \in K) \geq p.$$

Proof. After a change of variables $w \mapsto aw + b$ we may assume that $z = 0$ and $r = 1$, so that $D(z, r) = \mathbb{D}$. Suppose that $\gamma : [0, 1] \to \overline{\mathbb{D}}$ is as in Figure 22.2:

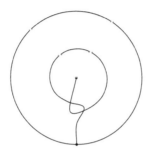

Figure 22.2

In particular $\gamma(0) = 0$, $|\gamma(1)| = 1$, and γ "includes" a closed loop about the origin. (Note that K is not included in the figure. This is intentional: The point is that K can be an arbitrarily wild connected set; any picture of K we might draw would be smoother than K actually needs to be, and we would not want to give the impression that the smoothness in our depiction of K was relevant to the proof.)

Let p be the probability that $B_t(0)$ traces γ within $1/100$; note that the previous proposition shows that $p > 0$. Now, if it happens that $B_t(0)$ does trace γ within $1/100$ then it follows that there exists an interval $I = [t_0, t_1]$ such that if we define $\phi : I \to \mathbb{C}$ by $\phi(t) = B_t(z)$ then ϕ is a closed curve in $\mathbb{D} \setminus \{0\}$. Further, Proposition 5.3 shows that

$$\mathrm{Ind}(\phi, 0) = 1$$

and

$$\mathrm{Ind}(\phi, w) = 0 \quad (|w| = 1).$$

Finally, it is clear that

$$D(0, 1/5) \cap \phi^* = \emptyset.$$

This shows that $\mathrm{Ind}(\phi, w_1) = 1$ and $\mathrm{Ind}(\phi, w_2) = 0$, and hence that ϕ^* must intersect K, since $\mathrm{Ind}(\phi, w)$ is constant for w in any connected subset of the complement of ϕ^*. □

Theorem 22.3. *Suppose that D is a bounded open set in $\mathbb{C}$ and every component of the boundary of D contains more than one point. Then D is a Dirichlet domain.*

Proof. The proof is essentially the same as the proof of Theorem 10.6.2, except using Lemma 22.2 instead of the argument about Brownian motion in squares to show that there is a fixed positive probability of success at each stage.

Suppose then that $\zeta \in \partial D$; we need to show that condition (∗∗∗) of Lemma 10.6.1 holds for every $\delta > 0$. If the condition holds for one value of δ then it certainly holds for all larger δ, so we may assume that δ is as small as we wish. Let K be the component of ∂D containing ζ. By hypothesis there exists $\zeta' \in K$ with $\zeta' \neq \zeta$; we will consider only δ with $\delta < |\zeta - \zeta'|$.

Let p be as in the conclusion of Lemma 22.2.

Assume first that $z \in D$ and $|z - \zeta| < \delta/11$. Let $r = 10|z - \zeta|$. Then $|z - \zeta| \leq r/10$ and there exists $\zeta' \in K$ with $|z - \zeta'| \geq r$, and so the lemma shows that
$$\Pr(\text{Exit}(D(z,r) \setminus K, z) \in K) \geq p.$$
But if $B_t(z)$ hits K before leaving $D(z,r)$ it follows that $B_t(z)$ hits ∂D at a point at distance no more than
$$|z - \zeta| + r \leq \delta$$
from ζ; hence $|z - \zeta| < \delta/11$ implies that
$$\Pr(|\text{Exit}(D, z) - \zeta| \leq \delta) \geq p.$$

Assume now that $|z - \zeta| < \delta/11^2$. Let $r = 10|z - \zeta|$ as before. Let $z_1 = \text{Exit}(D(z,r) \setminus K, z)$. If $z_1 \in K$ we're happy; this happens with probability at least p. On the other hand, if $z_1 \in \partial D(z,r)$ then it follows that
$$|z_1 - \zeta| \leq |z_1 - z| + |z - \zeta| = r + |z - \zeta| = 11|z - \zeta| \leq \delta/11,$$
and so the argument in the previous paragraph, with z_1 in place of z and $r_1 = 11r$ in place of r, shows that
$$\Pr(\text{Exit}(D(z_1, 11r) \setminus K, z_1) \in K) \geq p.$$
But the event $\text{Exit}(D(z_1, 11r) \setminus K, z_1) \in K$ implies that $B_t(z)$ hits ∂D at a point at distance no more than $|z_1 - \zeta| + 11r \leq (11 + 110)|z - \zeta| \leq \delta$ from ζ, which again counts as a success. As in the proof of Theorem 10.6.2 we see that if $|z - \zeta| < \delta/11^2$ then
$$\Pr(|\text{Exit}(D, z) - \zeta| \leq \delta) \geq 1 - (1 - p)^2.$$

It follows by induction that if $|z - \zeta| < \delta/11^N$ then
$$\Pr(|\text{Exit}(D, z) - \zeta| \leq \delta) \geq 1 - (1 - p)^N. \qquad \square$$

We now turn to the promised application to conformal mapping. For small $\delta > 0$ let I_δ be the arc
$$I_\delta = \{\, e^{it} : \delta \leq t \leq 2\pi - \delta \,\}.$$

We will assume that D_δ is a simply connected Dirichlet domain such that $\mathbb{D} \subset D_\delta$ and $I_\delta \subset \partial D_\delta$, as in Figure 22.3:

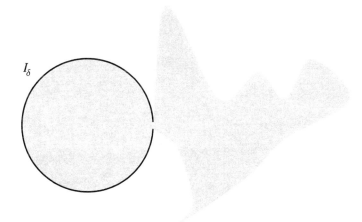

Figure 22.3

Suppose that $f \in C(\partial D_\delta)$ and let u be the harmonic function in D_δ with boundary values given by f. A basic principle is this: If $z \in \mathbb{D}$ then the values of f at points of $\partial D_\delta \setminus I_\delta$ should be largely irrelevant to the value of $u(z)$, at least if $|1 - z|$ is large relative to δ. This is clear if we think about the solution to the Dirichlet problem given by Brownian motion: If we start at a point of $\mathbb{D}$, other than a point close to 1, then it is very unlikely that a random path will make it through the "bottleneck" separating the two parts of D_δ before hitting I_δ. We will give a precise statement of this below, and prove it; then we will show that it follows that the shape of the blob constituting the right half of D_δ is also largely irrelevant to $\psi(z)$ for $z \in \mathbb{D}$, if $\psi : D_\delta \to \mathbb{D}$ is the conformal equivalence with $\psi(0) = 0$ and $\psi'(0) > 0$!

We begin with a simple estimate on the Poisson kernel:

Lemma 22.4. *Suppose $t, \theta \in \mathbb{R}$ and $0 \leq r < 1$. If $|1 - re^{it}| \geq 2\delta$ and $|\theta| \leq \delta$ then*

$$P_r(t - \theta) \leq \frac{8(1-r)}{|1 - re^{it}|^2}.$$

(Note that "P_r" is the Poisson kernel and "$P[f]$" is the Poisson integral of f, while the probability of the event A is "$\Pr(A)$". Alas P_r and Pr are both standard notations; luckily they don't look *quite* the same ...)

Proof. Exercise 10.3 shows that $|1 - e^{-i\theta}| \leq |\theta| \leq \delta$, so

$$\begin{aligned}
|1 - re^{i(t-\theta)}| &= |e^{-i\theta} - re^{i(t-\theta)} + 1 - e^{-i\theta}| \\
&\geq |e^{-i\theta} - re^{i(t-\theta)}| - |1 - e^{-i\theta}| \\
&= |1 - re^{it}| - |1 - e^{-i\theta}| \\
&\geq |1 - re^{it}| - \delta \\
&\geq |1 - re^{it}| - |1 - re^{it}|/2 \\
&= |1 - re^{it}|/2.
\end{aligned}$$

Hence

$$P_r(t - \theta) = \frac{1 - r^2}{|1 - re^{i(t-\theta)}|^2} \leq \frac{4(1-r)(1+r)}{|1 - re^{it}|^2}. \qquad \square$$

This allows us to prove the intuitively obvious fact that the probability that Brownian motion started at a point of $\mathbb{D}$ (other than a point "near" 1) exits D_δ at a point of $\partial D_\delta \setminus I_\delta$ is small:

Lemma 22.5. *If $z \in \mathbb{D}$ and $|1 - z| \geq 2\delta$ then*

$$\Pr(\mathrm{Exit}(D_\delta, z) \notin I_\delta) \leq \frac{8(1-|z|)}{\pi|1-z|^2}\delta.$$

Proof. The event $\mathrm{Exit}(D_\delta, z) \notin I_\delta$ implies the event $\mathrm{Exit}(\mathbb{D}, z) \notin I_\delta$, so

$$\Pr(\mathrm{Exit}(D_\delta, z) \notin I_\delta) \leq \Pr(\mathrm{Exit}(\mathbb{D}, z) \notin I_\delta).$$

Suppose $\epsilon > 0$, and choose $\phi \in C([-\pi, \pi])$ such that $0 \leq \phi \leq 1$ everywhere, $\phi = 1$ on $[-\delta, \delta]$, and $\phi = 0$ on $[-\pi, \pi] \setminus [-(\delta+\epsilon), \delta+\epsilon]$. Define $f \in C(\partial \mathbb{D})$ by

$$f(e^{it}) = \phi(t) \quad (-\pi \leq t \leq \pi),$$

and let $u = P[f]$. Let A be the event $\text{Exit}(\mathbb{D}, z) \notin I_\delta$. Then A implies $f(\text{Exit}(\mathbb{D}, z)) = 1$, and so, since $0 \leq f \leq 1$, if we set $z = re^{it}$ we have

$$\begin{aligned}
\Pr(A) &= \Pr(A) E(f(\text{Exit}(\mathbb{D}, z))|A) \\
&\leq \Pr(A) E(f(\text{Exit}(\mathbb{D}, z))|A) + (1 - \Pr(A)) E(f(\text{Exit}(\mathbb{D}, z))|A^c) \\
&= E(f(\text{Exit}(\mathbb{D}, z))) \\
&= u(z) \\
&= \frac{1}{2\pi} \int_{-\pi}^{\pi} f(e^{i\theta}) P_r(t - \theta)\, d\theta \\
&\leq \frac{1}{2\pi} \int_{-(\delta+\epsilon)}^{\delta+\epsilon} P_r(t - \theta)\, d\theta.
\end{aligned}$$

Now letting ϵ tend to 0 and applying the previous lemma we obtain

$$\Pr(A) \leq \frac{1}{2\pi} \int_{-\delta}^{\delta} P_r(t - \theta)\, d\theta \leq \frac{1}{2\pi} \int_{-\delta}^{\delta} \frac{8(1-r)}{|1 - re^{it}|^2}\, d\theta = \frac{8(1-r)\delta}{\pi |1 - re^{it}|^2}. \quad \square$$

And this allows us to show that if $z \in \mathbb{D}$ is not close to 1 then the values of f on $\partial D_\delta \setminus I_\delta$ are more or less irrelevant to the solution of the Dirichlet problem at z. There are various ways one could make this statement precise; in particular, if we define

$$K_\rho = \{z \in \mathbb{D} : |1 - z| \geq \rho\}$$

we have

Theorem 22.6. *Suppose that $f \in C(\partial \mathbb{D})$ and $|f| \leq M$ on $\partial \mathbb{D}$. Let $u = P[f]$. Suppose that for every $\delta \in (0, 1)$ we have $f_\delta \in C(\partial D_\delta)$ such that $|f_\delta| \leq M$ on ∂D_δ and such that $f_\delta = f$ on I_δ. Let u_δ be the harmonic function in D_δ with boundary values f_δ. Then for every $\rho > 0$ we have*

$$u_\delta \to u$$

uniformly on K_ρ as $\delta \to 0$.

In particular $u_\delta \to u$ uniformly on compact subsets of $\mathbb{D}$; in fact the theorem says more than that.

Proof. Suppose that $0 < \delta < \rho/2$ and $z \in K_\rho$, so the previous lemma may be applied. Let A be the event $\text{Exit}(\mathbb{D}, z) \in I_\delta$; then we have

$$u(z) = \Pr(A) E(f(\text{Exit}(\mathbb{D}, z)|A)) + (1 - \Pr(A)) E(f(\text{Exit}(\mathbb{D}, z))|A^c).$$

Now, *if A occurs* then it follows that $\text{Exit}(D_\delta, z) = \text{Exit}(\mathbb{D}, z)$ and also that $f_\delta(\text{Exit}(\mathbb{D}, z)) = f(\text{Exit}(\mathbb{D}, z))$. So we have

$$u_\delta(z) = \Pr(A)E(f_\delta(\text{Exit}(D_\delta, z))|A)) + (1 - \Pr(A))E(f_\delta(\text{Exit}(D_\delta, z))|A^c)$$
$$= \Pr(A)E(f(\text{Exit}(\mathbb{D}, z))|A)) + (1 - \Pr(A))E(f_\delta(\text{Exit}(D_\delta, z))|A^c).$$

Thus

$$u_\delta(z) - u(z) = (1 - \Pr(A))(E(f_\delta(\text{Exit}(D_\delta, z))|A^c) - E(f(\text{Exit}(\mathbb{D}, z))|A^c)).$$

Applying the estimate on $1 - \Pr(A)$ obtained in the previous lemma and using the fact that both f and f_δ are bounded by M it follows that

$$|u_\delta(z) - u(z)| \leq 2M(1 - \Pr(A)) \leq 2M \frac{8(1 - |z|)}{\pi|1 - z|^2} \delta \leq M \frac{16}{\pi \rho^2} \delta. \qquad \square$$

One interesting aspect of the theorem is that there is no bound on the size of the blob constituting the right half of D_δ; the result remains valid even if D_δ blows up as δ tends to 0! However, we do require that the functions f_δ be uniformly bounded; it is clear that extremely large values on parts of the boundary far away will become relevant again. This is the reason that the open sets are required to be uniformly bounded in the application to conformal mapping:

Theorem 22.7. *Suppose in addition to our previous hypotheses that there exists $R < \infty$ such that $D_\delta \subset D(0, R)$ for all $\delta \in (0, 1)$. Let $\psi_\delta : D_\delta \to \mathbb{D}$ be the conformal equivalence with $\psi_\delta(0) = 0$ and $\psi_\delta'(0) > 0$. Then*

$$\psi_\delta(z) \to z$$

as $\delta \to 0$, uniformly on compact subsets of $\mathbb{D}$.

This is a special case of a theorem of Carathéodory (that seems to happen a lot with results on conformal mapping). See for example Theorem 2.1 in [**M**].

Proof. Let u_δ be the harmonic function in D_δ with boundary values f_δ, where $f_\delta(z) = -\log(|z|)$, and let v_δ be the harmonic conjugate with $v_\delta(0) = 0$. Define $h_\delta \in H(D_\delta)$ by $h_\delta = u_\delta + iv_\delta$.

The proof of Theorem 10.5.3 shows that

$$\psi_\delta(z) = ze^{h_\delta(z)}.$$

The previous theorem shows that $u_\delta \to 0$ uniformly on compact subsets of $\mathbb{D}$ (since $\log(|z|) = 0$ for $z \in \partial\mathbb{D}$). It follows from Exercise 2 below that $h_\delta \to 0$ uniformly on compact subsets of $\mathbb{D}$. $\qquad \square$

22. More on Brownian Motion

Note. In fact the hypothesis that $D_\delta \subset D(0,R)$ is not really needed. Without this hypothesis the f_δ in the proof are no longer uniformly bounded, so we cannot apply the previous theorem as stated. However, a slightly more subtle analysis shows that the probability that $\text{Exit}(D_\delta, z)$ is large is small enough that one can still deduce the needed conclusion, since $\log(|z|)$ grows slowly.

Exercise 22.3 gives another strengthening of the theorem.

Exercises

22.1. Suppose that $f_n \in C(\partial \mathbb{D})$ is real-valued. Let $u_n = P[f_n]$ and let v_n be the harmonic conjugate of u_n with $v_n(0) = 0$; let $F_n = u_n + iv_n$. If $f_n \to 0$ uniformly on $\partial \mathbb{D}$ then $F_n \to 0$ uniformly on compact subsets of $\mathbb{D}$.

Hint: The proof of Lemma 10.1.3 shows that

$$F_n(z) = \frac{1}{2\pi} \int_0^{2\pi} f_n(e^{it}) \frac{1 + e^{-it}z}{1 - e^{-it}z} \, dt.$$

22.2. Suppose u_n is harmonic in $\mathbb{D}$, v_n is the harmonic conjugate of u_n with $v_n(0) = 0$, and $u_n \to 0$ uniformly on compact subsets of $\mathbb{D}$. Then $v_n \to 0$ uniformly on compact subsets of $\mathbb{D}$.

22.3. Show that in Theorem 22.7 we actually have $\psi_\delta(z) \to z$ uniformly on K_ρ for every $\rho > 0$. (Repeat the proof of the theorem, looking carefully at the formula for $h_\delta|_\mathbb{D}$ in terms of $u_\delta|_{\partial \mathbb{D}}$ (see the hint for Exercise 22.1) instead of just using the statement of Exercise 22.2. An inequality obtained in the proof of Lemma 22.4 may be useful.)

Chapter 23

More on the Maximum Modulus Theorem

23.0. Theorems of Hadamard and Phragmén-Lindelöf

One can often prove interesting and useful things about a holomorphic function f by applying the Maximum Modulus Theorem to $f\phi$, where ϕ is a suitably chosen auxiliary function. We give two examples: Hadamard's "Three-Lines Theorem" (with its corollary, the "Three-Circles Theorem") and a version of the Phragmén-Lindelöf Theorem. These results have applications in areas as diverse as partial differential equations, harmonic analysis and analytic number theory.

If D is an open subset of the plane we will let $A(D)$ denote the space of all functions continuous on $\overline{D}$ which are holomorphic in D. We begin with two versions of the Maximum Modulus Theorem for unbounded open sets.

Lemma 23.0.0. *Suppose that $\Omega \subset \mathbb{C}$ is an unbounded open set, $f \in A(\Omega)$, $|f(\zeta)| \leq M$ for all $\zeta \in \partial\Omega$, and $\lim_{z \to \infty} f(z) = 0$. Then $|f(z)| \leq M$ for all $z \in \Omega$.*

Proof. Fix $z \in \Omega$, and choose $R > |z|$ such that $|f(w)| \leq M$ for all $w \in \Omega$ with $|w| \geq R$. Then $|f(\zeta)| \leq M$ for all $\zeta \in \partial(\Omega \cup D(0, R))$, so the Maximum Modulus Theorem implies that $|f(w)| \leq M$ for all $w \in \Omega \cup D(0, R)$; in particular $|f(z)| \leq M$. □

Theorem 23.0.1. *Suppose that $\Omega \subset \mathbb{C}$ is an unbounded open set, $\Omega \ne \mathbb{C}$. If $f \in A(\Omega)$ is bounded and $|f(\zeta)| \le M$ for all $\zeta \in \partial\Omega$ then $|f(z)| \le M$ for all $z \in \Omega$.*

Proof. Let $\epsilon > 0$ and fix $z \in \Omega$. It is enough to show that $|f(z)| \le M + \epsilon$. Now, the hypotheses on Ω imply that $\partial\Omega \ne \emptyset$; if $\zeta \in \partial\Omega$ then $|f| \le M + \epsilon$ at every point of Ω sufficiently close to ζ. Hence there exist $p \in \Omega$ and $r > 0$ such that $\overline{D(p,r)} \subset \Omega$, $z \in \Omega \setminus \overline{D(p,r)}$, and $|f| \le M + \epsilon$ on $\partial D(p,r)$.

Let $V = \Omega \setminus \overline{D(p,r)}$, and for $n = 1, 2, \ldots$ define $g_n \in A(V)$ by

$$g_n(w) = \frac{r}{w - p} f(w)^n.$$

Then $|g_n| \le (M + \epsilon)^n$ at every point of ∂V and the fact that f is bounded shows that g_n tends to 0 at infinity; thus the previous lemma implies that $|g_n| \le (M + \epsilon)^n$ in V and in particular that $|g_n(z)| \le (M + \epsilon)^n$.

Thus

$$|f(z)| \le (M + \epsilon) \left(\frac{|z - p|}{r} \right)^{1/n}$$

for $n = 1, 2, \ldots$; hence $|f(z)| \le M + \epsilon$. $\square$

In that proof we applied the Maximum Modulus Theorem to $f^n \phi$ instead of $f\phi$. If we had used just $f\phi$ we would have obtained the inequality $|f(z)| \le (M + \epsilon)|z - p|/r$. Knowing this for all f satisfying the hypotheses of the theorem would actually be enough, by a little trick which we inserted into the proof. Recording that trick for future reference:

- **Exercise 23.1.** Suppose that X is a set, A is a set of complex-valued functions defined on X and A is closed under multiplication. Suppose that $K \subset X$, $x \in X$, and there exists a constant c such that

$$|f(x)| \le c \sup_{w \in K} |f(w)|$$

for all $f \in A$. Then in fact

$$|f(x)| \le \sup_{w \in K} |f(w)|$$

for all $f \in A$.

We can now prove the Hadamard Three-Lines Theorem. For $M > 0$ and $z \in \mathbb{C}$ we define

$$M^z = \exp(z \log(M))$$

(where $\log(M)$ is the real logarithm) and we note that

$$|M^z| = M^{\operatorname{Re}(z)}.$$

23.0. Theorems of Hadamard and Phragmén-Lindelöf

Theorem 23.0.2 (Three-Lines Theorem). *Suppose that $a < b$ and let $S = \{x + iy : x \in \mathbb{R}, a < y < b\}$. Suppose that $f \in A(S)$ is bounded. If $|f(x + ia)| \leq M_a$ and $|f(x + ib)| \leq M_b$ for all $x \in \mathbb{R}$ then*

$$|f(x + iy)| \leq M_y = M_a^{(b-y)/(b-a)} M_b^{(y-a)/(b-a)} \quad (x \in \mathbb{R}, \ a \leq y \leq b).$$

In other words, $\sup_{x \in \mathbb{R}} \log |f(x + iy)|$ is a convex function of y.

Proof. Define $g \in A(S)$ by

$$g(z) = f(z) M_a^{(b+iz)/(a-b)} M_b^{(iz+a)/(b-a)}.$$

Then g is bounded, $|g(x + ia)| \leq 1$ and $|g(x + ib)| \leq 1$. So Theorem 23.0.1 shows that $|g| \leq 1$ in S, and the result follows. $\square$

The Three-Circles Theorem is an immediate corollary:

Theorem 23.0.3 (Three-Circles Theorem). *Suppose that $0 < a < b$ and let Ω be the annulus $\Omega = \{re^{it} : a < r < b, t \in \mathbb{R}\}$. If $f \in A(\Omega)$, $|f(ae^{it})| \leq M_a$ and $|f(be^{it})| \leq M_b$ for all $t \in \mathbb{R}$ then*

$$|f(re^{it})| \leq M_r = M_a^{\log(b/r)/\log(b/a)} M_b^{\log(r/a)/\log(b/a)} \quad (a \leq r \leq b, \ t \in \mathbb{R}).$$

Proof. Let $S = \{x + iy : x \in \mathbb{R}, \log(a) < y < \log(b)\}$, and define $F \in A(S)$ by

$$F(z) = f(e^{-iz}).$$

Then $|F(x + i\log(a))| = |f(ae^{-ix})| \leq M_a$ and similarly $|F(x + i\log(b))| \leq M_b$, so the previous theorem shows that

$$|f(re^{it})| = |F(-t + i\log(r))|$$
$$\leq M_a^{(\log(b) - \log(r))/(\log(b) - \log(a))} M_b^{(\log(r) - \log(a))/(\log(b) - \log(a))}. \quad \square$$

We derived Theorem 23.0.3 from Theorem 23.0.2 by a trick. It's a good trick, well worth remembering, but one might also want a direct proof of Theorem 23.0.3, using the same technique as in the proof of Theorem 23.0.2, because the direct proof might work in situations where the trick was not available.

Unfortunately the direct proof does not seem to work. If one thinks about deriving Theorem 23.0.3 from Theorem 23.0.1, one is led to an argument that begins by defining

$$g(z) = f(z) z^\alpha$$

for a certain real number α (since this doesn't quite work anyway, we will leave it to you to determine what α is appropriate). Alas α is not in general an integer, so there simply is no function z^α holomorphic in all of the annulus Ω.

There *is* a simple way to give a direct proof of Theorem 23.0.3, but it involves a notion that we have not talked about, namely *subharmonic* functions. We are not going to bother defining the word "subharmonic" here, since a proper treatment of the subject would take some space (and would also require some real analysis). But in fact if f is holomorphic and not identically zero then $\log|f|$ is subharmonic (in this context one defines $\log(0) = -\infty$ so that $\log|f|$ is defined everywhere). Subharmonic functions satisfy a maximum principle, and even though z^α is "multi-valued" its absolute value is single-valued (at least when $\alpha \in \mathbb{R}$), so that $\log|z^\alpha|$ is a well defined subharmonic function in Ω. One could prove Theorem 23.0.3 by applying a maximum principle to the subharmonic function $\log|f| + \alpha \log|z|$.

Or one could if one knew something about subharmonic functions. This is a good reason to study the subject someday: They allow one to give arguments analogous to the arguments in this section, but they are more flexible. (Our theme here is deriving results about the holomorphic function f by applying the maximum principle to the holomorphic function fg. This is equivalent to applying a maximum principle to the subharmonic function $\log|f| + \log|g|$; now applying a maximum principle to $\log|f| + u$, where u is a suitable subharmonic function, is a more powerful technique, because there are a lot of subharmonic functions around which are not of the form $\log|g|$ for holomorphic g.)

- **Exercise 23.2.** Formulate and prove a "Three-Rays Theorem", for functions holomorphic in a sector $V = \{\, re^{it} : r > 0,\, a < t < b \,\}$.

We turn now to the Phragmén-Lindelöf Theorem. We should say that there are actually many theorems referred to by that name; probably you should think of it as the Phragmén-Lindelöf *technique*. In a typical Phragmén-Lindelöf theorem we have a function $f \in A(\Omega)$, where Ω is some unbounded open set, we are given that f satisfies some growth condition on $\partial \Omega$ and a much weaker condition in all of Ω, and we deduce that f satisfies a stronger growth condition in Ω. We give one example here and a few more in the exercises.

Theorem 23.0.4. *Suppose that $a < b \leq a + 2\pi$, and let $S_{a,b}$ be the truncated sector*

$$S_{a,b} = \{\, re^{it} : r > 1,\, a < t < b \,\}.$$

Suppose that $0 \leq \alpha < \pi/(b-a)$, $R \geq 0$, and $f \in A(S_{a,b})$ satisfies

$$|f(z)| \leq e^{|z|^\alpha}$$

for all $z \in S_{a,b}$ with $|z| > R$. If $|f(\zeta)| \leq M$ for all $\zeta \in \partial S_{a,b}$ then $|f| \leq M$ everywhere in $S_{a,b}$.

Proof. Rotating $S_{a,b}$ if necessary we may assume that $a = -b$ (so that $0 < b \leq \pi$ and $\alpha < \pi/(2b)$). Let log be the principal branch of the logarithm (so $\log(re^{it}) = \log(r) + it$ for $r > 0$ and $-\pi < t < \pi$) and define $z^w = \exp(w \log(z))$ for $z \in S_{-b,b}$ and $w \in \mathbb{C}$.

Now choose $\beta \in (\alpha, \pi/(2b))$ and for $c > 0$ define

$$g_c(z) = f(z) e^{-cz^\beta}.$$

Fix $z \in \overline{S_{-b,b}}$ for a moment and write $z = re^{it}$ with $r > 0$ and $-b \leq t \leq b$. Then $z^\beta = r^\beta e^{i\beta t}$, so that $\operatorname{Re}(z^\beta) = |z|^\beta \cos(\beta t)$. Note that $-b\beta \leq \beta t \leq b\beta$; since $b\beta < \pi/2$, it follows that there exists $\delta > 0$ such that $\operatorname{Re}(z^\beta) \geq \delta |z|^\beta$. Since $c > 0$, this shows that $|g_c(\zeta)| \leq |f(\zeta)| \leq M$ for all $\zeta \in \partial S_{-b,b}$. Furthermore, if $|z| > R$ then

$$|g_c(z)| \leq e^{|z|^\alpha - c\delta |z|^\beta}.$$

Since $\beta > \alpha$, it follows that $g_c(z) \to 0$ as $z \to \infty$. So Lemma 23.0.0 shows that $|g_c(z)| \leq M$, and hence $|f(z)| = \lim_{c \to 0} |g_c(z)| \leq M$. $\square$

Exercises

The next exercise arises in harmonic analysis, in connection with a result known as the "Paley-Wiener Theorem".

23.3. Suppose that f is an entire function, $|f(z)| \leq e^{|z|}$ for all $z \in \mathbb{C}$, and $|f(x)| \leq 1$ for all $x \in \mathbb{R}$. Show that $|f(x+iy)| \leq e^{|y|}$.

23.4. Formulate and prove stronger versions of the results in this section, in analogy with Lemma 10.5.2.

23.5. Suppose in addition to the hypotheses of Theorem 23.0.1 that Ω is not dense in the plane and that $\partial \Omega$ is bounded, which is to say that there exists $R > 0$ such that $\{z : |z| > R\} \subset \Omega$. Give a somewhat simpler proof of the theorem, using the fact that f has a removable singularity at ∞.

23.6. Suppose that p is a polynomial of degree no larger than n and
$$\sup_{|z|=1} |p(z)| = 1.$$
Show that
$$\sup_{|z|=r} |p(z)| \leq r^n$$
for all $r > 1$, and determine when equality holds. (Apply Theorem 23.0.1 to the function $q(z) = \ldots$)

23.1. An Application: The Hausdorff-Young Inequality

As mentioned previously, the results in the previous section have various applications in real analysis. Even giving a statement of most such results is beyond the scope of this book, since we are not assuming any prior knowledge of real analysis. We can however state and prove a version of the Hausdorff-Young inequality, a result in Fourier analysis. (The idea that this might be possible was suggested by the treatment in [**R**]; the details of the proof we give will necessarily be somewhat different.)

We will say that $f \in C(\mathbb{T})$ if $f : \mathbb{R} \to \mathbb{C}$ is continuous and 2π-periodic: $f(t + 2\pi) = f(t)$. If $f \in C(\mathbb{T})$ and $n \in \mathbb{Z}$ we define the *Fourier coefficient* $\hat{f}(n)$ by
$$\hat{f}(n) = \frac{1}{2\pi} \int_0^{2\pi} f(t) e^{-int} \, dt.$$

Theorem 23.1.0. (Hausdorff-Young Inequality). *Suppose that $1 < p < 2$ and $\frac{1}{p} + \frac{1}{q} = 1$. If $f \in C(\mathbb{T})$ then*
$$\left(\sum_{n=-\infty}^{\infty} |\hat{f}(n)|^q \right)^{1/q} \leq \left(\frac{1}{2\pi} \int_0^{2\pi} |f(t)|^p \, dt \right)^{1/p}.$$

The proof of the theorem is an example of a technique in real analysis known as "complex interpolation" — the proof proceeds by using the Three-Lines Theorem to "interpolate" between the cases $p = 1$ and $p = 2$. The fact that the theorem follows from those two cases is a special case of the Riesz-Thorin Theorem, originally proved by M. Riesz. Thorin found a much simpler proof using the Three-Lines Theorem; Littlewood called this application of complex analysis to real analysis "the most impudent idea in mathematics".

23.1. An Application: The Hausdorff-Young Inequality

When we said that the theorem follows from the two cases $p = 1$ and $p = 2$ the alert reader will have noticed that the theorem as stated makes no sense for $p = 1$. We will see that

$$\lim_{q \to \infty} \left(\sum_{n=-\infty}^{\infty} |\hat{f}(n)|^q \right)^{1/q} = \sup_{n \in \mathbb{Z}} |\hat{f}(n)|,$$

so for $p = 1$ the appropriate interpretation of the inequality in the theorem is the following:

Lemma 23.1.1. *If $f \in C(\mathbb{T})$ then*

$$\sup_{n \in \mathbb{Z}} |\hat{f}(n)| \leq \frac{1}{2\pi} \int_0^{2\pi} |f(t)|\, dt.$$

Of course the proof of Lemma 23.1.1 is trivial. For $p = 2$ we need a slightly more sophisticated version of the Parseval Formula than we have seen previously:

Lemma 23.1.2 (Parseval Formula). *If $f \in C(\mathbb{T})$ then*

$$\sum_{n=-\infty}^{\infty} |\hat{f}(n)|^2 = \frac{1}{2\pi} \int_0^{2\pi} |f(t)|^2\, dt.$$

Proof. For $0 \leq r < 1$ define

$$f_r(t) = u(re^{it}),$$

where u is the Poisson integral of the function g defined by $g(e^{it}) = f(t)$. Proposition 10.4.0 shows that

$$f_r(t) = \sum_{n=-\infty}^{\infty} \hat{f}(n) r^{|n|} e^{int}.$$

Since $\hat{f}(n)$ is bounded, this series converges absolutely and uniformly, which justifies the following manipulations:

$$|f_r(t)|^2 = f_r(t) \overline{f_r(t)}$$
$$= \sum_{n=-\infty}^{\infty} \hat{f}(n) r^{|n|} e^{int} \sum_{m=-\infty}^{\infty} \overline{\hat{f}(m)} r^{|m|} e^{-imt}$$
$$= \sum_{n,m=-\infty}^{\infty} \hat{f}(n) \overline{\hat{f}(m)} r^{|n|+|m|} e^{i(n-m)t},$$

hence

$$\frac{1}{2\pi}\int_0^{2\pi}|f_r(t)|^2\,dt = \sum_{n,m=-\infty}^{\infty}\hat{f}(n)\overline{\hat{f}(m)}r^{|n|+|m|}\frac{1}{2\pi}\int_0^{2\pi}e^{i(n-m)t}\,dt$$

$$= \sum_{n=-\infty}^{\infty}|\hat{f}(n)|^2 r^{2|n|}.$$

Now Theorem 10.1.3 shows that $f_r \to f$ uniformly as r increases to 1, so that

$$\frac{1}{2\pi}\int_0^{2\pi}|f_r(t)|^2\,dt \to \frac{1}{2\pi}\int_0^{2\pi}|f(t)|^2\,dt,$$

and we are done if we can show that $\sum_{n=-\infty}^{\infty}|\hat{f}(n)|^2 r^{2|n|} \to \sum_{n=-\infty}^{\infty}|\hat{f}(n)|^2$.

This is easy: Suppose that $\alpha < \sum_{n=-\infty}^{\infty}|\hat{f}(n)|^2$. There exists N such that $\sum_{n=-N}^{N}|\hat{f}(n)|^2 > \alpha$, and now there exists $R \in (0,1)$ such that $\sum_{n=-N}^{N}|\hat{f}(n)|^2 R^{2|n|} > \alpha$; it follows that

$$\alpha < \sum_{n=-\infty}^{\infty}|\hat{f}(n)|^2 r^{2|n|} \leq \sum_{n=-\infty}^{\infty}|\hat{f}(n)|^2$$

for all $r \in (R,1)$. □

Although we do not need to do so for the proof of the theorem, we feel we should show that the conclusion of the theorem does in fact tend to Lemma 23.1.1 as p tends to 1, in order to justify our claim that the theorem is proved by interpolation between the cases $p = 1$ and $p = 2$:

Lemma 23.1.3. *Suppose that $a_n \geq 0$ and $\sum_{n=1}^{\infty} a_n^2 < \infty$. Then*

$$\lim_{q\to\infty}\left(\sum_{n=1}^{\infty} a_n^q\right)^{1/q} = \sup_{n\geq 1} a_n.$$

Proof. Set $\alpha = \sup_n a_n$ and $\beta = \sum_n a_n^2$, and assume that $\alpha > 0$ (if $\alpha = 0$ there is nothing to prove). For $q > 2$ we have

$$\alpha \leq \left(\sum_{n=1}^{\infty} a_n^q\right)^{1/q} \leq (\alpha^{q-2}\beta)^{1/q} = \alpha(\beta/\alpha^2)^{1/q};$$

note that $(\beta/\alpha^2)^{1/q} \to 1$ as $q \to \infty$. □

We need to give two other results that would be clear if we had studied some reals:

23.1. An Application: The Hausdorff-Young Inequality

Lemma 23.1.4 (Cauchy-Schwarz Inequality). *If $z_j, w_j \in \mathbb{C}$ for $1 \le j \le n$ then*
$$\left|\sum_{j=1}^n z_j \overline{w_j}\right| \le \left(\sum_{j=1}^n |z_j|^2\right)^{1/2} \left(\sum_{j=1}^n |w_j|^2\right)^{1/2}.$$

We leave the proof as an exercise; the hint is that
$$\sum_{j=1}^n (z_j - \alpha w_j)\overline{(z_j - \alpha w_j)} = \sum_{j=1}^n |z_j - \alpha w_j|^2 \ge 0$$

for every $\alpha \in \mathbb{C}$.

Lemma 23.1.5. *Suppose that $q > 1$ and $\frac{1}{p} + \frac{1}{q} = 1$. If $a_1, \ldots, a_n \in \mathbb{C}$ then there exist $b_1, \ldots, b_n \in \mathbb{C}$ such that $\sum_{j=1}^n |b_j|^p = 1$ and*
$$\sum_{j=1}^n a_j b_j = \left(\sum_{j=1}^n |a_j|^q\right)^{1/q}.$$

Proof. We may assume that $a_j \ge 0$. Let $\alpha = \left(\sum_{j=1}^n a_j^q\right)^{1/q}$ and assume that $\alpha > 0$. Set
$$b_j = a_j^{q-1}/\alpha^{q-1}$$
(and note that $pq - p = q$). □

Our final lemma is just a technicality. You should note however that it is false for real-valued functions:

Lemma 23.1.6. *If $f \in C(\mathbb{T})$ then there exists a sequence $(f_n) \subset C(\mathbb{T})$ such that f_n has no zero on $\mathbb{R}$ and $f_n \to f$ uniformly as $n \to \infty$.*

Proof. Exercise 10.8 shows that we may assume that f is a trigonometric polynomial:
$$f(t) = \sum_{n=-N}^N a_n e^{int}.$$

Define a rational function F by
$$F(z) = \sum_{n=-N}^N a_n z^n$$

and for $r > 0$ set
$$g_r(t) = F(re^{it}),$$

so that $g_1 = f$. Now $g_r \in C(\mathbb{T})$ for every $r > 0$, $g_r \to f$ uniformly as $r \to 1$, and, with the exception of finitely many values of r, g_r has no zero. □

That was definitely a "trick proof". It is possible to give more elementary proofs of the lemma, although as far as I can see they take slightly more space to write out in detail. A few exercises in that regard:

- **Exercise 23.7.** Suppose that $\phi : \mathbb{C} \to \mathbb{C}$ is a continuous function such that $|\phi(z)| \geq 1$ for all $z \in \mathbb{C}$ and $\phi(z) = z$ whenever $|z| \geq 1$. Use ϕ to give a simple proof of Lemma 23.1.6.

- **Exercise 23.8.** Show that the proof of the lemma suggested in the previous exercise does not work, because in fact there does not exist a function ϕ with all the properties assumed in the previous exercise! (Let $\gamma_r(t) = \phi(re^{it})$ for $r \geq 0$ and $t \in [0, 2\pi]$. Consider $\operatorname{Ind}(\gamma_r, 0)$.)

The previous exercise contains a special case of something known as the "no-retraction theorem". The next exercise gives a hint towards an elementary proof of the lemma that actually does work:

- **Exercise 23.9.** Prove Lemma 23.1.6 as follows: Suppose that $f \in C(\mathbb{T})$. Let Z denote the set of all $t \in [0, 2\pi]$ with $f(t) = 0$. Let $\epsilon > 0$. Show that Z is contained in the union of finitely many intervals I_j such that $|f| < \epsilon$ on each I_j. Now ...

We can now prove the Hausdorff-Young inequality:

Proof (Theorem 23.1.0). We fix a positive integer N and note that it is enough to show that

$$\left(\sum_{n=-N}^{N} |\hat{f}(n)|^q \right)^{1/q} \leq \left(\frac{1}{2\pi} \int_0^{2\pi} |f(t)|^p \, dt \right)^{1/p}.$$

If $(f_j) \in C(\mathbb{T})$ and $f_j \to f$ uniformly then

$$\left(\frac{1}{2\pi} \int_0^{2\pi} |f_j(t)|^p \, dt \right)^{1/p} \to \left(\frac{1}{2\pi} \int_0^{2\pi} |f(t)|^p \, dt \right)^{1/p}$$

and

$$\left(\sum_{n=-N}^{N} |\hat{f_j}(n)|^q \right)^{1/q} \to \left(\sum_{n=-N}^{N} |\hat{f}(n)|^q \right)^{1/q};$$

thus Lemma 23.1.6 shows that we may assume that f has no zero on $\mathbb{R}$.

23.1. An Application: The Hausdorff-Young Inequality

Now we fix $b_{-N}, \ldots, b_N \in \mathbb{C}$ with

$$\sum_{n=-N}^{N} |b_n|^p = 1;$$

if we can show that

$$\left| \sum_{n=-N}^{N} b_n \hat{f}(n) \right| \leq \left(\frac{1}{2\pi} \int_0^{2\pi} |f(t)|^p \, dt \right)^{1/p}$$

then we are done, by Lemma 23.1.5.

Since f has no zero, there is a continuous function $\theta : [0, 2\pi] \to \mathbb{R}$ with

$$f(t) = |f(t)|e^{i\theta(t)} \quad (0 \leq t \leq 2\pi);$$

similarly there exist $\phi_{-N}, \ldots, \phi_N \in \mathbb{R}$ with

$$b_n = |b_n|e^{i\phi_n} \quad (n = -N, \ldots, N).$$

Now we define $F : \mathbb{C} \to \mathbb{C}$ by

$$F(z) = \frac{1}{2\pi} \int_0^{2\pi} \left(\sum_{n=-N}^{N} |b_n|^{pz} e^{i\phi_n} e^{-int} \right) |f(t)|^{pz} e^{i\theta(t)} \, dt.$$

(As always, if $a > 0$ and $z \in \mathbb{C}$ we define $a^z = \exp(z \ln(a))$; in case $b_n = 0$ for some n we set $a^z = 0$ for $a = 0$.)

It will be easy to show (using Lemma 23.1.1) that F is an entire function and that F is bounded in every vertical strip. Now comes the promised "interpolation": Lemma 23.1.2 will give a sharper bound on $|F(1/2 + iy)|$, and then the bound that the Three-Lines Theorem gives on $|F(1/p)|$ will turn out to be exactly the Hausdorff-Young inequality.

The fact that F is an entire function should be an easy exercise by now (see the proof of Theorem 4.9 if you need a hint). Now note that

$$F(z) = \sum_{n=-N}^{N} |b_n|^{pz} e^{i\phi_n} \hat{g}_z(n),$$

where

$$g_z(t) = |f(t)|^{pz} e^{i\theta(t)}.$$

(Note that while θ need not be an element of $C(\mathbb{T})$, or more precisely need not be the restriction to $[0, 2\pi]$ of an element of $C(\mathbb{T})$ (why not?), g_z is nonetheless in $C(\mathbb{T})$.)

It follows from Lemma 23.1.1 that

$$|F(x+iy)| \le \sum_{n=-N}^{N} |b_n|^{px} \frac{1}{2\pi} \int_0^{2\pi} |f(t)|^{px}\, dt,$$

which shows that F is bounded in any vertical strip $a < x < b$. Since $\sum |b_n|^p = 1$, it also shows that

$$|F(1+iy)| \le \frac{1}{2\pi} \int_0^{2\pi} |f(t)|^p\, dt$$

for every $y \in \mathbb{R}$.

Now note that if $z = 1/2 + iy$ then $|g_z(t)|^2 = |f(t)|^p$. Considering Lemma 23.1.2 and using Lemma 23.1.4 we see that

$$|F(1/2+iy)| \le \sum_{n=-N}^{N} |b_n|^{p/2} |\widehat{g_{1/2+iy}}(n)|$$

$$\le \left(\sum_{n=-N}^{N} |b_n|^p\right)^{1/2} \left(\sum_{n=-N}^{N} |\widehat{g_{1/2+iy}}(n)|^2\right)^{1/2}$$

$$\le \left(\frac{1}{2\pi} \int_0^{2\pi} |f(t)|^p\, dt\right)^{1/2}.$$

Now, since $g_{1/p} = f$ and $|b_n|e^{i\phi_n} = b_n$, the Three-Lines Theorem (Theorem 23.0.2) shows that

$$\left|\sum_{-N}^{N} b_n \hat{f}(n)\right| = |F(1/p)|$$

$$\le \left(\frac{1}{2\pi} \int_0^{2\pi} |f(t)|^p\right)^{1(1/p-1/2)/(1/2)+(1/2)(1-1/p)/(1/2)}$$

$$= \left(\frac{1}{2\pi} \int_0^{2\pi} |f(t)|^p\right)^{1/p},$$

as required. □

We might remark in conclusion that there is nothing corresponding to the Hausdorff-Young inequality for $p > 2$; in fact there exists $f \in C(\mathbb{T})$ such that $\sum_{n=-\infty}^{\infty} |\hat{f}(n)|^q = \infty$ for all $q \in (0, 2)$. This has little to do with complex analysis, but we sketch a proof for the benefit of interested readers.

23.1. An Application: The Hausdorff-Young Inequality

The key is a clever construction of the so-called *Rudin-Shapiro polynomials*. We define two sequences (P_n) and (Q_n) of trigonometric polynomials by $P_0 = 1$, $Q_0 = 1$, and

$$P_{n+1}(t) = P_n(t) + e^{i2^n t} Q_n(t),$$

$$Q_{n+1}(t) = P_n(t) - e^{i2^n t} Q_n(t).$$

Since $|z|^2 = z\bar{z}$, it follows that

$$|P_{n+1}|^2 + |Q_{n+1}|^2 = 2(|P_n|^2 + |Q_n|^2),$$

hence

$$|P_n|^2 + |Q_n|^2 = 2^{n+1}$$

and so

$$|P_n| \le 2^{(n+1)/2}.$$

It is clear that

$$\hat{P}_n(k) = \begin{cases} \pm 1 & (0 \le k < 2^n), \\ 0 & \text{otherwise}; \end{cases}$$

in particular

$$\sum_{k=0}^{2^n - 1} |\hat{P}_n(k)|^q = 2^n.$$

These polynomials are useful in constructing examples of various phenomena. Define

$$c_n = \frac{2^{-n/2}}{n^2}$$

and set

$$f(t) = \sum_{n=1}^{\infty} c_n e^{i2^n t} P_n(t).$$

Since

$$|c_n e^{i2^n t} P_n(t)| \le \frac{2^{1/2}}{n^2},$$

the series converges uniformly, so that $f \in C(\mathbb{T})$. On the other hand

$$\sum_{k=0}^{\infty} |\hat{f}(k)|^q = \sum_{n=1}^{\infty} \frac{2^{n(1-q/2)}}{n^{2q}},$$

which is infinite for $q \in (0, 2)$ since $1 - q/2 > 0$.

Chapter 24

The Gamma Function

In this chapter we will develop some of the basic properties of the Gamma function $\Gamma(z)$. This is probably the most important of the classical "special functions", because it seems to arise in many places in mathematics. We mention just three examples: The "volume" of the unit ball in $\mathbb{R}^n$ is $\pi^{n/2}/\Gamma((n/2)+1)$, one proof of Littlewood's Theorem (mentioned in Chapter 21) depends on the fact that $\Gamma(1+i\xi) \ne 0$ for $\xi \in \mathbb{R}$, and the Gamma function is very important in analytic number theory, in large part because of various relations between the Gamma function and the "zeta function" (we will say a little bit about this at the end of the chapter).

The proofs will involve a certain amount of calculus as well as complex analysis; we will leave some details of the calculus to the reader. First we note for the record that although it is good to be very careful with expressions of the form z^w for complex z and w (because various "laws of exponents", etc., don't always work), there is no problem with t^z for $t > 0$; in that case we define

$$t^z = \exp(z \ln(t)),$$

and everything works the way one would hope. In particular you may easily verify that the following hold for $s, t > 0$ and $z \in \mathbb{C}$:

(i) $(st)^z = s^z t^z$;

(ii) $|t^z| = t^{\operatorname{Re}(z)}$ (so in particular $|t^{i\xi}| = 1$ for $\xi \in \mathbb{R}$);

(iii) if $t > 0$ is fixed and $f(z) = t^z$ then f is entire with $f'(z) = \ln(t) f(z)$;

(iv) if z is fixed and $f(t) = t^z$ then f is differentiable on $(0, \infty)$ with $f'(t) = z t^{z-1}$.

It is easy to see that
$$\int_0^\infty e^{-t} t^x \, dt < \infty$$
for $x > -1$; since $|t^z| = t^{\operatorname{Re}(z)}$, it follows that we may define $\Gamma(z)$ for $\operatorname{Re}(z) > 0$ by
$$\Gamma(z) = \int_0^\infty e^{-t} t^{z-1} \, dt.$$

(We will see shortly that Γ can be analytically continued to a larger region.) It follows from Fubini's Theorem plus Morera's Theorem that Γ is holomorphic in the right half-plane. A simple way to deduce this from the version of Fubini's Theorem that you may have learned in Advanced Calculus is as follows: For $0 < A < \infty$ define
$$\Gamma_A(z) = \int_0^A e^{-t} t^{z-1} \, dt.$$

Use Fubini's Theorem and Morera's Theorem to show that Γ_A is holomorphic in the right half-plane (as in the proof that $F_1 \in H(V)$ in the proof of Theorem 4.9) and show that $\Gamma_A \to \Gamma$ uniformly on compact subsets of the right half-plane as $A \to \infty$.

The following lemma gives the so-called "functional equation" satisfied by the Gamma function; it may be proved simply by integration by parts (with a little care, making certain that the "boundary terms" tend to 0):

Lemma 24.0. *If $\operatorname{Re}(z) > 0$ then*
$$\Gamma(z+1) = z\Gamma(z).$$

One finds a calculus student to verify that $\Gamma(1) = 1$, and then the lemma shows by induction that the Gamma function is related to the factorial:

Corollary 24.1. $n! = \Gamma(n+1)$ $(n = 0, 1, \dots)$.

As mentioned in the introduction to Chapter 16, the functional equation also shows that the Gamma function allows analytic continuation to a region larger than the right half-plane:

Corollary 24.2. *The Gamma function extends to a function meromorphic in the entire plane, with simple poles at the nonpositive integers (and no poles elsewhere).*

24. The Gamma Function

Proof. For $\operatorname{Re}(z) > 0$ we have
$$\Gamma(z) = \frac{\Gamma(z+1)}{z}.$$
But $\Gamma(z+1)$ is holomorphic in the region $\operatorname{Re}(z) > -1$; since $\Gamma(1) \neq 0$, it follows that Γ may be extended to a function meromorphic in the region $\operatorname{Re}(z) > -1$, with a simple pole at the origin.

The same argument now shows that Γ extends to a function meromorphic for $\operatorname{Re}(z) > -2$, with simple poles at -1 and 0. Etc. □

The first nontrivial result we shall prove about the Gamma function is that it can also be defined by a certain infinite product; we will use an argument analogous to one of the proofs in Chapter 6. To begin, we note that Theorem 13.1 shows that the expression
$$h(z) = \frac{1}{z} \prod_{n=1}^{\infty} e^{z/n} \left(1 + \frac{z}{n}\right)^{-1}$$
defines a function meromorphic in the plane with simple poles at the nonpositive integers. In order to investigate how close h comes to satisfying the functional equation satisfied by the Gamma function we set
$$h_N(z) = \frac{1}{z} \prod_{n=1}^{N} e^{z/n} \left(1 + \frac{z}{n}\right)^{-1}.$$
It follows that
$$h_N(z+1) = \frac{1}{z+1} \prod_{n=1}^{N} e^{(z+1)/n} \left(1 + \frac{z+1}{n}\right)^{-1}$$
$$= \frac{1}{z+1} \exp\left(\sum_{n=1}^{N} 1/n\right) \prod_{n=1}^{N} e^{z/n} \prod_{n=1}^{N} \frac{n}{z+1+n}$$
$$= \frac{z}{z+1+N} \exp\left(\sum_{n=1}^{N} 1/n\right) \frac{1}{z} \prod_{n=1}^{N} e^{z/n} \prod_{n=1}^{N} \frac{n}{z+n}$$
$$= z \frac{N}{z+1+N} \exp\left(-\log(N) + \sum_{n=1}^{N} 1/n\right) h_N(z).$$

The reader is probably aware that $-\log(N) + \sum_{n=1}^{N} 1/n$ tends to a positive constant γ, known as the *Euler-Mascheroni constant*, as $N \to \infty$. (A hint if you want to prove this yourself: The inequality
$$\frac{1}{n} - (\log(n+1) - \log(n)) = \int_n^{n+1} \left(\frac{1}{n} - \frac{1}{t}\right) dt < \int_n^{n+1} \left(\frac{1}{n} - \frac{1}{n+1}\right) dt$$

shows that the limit exists, and in fact shows that $0 < \gamma < 1$.) Hence, letting $N \to \infty$ above, we see that
$$h(z+1) = e^\gamma z h(z);$$
now if we define
$$g(z) = \frac{e^{-\gamma z}}{z} \prod_{n=1}^{\infty} e^{z/n} \left(1 + \frac{z}{n}\right)^{-1}$$
then g is meromorphic in the plane with simple poles at the nonpositive integers, and g also satisfies
$$g(z+1) = zg(z)$$
if z is not a pole of g. It is easy to see that the residue of g at the origin is 1, while the fact that $\Gamma(z+1) = z\Gamma(z)$ and $\Gamma(1) = 1$ shows that the residue of Γ at the origin is also 1. The function g is starting to look a lot like the Gamma function.

The function
$$f(z) = \frac{\Gamma(z)}{g(z)}$$
is entire and equals 1 at the origin; if we could show f was bounded then Liouville's Theorem would show that $f(z) = 1$ for all z, so that $\Gamma = g$. Since $f(z+1) = f(z)$, we need only obtain a bound on $|f(x+iy)|$ for $1 \leq x \leq 2$.

Now, since $|t^{x+iy}| = t^x$ for $t > 0$ and $x, y \in \mathbb{R}$, it is clear from the original definition that
$$|\Gamma(x+iy)| \leq \Gamma(x) \leq c \quad (1 \leq x \leq 2).$$

Note. In this chapter we have adopted the common convention that the letter "c" denotes some constant, the value of which may vary from occurrence to occurrence.

A suitable bound on $1/|g(x+iy)|$ is not so obvious, but can be obtained by a little trick:

Lemma 24.3. *If $z = x + iy$ with $x > 0$ then*
$$-x + \log|1 + z| \leq \log\left(\left(1 + |z|^2\right)^{1/2}\right).$$

Proof. Multiplying both sides of the inequality by 2, we need to prove that
$$-2x + \log((1+x)^2 + y^2) \leq \log(1 + x^2 + y^2).$$
But it is clear from the Mean Value Theorem that if $1 \leq a < b$ then $\log(b) - \log(a) \leq b - a$. □

24. The Gamma Function

Lemma 24.4. *If g is as above and $z = x + iy$ with $x > 0$ then*
$$\frac{1}{|g(z)|} \leq \sqrt{|z|/2\pi}\, e^{\gamma x} e^{\pi|z|/2}.$$

Proof. The definition of g and the previous lemma show that
$$\frac{1}{|g(z)|} = |z| e^{\gamma x} \prod_{n=1}^{\infty} e^{-x/n} \left|1 + \frac{z}{n}\right| \leq |z| e^{\gamma x} \prod_{n=1}^{\infty} \left(1 + \frac{|z|^2}{n^2}\right)^{1/2}.$$

We could estimate that product as in Chapter 6, but we don't need to; in fact Euler's formula for the sine shows that
$$\pi|z| \prod_{n=1}^{\infty} \left(1 + \frac{|z|^2}{n^2}\right) = -i\sin(i\pi|z|) = (e^{\pi|z|} - e^{-\pi|z|})/2 < e^{\pi|z|}/2,$$

which gives the stated inequality. $\square$

Now if $f = \Gamma/g$ as above and $z = x + iy$ with $1 \leq x \leq 2$ we have shown that
$$|f(z)| \leq c\sqrt{|z|}\, e^{\pi|z|/2} \leq c\sqrt{1 + |y|}\, e^{\pi|y|/2}.$$

(Recall that different instances of the letter "c" may denote different constants.) Since f is an entire function of period 1, it follows as in Corollary 19.0.10 that there exists a function F holomorphic in the punctured plane $\mathbb{C} \setminus \{0\}$ with
$$f(z) = F(e^{2\pi i z}).$$

Now if $z = x + iy$ and $w = e^{2\pi i z}$ then $|y| = |\log|w||/2\pi$, so the estimate above shows that
$$|F(w)| \leq c\sqrt{1 + |\log|w||}\, e^{|\log|w||/4}$$
$$\leq c e^{|\log|w||/2} \leq c\left(e^{\log|w|/2} + c e^{-\log|w|/2}\right).$$

In other words, $|F(w)| \leq c(|w|^{-1/2} + |w|^{1/2})$. This implies that F is bounded, for example by Exercise 3.17 in Chapter 3 (applied to F and to $F(1/w)$). Thus f is bounded and hence constant, and so we have proved

Theorem 24.5. *If z is not a nonpositive integer then*
$$\Gamma(z) = \frac{e^{-\gamma z}}{z} \prod_{n=1}^{\infty} e^{z/n} \left(1 + \frac{z}{n}\right)^{-1}.$$

Theorem 24.5 shows that the Gamma function has no zero. A slightly less trivial corollary follows:

Corollary 24.6. *If z is not an integer then*

$$\Gamma(z)\Gamma(1-z) = \frac{\pi}{\sin(\pi z)}.$$

- **Exercise 24.1.** Show that Corollary 24.6 follows from Theorem 24.5. (Use the identity $\Gamma(1-z) = -z\Gamma(-z)$ and then apply the theorem to $\Gamma(z)\Gamma(-z)$.)

In particular, Corollary 24.6 shows that

$$\Gamma\left(\frac{1}{2}\right) = \sqrt{\pi}.$$

This allows one to find the value of $\Gamma(n+1/2)$ for any $n \in \mathbb{Z}$, using the functional equation.

The product formula for the Gamma function is more or less equivalent to a result due to Gauss:

Theorem 24.7 (Gauss). *If $z \in \mathbb{C}$ is not a nonpositive integer then*

$$\Gamma(z) = \lim_{n\to\infty} \frac{n!n^z}{z(z+1)\cdots(z+n)}.$$

Proof. Theorem 24.5 shows that

$$\Gamma(z) = \lim_{n\to\infty} \frac{e^{z(-\gamma+\sum_{j=1}^{n}\frac{1}{j})}}{z} \prod_{j=1}^{n}\left(1+\frac{z}{j}\right)^{-1}$$

$$= \lim_{n\to\infty} \frac{e^{z\log(n)}}{z} \prod_{j=1}^{n}\left(\frac{j}{z+j}\right)$$

$$= \lim_{n\to\infty} \frac{n!n^z}{z(z+1)\cdots(z+n)}. \qquad \square$$

- **Exercise 24.2.** Show that the functional equation $\Gamma(z+1) = z\Gamma(z)$ follows immediately from Theorem 24.7.

Our next goal is Stirling's formula, which gives an asymptotic expression for $\Gamma(z)$ as $z \to +\infty$ within certain sectors of the plane. In general the notation

$$f(z) \sim g(z),$$

24. The Gamma Function

which might be read "$f(z)$ is asymptotic to $g(z)$", means that

$$\frac{f(z)}{g(z)} \to 1$$

(as z tends to something, within some set).

We should say for the record that for the rest of this chapter "log" will be the principal-value logarithm, i.e. the function $\log : \mathbb{C} \setminus (-\infty, 0] \to \mathbb{C}$ defined by

$$\log(re^{it}) = \log(r) + it \quad (r > 0, |t| < \pi),$$

and if $z \in \mathbb{C} \setminus (-\infty, 0]$ and $w \in \mathbb{C}$ then z^w will be defined by

$$z^w = \exp(w \log(z)).$$

(We will write $\sqrt{z}$ for $z^{1/2}$, $z \in \mathbb{C} \setminus (-\infty, 0]$.) Note that although $(z_1 z_2)^w \ne z_1^w z_2^w$ in general we do have $(tz)^w = t^w z^w$ if $z \in \mathbb{C} \setminus (-\infty, 0]$ and $t > 0$.

For $0 < a < \pi$ we will let S_a denote the sector

$$S_a = \{\, re^{it} : r > 0,\ -a < t < a \,\}.$$

We can now state

Theorem 24.8 (Stirling's Formula). *If $0 < a < \pi$ then*

$$\Gamma(z+1) \sim \sqrt{2\pi z}\left(\frac{z}{e}\right)^z$$

as $z \to \infty$ within S_a.

Since Γ has no zero and $\mathbb{C} \setminus (-\infty, 0]$ is simply connected, Γ has a holomorphic logarithm in $\mathbb{C} \setminus (-\infty, 0]$. Since $\Gamma(1) > 0$, we may choose this logarithm so that it is real at 1, and now continuity shows that it is real on $(0, \infty)$. So: We will let λ denote a function holomorphic in $\mathbb{C} \setminus (-\infty, 0]$ such that $e^\lambda = \Gamma$ and $\lambda(x) \in \mathbb{R}$ for all $x \in (0, \infty)$.

As always one needs to be careful in applying various identities involving complex logarithms. Note here that

$$\lambda(z+1) = \lambda(z) + \log(z),$$

exactly as one would hope; this is because both sides are holomorphic and they agree for $z \in (0, \infty)$. The same argument shows that if $t > 0$ then

$$\log(z+t) - \log(z) = \log(1 - t/z)$$

for all $z \in \mathbb{C} \setminus (-\infty, 0]$.

We begin with the most technical part of the proof:

Lemma 24.9. *If $0 < a < \pi$ then there exists a constant c such that*

$$|\lambda''(z)| \leq c/|z|$$

for all $z \in S_a$ with $|z| \geq 2$.

Proof. Note first that Theorem 24.5 shows that

$$\lambda''(z) = \sum_{n=0}^{\infty} \frac{1}{(z+n)^2}.$$

(See the comments on the "logarithmic derivative" in Chapter 6.) In estimating this sum we will use the fact that there exists $c > 0$ (depending on a) such that

$$|z + n| \geq cn$$

for all $z \in S_a$ and $n > 0$. (If $0 < a \leq \pi/2$ this is clear, since $\operatorname{Re}(z) \geq 0$. If $\pi/2 < a < \pi$ let z_0 be one of the points of $\overline{S}_a$ closest to $-n$; then it is clear that $|z_0 - (-n)| = cn$.)

Choose a positive integer N with

$$N \leq |z|/2 < N + 1.$$

If $0 \leq n \leq |z|/2$ then $|z + n| \geq |z|/2$; hence

$$\sum_{n=0}^{N} \frac{1}{|z+n|^2} \leq 4 \sum_{n=0}^{N} \frac{1}{|z|^2} = 4\frac{N+1}{|z|^2} \leq c\frac{N}{|z|^2} \leq \frac{c}{|z|}.$$

On the other hand,

$$\sum_{n=N+1}^{\infty} \frac{1}{|z+n|^2} \leq c \sum_{N+1}^{\infty} \frac{1}{n^2} \leq c \int_N^{\infty} \frac{1}{t^2}\, dt = \frac{c}{N} \leq \frac{c}{|z|}. \quad \square$$

Stirling's formula will now follow easily from a certain identity, which might be regarded as an "averaged" version of Stirling's formula:

Theorem 24.10. *There exists a constant $\beta \in \mathbb{R}$ such that*

$$\int_0^1 \lambda(z+t)\, dt = z \log(z) - z + \beta$$

for all $z \in \mathbb{C} \setminus (-\infty, 0]$.

24. The Gamma Function

Proof. If $F(z) = \int_0^1 \lambda(z+t)\,dt$ then F is holomorphic and the fundamental theorem of calculus shows that

$$F'(z) = \lambda(z+1) - \lambda(z) = \log(z).$$

This is the same as the derivative of $z\log(z) - z$, so the two functions must differ by a constant β; now the case $z \in (0, \infty)$ shows that $\beta \in \mathbb{R}$. $\square$

For the next few pages the letter β will denote the constant in the statement of the previous theorem. The following corollary might be regarded as a version of Stirling's formula, except that the formula is not written the way it usually is and the value of β has not yet been determined:

Corollary 24.11. *If $0 < a < \pi$ then*

$$\lambda(z+1) - ((z+1/2)\log(z+1/2) - (z+1/2) + \beta) \to 0$$

as $z \to \infty$ within S_a.

Proof. Replacing z by $z + 1/2$ in the previous theorem shows that

$$\int_{-1/2}^{1/2} \lambda(z+1+t)\,dt = \int_0^1 \lambda(z+1/2+t)\,dt$$
$$= (z+1/2)\log(z+1/2) - (z+1/2) + \beta.$$

Since λ'' tends to 0 at infinity, Exercise 6.3 shows that

$$\lambda(z+1) - \int_{-1/2}^{1/2} \lambda(z+1+t)\,dt \to 0. \quad \square$$

Now we note that the expression in Corollary 24.11 is equivalent to the expression in the usual statement of Stirling's formula, a small miracle happens which gives us the value of β, and we are done:

Proof of Theorem 24.8. First we note that

$$(z+1/2)\log(z+1/2) - (z+1/2) - (z\log(z) - z + \log(z)/2)$$
$$= (z+1/2)(\log(z+1/2) - \log(z)) - 1/2$$
$$= (z+1/2)\log(1 + 1/(2z)) - 1/2$$
$$= (z+1/2)(1/(2z) + O(1/|z|^2)) - 1/2$$
$$= O(1/|z|)$$
$$\to 0;$$

thus the previous corollary implies that
$$\lambda(z+1) - (z\log(z) - z + \log(z)/2 + \beta) \to 0,$$
and hence
$$\Gamma(z+1) \sim e^\beta \sqrt{z} \left(\frac{z}{e}\right)^z,$$
as $z \to \infty$ within S_a.

We can determine the value of e^β using the curious fact that it is very easy to determine the value of $|\Gamma(1+iy)|$ exactly. If $y > 0$ then the identity $\Gamma(1+iy) = iy\Gamma(iy)$ shows that
$$|\Gamma(1+iy)|^2 = \prod_{n=1}^\infty \left(1 + \frac{y^2}{n^2}\right)^{-1} = \frac{\pi i y}{\sin(\pi i y)} = \frac{2\pi y}{e^{\pi y} - e^{-\pi y}},$$
so that
$$|\Gamma(1+iy)| \sim \sqrt{2\pi y}\, e^{-\pi y/2}$$
as $y \to +\infty$. Now, if $y > 0$ then $iy = ye^{i\pi/2}$; since $-\pi < \pi/2 < \pi$ it follows that $\log(iy) = \log(y) + i\pi/2$, so that
$$\left(\frac{iy}{e}\right)^{iy} = e^{iy(\log(y) - 1 + i\pi/2)} = e^{-\pi y/2} e^{iy(\log(y)-1)}.$$
Thus
$$\left|\left(\frac{iy}{e}\right)^{iy}\right| = e^{-\pi y/2},$$
so that
$$|\Gamma(1+iy)| \sim e^\beta \sqrt{y}\, e^{-\pi y/2}$$
as $y \to +\infty$. Comparing the two asymptotic expressions for $|\Gamma(1+iy)|$ shows that $e^\beta = \sqrt{2\pi}$. $\square$

Now that we know the value of β we should rephrase Lemma 24.10:

Corollary 24.12. *If $z \in \mathbb{C} \setminus (-\infty, 0]$ then*
$$e^{\int_0^1 \log(\Gamma(z+t))\, dt} = \sqrt{2\pi} \left(\frac{z}{e}\right)^z.$$

(The statement of Corollary 24.12 is valid for any (continuous) choice of $\log(\Gamma(z+t))$; why is that?)

We should note that our proof of Stirling's formula contains implicit information on the rate of convergence; the reader may verify that for every $a \in (0, \pi)$ there exists c such that
$$\left|1 - \frac{\Gamma(z+1)}{\sqrt{2\pi z}\, (z/e)^z}\right| \leq \frac{c}{|z|}$$
for all $z \in S_a$ with $|z| \geq 1$.

- **Exercise 24.3.** Prove Theorem 24.7 by starting with the identity

$$\Gamma(z) = \frac{\Gamma(z+n+1)}{z(z+1)\cdots(z+n)}$$

and applying Stirling's formula. (Of course this is much less straightforward than the proof in the text; the point is practice dealing with Stirling's formula.)

A more substantial application of Stirling's formula is given by the so-called duplication formula for $\Gamma(2z)$.

We consider $(2n)!$ first, for motivation. Since $2j < 2j+1 < 2(j+1)$, it is easy to see from the definition of $n!$ that

$$2^{2n-1}n!(n-1)! < (2n)! < 2^{2n}(n!)^2.$$

Now, $2j+1$ is presumably better approximated by the geometric mean $2\sqrt{j(j+1)}$ than by either the lower bound $2j$ or the upper bound $2j+2$, so one might conjecture that $2^{2n}(n!)^2/\sqrt{2n}$, the geometric mean of the resulting upper and lower bounds on $(2n)!$, should give a better approximation to $(2n)!$. Indeed it does; we give two proofs of this to illustrate various techniques:

- **Exercise 24.4.** Show, using only Theorem 6.1.4, that the product

$$\prod_{j=1}^{\infty} \frac{2j+1}{2\sqrt{j(j+1)}}$$

converges (to a strictly positive finite number). Note that the algebra may work out better if instead you consider the squared product

$$\prod_{j=1}^{\infty} \frac{(2j+1)^2}{4j(j+1)}.$$

- **Exercise 24.5.** Deduce from the previous exercise that there exists a constant c such that

$$(2n)! \sim \frac{c}{\sqrt{n}} 2^{2n}(n!)^2$$

as $n \to \infty$.

That is a very precise result considering that it used such elementary methods. We can easily do better:

- **Exercise 24.6.** Deduce from Stirling's formula that in fact

$$(2n)! \sim \frac{2^{2n}(n!)^2}{\sqrt{\pi n}}$$

as $n \to \infty$.

One might conjecture on this basis that $\Gamma(2z)$ should be at least somewhat like $2^{2z}\Gamma(z)^2$. Of course the two canot be equal, since for example they have poles in different places. We can easily fix that: The functions $\Gamma(2z)$ and

$$2^{2z}\Gamma(z)\Gamma(z+1/2)$$

both have simple poles precisely at $-n/2$, $n = 0, 1, \ldots$. So all the singularities of the quotient of the two are removable. Who knows, maybe the quotient is bounded?

So we define

$$f(z) = \frac{\Gamma(2z)}{2^{2z}\Gamma(z)\Gamma\left(z + \frac{1}{2}\right)}.$$

As noted above, f is (or, strictly speaking, can be extended so as to be) an entire function. We note that

$$f\left(z + \frac{1}{2}\right) = \frac{\Gamma(2z+1)}{2^{2z+1}\Gamma\left(z + \frac{1}{2}\right)\Gamma(z+1)} = \frac{2z\Gamma(2z)}{2^{2z+1}\Gamma\left(z + \frac{1}{2}\right)z\Gamma(z)} = f(z),$$

so if we want to show that f is bounded we need only obtain a bound on $f(z)$ for, say, $0 \leq \mathrm{Re}(z) \leq 1/2$; this is straightforward using Stirling's formula.

In fact the duplication formula gives another way to find the constant in Stirling's formula — we shall use Stirling's formula in the form

$$\Gamma(z+1) \sim \alpha\sqrt{z}\left(\frac{z}{e}\right)^z$$

to illustrate this. As $z \to \infty$ in the right half-plane we have

$$f(z) \sim \frac{\alpha\sqrt{2z-1}\left(\frac{2z-1}{e}\right)^{2z-1}}{2^{2z}\alpha\sqrt{z-1}\left(\frac{z-1}{e}\right)^{z-1}\alpha\sqrt{z-1/2}\left(\frac{z-1/2}{e}\right)^{z-1/2}}$$

$$= \frac{e^{-1/2}}{2\alpha}\left(\frac{2z-1}{z-1}\right)^{1/2}\left(\left(1 + \frac{1}{2(z-1)}\right)^{2(z-1)}\right)^{1/2}$$

$$\sim \frac{1}{\sqrt{2}\alpha}.$$

24. The Gamma Function

Since f is periodic, it follows that f is bounded, hence constant, and in fact $f(z) = 1/(\sqrt{2}\alpha)$ for all z:

$$\Gamma(2z) = \frac{2^{2z-1/2}}{\alpha}\Gamma(z)\Gamma\left(z+\frac{1}{2}\right).$$

Now setting $z = 1/2$ shows that $\alpha = \sqrt{2\pi}$, and so we have proved the duplication formula:

Theorem 24.13.

$$\Gamma(2z) = \frac{2^{2z-1}}{\sqrt{\pi}}\Gamma(z)\Gamma\left(z+\frac{1}{2}\right).$$

- **Exercise 24.7.** Deduce the duplication formula directly from Theorem 24.7, using Exercise 24.6 (or Exercise 24.5).

Readers possessing a sufficiently wide sheet of paper can easily adapt either proof of the duplication formula to prove a more general result known as the multiplication formula:

- **Exercise 24.8.** If m is a positive integer then

$$\Gamma(z)\Gamma\left(z+\frac{1}{m}\right)\cdots\Gamma\left(z+\frac{m-1}{m}\right) = (2\pi)^{(m-1)/2}m^{\frac{1}{2}-mz}\Gamma(mz).$$

Something very interesting happens if we take m-th roots on both sides of the multiplication formula and then let m tend to infinity — we obtain Corollary 24.12:

- **Exercise 24.9.** Show that

$$\lim_{m\to\infty}\left(\Gamma(z)\Gamma\left(z+\frac{1}{m}\right)\cdots\Gamma\left(z+\frac{m-1}{m}\right)\right)^{1/m} = e^{\int_0^1 \log(\Gamma(z+t))\,dt}$$

and

$$\lim_{m\to\infty}\left((2\pi)^{(m-1)/2}m^{\frac{1}{2}-mz}\Gamma(mz)\right)^{1/m} = \sqrt{2\pi}\left(\frac{z}{e}\right)^z.$$

Of course a complex number has more than one m-th root; you should specify which root you are using in your solution. (There is an obvious choice, based on the preceding material, which turns out to work.)

The exercise illustrates why $\exp\left(\int_0^1 \log(f(t))\,dt\right)$ is known as the *geometric mean* of $f(t)$ on $[0,1]$.

The Bohr-Mollerup Theorem is a very interesting uniqueness result for the Gamma function. Note that $\log(\Gamma(x))$ is convex on $(0, \infty)$; this is very easy to see from the definition as an integral, using a result from real analysis called "Hölder's inequality". It is also clear from Theorem 24.5 above, which shows that if $x > 0$ then

$$\log(\Gamma(x)) = -\gamma x - \log(x) + \sum_{n=1}^{\infty} \left(\frac{x}{n} - \log\left(1 + \frac{x}{n}\right)\right).$$

Theorem 24.14 (Bohr-Mollerup). *Suppose that $f : (0, \infty) \to (0, \infty)$ satisfies $f(x + 1) = xf(x)$, $f(1) = 1$, and suppose that $\log(f)$ is convex. Then $f(x) = \Gamma(x)$ for all $x > 0$.*

Proof. If $n \in \mathbb{N}$ then

$$f(x + n) = x(x + 1) \cdots (x + n - 1)f(x);$$

in particular $f(n + 1) = n!$.

Since $f(x + 1) = xf(x)$ and $f(1) = 1$, we may assume that $x \in (0, 1)$; hence the convexity of $\log(f)$ shows that

$$f(xa + (1 - x)b) \leq f(a)^x f(b)^{1-x}$$

for $a, b > 0$. Now for $n \in \mathbb{N}$ we have

$$x + n = (1 - x)n + x(n + 1),$$

so the convexity of $\log(f)$ shows that

$$x(x - 1) \cdots (x + n - 1)f(x) = f(x + n) \leq f(n)^{1-x} f(n + 1)^x$$
$$= ((n - 1)!)^{1-x}(n!)^x = n^x(n - 1)!.$$

Similarly

$$n + 1 = x(x + n) + (1 - x)(x + n + 1)$$

shows that

$$n! = f(n + 1) \leq f(x + n)^x f(x + n + 1)^{1-x}$$
$$= (x + n)^{1-x} x(x + 1) \cdots (x + n - 1)f(x).$$

24. The Gamma Function

Rearranging the two inequalities we have proved shows that

$$\frac{n!n^x}{x(x+1)\cdots(x+n)}\left(\frac{x+n}{n}\right)^x \le f(x) \le \frac{n!n^x}{x(x+1)\cdots(x+n)}\left(\frac{x+n}{n}\right);$$

now Theorem 24.7 shows that $f(x) = \Gamma(x)$. □

We should note that the application of Theorem 24.7 in the proof of the Bohr-Mollerup Theorem was not really essential: The proof shows that $0 < a_n \le f(x) \le b_n$ where $b_n/a_n \to 1$, and hence it shows that there is at most one function f satisfying the hypotheses. Indeed, looking at the argument this way gives a proof of Theorem 24.7 in the special case $x > 0$. One may use Theorem 24.13 to prove various results about the Gamma function; for example if we used Hölder's inequality to show that $\log(\Gamma)$ is convex on $(0, \infty)$ then we could use Theorem 24.13 to prove Theorem 24.5.

Next we give an integral formula for the Beta function, which is defined (for appropriate $z, w \in \mathbb{C}$) by

$$B(z, w) = \frac{\Gamma(z)\Gamma(w)}{\Gamma(z+w)}.$$

We will use the following fact from calculus: If $f : (0, \infty)^2 \to \mathbb{C}$ is continuous and

$$\int_0^\infty \left(\int_0^\infty |f(x, y)|\, dx\right) dy < \infty$$

then

$$\int_0^\infty \int_0^\infty f(x, y)\, dx\, dy = \int_0^{\pi/2} \int_0^\infty f(r\cos(\theta), r\sin(\theta))\, r\, dr\, d\theta.$$

Theorem 24.15. *If* $\operatorname{Re}(z), \operatorname{Re}(w) > 0$ *then*

$$B(z, w) = \frac{\Gamma(z)\Gamma(w)}{\Gamma(z+w)}$$

$$= 2\int_0^{\pi/2} (\cos(\theta))^{2z-1}(\sin(\theta))^{2w-1}\, d\theta$$

$$= \int_0^1 t^{z-1}(1-t)^{w-1}\, dt.$$

Proof. A change of variable shows that

$$\Gamma(z) = 2 \int_0^\infty e^{-t^2} t^{2z-1} \, dt;$$

thus

$$\Gamma(z)\Gamma(w) = 4 \int_0^\infty \int_0^\infty e^{-(t^2+s^2)} t^{2z-1} s^{2w-1} \, dt \, ds$$

$$= 4 \int_0^{\pi/2} \int_0^\infty e^{-r^2} r^{2z+2w-2} (\cos(\theta))^{2z-1} (\sin(\theta))^{2w-1} r \, dr \, d\theta$$

$$= \left(2 \int_0^\infty e^{-r^2} r^{2(z+w)-1} \, dr \right)$$

$$\times \left(2 \int_0^{\pi/2} (\cos(\theta))^{2z-1} (\sin(\theta))^{2w-1} \, d\theta \right)$$

$$= \Gamma(z+w) \left(2 \int_0^{\pi/2} (\cos^2(\theta))^{z-1} (\sin^2(\theta))^{w-1} \cos(\theta) \sin(\theta) \, d\theta \right)$$

$$= \Gamma(z+w) \int_0^1 t^{z-1} (1-t)^{w-1} \, dt. \qquad \square$$

This gives another proof that

$$\Gamma\left(\frac{1}{2}\right) = \sqrt{\pi}.$$

We conclude with a hint regarding connections between the Gamma function and the zeta function. The zeta function $\zeta(z)$ is initially defined for $\operatorname{Re}(z) > 1$ by

$$\zeta(z) = \sum_{n=1}^\infty \frac{1}{n^z}.$$

You can consult any text on analytic number theory for many fascinating facts about the zeta function, many applications to number theory, and many connections between the zeta function and the Gamma function; here we will give just one of each.

We will show that the zeta function can be analytically continued to a function meromorphic in the plane, with a simple pole at 1 (with residue 1) and no other poles. This will follow from a simple identity relating the zeta function and the Gamma function:

24. The Gamma Function

Theorem 24.16. *If* $\operatorname{Re}(z) > 1$ *then*

$$\zeta(z)\Gamma(z) = \int_0^\infty \frac{t^{z-1}}{e^t - 1}\, dt.$$

Proof. The formula for the sum of a geometric series shows that if $t > 0$ then

$$\frac{1}{e^t - 1} = \sum_{n=1}^\infty e^{-nt};$$

the series converges uniformly on compact subsets of $(0, \infty)$. Suppose that $z = x + iy$ with $x > 1$ and $y \in \mathbb{R}$. Since $1/(e^t - 1) \leq c/t$ for small $t > 0$ and $1/(e^t - 1) \leq c/e^t$ for large $t > 0$, it follows that

$$\int_0^\infty g(t)\, dt < \infty,$$

if $g(t) = t^{x-1}/(e^t - 1)$. But

$$\left| t^{z-1} \sum_{n=1}^N e^{-nt} \right| \leq g(t)$$

for all $t > 0$; hence we may use Lemma 17.7 (or rather a similar result for functions defined on $(0, \infty)$) to show that the integral of the sum equals the sum of the integrals in the following computation:

$$\int_0^\infty \frac{t^{z-1}}{e^t - 1}\, dt = \int_0^\infty t^{z-1} \sum_{n=1}^\infty e^{-nt}\, dt$$

$$= \sum_{n=1}^\infty \int_0^\infty t^{z-1} e^{-nt}\, dt$$

$$= \sum_{n=1}^\infty \frac{1}{n} \int_0^\infty \left(\frac{t}{n}\right)^{z-1} e^{-t}\, dt$$

$$= \sum_{n=1}^\infty \frac{1}{n^z} \int_0^\infty t^{z-1} e^{-t}\, dt$$

$$= \zeta(z)\Gamma(z). \qquad \square$$

Thus for $\operatorname{Re}(z) > 1$ we have

$$\zeta(z)\Gamma(z) = \int_0^1 \frac{t^{z-1}}{e^t - 1}\, dt + \int_1^\infty \frac{t^{z-1}}{e^t - 1}\, dt = \int_0^1 \frac{t^{z-1}}{e^t - 1}\, dt + F(z).$$

It follows from Fubini's Theorem and Morera's Theorem that F is actually entire (see the proof at the start of this chapter that $\Gamma(z)$ is holomorphic for $\operatorname{Re}(z) > 0$); we turn our attention to the integral from 0 to 1.

Let
$$\phi(t) = \frac{1}{e^t - 1}.$$

Then ϕ is holomorphic in the punctured disk $0 < |t| < 2\pi$ and has a simple pole with residue 1 at the origin; thus
$$\phi(t) = \frac{1}{t} + \sum_{n=0}^{\infty} a_n t^n,$$

where the power series has radius of convergence 2π; in particular $|a_n| \leq c\, 2^{-n}$. The definition of ϕ shows that $\phi(t) + \phi(-t) = -1$; hence $a_0 = -1/2$ and $a_{2k} = 0$ for $k = 1, 2, \ldots$, so that
$$\phi(t) = \frac{1}{t} - \frac{1}{2} + \sum_{k=0}^{\infty} b_k t^{2k+1},$$

where $b_k = a_{2k+1}$. Now if $\operatorname{Re}(z) > 1$ then the series
$$\sum_{k=0}^{\infty} t^{z-1} b_k t^{2k+1}$$

converges uniformly on $(0,1)$, since $|b_k| \leq c\, 2^{-2k}$, so we obtain
$$\int_0^1 t^{z-1} \phi(t)\, dt = \int_0^1 t^{z-2}\, dt - \frac{1}{2} \int_0^1 t^{z-1}\, dt + \sum_{k=0}^{\infty} b_k \int_0^1 t^{z+2k}$$
$$= \frac{1}{z-1} - \frac{1}{2z} + \sum_{k=0}^{\infty} \frac{b_k}{z + (2k+1)}$$
$$= G(z).$$

Since $|b_k| \leq c2^{-2k}$, the series defining G converges uniformly on compact subsets of $\mathbb{C} \setminus (\{0, 1\} \cup \{-2k - 1 : k = 0, 1, \ldots\})$ to a function meromorphic in $\mathbb{C}$, with simple poles at 0 and 1, possible poles at $-2k - 1$ ("possible" poles because we have not shown that $b_k \neq 0$) and no other poles.

Since $1/\Gamma$ is entire, it follows that
$$\zeta(z) = (G(z) + F(z))/\Gamma(z)$$

is meromorphic in $\mathbb{C}$. Since $\Gamma(1) = 1$, it follows that ζ has a pole at 1 with residue 1; all the other possible poles of $G + F$ are cancelled by zeroes of $1/\Gamma$, so ζ is holomorphic in the entire plane except for the pole at 1. Considering the other zeroes of $1/\Gamma$ that are not cancelled by poles of $G + F$ we see that we have proved the following result:

Theorem 24.17. *The zeta function extends to a function holomorphic in* $\mathbb{C}\setminus\{1\}$ *with a simple pole at 1 with residue 1 and zeroes at* $-2k$, $k = 1, 2, \ldots$.

The zeroes of the zeta function at the negative even integers are known as the "trivial" zeroes. It is relatively easy to show that the trivial zeroes are the only zeroes outside of the "critical strip" defined by $0 < \mathrm{Re}\,(z) < 1$. The location of the zeroes in the critical strip is of great importance in number theory, for reasons we cannot begin to explain in detail here. It is known that there are infinitely many zeroes in the critical strip, and the *Riemann Hypothesis*, one of the most famous and important unsolved problems in mathematics, conjectures that all the zeroes in the critical strip lie on the line $\mathrm{Re}\,(z) = 1/2$.

Indeed, the Prime Number Theorem states that if $\pi(x)$ is the number of primes p with $p \leq x$ then

$$\pi(x) \sim \frac{x}{\log(x)}$$

as $x \to +\infty$. Attempts to prove this were in fact the motivation for the development of a great deal of complex analysis towards the end of the 19th century — the fact that $\zeta(1 + iy) \neq 0$ is an important part of the proof, and the truth of the Riemann Hypothesis would imply improved versions of the Prime Number Theorem (with better information on the rate of convergence).

The reader can find much more about all of this in a text on analytic number theory. Here we give a trivial illustration of how the zeta function might have something to do with distribution of primes:

The fact that a positive integer has a unique prime factorization shows that

$$\zeta(z) = \prod_p \left(1 + \frac{1}{p^z} + \frac{1}{p^{2z}} + \cdots\right) = \prod_p \left(1 + \frac{1}{p^z - 1}\right) \quad (\mathrm{Re}\,(z) > 1);$$

here $\prod_p$ denotes the product over all primes p. Since $\lim_{x \to 1^+} \zeta(x) = +\infty$, it follows that there are infinitely many primes.

We give a brief explanation of why the zeta function satisfies that product formula, omitting details like convergence. First, note that

$$\left(\sum_{j=0}^N a_j\right)\left(\sum_{k=0}^N b_k\right) = \sum_{j,k=0}^N a_j b_k.$$

Now if $\sum_{j=0}^\infty a_j$ and $\sum_{k=0}^\infty b_k$ converge absolutely then letting $N \to \infty$ shows that

$$\left(\sum_{j=0}^\infty a_j\right)\left(\sum_{k=0}^\infty b_k\right) = \sum_{j,k=0}^\infty a_j b_k.$$

(The order of summation of the terms in the final double sum is ambiguous, but it doesn't matter because the sum converges absolutely.) It follows by induction on N that if $\sum_{j=0}^{\infty} a_{j,k}$ converges absolutely for $1 \leq k \leq N$ then

$$\prod_{k=1}^{N}\left(\sum_{j=0}^{\infty} a_{j,k}\right) = \sum_{j_1,j_2,\ldots,j_N=0}^{\infty} a_{j_1,1} a_{j_2,2} \cdots a_{j_N,N}.$$

In particular, if we list the prime numbers $p_1 < p_2 < \cdots$, then unique prime factorization shows that

$$\prod_{k=1}^{N}\left(\sum_{j=0}^{\infty} p_k^{-jz}\right) = \sum_{j_1,j_2,\ldots,j_N=0}^{\infty} \left(p_1^{j_1} p_2^{j_2} \cdots p_N^{j_N}\right)^{-z} = \sum_{n \in \mathbb{N}_N} n^{-z},$$

where $\mathbb{N}_N$ is the set of all natural numbers (including 1, excluding 0) which use no primes other than $p_1, p_2, \ldots, p_N$ in their factorizations. If there are only N primes then $\mathbb{N}_N = \mathbb{N}$ and we are done; if there are infinitely many primes we let $N \to \infty$ to obtain

$$\prod_{k=1}^{\infty}\left(\sum_{j=0}^{\infty} p_k^{-jz}\right) = \lim_{N \to \infty} \sum_{n \in \mathbb{N}_N} n^{-z} = \sum_{n \in \mathbb{N}} n^{-z}.$$

Of course there are easier ways to prove the existence of infinitely many primes. We conclude with an exercise designed for readers who feel that the rest of the exercises in the book are too easy:

- **Exercise 24.8.** Assume that $\lim_{x \to +\infty} \pi(x) \log(x)/x$ exists, and show that the limit equals 1, by filling in the details in the following outline.

(We should note that the exercise should definitely not be thought of as bringing us close to a proof of the Prime Number Theorem; showing that the limit *exists* is the hard part.)

Assume that $1 < x < 2$. Recall that $O(1)$ denotes "something bounded". It is not too hard to show that we introduce a bounded error at each step

24. The Gamma Function

in the following calculation:

$$\log(\zeta(x)) = \sum_p \log\left(1 + \frac{1}{p^x - 1}\right)$$

$$= O(1) + \sum_p \frac{1}{p^x - 1}$$

$$= O(1) + \sum_p \frac{1}{p^x}$$

$$= O(1) + \sum_{n=2}^{\infty} \frac{\pi(n) - \pi(n-1)}{n^x}$$

$$= O(1) + \sum_{n=2}^{\infty} \pi(n) \left(\frac{1}{n^x} - \frac{1}{(n+1)^x}\right)$$

$$= O(1) + x \sum_{n=2}^{\infty} \frac{\pi(n)}{n^{x+1}}.$$

(Most of this can be justified using techniques we have seen; the one step that may look mysterious is in fact trivial, because $\pi(n) - \pi(n-1)$ is equal to 1 if n is prime and 0 otherwise.)

Define a sequence (α_n) by

$$\pi(n) = \alpha_n \frac{n}{\log(n)};$$

we need to show that $\alpha_n \to 1$ as $n \to \infty$. Since ζ has a simple pole at 1 with residue 1, the calculation above shows that

$$x \sum_{n=2}^{\infty} \frac{\alpha_n}{n^x \log(n)} = O(1) + \log\left(\frac{1}{x-1}\right) \quad (1 < x < 2).$$

On the other hand, since $1 < x < 2$ we see that

$$x \sum_{n=2}^{\infty} \frac{1}{n^x \log(n)} = O(1) + x \int_2^{\infty} \frac{dt}{t^x \log(t)}$$

$$= O(1) + x \int_{\log(2)}^{\infty} \frac{e^{-(x-1)s}}{s} ds$$

$$= O(1) + x \int_{(x-1)\log(2)}^{\infty} \frac{e^{-s}}{s} ds$$

$$= O(1) + x \int_{(x-1)\log(2)}^{1} \frac{ds}{s}$$

$$+ x \int_{(x-1)\log(2)}^{1} \frac{e^{-s}-1}{s} ds + x \int_{1}^{\infty} \frac{e^{-s}}{s} ds$$

$$= O(1) + x \log\left(\frac{1}{x-1}\right)$$

$$= O(1) + (x-1)\log\left(\frac{1}{x-1}\right) + \log\left(\frac{1}{x-1}\right)$$

$$= O(1) + \log\left(\frac{1}{x-1}\right).$$

This shows that if $c_n \to 0$ as $n \to \infty$ then

$$x \sum_{n=2}^{\infty} \frac{c_n}{n^x \log(x)} = o(1) \log\left(\frac{1}{x-1}\right) \quad (x \to 1^+).$$

(Recall that $o(1)$ denotes something that tends to 0, here as x decreases to 1.) To prove this let $\epsilon > 0$, choose N so that $|c_n| < \epsilon$ for all $n > N$, and note that

$$x \sum_{n=2}^{N} \frac{1}{n^x \log(n)} \leq 2 \sum_{n=2}^{N} \frac{1}{n \log(n)}.$$

Now let $\alpha = \lim_{n \to \infty} \alpha_n$, so that $\alpha_n = \alpha + o(1)$. The above shows that

$$x \sum_{n=2}^{\infty} \frac{\alpha_n}{n^x \log(n)} = O(1) + (\alpha + o(1)) \log\left(\frac{1}{x-1}\right).$$

With our previous estimate this shows that

$$O(1) + \log\left(\frac{1}{x-1}\right) = O(1) + (\alpha + o(1)) \log\left(\frac{1}{x-1}\right),$$

hence $\alpha = 1$.

That finishes the outline of the solution to Exercise 24.8. Regarding why this is nothing close to a proof of the Prime Number Theorem we should note the following: The solution showed that if $\lim_{n \to \infty} c_n$ exists and

$$x \sum_{n=2}^{\infty} \frac{c_n}{n^x \log(x)} = O(1) + \log\left(\frac{1}{x-1}\right)$$

for $1 < x < 2$ then $c_n \to 1$. To show that we cannot remove the assumption that $\lim_{n \to \infty} c_n$ exists, consider a sequence like $c_n = (1 + (-1)^n)/2$. Thus the solution to the exercise tells us nothing about the *existence* of the limit in the Prime Number Theorem.

Chapter 25

Universal Covering Spaces

This chapter is a continuation of Chapter 19; our topic here is both more technical and more abstract than the results in that chapter. We will again assume that the reader has some familiarity with general topology, as we did in Section 16.4. In particular we will need the notion of "one-dimensional complex manifold" which was defined in that section (we may omit the words "one-dimensional", since these are the only complex manifolds we will be talking about). The results in this chapter will not be used elsewhere, but they're too interesting to omit.

We said a lot in Chapter 19 about holomorphic covering maps $p : \Omega \to D$, where Ω is simply connected. The reader may have wondered whether every connected open subset of the plane *has* a simply connected holomorphic covering space. We will show here that the answer is yes: We will show that there exists a simply connected complex manifold M and a holomorphic covering map from M onto D (Theorem 25.0), and then we will show that in fact M is conformally equivalent to $\mathbb{C}$ or Π^+ (see Theorem 25.3 and the comment following the statement of the theorem).

Theorem 25.0. *If D is a connected open subset of the plane then there exists a simply connected one-dimensional complex manifold M such that there exists a holomorphic covering map $p : M \to D$.*

The proof is not too hard, but it is slightly tricky and somewhat tedious, so we defer it to the end of the chapter.

Note that essentially all of our previous results on holomorphic covering maps remain true, with essentially the same proofs, in the context of complex manifolds. In particular Theorem 19.4.0 shows that M is unique up to conformal equivalence.

The manifold M is known as the *universal covering space* of D. This is because M in some sense "contains" any other covering space for D. Theorem 25.1 and Corollary 25.2 explain what we mean by this:

Theorem 25.1. *Suppose that M and N are one-dimensional complex manifolds, D is a connected open set in the plane, and that $p : M \to D$ and $q : N \to D$ are holomorphic covering maps. If M is simply connected then there exists a holomorphic covering map $r : M \to N$ such that*

$$p = q \circ r$$

and $\mathrm{Aut}(M, r) \subset \mathrm{Aut}(M, p)$. *If $a \in D$, $b \in M$ and $c \in N$ satisfy $p(b) = a$ and $q(c) = a$ then there is a unique choice of r which satisfies $r(p) = q$.*

Proof. The existence of a holomorphic map r with $p = q \circ r$ and the uniqueness statement are immediate from Theorem 19.0.5. We leave it to you to show that r is a covering map; see the exercises at the end of the chapter. □

It follows that any covering space for D is a "quotient space" of the simply connected covering space. We are not going to prove this in detail; a brief explanation follows:

In general, if X is simply connected and $p : X \to Y$ is a covering map one defines $X/\mathrm{Aut}(X, p)$ to be the set of equivalence classes of elements of X, where $x_1, x_2 \in X$ are equivalent if $x_1 = \phi(x_2)$ for some $\phi \in \mathrm{Aut}(X, p)$. Say $Q : X \to X/\mathrm{Aut}(X, p)$ is the quotient map, taking an element of X to its equivalence class: $Q(x) = [x]$. One defines a topology on $X/\mathrm{Aut}(X, p)$ by saying that $E \subset X/\mathrm{Aut}(X, p)$ is open if and only if $Q^{-1}(E)$ is open. It is very easy to see that there is a unique map $\tilde{p} : X/\mathrm{Aut}(X, p) \to Y$ such that $p = \tilde{p} \circ Q$, and then it is not too hard to show, using the fact that p is a covering map, that $\tilde{p}$ is actually a homeomorphism. If X is a one-dimensional complex manifold and p is a holomorphic covering map then $X/\mathrm{Aut}(X, p)$ is also a one-dimensional complex manifold in a natural way, and $\tilde{p}$ turns out to be holomorphic. If one worked out the details in all this then one would obtain the following:

Corollary 25.2. *If D is a connected open subset of the plane and (M, p) is the universal covering space of D then every holomorphic covering space of D is conformally equivalent to M/G for some subgroup G of $\mathrm{Aut}(M, p)$.*

Hence the word "universal".

The Uniformization Theorem is a fundamentally important result here. It is also much deeper than the results we have proved about covering maps and manifolds; we shall not attempt to give a proof:

25. Universal Covering Spaces

Uniformization Theorem. *If M is a simply connected one-dimensional complex manifold then M is conformally equivalent to $\mathbb{C}_\infty$, $\mathbb{C}$, or Π^+.*

It is not hard to deduce from this that for most connected open sets D in the plane there exists a holomorphic covering map $p : \Pi^+ \to D$:

Theorem 25.3. *If D is a connected open subset of $\mathbb{C}$ and $\mathbb{C} \setminus D$ contains more than one point then there is a holomorphic covering map $p : \Pi^+ \to D$.*

Note. The identity is a holomorphic covering map from $\mathbb{C}$ to $\mathbb{C}$ and the exponential is a holomorphic covering map from $\mathbb{C}$ onto $\mathbb{C} \setminus \{0\}$, so it follows from Theorem 25.3 that every connected open set in the plane has either $\mathbb{C}$ or Π^+ for its universal covering space.

One can think of Theorem 25.3 as a massive extension of the Riemann Mapping Theorem, which is to say that the latter theorem is a trivial corollary of the former: Note first that if $D \subset \mathbb{C}$ is simply connected and $D \neq \mathbb{C}$ then $\mathbb{C} \setminus D$ contains at least two points. Hence Theorem 25.3 shows that there exists a holomorphic covering map $p : \Pi^+ \to D$. Since D is simply connected, Theorem 19.4.0 now shows that p is actually a conformal equivalence.

Theorem 25.3 is itself a trivial corollary of the Uniformization Theorem: By Theorem 25.0 we know that there exists a holomorphic covering map $p : \Omega \to D$, where Ω is either $\mathbb{C}_\infty$, $\mathbb{C}$ or Π^+. We need to show that the first two cases are impossible. It is clear that $\Omega \neq \mathbb{C}_\infty$, since D is not compact. And if $\Omega = \mathbb{C}$ then p is an entire function such that $\mathbb{C} \setminus p(\mathbb{C})$ contains two points; the Little Picard Theorem (Theorem 20.0) shows that p is constant, a contradiction. See also Exercise 19.12 at the end of Chapter 19.

So the Riemann Mapping Theorem is contained in Theorem 25.3; in fact the proof of Theorem 25.3 below (without using the Uniformization Theorem) is inspired by the proof of the Riemann Mapping Theorem in Chapter 9. (Exercise 19.3 in Section 19.0 gives a hint that the main idea in the proof of the Riemann Mapping Theorem might extend to the present context.)

A final comment before we get to work: The results in this chapter show that Π^+ is the universal covering space for "almost all" connected open sets in the plane, the only exceptions being $\mathbb{C}$ and $\mathbb{C}\setminus\{z_0\}$, which have $\mathbb{C}$ for their universal covering space. There do exist holomorphic covering maps from $\mathbb{C}$ onto various compact one-dimensional complex manifolds; for example for each $a \in \mathbb{C} \setminus \mathbb{R}$ there exists such a map p with $\mathrm{Aut}(\mathbb{C}, p)$ equal to the set of all translations by elements of the additive subgroup of $\mathbb{C}$ generated by 1 and a. On the other hand, $\mathbb{C}_\infty$ is quite boring, as simply connected covering spaces go:

- **Exercise 25.1.** Suppose that N is a one-dimensional complex manifold and $p : \mathbb{C}_\infty \to N$ is a holomorphic covering map. Show that N is conformally equivalent to $\mathbb{C}_\infty$. (It is enough (why?) to show that $\mathrm{Aut}(\mathbb{C}_\infty, p)$ is trivial (that is, that it contains just one element). Consider Corollary 19.0.2.1 and Theorem 8.1.1.)

We begin the proof of Theorem 25.3 with three lemmas, of some interest in themselves; the first gives a way to recognize a holomorphic covering map, the second says that under certain conditions limits of covering maps are covering maps, and the third says that square roots of covering maps are covering maps:

Lemma 25.4. *Suppose that M is a simply connected complex manifold, Ω is a connected open subset of $\mathbb{C}$, and $p : M \to \Omega$ is a holomorphic covering map; suppose that G is a subgroup of $\mathrm{Aut}(M, p)$. If $f : M \to \mathbb{C}$ has the property that $f(z') = f(z)$ if and only if $z' = g(z)$ for some $g \in G$ then f is a covering map onto $D = f(M)$, with $\mathrm{Aut}(M, f) = G$.*

Proof. Suppose that $w \in D$; we need to show that w has a neighborhood which is elementary for f. Choose $z \in M$ with $w = f(z)$.

Now let V be a neighborhood of $p(z)$ which is elementary for p. This says that z has a neighborhood O such that $p|_O$ is a homeomorphism of O onto V and

$$p^{-1}(V) = \bigcup_{g \in \mathrm{Aut}(M,p)} g(O),$$

as in the definition of "elementary for p". In particular $O \cap g(O) = \emptyset$ for every $g \in \mathrm{Aut}(M, p)$ except the identity; this shows that $f|_O$ is one-to-one. Theorem 7.5 now shows that $f|_O$ is a homeomorphism of O onto $W = f(O)$, and it is easy to verify that

$$f^{-1}(W) = \bigcup_{g \in G} g(O);$$

hence W is elementary for f. $\square$

Lemma 25.5. *Suppose that M is a simply connected complex manifold, Ω is a connected open subset of $\mathbb{C}$, and $p : M \to \Omega$ is a holomorphic covering map; suppose $z_0 \in M$. Suppose that D_n is a connected open subset of $\mathbb{C}$ and $f_n : M \to D_n$ is a holomorphic covering map with $\mathrm{Aut}(M, f_n) = G_n \subset \mathrm{Aut}(M, p)$ and $f_n(z_0) = 0$ for $n = 1, 2, \ldots$. If $f_n \to f$ uniformly on compact subsets of M and f is not constant then f is a covering map from M onto $D = f(M)$, with $\mathrm{Aut}(M, f) \subset \mathrm{Aut}(M, p)$.*

25. Universal Covering Spaces

Proof. Suppose that $z \in M$ and $f(z) = 0$. Since f is nonconstant and $f_n(z_0) = 0$ for all n, it follows from Exercise 5.7 in Chapter 5 that there exists a sequence $(z_n) \subset M$ with $f_n(z_n) = 0$ and $z_n \to z$. Now there exists $g_n \in G_n$ with $z_n = g_n(z_0)$. Since $g_n(z_0) \to z$ and $\{g(z_0) : g \in \mathrm{Aut}(M, p)\}$ is discrete, there must exist N and $g \in \mathrm{Aut}(M, p)$ such that $g_n = g$ for all $n > N$. In particular $z = g(z_0)$.

Now suppose that $g \in \mathrm{Aut}(M, p)$ and $f(g(z_0)) \neq 0$. Then there exists N such that $f_n(g(z_0)) \neq 0$ for all $n > N$, so that $g \notin G_n$ for all $n > N$.

So for every $g \in \mathrm{Aut}(M, p)$ either there exists N such that $g \in G_n$ for all $n > N$ or there exists N such that $g \notin G_n$ for all $n > N$. Let G be the set of elements of $\mathrm{Aut}(M, p)$ satisfying the first condition. Then G is a subgroup of $\mathrm{Aut}(M, p)$, and it is clear that $f \circ g = f$ for all $g \in G$.

Suppose now that $f(z) = f(z')$. As before we obtain $z_n \to z$ and $z'_n \to z'$ such that $f_n(z_n) = f(z) = f_n(z'_n)$ for all sufficiently large n; it follows that $z'_n = g_n(z_n)$ for some $g_n \in G_n$; it follows that there exists N such that $g_n = g$ for all $n > N$, and hence that $g \in G$ and $z' = g(z)$.

So we have $f(z') = f(z)$ if and only if $z' = g(z)$ for some $g \in G$; thus the previous lemma shows that f is a covering map with $\mathrm{Aut}(M, f) = G$. □

Lemma 25.6. *Suppose that M is a simply connected complex manifold, D is a connected open subset of $\mathbb{C} \setminus \{0\}$, and $f : M \to D$ is a holomorphic covering map. If $g : M \to \mathbb{C}$ is holomorphic and $g^2 = f$ then g is a covering map onto $g(M)$ and $\mathrm{Aut}(M, g) \subset \mathrm{Aut}(M, f)$.*

Proof. Let $S = \{w \in \mathbb{C} : w^2 \in D\}$, suppose $z \in M$ and let N be the connected component of S containing $g(z)$. Then $s : N \to D$ is a covering map if $s(w) = w^2$; the result now follows from Theorem 25.1 together with the uniqueness statement in Theorem 19.0.5. □

Proof of Theorem 25.3. We may assume that $D \subset \mathbb{C} \setminus \{0, 1\}$. Theorem 25.0, to be proved below, shows that there exists a simply connected complex manifold M and a holomorphic covering map $p : M \to D$; it is enough to show that M is conformally equivalent to $\mathbb{D}$ (and hence to Π^+).

Fix $z_0 \in M$, and let $\mathcal{F}$ denote the family of all holomorphic covering maps f from M onto an open subset of $\mathbb{D}$ such that $f(z_0) = 0$ and $\mathrm{Aut}(M, f) \subset \mathrm{Aut}(M, p)$. As in the proof of the Riemann Mapping Theorem (Theorem 9.4.2) we want to show that there exists $f \in \mathcal{F}$ which maximizes $|f'(z_0)|$ and that this f is actually a conformal equivalence of M and $\mathbb{D}$.

Except of course that in the present context there is no such thing as $|f'(z_0)|$ for $f \in \mathcal{F}$. As a substitute, we note that the definition of "complex manifold" shows that there exists a holomorphic map $c : \mathbb{D} \to M$ such that

$c(0) = z_0$ and c is a homeomorphism of $\mathbb{D}$ onto some neighborhood of z_0; we fix such a c and now look for $f \in \mathcal{F}$ which maximizes $\rho(f) = |(f \circ c)'(0)|$.

First we need to show that $\mathcal{F}$ is nonempty. We saw in Theorem 18.10 that there exists a holomorphic covering map from Π^+ onto $\mathbb{C} \setminus \{0, 1\}$; hence there is a holomorphic covering map $\Lambda : \mathbb{D} \to \mathbb{C} \setminus \{0, 1\}$. We may assume that $\Lambda(0) = p(z_0)$. Let N be the connected component of $\Lambda^{-1}(D)$ which contains 0 and set $q = \Lambda|_N$; then $q : N \to D$ is a covering map. Theorem 25.1 now shows that there is a holomorphic covering map $F : M \to N$ so that $p = q \circ F$, $F(z_0) = 0$ and $\mathrm{Aut}(M, F) \subset \mathrm{Aut}(M, p)$; thus $F \in \mathcal{F}$.

The Schwarz Lemma shows that $\rho(f) \leq 1$ for all $f \in \mathcal{F}$, so that $m = \sup\{\rho(f) : f \in \mathcal{F}\}$ is finite. The fact that F is a covering map shows that F is one-to-one in some neighborhood of z_0; hence $(F \circ c)'(0) \neq 0$, so that $m > 0$. By Montel's Theorem there exists a sequence $(f_n) \subset \mathcal{F}$ such that $\rho(f_n) \to m$ and (f_n) is uniformly convergent on compact subsets of M; say $f_n \to f$. It follows that $|(f \circ c)'(0)| = m > 0$, so that f is nonconstant; now Lemma 25.5 above shows that f is a covering map onto some open set in $\mathbb{C}$ and $\mathrm{Aut}(M, f) \subset \mathrm{Aut}(M, p)$. The Maximum Modulus Theorem shows that in fact $f(M) \subset \mathbb{D}$, so $f \in \mathcal{F}$.

The same argument as in the proof of the Riemann Mapping Theorem shows that in fact $f(M) = \mathbb{D}$: If, on the other hand, $\alpha \in \mathbb{D} \setminus f(M)$ then $\phi_\alpha \circ f$ has no zero; since M is simply connected, there exists a holomorphic square root $(\phi_\alpha \circ f)^{1/2}$. Note that Lemma 25.6 shows that $(\phi_\alpha \circ f)^{1/2}$ is a covering map onto its range, and that $\mathrm{Aut}(M, (\phi_\alpha \circ f)^{1/2}) \subset \mathrm{Aut}(M, f) \subset \mathrm{Aut}(M, p)$. Thus there exists β such that if $g = \phi_\beta \circ (\phi_\alpha \circ f)^{1/2}$ then $g \in \mathcal{F}$. As in the proof of Theorem 9.4.2, Lemma 9.4.1 shows that $H_{g \circ c} > H_{f \circ c}$, and in particular that $\rho(g) > \rho(f)$, contradicting the maximality of $\rho(f)$.

So $f : M \to \mathbb{D}$ is a covering map; since $\mathbb{D}$ is simply connected, it follows from Exercise 19.12 that f is in fact a conformal equivalence. $\square$

Note. We have been somewhat glib about complex manifolds, assuming without proof that various results for holomorphic functions in an open subset of the plane are still valid for holomorphic functions on a manifold. In most cases the result in question is indeed a trivial consequence of the definitions. For example:

Proposition 25.7. *Suppose that M is a one-dimensional complex manifold and $f_n : M \to \mathbb{C}$ is holomorphic for $n = 1, 2, \ldots$. If $f_n \to f$ uniformly on compact subsets of M then f is holomorphic.*

Proof. By definition (see Section 16.4) we may assume that (O, c) is a chart on M and we need to show that $f \circ c^{-1}$ is holomorphic in $V = c(O)$. But we know that $f_n \circ c^{-1}$ is holomorphic in V and the fact that $c : O \to V$

is a homeomorphism shows that $f_n \circ c^{-1} \to f \circ c^{-1}$ uniformly on compact subsets of V. □

There is however one bit of complex analysis on M used in the proof such that deducing it from the corresponding bit of complex analysis in the plane is not *quite* this simple. See if you can identify the gap (and if so see whether you can fix it) before looking at the exercises at the end of the chapter.

We now turn to the proof of Theorem 25.0. The construction below may seem somewhat curious at first glance; by way of motivation, let us assume that we *have* a simply connected M and a covering map $p : M \to D$ and see if we can somehow say what the points of M are, or what they correspond to in some manner, in terms of D.

Of course the first thing we notice is that the elements of M are not going to be the same as the points of D; there will typically be many elements of M mapped to a given point of D by p. Is there anything to do with M that *does* correspond to something to do with D in a one-to-one fashion? Yes! By now we realize that sort of the whole point to covering maps is that we can "lift" a curve in D uniquely to a curve in M, at least if we specify the initial points of the curves: If $a \in D$ and $\tilde{a} \in M$ satisfy $p(\tilde{a}) = a$ then there is a one-to-one correspondence between curves γ in D starting at a and curves $\tilde{\gamma}$ in M starting at $\tilde{a}$, given by $p \circ \tilde{\gamma} = \gamma$.

This is progress. But we want to figure out what the elements of M should be, not what the curves in M are. A point in M is not the same thing as a curve in M starting at $\tilde{a}$. But the points in M *are* in a natural one-to-one correspondence with equivalence classes of curves in M starting at $\tilde{a}$, if we say that two curves are equivalent if they end at the same point. And, since M is supposed to be simply connected, we know that two curves in M starting at $\tilde{a}$ will end at the same point if and only if their images in D under our covering map are path-homotopic! So the elements of M are in a natural one-to-one correspondence with the path-homotopy equivalence classes of curves in D starting at a. Since it doesn't matter what the elements of M actually *are* and they're in one-to-one correspondence with these equivalence classes, we may as well define M to *be* the set of all these equivalence classes, and then see whether we can make everything work. (This is why we need to use topological spaces in this chapter: When we define M to be the set of those equivalence classes it is not at all clear how to define a metric on M, but it turns out that there is a simple definition of a suitable topology.)

The actual proof of the theorem starts here (timid readers may want to skip the rest of the chapter):

Proof (Theorem 25.0).

Suppose that D is a connected open subset of the plane. Fix $a \in D$, and let C be the collection of all paths in D from a to some point of D. That is, C is the collection of all continuous $\gamma : [0,1] \to D$ with $\gamma(0) = a$. Define an equivalence relation $\equiv$ on C by saying that $\gamma_1 \equiv \gamma_2$ if and only if γ_1 and γ_2 are path-homotopic. Let $M = C/\equiv$ (that is, M is the set of all $\equiv$-equivalence classes). If $\gamma \in C$ we will let $[\gamma]$ denote the element of M with $\gamma \in [\gamma]$.

It is clear how $p : M \to D$ should be defined:
$$p([\gamma]) = \gamma(1).$$

We need to show that p is well defined, but this is trivial: If $[\gamma_1] = [\gamma_2]$ then $\gamma_1 \equiv \gamma_2$, and by definition this implies that $\gamma_1(1) = \gamma_2(1)$.

Suppose that $\gamma_j : [0,1] \to D$ is continuous for $j = 1, 2$ and $\gamma_1(1) = \gamma_2(0)$. Then we define a curve $\gamma_1 + \gamma_2 : [0,1] \to D$ by

$$(\gamma_1 + \gamma_2)(s) = \begin{cases} \gamma_1(2s) & (0 \leq s \leq 1/2), \\ \gamma_2(2s-1) & (1/2 < s \leq 1). \end{cases}$$

So $\gamma_1 + \gamma_2$ is just the curve obtained by joining γ_1 and γ_2 at the point $\gamma_1(1) = \gamma_2(0)$, parametrized so as to have parameter interval $[0,1]$. (Note that this is not quite the same thing as the cycle $\gamma_1 \dotplus \gamma_2$, although the two objects are certainly related. If you think they really *are* the same thing you might consider the fact that the definitions imply that $\dotplus$ is commutative, while $+$ is not; in fact if γ_1 and γ_2 are as above then $\gamma_2 + \gamma_1$ is typically undefined.)

Now if $[\gamma] \in M$ and $\gamma_2 : [0,1] \to D$ satisfies $\gamma_2(0) = \gamma(1)$ we define
$$[\gamma] + \gamma_2 = [\gamma + \gamma_2].$$

- **Exercise 25.2.** Show that $[\gamma] + \gamma_2$ is well defined. (Make certain to begin with a statement of exactly what needs to be proved!)

- **Exercise 25.3.** Suppose that $\gamma_1 \in C$ and define $\gamma_2 : [0,1] \to D$ by $\gamma_2(s) = \gamma_1(1)$ for all s. Show that $\gamma_1 + \gamma_2 \equiv \gamma_1$.

Now we can define the topology on M. Suppose that $[\gamma] \in M$, so that γ is a path in D from a to some point $b \in D$. For every $r > 0$ such that $D(b,r) \subset D$ and every $z \in D(b,r)$ let γ_z be the path
$$\gamma_z = \gamma + [b,z],$$
where $[b,z]$ is the straight line segment from b to z, as usual, and let
$$O([\gamma], r) = \{\, [\gamma_z] : z \in D(b,r) \,\}.$$

25. Universal Covering Spaces

- **Exercise 25.4.** Show that $O([\gamma], r)$ is well defined. (State clearly exactly what needs to be proved! Once you have figured that out, actually proving it should be easy.)

- **Exercise 25.5.** Show that the collection of all $O([\gamma], r)$, for $[\gamma] \in M$ and $r > 0$ such that $D(\gamma(1), r) \subset D$, forms a basis for a topology on M. (One of the previous exercises is needed to show that $[\gamma] \in O([\gamma], r)$.)

When discussing M we will assume the topology is as defined in the previous exercise. The next few exercises enumerate some basic properties of this topology; the proofs should be straightforward:

- **Exercise 25.6.** Show that $p : M \to D$ is continuous.

- **Exercise 25.7.** Show that $p : M \to D$ is a covering map.

- **Exercise 25.8.** Suppose that $[\gamma] \in M$, $b = \gamma(1)$, and $D(b, r) \subset D$. Suppose that $\gamma_2 : [0, 1] \to D(b, r)$ is continuous and $\gamma_2(0) = b$. Show that $[\gamma] + \gamma_2 \in O([\gamma], r)$.

- **Exercise 25.9.** Suppose that X is a topological space, and for each $x \in X$ we are given $\gamma_x \in C$. Suppose that $\gamma_x(s)$ depends continuously on $(x, s) \in X \times [0, 1]$. Define $G : X \to M$ by $G(x) = [\gamma_x]$. Show that G is continuous.

And now we can make M into a complex manifold: Suppose that $[\gamma] \in M$, $\gamma(1) = b$, $D(b, r) \subset D$, and as before define $\gamma_z = \gamma + [b, z]$ for $z \in D(b, r)$. Define $c : O([\gamma], r) \to \mathbb{C}$ by

$$c([\gamma_z]) = z.$$

- **Exercise 25.10.** Show that each $(O([\gamma], r), c)$ is a chart on M, and that the collection of all such charts is an atlas on M which makes M into a complex manifold. (See Section 16.4 for the definitions.)

When we speak of M as a complex manifold we will be referring to the atlas defined above.

- **Exercise 25.11.** Show that $p : M \to D$ is holomorphic.

So far we have shown that M is a one-dimensional complex manifold and that $p : M \to D$ is a holomorphic covering map. To finish the proof of Theorem 25.0 we need to show that M is simply connected. There is an obvious approach to this, but it doesn't quite work without a little trick.

First some more notation: If $\gamma \in C$ and $0 \leq t \leq 1$ then $\gamma|_{[0,t]} \in C$ is the curve defined by

$$\gamma|_{[0,t]}(s) = \gamma(ts).$$

So $\gamma|_{[0,t]}$ is not quite the restriction of γ to the interval $[0,t]$; rather it is this restriction reparametrized to have parameter interval $[0,1]$.

There is some possible confusion when we talk about curves in M, since the points of M are themselves (equivalence classes of) curves. We will attempt to minimize the confusion by using the letter Γ for curves in M and γ for curves in D.

Now the obvious way to show that M is simply connected is this: Suppose that $\Gamma : [0,1] \to M$ is a closed curve. For every $s \in [0,1]$ choose $\gamma_s \in C$ so that $\gamma_s \in \Gamma(s)$. Now for $0 \leq t \leq 1$ define

$$\Gamma_t(s) = [\gamma_s|_{[0,t]}].$$

At first blush it may seem that the family (Γ_t) shows that Γ is null-homotopic, as required: We have $\Gamma_1 = \Gamma$ and $\Gamma_0(s) = \Gamma_0(0)$ for every s (in fact $\Gamma_0(s) = [\gamma]$, where $\gamma(\alpha) = a$ for all α).

The problem with the simple solution above is that Γ_t does not depend continuously on t, and in fact Γ_t itself is not even continuous for $t \in (0,1)$. We chose $\gamma_s \in \Gamma(s)$ "at random". So γ_s is only determined up to path homotopy, and it would be an amazing coincidence if γ_s turned out to depend continuously on s.

It seems that if we are going to have a chance of showing that γ_s depends continuously on s we should probably choose a specific $\gamma_s \in \Gamma(s)$ somehow, instead of choosing γ_s "at random". Now, however we choose $\gamma_s \in \Gamma(s)$, it is going to be a curve in D from a to $p(\Gamma(s))$. There is a simple way to define a specific family of curves (γ_s) such that each γ_s is a curve in D from a to $p(\Gamma(s))$: Fix $\gamma \in \Gamma(0)$, define $c : [0,1] \to D$ by

$$c(s) = p(\Gamma(s))$$

(so that c is just the curve traced by the "endpoint" of Γ), and define

$$\gamma_s = \gamma + c|_{[0,s]}.$$

That is, γ_s follows γ from a to $p(\Gamma(0))$ and then follows c from there to $p(\Gamma(s))$. The miracle is that if we define γ_s this way then it turns out that $\gamma_s \in \Gamma(s)$. This is not obvious, or at least not obvious to me, but it is not hard to prove:

Lemma 25.4. *If $\Gamma : [0,1] \to M$ is continuous then there exists $\gamma_s \in \Gamma(s)$ ($s \in [0,1]$) such that $\gamma_s(\alpha)$ depends continuously on $(s, \alpha) \in [0,1] \times [0,1]$.*

Proof. Fix $\gamma \in \Gamma(0)$ and define c and γ_s as above. The fact that p is continuous shows that c is continuous, and hence that $\gamma_s(\alpha)$ depends continuously on (s, α). To show that $\gamma_s \in \Gamma(s)$ we use the fact that p is a covering map:

25. Universal Covering Spaces

Define $\phi : [0,1] \to M$ by $\phi(s) = [\gamma_s]$. Note that Exercise 25.9 shows that ϕ is continuous. Now, although γ_0 is not quite equal to γ (why not?), Exercise 25.3 shows that $\gamma_0 \equiv \gamma$, and hence that $\phi(0) = \Gamma(0)$. Also for every s we have
$$p(\phi(s)) = \gamma_s(1) = p(\Gamma(s)).$$
So the uniqueness assertion in Theorem 19.0.1(i) shows that $\phi(s) = \Gamma(s)$ for all s; hence $\gamma_s \in \Gamma(s)$. □

Now we can show that M is simply connected. Suppose that $\Gamma : [0,1] \to M$ is continuous and $\Gamma(1) = \Gamma(0)$. Choose $\gamma_s \in \Gamma(s)$ as in the lemma.

The obvious proof mentioned above that Γ can be shrunk to a point still does not quite work, because $\gamma_1 \neq \gamma_0$, so if we attempt to apply the obvious argument we will get continuous but nonclosed curves Γ_t. This is not hard to fix. Although $\gamma_1 \neq \gamma_0$, the fact that $\gamma_1 \in \Gamma(1) = \Gamma(0)$ and $\gamma_0 \in \Gamma(0)$ shows that $\gamma_1 \equiv \gamma_0$. Let $(\tilde{\gamma}_s)_{0 \le s \le 1}$ be a path homotopy of γ_1 and γ_0, with $\tilde{\gamma}_0 = \gamma_1$ and $\tilde{\gamma}_1 = \gamma_0$. Now patch the two families together: For $0 \le s \le 1$ define
$$\gamma_s^* = \begin{cases} \gamma_{2s} & (0 \le s \le 1/2), \\ \tilde{\gamma}_{2s-1} & (1/2 < s \le 1). \end{cases}$$
Then $\gamma_1^* = \gamma_0^*$, and it is clear that $\gamma_s^*|_{[0,t]}(\alpha)$ depends continuously on $(s, t, \alpha) \in [0,1]^3$, so that if we define $\Gamma_t : [0,1] \to M$ for $t \in [0,1]$ by
$$\Gamma_t(s) = [\gamma_s^*|_{[0,t]}]$$
then Exercise 25.9 shows that $\Gamma_t(s)$ depends continuously on $(s,t) \in [0,1]^2$. Each Γ_t is a closed curve and $\Gamma_0(s) = \Gamma_0(0)$ for all s, so we have shown that Γ_1 is null-homotopic.

No, we are not *quite* done, because $\Gamma_1 \neq \Gamma$. But the fact that $\gamma_s^* \equiv \gamma_1^*$ for $1/2 \le s \le 1$ shows that
$$\Gamma_1(s) = \begin{cases} \Gamma(2s) & (0 \le s \le 1/2), \\ \Gamma(1) & (1/2 < s \le 1). \end{cases}$$
Hence an argument as in Exercise 25.3 shows that Γ is homotopic to Γ_1, and hence Γ is null-homotopic.

Of course, having shown that every closed curve in M is null-homotopic, we also need to show that M is path-connected. Suppose that $[\gamma] \in M$. Define $\alpha \in C$ by $\alpha(s) = a$ for all s. We leave it to you to show that there is a curve in M from $[\alpha]$ to $[\gamma]$ (this is very simple, using one of the constructions above). □

Exercises

We begin with some exercises regarding Theorem 25.1 (in the exercises you should assume the notation is as in the statement of the theorem, and in particular that $a \in D$, $b \in M$ and $c \in N$ satisfy $p(b) = a$ and $r(b) = c$, so that $q(c) = a$):

25.12. Show that $r(M) = N$.

One possibility is this: Suppose that $z \in N$. Let γ be a curve in N from c to z. Then $q \circ \gamma$ is a curve in D from a to $q(z)$, and hence there exists a unique curve $\tilde{\gamma}$ in M from b to some point w such that $p \circ \tilde{\gamma} = q \circ \gamma$. Hence $q \circ r \circ \tilde{\gamma} = q \circ \gamma$; use one of our results on covering maps to conclude that $r(w) = z$.

25.13. Show that every point of D has a (connected!) neighborhood which is elementary for p and also elementary for q.

In fact it is very easy to show that if $w \in D$ and O is a neighborhood of w which is elementary for p then O contains a neighborhood of w which is elementary for p and also elementary for q. If that's too easy you could prove the following statement, from which the exercise follows: If $\pi : S \to D$ is a covering map and O is a simply connected neighborhood of $w \in D$ then O is elementary for π.

25.14. Show that r is a covering map.

You could do this: Fix $z \in N$. You need to show that z has a neighborhood which is elementary for r. Using the previous exercise, let O be a neighborhood of $q(z)$ which is elementary for p and elementary for q. Write

$$p^{-1}(O) = \bigcup_{\alpha \in A} U_\alpha, \quad q^{-1}(O) = \bigcup_{\beta \in B} V_\beta,$$

as in the definition of "elementary neighborhood". Fix $\beta \in B$ with $z \in V_\beta$.

Now you can show that V_β is an elementary neighborhood of z, by showing that there exists $A' \subset A$ such that

$$r^{-1}(V_\beta) = \bigcup_{\alpha \in A'} U_\alpha.$$

Note first that $p^{-1}(O) = r^{-1}(q^{-1}(O))$, so that $r^{-1}(V_\beta) \subset p^{-1}(O)$. Hence you need only show that $U_\alpha \subset r^{-1}(V_\beta)$ for every $\alpha \in A$ such that $U_\alpha \cap r^{-1}(V_\beta) \neq \emptyset$. This follows (how?) from the fact that $r(U_\alpha)$ is connected.

25.15. Show that $\text{Aut}(M, r) \subset \text{Aut}(M, p)$. (This is probably the easiest exercise in the book.)

We stated above that there was a gap in the proof of Theorem 25.3. The problem is this: We used Montel's Theorem on M; in order for the proof of Montel's Theorem to work we need to know that M has a countable dense subset (or what amounts to the same thing here, that M has an exhaustion by a *sequence* of compact sets; see the two proofs of Montel's Theorem in Chapter 9).

There are various ways to fix this. The most elegant would take some explanation: We could define a topology on $H(M)$ (the "compact-open" topology, also known as "the topology of uniform convergence on compact sets") without worrying about whether this topology was metrizable, give a version of Montel's Theorem characterizing the compact subsets of $H(M)$ in this topology, and then use that result, perhaps talking about "nets" instead of sequences.

Or we could simply show that M does in fact have a countable dense subset:

25.16. Suppose that M is a (connected!) one-dimensional complex manifold, Ω is an open subset of the plane, and $p : M \to \Omega$ is a covering map. Then M has a countable dense subset

Hint: Fix $a \in M$ and let $b = p(a)$. Let Q be a countable dense subset of Ω with $b \in Q$. Let C_Q be the collection of all *polygonal* paths $\gamma : [0, 1] \to \Omega$ such that $\gamma(0) = b$ and every vertex of γ lies in Q.

Show that C_Q is countable. Now for every $\gamma \in C_Q$ let $\tilde{\gamma} : [0, 1] \to M$ be the unique curve with $\tilde{\gamma}(0) = a$ and $p \circ \tilde{\gamma} = \gamma$. Define $\tilde{Q} = \{\tilde{\gamma}(1) : \gamma \in C_Q\}$. Show that $\tilde{Q}$ is dense in M.

The next exercise gives a sketch of a proof of Theorem 25.3 which is somewhat similar to the proof in the text, but which does not require Theorem 25.0. It is probably harder than most exercises in the book, since a large part of the argument is left to you. Note that you will need to be familiar with various results in Chapter 16, including results in Section 16.4 for which most of the proof was left to the reader:

25.17. Give a proof of Theorem 25.3 without using Theorem 25.0, as follows: Fix $z_0 \in D$ and choose $r > 0$ so that $D_0 = D(z_0, r) \subset D$. If (f, D_0) is a function element which admits unrestricted continuation in D, define the "extended range" of (f, D_0) to be the union of the sets $g(\Omega)$, where (g, Ω) is a function element which arises in the continuation of (f, D_0) along some curve in D. Let $\mathcal{F}$ be the family of all function elements (f, D_0) that admit unrestricted continuation in D, have the property that the extended range of (f, D_0) is a subset of $\mathbb{D}$ and $f(z_0) = 0$, and are one-to-one in the following sense: If (g, Ω) and $(\tilde{g}, \tilde{\Omega})$ are two function elements which arise by continuation of (f, D_0) along curves, $z \in \Omega$, $\tilde{z} \in \tilde{\Omega}$, and $g(z) = \tilde{g}(\tilde{z})$, then $z = \tilde{z}$.

Since there exists a holomorphic covering map from $\mathbb{D}$ onto $\mathbb{C} \setminus \{0, 1\}$ and local inverses for holomorphic covering maps admit unrestricted continuation, we see that $\mathcal{F}$ contains nonconstant functions, so that $m = \sup\{|f'(z_0)| : (f, D_0) \in \mathcal{F}\} > 0$. (Of course you need to explain why a local inverse for that covering map is one-to-one in the required sense.) It is easy to show that m is finite; now it follows from Montel's Theorem that there exists $(F, D_0) \in \mathcal{F}$ with $|F'(z_0)| = m$. (This requires some argument: It is easy to see that there exists a sequence $((f_n, D_0)) \subset \mathcal{F}$ such that $f_n \to F$ in $H(D_0)$, where $|F'(z_0)| = m$. Now showing that $(F, D_0) \in \mathcal{F}$ may take some thought — this follows from an argument which is at least somewhat analogous to Exercise 9.4 in Section 9.3.)

Now show that $|F'(z_0)| = m$ implies that the extended range of (F, D_0) is all of $\mathbb{D}$, and conclude from this that there exists a covering map from $\mathbb{D}$ onto D (you could imitate the proof of Theorem 16.4.3 or perhaps use Theorem 16.4.3 itself; in either case you probably need to *give* a proof of that theorem).

Chapter 26

Cauchy's Theorem for Nonholomorphic Functions

In this chapter we give versions of Cauchy's Theorem and the Cauchy Integral Formula that are valid for functions which are not holomorphic (the formulas contain "error terms" measuring how far from holomorphic the function is).

We will use Green's Theorem. Suppose that $\gamma : [a,b] \to \mathbb{R}^2$ is smooth and write $\gamma(t) = (\gamma_1(t), \gamma_2(t))$. Recall the definitions

$$\int_\gamma M(x,y)\, dx = \int_a^b M(\gamma_1(t), \gamma_2(t))\gamma_1'(t)\, dt$$

and

$$\int_\gamma N(x,y)\, dy = \int_a^b N(\gamma_1(t), \gamma_2(t))\gamma_2'(t)\, dt,$$

valid for $M, N \in C(\gamma^*)$. Supposing that D is an open set bounded by finitely many smooth closed curves, we say that the boundary ∂D is positively oriented if $\text{Ind}(\partial D, z) = 1$ for all $z \in D$ and $\text{Ind}(\partial D, z) = 0$ for all $z \in \mathbb{R}^2 \setminus \overline{D}$. (Below we will assume without comment that the boundary is positively oriented.)

If D is an open set in the plane we say that $f \in C^1(D)$ if f, $\partial f/\partial x$, $\partial f/\partial y \in C(D)$, and we say that $f \in C^1(\overline{D})$ if f and its first-order partial derivatives extend continuously to $\overline{D}$.

435

Various calculus books give versions of Green's Theorem with various hypotheses on the smoothness of the boundary. We shall refer to the following version as "Green's Theorem":

Green's Theorem. *Suppose that D is a bounded open set in the plane and ∂D consists of finitely many smooth closed curves. If M, $N \in C^1(\overline{D})$ then*

$$\int_{\partial D}(M\,dx + N\,dy) = \iint_D \left(\frac{\partial N}{\partial x} - \frac{\partial M}{\partial y}\right) dx\,dy.$$

Note that although this is typically stated only for real-valued functions M and N in calculus books, the extension to complex-valued functions is trivial by linearity.

Now, informally we might say that $z = x + iy$, hence $dz = dx + i\,dy$, and so if f is continuous on γ^* we have

$$\int_\gamma f\,dz = \int_\gamma (f\,dx + if\,dy).$$

We call that argument "informal" because we have not assigned any independent meaning to the symbols dz, dx, and dy, so the equation $dz = dx + i\,dy$ is officially meaningless in the present context (if we introduced "differential forms" then the argument above would be perfectly valid). It is easy to show from the definitions that the formula is correct, recalling that by definition $\gamma_1 + i\gamma_2 = (\gamma_1, \gamma_2)$ (see Appendix 1):

$$\int_\gamma f\,dz = \int_a^b f(\gamma(t))\gamma'(t)\,dt$$
$$= \int_a^b f(\gamma(t))(\gamma_1'(t) + i\gamma_2'(t))\,dt$$
$$= \int_a^b (f(\gamma(t))\gamma_1'(t) + if(\gamma(t))\gamma_2'(t))\,dt$$
$$= \int_\gamma (f\,dx + if\,dy).$$

This leads to a generalization of Cauchy's Theorem to nonholomorphic functions. Using the notation $\partial f/\partial \overline{z}$ introduced in Appendix 7 we have

Theorem 26.0 (Extended Cauchy's Theorem). *If $f \in C^1(\overline{D})$ then*

$$\frac{1}{2\pi i}\int_{\partial D} f\,dz = \frac{1}{\pi}\iint_D \frac{\partial f}{\partial \overline{z}}\,dx\,dy.$$

26. Cauchy's Theorem for Nonholomorphic Functions

Proof. Green's Theorem shows that

$$\frac{1}{2\pi i}\int_{\partial D} f\,dz = \frac{1}{2\pi i}\int_{\partial D}(f\,dx + if\,dy)$$
$$= \frac{1}{2\pi i}\iint_D \left(i\frac{\partial f}{\partial x} - \frac{\partial f}{\partial y}\right)dx\,dy$$
$$= \frac{1}{2\pi}\iint_D \left(\frac{\partial f}{\partial x} + i\frac{\partial f}{\partial y}\right)dx\,dy$$
$$= \frac{1}{\pi}\iint_D \frac{\partial f}{\partial \bar{z}}\,dx\,dy. \qquad \square$$

If f is holomorphic this reduces to Cauchy's Theorem:

Theorem 26.1 (Cauchy's Theorem). *If $f \in C^1(\overline{D})$ and $\partial f/\partial \bar{z} = 0$ then*

$$\frac{1}{2\pi i}\int_{\partial D} f\,dz = 0.$$

(The reader may be wondering why we didn't just do this in Chapter 2 instead of the more complicated arguments we gave there. Part of the answer is that if we included a complete proof of Green's Theorem, at the level of rigor of Chapter 2, then the present argument would not appear quite so simple. A more important reason is that the versions of Cauchy's Theorem in Chapter 2 are true under much weaker hypotheses on f! As we noted then, if we assume that $f \in C^1(\overline{D})$ the proof is simpler. A related point is that Cauchy's Theorem remains valid under weaker hypotheses on smoothness of the boundary of the region than Green's Theorem; this is less important for our purposes, since we have only discussed contour integrals for smooth curves.)

Another amusing and sometimes useful corollary is this:

Corollary 26.2. *If D is bounded by finitely many smooth closed curves then the area $A(D)$ of D is given by*

$$A(D) = \frac{1}{2i}\int_{\partial D} \bar{z}\,dz.$$

Our Extended Cauchy's Theorem leads to an Extended Cauchy Integral Formula, by techniques that should be familiar by now:

Theorem 26.3 (Extended Cauchy Integral Formula). *If $f \in C^1(\overline{D})$ and $z_0 \in D$ then*

$$\frac{1}{2\pi i} \int_{\partial D} \frac{f(z)\,dz}{z - z_0} = f(z_0) + \frac{1}{\pi} \iint_D \frac{1}{z - z_0} \frac{\partial f}{\partial \bar{z}}\, dx\, dy \quad (z = x + iy).$$

(If that looks wrong recall our standing assumptions on the winding number of ∂D about various points.)

Note. The last integral in the theorem does not exist, at least not as a proper Riemann integral, since the integrand is unbounded. It must be interpreted as an "improper" integral: We set $D_r = \{z \in D : |z - z_0| > r\}$ for $r > 0$ and then we *define*

$$\iint_D \frac{1}{z - z_0} \frac{\partial f}{\partial \bar{z}}\, dx\, dy = \lim_{r \to 0^+} \iint_{D_r} \frac{1}{z - z_0} \frac{\partial f}{\partial \bar{z}}\, dx\, dy.$$

The proof of the theorem will show that this limit exists.

Proof. Define $g(z) = f(z)/(z - z_0)$. Then $g \in C^1(\overline{D_r})$, and, since $1/(z - z_0)$ is a holomorphic function of z (for $z \neq z_0$), we have

$$\frac{\partial g(z)}{\partial \bar{z}} = \frac{1}{z - z_0} \frac{\partial f(z)}{\partial \bar{z}}$$

in D_r. If $r > 0$ is small enough that $\overline{D(z_0, r)} \subset D$ then $\partial D_r = \partial D \dot{-} \partial D(z_0, r)$, so Theorem 26.0 shows that

$$\frac{1}{2\pi i} \int_{\partial D} \frac{f(z)\,dz}{z - z_0} - \frac{1}{2\pi i} \int_{\partial D(z_0, r)} \frac{f(z)\,dz}{z - z_0} = \frac{1}{\pi} \iint_{D_r} \frac{1}{z - z_0} \frac{\partial f(z)}{\partial \bar{z}}\, dx\, dy.$$

If we parametrize $\partial D(z_0, r)$ in the usual way we obtain

$$\frac{1}{2\pi i} \int_{\partial D(z_0, r)} \frac{f(z)\,dw}{z - z_0} = \frac{1}{2\pi} \int_0^{2\pi} f(z_0 + re^{i\theta})\, d\theta;$$

since $f(z_0 + re^{i\theta}) \to f(z_0)$ as $r \to 0$, uniformly for $\theta \in [0, 2\pi]$, this shows that

$$\lim_{r \to 0} \frac{1}{2\pi i} \int_{\partial D(z_0, r)} \frac{f(z)\,dz}{z - z_0} = f(z_0). \qquad \square$$

One consequence of Theorem 26.3 is this: *A function which is close to holomorphic is close to a holomorphic function.*

Or, less flippantly: *If $\partial f / \partial \bar{z}$ is small then there exists a holomorphic function g such that $f - g$ is small.* Here "small" can be interpreted in various ways; a precise version of this principle is the following:

26. Cauchy's Theorem for Nonholomorphic Functions

Corollary 26.4. *There exists a constant c, depending only on D, such that if $f \in C^1(\overline{D})$ and $|\partial f(z)/\partial \bar{z}| \leq \epsilon$ for all $z \in D$ then there exists $g \in H(D)$ such that $|f(z) - g(z)| \leq c\epsilon$ for all $z \in D$.*

If R is large enough that $D \subset D(z, R)$ for every $z \in D$ we may take $c = 2R$.

Proof. Define
$$g(z) = \frac{1}{2\pi i} \int_{\partial D} \frac{f(w)\, dw}{w - z}.$$
An argument using Fubini's Theorem and Morera's Theorem that we have seen several times shows that $g \in H(D)$. And for any $z \in D$, Theorem 26.3 and an integration in polar coordinates show that
$$|f(z) - g(z)| \leq \frac{\epsilon}{\pi} \iint_D \frac{ds\, dt}{|s + it - z|}$$
$$\leq \frac{\epsilon}{\pi} \iint_{D(z,R)} \frac{ds\, dt}{|s + it - z|}$$
$$= \frac{\epsilon}{\pi} \int_0^{2\pi} \int_0^R \frac{1}{\rho} \rho\, d\rho\, d\theta = 2R\epsilon. \qquad \square$$

This can sometimes be useful in constructing holomorphic functions. For example:

- **Exercise 26.1.** Suppose that $f_n \in C^1(D)$, $f_n \to g$ uniformly on compact subsets of D, and $\partial f_n / \partial \bar{z} \to 0$ uniformly on compact subsets of D. Then $g \in H(D)$.

Results analogous to this, otherwise known as "$\bar{\partial}$ techniques", are used a great deal in several complex variables to construct holomorphic functions. Such techniques are sometimes useful even in one complex variable; see for example the proof of Mergelyan's Theorem in [**R**].

We need to point out that Exercise 26.1 is amazing; in fact if anything it's even more amazing than the amazing Corollary 3.0 (which is of course a special case of Exercise 26.1). It would be obvious if we assumed that $f_n \to g$ "in $C^1(D)$", meaning that $f_n \to g$, $\partial f_n / \partial x \to \partial g / \partial x$ and $\partial f_n / \partial y \to \partial g / \partial y$ uniformly on compact sets; it would be clear from just "advanced calculus", if not quite obvious, if we assumed that $f_n \to g$ uniformly on compact sets and that the sequences $(\partial f_n / \partial x)$ and $(\partial f_n / \partial y)$ were uniformly Cauchy on compact sets. But there is no assumption about individual partial derivatives of f_n in the exercise, only an assumption about a certain linear combination of the two partials.

For readers who don't see what's so surprising about all this we give an example showing that the corollary fails with another differential operator in place of $\partial/\partial\bar{z}$. Define
$$Tf = \frac{\partial f}{\partial x} - \frac{\partial f}{\partial y}.$$

Recall from advanced calculus that a theorem of Weierstrass shows that there exists a sequence of (real) polynomials P_n such that $P_n(t) \to |t|$ uniformly on compact subsets of $\mathbb{R}$. Define $f_n \in C^1(\mathbb{C})$ by

$$f_n(x+iy) = P_n(x+y).$$

Then $f_n \to g$ uniformly on compact subsets of $\mathbb{C}$, where $g(x+iy) = |x+y|$. And it is immediate from the chain rule that $Tf_n = 0$, but Tg does not even exist.

Chapter 27

Harmonic Conjugates

We conclude with some results on harmonic conjugates. We leave all the proofs as exercises for the reader.

Some of the things you are asked to prove are immediate consequences of results proved earlier in the text; in *this* chapter I don't tell you *which* previous results to use. Some other things are not such immediate consequences of previous results, although in every case some of the previous material is at least relevant.

Note. Of course some of the assertions below are most easily proved using another one of the assertions below. Don't jump to conclusions: It may happen that the easiest proof of something uses a fact stated *later*. In fact of course there is more than one possible path through the material; it's up to you to find such a path (you should be careful that you don't prove A using B and then prove B using A).

You might regard this chapter as a (lengthy) final exam. Ready... pick up your pencils... you may begin:

Exercise

Suppose that $D \subset \mathbb{C}$ is open and $u : D \to \mathbb{R}$ is harmonic. Recall that we say that u has a *harmonic conjugate* if there exists $f \in H(D)$ with $u = \operatorname{Re}(f)$. Show that u has a harmonic conjugate if and only if

$$\int_\gamma \frac{\partial u}{\partial z} \, dz = 0$$

for every smooth closed curve γ in D. (See Appendix 7 for the definition of $\partial u/\partial z$.)

Assume that u is harmonic in $\mathbb{D}' = \{\, z \in \mathbb{C} : 0 < |z| < 1 \,\}$. Define

$$u^{\#}(z) = \frac{1}{2\pi} \int_0^{2\pi} u(e^{it} z) \, dt.$$

Show that $u^{\#}$ is harmonic and $u^{\#}(e^{it} z) = u^{\#}(z)$. Show that there exist constants a and b such that

$$u^{\#}(z) = a + b \log(|z|).$$

The notation $b(u)$ below will refer to the coefficient of $\log(|z|)$ in $u^{\#}$.

We know that functions holomorphic in $\mathbb{D}$ are given by power series, and we have seen an analogous series representation for functions harmonic in $\mathbb{D}$. We also know that functions holomorphic in $\mathbb{D}'$ are given by Laurent series. Give the analogous series representation for functions harmonic in $\mathbb{D}'$. Show that if u is harmonic in $\mathbb{D}'$ then there exist $f, g \in H(\mathbb{D}')$ such that

$$u(z) = b(u) \log(|z|) + f(z) + \overline{g(z)}.$$

Now assume that u is a real-valued harmonic function in $\mathbb{D}'$. Show that that following are equivalent:

(i) u has a harmonic conjugate.

(ii) $u^{\#}$ is constant.

(iii) $b(u) = 0$.

(iv) $\int_{\partial D(0,1/2)} \partial u/\partial z \, dz = 0$.

Note. The previous results in this chapter suffice to give fairly easy proofs of enough implications among those four conditions to show they are all equivalent. For extra credit you should try to give "direct" proofs of all twelve implications! The analogous suggestion at the end of Chapter 11 was slightly silly, but here you should really try to do this. In particular, giving a direct proof that (ii) and (iv) are equivalent (meaning a proof without using the results above and without even mentioning harmonic conjugates, but using certain key results in earlier chapters in the book) would be instructive.

Part 3

Appendices

Appendix 1

Complex Numbers

In this appendix we will construct the complex numbers.

What do I mean, "construct"? Well, this complex-number business started some centuries ago, with some Italian mathematicians who were studying the problem of solving polynomial equations. Someone found a "cubic formula", which gives the three solutions to an arbitrary cubic, just as the familiar quadratic formula gives the two solutions to a quadratic equation, in terms of roots and algebraic operations.

We should probably point out that at first complex numbers were regarded as totally "imaginary" things, with no real meaning at all. That is, even after people started using complex numbers, for some time they would have said that the equation $x^2 + 1 = 0$ had no solutions, instead of saying that it has two complex roots i and $-i$. But it turned out that the formula for the three roots of a cubic involved complex numbers, even when the actual solutions to the equation were all real. Those complex numbers were "imaginary", just formal devices used on the way to finding the real roots of the equation.

They couldn't be "real", because it is obvious that an equation like $x^2 + 1 = 0$ has no solutions: $x^2 + 1 \geq 1$ for all x, and hence $x^2 + 1 \neq 0$. While we're solving some cubic equation we may pretend that it has a solution and call it i, but in reality there is no such thing — if someone talked about i as something that actually existed he would have to answer the obvious question of exactly what it *was,* and they didn't have an answer to that. (Someone else might point out that we could just as well assume that the equation $x^2 + 1 = x^2$ has a solution and call it j. But this leads to the conclusion that $0 = 1$ and the universe explodes... why then is it ok to just

445

assume that $x^2 + 1 = 0$ has a solution, if we cannot say what the solution actually *is*?)

But we are going to actually *construct* the complex numbers; this means we are going to say exactly what a complex number is, in a perfectly straightforward way with nothing mystical or "imaginary" involved. Here we go:

By definition, a complex number is an ordered pair of real numbers: $\mathbb{C} = \mathbb{R}^2$. (As usual, $\mathbb{R}$ denotes the set of all real numbers, $A^2 = \{(x, y) : x, y \in A\}$, and $\mathbb{C}$ denotes the set of all complex numbers.)

We also have to say how to add and multiply two complex numbers. The sum of two complex numbers is the same as their sum when regarded as elements of $\mathbb{R}^2$, but the product is defined by a curious formula:

$$(x_1, y_1) + (x_2, y_2) = (x_1 + x_2, y_1 + y_2),$$

$$(x_1, y_1)(x_2, y_2) = (x_1 x_2 - y_1 y_2, x_1 y_2 + x_2 y_1).$$

The first thing we need to prove is that this definition "works", meaning roughly that the basic rules of algebra are the same in $\mathbb{C}$ as in $\mathbb{R}$. A precise statement of what we mean by that is given by the following exercise:

- **Exercise A1.1.** The operations defined above make $\mathbb{C}$ into a field (in the sense of abstract algebra). That is, for every $a, b, c \in \mathbb{C}$ we have

$$a + b = b + a,$$
$$(a + b) + c = a + (b + c),$$
$$ab = ba,$$
$$(ab)c = a(bc),$$
$$(a + b)c = ac + bc,$$

there exists $0_\mathbb{C} \in \mathbb{C}$ such that

$$a + 0_\mathbb{C} = a \quad (a \in \mathbb{C}),$$

for every $a \in \mathbb{C}$ there exists $-a \in \mathbb{C}$ such that

$$a + (-a) = 0_\mathbb{C},$$

there exists $1_\mathbb{C} \in \mathbb{C}$ such that $1_\mathbb{C} \neq 0_\mathbb{C}$ and

$$a(1_\mathbb{C}) = a \quad (a \in \mathbb{C}),$$

and finally for all $a \in \mathbb{C}$ other than $a = 0_\mathbb{C}$ there exists $a^{-1} \in \mathbb{C}$ with

$$a(a^{-1}) = 1_\mathbb{C}.$$

Appendix 1. Complex Numbers

We use the notation "$0_\mathbb{C}$" in the exercise because "0" already denotes the real number 0, and the complex number $0_\mathbb{C}$ is not the same thing (at least for now); similarly for $1_\mathbb{C}$. Most of the exercise is utterly trivial. It is probably obvious that one should define $0_\mathbb{C} = (0,0)$, and it is almost as obvious that if we set $1_\mathbb{C} = (1,0)$ then we will have $a(1_\mathbb{C}) = a$ for all a. The existence of multiplicative inverses is the trickiest part; in fact if $a = (x,y) \in \mathbb{C}$ and $a \neq 0_\mathbb{C}$ then one can define

$$a^{-1} = \left(\frac{x}{x^2 + y^2}, \frac{-y}{x^2 + y^2} \right).$$

(In your solution you should make certain to explain why this formula makes sense for all $a \in \mathbb{C}$ other than $0_\mathbb{C}$.)

So $\mathbb{C}$ is a field, great. Roughly speaking, it follows that (as long as we are only talking about addition and multiplication) algebra in $\mathbb{C}$ works the same as in $\mathbb{R}$. We will use traditional notation for inverses from now on, writing $z - w$ in place of $z + (-w)$, z/w in place of $z(w^{-1})$, etc.

This all started because we wanted the equation $z^2 + 1 = 0$ to have a solution $z \in \mathbb{C}$. Does it? Well, strictly speaking the question makes no sense: If $z \in \mathbb{C}$ then $z^2 = zz \in \mathbb{C}$, while $1 \in \mathbb{R}$, so there is no such thing as $z^2 + 1$ according to our definitions. But we can ask whether the equation $z^2 + 1_\mathbb{C} = 0_\mathbb{C}$ has a solution $z \in \mathbb{C}$, and the answer is yes: We define

$$i = (0, 1),$$

and then it is easy to verify that $i^2 + 1_\mathbb{C} = 0_\mathbb{C}$. Great.

But of course we wanted the complex numbers to *contain* the real numbers, and they don't; $\mathbb{R}$ is not a subset of $\mathbb{R}^2$. This is a problem just because of the funny definition of complex numbers we gave; while $\mathbb{R}$ is not literally a subset of $\mathbb{C}$, it turns out that $\mathbb{R}$ is isomorphic to a subfield of $\mathbb{C}$. We define $\phi : \mathbb{R} \to \mathbb{C}$ by

$$\phi(x) = (x, 0),$$

and it turns out that ϕ is an isomorphism from $\mathbb{R}$ onto a subfield of $\mathbb{C}$. This follows from the next exercise:

- **Exercise A1.2.** Show that ϕ is one-to-one, that $\phi(0) = 0_\mathbb{C}$, $\phi(1) = 1_\mathbb{C}$, and that

$$\phi(x+y) = \phi(x) + \phi(y),$$
$$\phi(xy) = \phi(x)\phi(y)$$

for all $x, y \in \mathbb{R}$.

It turns out that $\mathbb{C}$ is generated by $\{i\} \cup \phi(\mathbb{R})$; in fact you can easily verify that if $z = (x, y) \in \mathbb{C}$ then

$$z = (x, 0) + (0, 1)(y, 0) = \phi(x) + i\phi(y).$$

As of now, and for the rest of our mathematical lives, we "identify" the real number x with the complex number $\phi(x)$. This means that although they are not literally the same thing we are going to pretend they *are* the same thing; the fact that ϕ is an isomorphism from $\mathbb{R}$ onto a subfield of $\mathbb{C}$ shows that the arithmetic is going to come out right. Now we can simply write 0 instead of $0_\mathbb{C}$, and the previous equation $z = \phi(x) + i\phi(y)$ becomes

$$z = x + iy.$$

What all this amounts to is that we can now take the following as the "definition" of the complex numbers: "A complex number is a quantity $z = x + iy$, where $x, y \in \mathbb{R}$ and i is the square root of -1."

That cannot be the "official" definition of what a complex number is, because it raises as many questions as it answers, as mentioned at the start of this appendix. But now that we have answers to those questions we can pretend that that's the definition; we shall do so from now on (with rare exceptions, in places where the fact that a complex number is literally an ordered pair of reals is important for some reason).

Appendix 2

Complex Numbers, Continued

In the previous appendix we constructed the complex numbers, making sense of the statement "A complex number is a quantity $z = x + iy$, where x, $y \in \mathbb{R}$ and i is the square root of -1." Here we will say a little bit about the "arithmetic" of $\mathbb{C}$.

Suppose that $z = x + iy \in \mathbb{C}$, where $x, y \in \mathbb{R}$. The *real part* of z is $\operatorname{Re} z = x$, and the *imaginary part* of z is $\operatorname{Im} z = y$. We say that z is *real* if $\operatorname{Im} z = 0$, in which case $z = \operatorname{Re} z$, and that z is *(purely) imaginary* if $\operatorname{Re} z = 0$, in which case $z = i\operatorname{Im} z$. The *modulus* or *absolute value* of z is $|z| = (x^2 + y^2)^{1/2}$ (the *non-negative* square root of $x^2 + y^2$) and the *complex conjugate* of z is $\overline{z} = x - iy$. (Note that if z is real then $|z|$ is the same as the absolute value of z regarded as a real number.)

It is easy to verify that the map $z \mapsto \overline{z}$ is an automorphism of $\mathbb{C}$, which is to say that

$$\overline{z + w} = \overline{z} + \overline{w},$$
$$\overline{zw} = \overline{z}\,\overline{w},$$

for all $z, w \in \mathbb{C}$. There is an important relation between complex conjugation and the modulus, given by

$$|z|^2 = z\overline{z}.$$

This fact is often useful in changing a problem involving $|z|$ into a problem involving $\overline{z}$; the latter is often simpler because, for example, $\overline{z + w} = \overline{z} + \overline{w}$, while it is certainly not true in general that $|z + w| = |z| + |w|$ (see the

triangle inequality below). As a trivial example of this technique we show that
$$|zw| = |z||w|$$
for all $z, w \in \mathbb{C}$:
$$|zw|^2 = (zw)\overline{(zw)} = zw\overline{z}\overline{w} = (z\overline{z})(w\overline{w}) = |z|^2|w|^2.$$

As a slightly less trivial example, we note that the fact that $|z|^2 = z\overline{z}$ makes the proof of the so-called *triangle inequality*
$$|z + w| \leq |z| + |w| \quad (z, w \in \mathbb{C})$$
essentially automatic:
$$\begin{aligned}
|z + w|^2 &= \operatorname{Re}|z + w|^2 \\
&= \operatorname{Re}(z + w)\overline{(z + w)} \\
&= \operatorname{Re}(z + w)(\overline{z} + \overline{w}) \\
&= \operatorname{Re}(z\overline{z} + z\overline{w} + w\overline{z} + w\overline{w}) \\
&= \operatorname{Re}|z|^2 + \operatorname{Re}(z\overline{w}) + \operatorname{Re}(w\overline{z}) + \operatorname{Re}|w|^2 \\
&= |z|^2 + \operatorname{Re}(z\overline{w}) + \operatorname{Re}(w\overline{z}) + |w|^2 \\
&\leq |z|^2 + |z\overline{w}| + |w\overline{z}| + |w|^2 \\
&= |z|^2 + |z||\overline{w}| + |w||\overline{z}| + |w|^2 \\
&= |z|^2 + |z||w| + |w||z| + |w|^2 \\
&= |z|^2 + 2|z||w| + |w|^2 \\
&= (|z| + |w|)^2.
\end{aligned}$$

Ok, that took a few lines, but only because we wrote out every detail; once we decided to look at $|z + w|^2$ instead of $|z + w|$, so that we could apply the identity $|z|^2 = z\overline{z}$, every step really was more or less obvious.

Note that we used the easily verified inequality $\operatorname{Re} z \leq |z|$ above. For the following proposition we will use the facts that $|-z| = |z|$ and that $|z| = 0$ if and only if $z = 0$, which are also very easy to check.

Proposition. *If $d : \mathbb{C} \times \mathbb{C} \to \mathbb{C}$ is defined by*
$$d(z, w) = |z - w|$$
then d is a metric on $\mathbb{C}$.

(See Appendix 4 if you are not familiar with metric spaces.)

Proof. It is clear that $d(z, w) \geq 0$ for all $z, w \in \mathbb{C}$; also $d(z, w) = 0$ if and only if $z - w = 0$, or $z = w$. It is also clear that $d(w, z) = |w - z| = |-(z - w)| = |z - w| = d(z, w)$. Finally, the triangle inequality for d follows from the inequality called the "triangle inequality" above, by a little trick:

$$d(z, v) = |z - v| = |(z - w) + (w - v)| \leq |z - w| + |w - v| = d(z, w) + d(w, v). \quad \square$$

In the rest of the book, metric-space concepts (convergence, continuity, open and closed set, etc.) in the context of the complex numbers will refer to the metric defined in the proposition above; it is the *euclidean* metric on $\mathbb{C}$, also known as the *standard* metric. In particular:

If (z_n) is a sequence of complex numbers and $z \in \mathbb{C}$, then we say that $\lim_{n \to \infty} z_n = z$, or just $z_n \to z$, if $|z_n - z| \to 0$. If $S \subset \mathbb{C}$, $f : S \to \mathbb{C}$, and $z \in S$, then f is *continuous at* z if $f(z_n) \to f(z)$ for every sequence (z_n) in S with $z_n \to z$. (As in any metric space, f is continuous at z if and only if for every $\epsilon > 0$ there exists $\delta > 0$ such that $|f(w) - f(z)| < \epsilon$ for all $w \in S$ with $|w - z| < \delta$.)

We are going to refer to the balls in this metric space as "disks" instead of "balls", because that's what they are; we will also use the notation $D(z, r)$ instead of $B(z, r)$. So: If $z \in \mathbb{C}$ and $r > 0$ then the open disk about z of radius r is

$$D(z, r) = \{ w \in \mathbb{C} : |w - z| < r \}.$$

The set $S \subset \mathbb{C}$ is *open* if for every $z \in S$ there exists $r > 0$ such that $D(z, r) \subset S$. The set $S \subset \mathbb{C}$ is *closed* if its complement $\mathbb{C} \setminus S = \{ z \in \mathbb{C} : z \notin S \}$ is open; as in any metric space it follows that S is closed if $z \in S$ whenever there exists a sequence (z_n) in S which converges to z.

- **Exercise A2.1.** Suppose that (z_n) is a sequence of complex numbers and $z \in \mathbb{C}$. Show that $z_n \to z$ if and only if $\operatorname{Re} z_n \to \operatorname{Re} z$ and $\operatorname{Im} z_n \to \operatorname{Im} z$.

- **Exercise A2.2.** Show that $\mathbb{C}$ is a *complete* metric space (in the standard metric).

   Hint: Sometimes the result of one exercise is useful in another exercise ...

The completeness of the complex numbers (together with the triangle inequality) allows one to show that an absolutely convergent series is convergent, exactly as is done in calculus for series of real numbers. (We recall the relevant definitions: The series $\sum_{j=0}^{\infty} z_j$ is said to *converge* if there exists $z \in \mathbb{C}$ such that $s_n \to z$, where $s_n = \sum_{j=0}^{n} z_j$; in this case we write

$\sum_{j=0}^{\infty} z_j = z$. The series $\sum_{j=0}^{\infty} z_j$ is said to *converge absolutely* if the series $\sum_{j=0}^{\infty} |z_j|$ converges.)

Similarly, the usual proof for real-valued functions suffices to prove the "Weierstrass M-test" for complex-valued functions: Suppose that $f_n : S \to \mathbb{C}$, $|f_n(x)| \le M_n$ for all $x \in S$, and the sum $\sum_{n=0}^{\infty} M_n$ converges. Then the sum $\sum_{n=0}^{\infty} f_n$ converges uniformly on S (which is to say the sequence (s_n) converges uniformly on S, where $s_n = \sum_{j=0}^{n} f_j$).

We turn next to what one might refer to as the representation of a complex number in polar coordinates (where $z = x + iy$ is the representation in rectangular coordinates). Recall that any vector in $\mathbb{R}^2$ can be written in the form

$$(x, y) = (r\cos(t), r\sin(t))$$

for some $r \ge 0$ and $t \in \mathbb{R}$; in fact $r = \sqrt{x^2 + y^2}$ and $\tan(t) = y/x$ (at least when $x \ne 0$). This is the same as saying that

$$z = x + iy = re^{it},$$

where for now we simply define $e^{it} = \cos(t) + i\sin(t)$ for $t \in \mathbb{R}$ (we will see in Appendix 3 why this definition makes sense); here $t = \arg(z)$ is an *argument* of z.

We said "an" argument, not "the" argument: If t is an argument of z then so is $t + 2\pi k$ for any integer k. Thus arg is not a function, the function-like appearance of the notation $\arg(z)$ notwithstanding.

- **Exercise A2.3.** Show that $e^{it}e^{is} = e^{i(t+s)}$ for all $t, s \in \mathbb{R}$, using the familiar addition formulas for sine and cosine.

It follows that
$$r_1 e^{it_1} r_2 e^{it_2} = r_1 r_2 e^{i(t_1 + t_2)},$$

or equivalently
$$|z_1 z_2| = |z_1||z_2|$$

and
$$\arg(z_1 z_2) = \arg(z_1) + \arg(z_2),$$

although we need to be a little careful about what that last equation means, since arg is not a function — it means that if $\arg(z_1)$ is any argument of z_1 and $\arg(z_2)$ is any argument of z_2 then $\arg(z_1) + \arg(z_2)$ is *an* argument of $z_1 z_2$. This is why we are interested in the polar representation of complex numbers: It makes multiplication easy to understand. The next exercise is an elementary example of a problem that becomes much simpler in polar coordinates:

Appendix 2. Complex Numbers, Continued

- **Exercise A2.4.** Find all four complex solutions to the equation $z^4 = i$. For extra credit write the solutions in rectangular coordinates (in terms of radicals, not inverse trig functions!).

Actually the elegant approach would be to define $\arg(z)$ (for $z \neq 0$) to be an element of the additive quotient group $\mathbb{R}/2\pi\mathbb{Z}$; then arg becomes an actual function, the equation $\arg(z_1 z_2) = \arg(z_1) + \arg(z_2)$ no longer requires any special interpretation, etc. But people do not typically do that so we won't.

There *is* an actual function that is closely related to arg: The *principal-value argument* of $z \neq 0$, written $\mathrm{Arg}(z)$, is defined to be the unique argument of z which satisfies

$$-\pi < \mathrm{Arg}(z) \leq \pi.$$

Alas Arg does not do various things one might want it to do:

- **Exercise A2.5.** Show that $\mathrm{Arg} : \mathbb{C} \setminus \{0\} \to (-\pi, \pi]$ is not continuous.

- **Exercise A2.6.** Give an example of $z, w \neq 0$ such that $\mathrm{Arg}(zw) \neq \mathrm{Arg}(z) + \mathrm{Arg}(w)$.

There is no perfect solution; any time you see a statement about arg you have to be careful about the fact that a complex number has more than one argument.

Appendix 3

Sin, Cos and Exp

In the previous appendix we used a few facts that "everybody knows" about trigonometric functions. The proofs one typically sees for these things in elementary courses are far from rigorous, relying on the way a picture looks, etc. Just for fun we are going to show here how one can derive everything there is to know about trig functions and the exponential function in a completely rigorous way, starting with the power series for the definitions. We assume that the reader is familiar with the material in Chapters 0 and 1. This appendix is inspired by the Prologue to [**R**]; I gather that Weierstrass was the first person to actually do more or less what we do here (there were giants in the earth in those days ...).

Fix $r > 0$ for a moment, and for $n = 0, 1, \ldots$ let $c_n = r^n/n!$. Since $c_{n+1}/c_n = r/(n+1)$, it follows that $0 < c_{n+1} < c_n$ for all $n > r - 1$, which implies that the sequence (c_n) is bounded. Hence the three power series in the following definitions have infinite radius of convergence, which is to say they converge uniformly on every compact subset of the plane (see Lemma 1.0): For $z \in \mathbb{C}$ we define

$$\exp(z) = \sum_{n=0}^{\infty} \frac{z^n}{n!},$$

$$\sin(z) = \sum_{n=0}^{\infty} \frac{(-1)^n z^{2n+1}}{(2n+1)!},$$

$$\cos(z) = \sum_{n=0}^{\infty} \frac{(-1)^n z^{2n}}{(2n)!}.$$

Now Proposition 1.1 shows that these functions are differentiable in the entire plane, and that the derivatives can be calculated by termwise differentiation of the power series; if one does that one sees that

$$\exp' = \exp,$$
$$\sin' = \cos,$$
$$\cos' = -\sin.$$

The absolute convergence of the series shows that the terms can be rearranged as desired, and it follows that

$$\exp(iz) = \cos(z) + i\sin(z)$$

for all $z \in \mathbb{C}$, which implies in turn that

$$\cos(z) = \frac{\exp(iz) + \exp(-iz)}{2},$$
$$\sin(z) = \frac{\exp(iz) - \exp(-iz)}{2i}.$$

At this point we need a lemma. If we knew just a little bit of complex analysis we could give a much stronger statement, with a proof that would be much more satisfactory in certain respects; we settle for a somewhat lame version that has the advantage that it follows immediately from just the results in Chapter 1:

Lemma. *Suppose that $r > 0$ and f is defined in $D(z_0, r)$ by a convergent power series. If $f' = 0$ at every point of this disk then f is constant.*

Proof. Say $f(z) = \sum_{n=0}^{\infty} c_n (z - z_0)^n$. Proposition 1.1 shows that $f'(z) = \sum_{n=0}^{\infty} (n+1) c_{n+1} (z - z_0)^n$. Now Corollary 1.3 shows that the coefficients of the series defining f' must vanish. This shows that $c_n = 0$ for all $n \geq 1$, so that $f(z) = c_0$. □

We use the lemma to give a simple proof that exp does the things it is supposed to do. To start, define $f(z) = \exp(z)\exp(-z)$. Then the product rule and the chain rule (which can be proved just as in ordinary calculus) show that f is differentiable in $\mathbb{C}$ and that in fact $f' = 0$. Hence f is constant; since $f(0) = 1$, we have proved that

$$\exp(z)\exp(-z) = 1 \quad (z \in \mathbb{C}).$$

This has a few consequences; for example it shows that exp never vanishes. It is clear that $\exp(x) \geq 1 + x$ for $x \geq 0$, so that

$$\lim_{\mathbb{R} \ni x \to +\infty} \exp(x) = +\infty;$$

Appendix 3. Sin, Cos and Exp

since $\exp(-z) = 1/\exp(z)$, it follows that

$$\lim_{\mathbb{R} \ni x \to -\infty} \exp(x) = 0.$$

Since $\exp'(x) = \exp(x) > 0$ for $x \in \mathbb{R}$, it follows that $\exp|_\mathbb{R}$ is strictly increasing; since $\exp$ is continuous, this implies that $\exp$ maps $\mathbb{R}$ bijectively onto the interval $(0, \infty)$. So $\exp|_\mathbb{R}$ has an inverse

$$\ln : (0, \infty) \to \mathbb{R}.$$

Now fix $w \in \mathbb{C}$ and define $f(z) = \exp(z)\exp(w)\exp(-(z+w))$. As before one sees that $f' = 0$, so that $f(z) = f(0) = 1$; this shows that

$$\exp(z+w) = \exp(z)\exp(w).$$

We define $e = \exp(1)$ and it follows that

$$e^n = \exp(n) \quad (n \in \mathbb{Z});$$

at this point we decide that $\exp$ is looking somewhat exponential so we *define*

$$e^z = \exp(z) \quad (z \in \mathbb{C}).$$

(If $a > 0$ we define

$$a^z = e^{z \ln(a)} \quad (z \in \mathbb{C});$$

we will talk about a^z for other values of a later.)

We turn now to sine and cosine. The first thing we note is that

$$\sin^2(z) + \cos^2(z) = 1 \quad (z \in \mathbb{C}).$$

- **Exercise A3.1.** Give two proofs that $\sin^2(z) + \cos^2(z) = 1$ for all $z \in \mathbb{C}$, one by writing $\sin$ and $\cos$ in terms of $\exp$ and one by using the facts that $\sin' = \cos$ and $\cos' = -\sin$.

This has a few consequences. First, it follows that

$$|\sin(x)| \leq 1 \quad (x \in \mathbb{R})$$

and

$$|\cos(x)| \leq 1 \quad (x \in \mathbb{R});$$

second, since $e^{iz} = \cos(z) + i\sin(z)$, it also follows that

$$|e^{ix}| = 1 \quad (x \in \mathbb{R}).$$

You should note that the last three statements hold only for real x:

- **Exercise A3.2.** Note that the identities $e^{iz} = \cos(z) + i\sin(z)$ and $\sin^2(z) + \cos^2(z) = 1$ are valid for all complex z. Why does it not follow that $|e^{iz}| = 1$ for all complex z? (In other words, how is the fact that x is real used in the proof that $|e^{ix}| = 1$?)

- **Exercise A3.3.** If $z = x + iy$ ($x, y \in \mathbb{R}$) what is $|e^z|$? Explain.

We might note at this point that we have proved a meta-theorem: *knowing that a function satisfies a differential equation can be very useful in proving things about that function!* Indeed, what's actually happened here is that we used the fact that $f(x) = \cos(x)$ satisfies $f'' + f = 0$ to show that $|\cos(x)| \leq 1$; this is far from obvious from the definition of cos as a power series.

Now comes the interesting part: It is clear from the power series that $\cos(3) < 0$. Indeed, the fact that $9 < 7 \cdot 8$ shows that $3^8/8! < 3^6/6!$. Similarly $3^{12}/12! < 3^{10}/10!$, etc., so

$$\cos(3) = 1 - \frac{3^2}{2!} + \frac{3^4}{4!} - \left(\frac{3^6}{6!} - \frac{3^8}{8!}\right) - \cdots < 1 - \frac{3^2}{2!} + \frac{3^4}{4!} = \frac{-3}{4!} < 0.$$

Since $\cos|_\mathbb{R}$ is continuous and real-valued, $\cos(0) = 1$ and $\cos(3) < 0$ the Intermediate Value Theorem shows that cos has at least one zero in the interval $(0, 3)$. The fact that cos is continuous shows that its zero set is closed; hence cos has a smallest positive zero. We define

$$\pi = 2\alpha,$$

where α is the smallest positive zero of cos.

Now $\cos > 0$ on the interval $(0, \pi/2)$, so sin is increasing on this interval, since $\sin' = \cos$; hence $\sin(\pi/2) > 0$. Since $\sin^2 + \cos^2 = 1$, it follows that $\sin(\pi/2) = 1$. This shows that $e^{i\pi/2} = i$, and hence

$$e^{2\pi i} = i^4 = 1.$$

So exp has period $2\pi i$:

$$e^{z+2\pi i} = e^z e^{2\pi i} = e^z \quad (z \in \mathbb{C}),$$

which implies that sin and cos have period 2π.

The fact that sin is increasing on the interval $(0, \pi/2)$ shows that $\sin > 0$ on the same interval, and hence cos is strictly decreasing on $(0, \pi/2)$. Since cos is continuous, it follows that cos maps the interval $[0, \pi/2]$ bijectively onto $[0, 1]$. Finally, since $\sin > 0$ on $(0, \pi/2)$, it follows that if we set

$$\gamma(t) = e^{it}$$

Appendix 3. Sin, Cos and Exp

then γ maps the interval $[0, \pi/2]$ bijectively onto $T_1 = \{x + iy : x, y \geq 0, x^2 + y^2 = 1\}$, that is, the intersection of the unit circle with the closed first quadrant.

Now, since $\gamma(t + \pi/2) = e^{i\pi/2}\gamma(t) = i\gamma(t)$, this shows that γ maps the interval $[\pi/2, \pi]$ bijectively onto the intersection of the unit circle with the closed second quadrant. Repeating this twice more we see we have proved the following:

Theorem. *Define* $\gamma : \mathbb{R} \to \mathbb{C}$ *by* $\gamma(t) = e^{it}$. *Then* $|\gamma(t)| = 1$ *for all* t, γ *has period* 2π, *and for every* $z \in \mathbb{C}$ *with* $|z| = 1$ *there exists exactly one* $t \subset [0, 2\pi)$ *with* $\gamma(t) = z$.

Excellent. As promised, the theorem gives a rigorous proof that a complex number can be written in polar coordinates, as in the previous appendix:

- **Exercise A3.4.** Suppose $z \in \mathbb{C}$ and $z \neq 0$. Show that there exist a unique $r > 0$ and $t \in [0, 2\pi)$ such that $z = re^{it}$.

- **Exercise A3.5.** Show that $\sin(z) = 0$ if and only if $z = n\pi$ for some $n \in \mathbb{Z}$, and that all the zeroes of sin are simple. (The complex number z is a *simple zero* of the differentiable function f if $f(z) = 0$ and $f'(z) \neq 0$.)

    Hint: If $\operatorname{Re}(z) \neq 0$ then $|e^{iz}| \neq |e^{-iz}|$, so that $\sin(z) \neq 0$. So all the zeroes of sin are real. Now it follows from the theorem above that sin has exactly two zeroes in the interval $[0, 2\pi)$, and various other facts above show that $\sin(0) = \sin(\pi) = 0$ while $\cos(0)$ and $\cos(\pi)$ are nonzero.

We conclude with an informal demonstration that our trig functions are the same as those you saw defined long ago in terms of triangles (this part has to be informal, since the definitions you saw long ago were somewhat informal). Suppose $0 < \theta < \pi/2$ and consider the triangle ABC, where $A = (0,0)$, $B = (\cos(\theta), 0)$ and $C = (\cos(\theta), \sin(\theta))$. This is a right triangle with base $\cos(\theta)$, height $\sin(\theta)$ and hypotenuse 1, so we're done if we can show that angle A is equal to θ. Recalling the definition of the radian measure of an angle, we need to show that the arc on the unit circle from B to C has length θ. This is easy: If γ is as above then $\gamma|_{[0,\theta]}$ is a smooth parametrization of this arc, so the formula for the length of a parametrized curve shows that the arc has length

$$\int_0^\theta |\gamma'(t)|\, dt = \int_0^\theta |ie^{it}|\, dt = \int_0^\theta 1\, dt = \theta.$$

The same argument shows that the circumference of the unit circle is 2π, which is reassuring.

Appendix 4

Metric Spaces

Suppose X is a set. A *metric on X* is a function $d : X \times X \to [0, \infty)$ such that for all $x, y, z \in X$ we have

(i) $d(x, y) = 0$ if and only if $x = y$,

(ii) $d(x, y) = d(y, x)$,

(iii) $d(x, y) \leq d(x, z) + d(z, y)$.

A *metric space* is an ordered pair (X, d) where d is a metric on X. (Usually one refers to X itself as a metric space.)

In this appendix we summarize some basic facts about metric spaces used in the text, often without comment. Most of the proofs, and indeed the statements of most of the results, are left as exercises, so the reader should make certain not to skip any of them.

We begin with a few examples of metric spaces:

If X is any set and we define

$$d(x, y) = \begin{cases} 1 & (x \neq y), \\ 0 & (x = y) \end{cases}$$

for $x, y \in X$ then d is a metric on X, known as the *discrete* metric.

The formula $d(x, y) = |x - y|$ defines a metric on $\mathbb{R}$.

The formula $d(x, y) = \left(\sum_{j=1}^{n} |x_j - y_j|^2 \right)^{1/2}$ defines a metric on $\mathbb{R}^n$; this is the *standard* or *euclidean* metric. (When $n = 2$ this is the metric on $\mathbb{C}$ defined in Appendix 2.)

The formula $d(x, y) = \max_{1 \leq j \leq n} |x_j - y_j|$ defines a metric on $\mathbb{R}^n$, as does the formula $d(x, y) = \sum_{j=1}^{n} |x_j - y_j|$.

The formula $d(x,y) = |x-y|$ defines a metric on $\mathbb{Q}$, the set of rational numbers.

For the rest of this appendix we will assume that (X, d) and (Y, ρ) are metric spaces.

Suppose that (x_n) is a sequence in X and $x \in X$. We say that $x_n \to x$, or $\lim_{n\to\infty} x_n = x$, if $d(x_n, x) \to 0$ as $n \to \infty$. If there exists $x \in X$ such that $x_n \to x$ the sequence (x_n) is said to be *convergent*. (Note that "*convergent in X*" would be better terminology, because whether a sequence converges depends on what metric space we are considering! For example, if (r_n) is a sequence of rationals and $\lim r_n = \sqrt{2}$ then (r_n) is convergent in $\mathbb{R}$ but not in $\mathbb{Q}$.) The sequence (x_n) is a *Cauchy sequence* if $\lim_{n,m\to\infty} d(x_n, x_m) = 0$, meaning that for every $\epsilon > 0$ there exists N such that $d(x_n, x_m) < \epsilon$ for all $n, m > N$.

- **Exercise A4.1.** Show that a convergent sequence must be Cauchy.

A metric space is *complete* if every Cauchy sequence converges. All the metric spaces listed above are complete except for $\mathbb{Q}$. This follows easily from the completeness of $\mathbb{R}$, as in the exercises for Appendix 2. (We shall not prove the completeness of $\mathbb{R}$; how one proves this depends on the details of how the reals were constructed.)

The set $S \subset X$ is *open* if for every $x \in S$ there exists $r > 0$ such that $B(x, r) \subset S$; here

$$B(x, r) = \{ y \in X : d(x, y) < r \}$$

is the *(open) ball about x* with *radius r*.

- **Exercise A4.2.** Show that X and $\emptyset$ are open.

- **Exercise A4.3.** Show that an arbitrary union of open sets is open. That is, assume that $S_j \subset X$ is open for every $j \in J$ and show that $S = \bigcup_{j \in J} S_j$ is open.

- **Exercise A4.4.** Show that the intersection of two open sets is open. (Note that it follows by induction that the intersection of finitely many open sets is open.)

- **Exercise A4.5.** Give an example of a sequence of open sets $S_n \subset \mathbb{R}$ ($n = 1, 2, \dots$) such that the intersection $S = \bigcap_{n=1}^{\infty} S_n$ is not open.

We note in passing that Exercises A4.2, A4.3, and A4.4 show that the collection of open subsets of X is a *topology* on X. More about that in some other course.

Appendix 4. Metric Spaces 463

The set $S \subset X$ is *closed* if its complement $X \setminus S$ is open. (Note again that whether S is closed depends both on S and on X; for example $\mathbb{Q}$ is a closed subset of $\mathbb{Q}$ but not a closed subset of $\mathbb{R}$. So one should really say "closed in X" instead of just "closed".)

- **Exercise A4.6.** Show that $S \subset X$ is closed if and only if $x \in S$ whenever (x_n) is a sequence in S, $x \in X$, and $x_n \to x$.

- **Exercise A4.7.** Show that "closed" is not the same as "not open" by giving an example of a subset of $\mathbb{R}$ which is neither open nor closed.

Note that the next four exercises are immediate consequences of the corresponding exercises for open sets above:

- **Exercise A4.8.** Show that X and $\emptyset$ are closed.

- **Exercise A4.9.** Show that an arbitrary intersection of closed sets is closed. That is, assume that $S_j \subset X$ is closed for every $j \in J$ and show that $S = \bigcap_{j \in J} S_j$ is closed.

- **Exercise A4.10.** Show that the union of two closed sets is closed. (Note that it follows by induction that the union of finitely many closed sets is closed.)

- **Exercise A4.11.** Give an example of a sequence of closed sets $S_n \subset \mathbb{R}$ ($n = 1, 2, \ldots$) such that the union $S = \bigcup_{n=1}^{\infty} S_n$ is not closed.

The point $x \in X$ is a *limit point* of $E \subset X$ if there exists a sequence $(x_n) \subset E$ such that $x_n \neq x$ for $n = 1, 2, \ldots$ and $x_n \to x$.

- **Exercise A4.12.** Show that $x \in X$ is a limit point of $E \subset X$ if and only if for every $r > 0$ there exists $y \in E$ with $0 < d(x, y) < r$.

Suppose $S \subset X$. The *closure* of S (in X), denoted $\overline{S}$, is defined to be the intersection of all the closed subsets of X containing S; note that Exercise A4.9 shows that $\overline{S}$ is closed, so it is the smallest closed subset of X containing S. The *interior* of S, denoted S°, is the union of all the open subsets of S; Exercise A4.3 shows that S° is the largest open subset of S. The *boundary* of S, denoted ∂S, is $\overline{S} \cap \overline{(X \setminus S)}$.

- **Exercise A4.13.** Show that $\overline{S}$ is the union of S and the set of limit points of S.

If $S \subset X$ the *diameter* of S is

$$\mathrm{diam}(S) = \sup_{x, y \in S} d(x, y).$$

It is often useful to know that in a complete metric space the intersection of a nested sequence of closed sets with diameter tending to 0 consists of exactly one point:

- **Exercise A4.14.** Suppose X is complete, $E_n \subset X$ is closed ($n = 1, 2, \ldots$), each E_n is nonempty, $E_{n+1} \subset E_n$ for all n, and $\operatorname{diam}(E_n) \to 0$. Show that there exists $x_0 \in X$ such that
$$\bigcap_{n=1}^{\infty} E_n = \{x_0\}.$$

  Hint: Let $x_n \in E_n$ and show that (x_n) is a Cauchy sequence. So there exists x_0 with $x_n \to x_0$. Now $x_j \in E_n$ for all $j \geq n$ (why?), so $x_0 \in E_n$ for all n (why?). So $x_0 \in \bigcap_{n=1}^{\infty} E_n$. On the other hand, if y is another element of $\bigcap_{n=1}^{\infty} E_n$ then $d(x_0, y) = 0$ (why?), hence $y = x_0$.

- **Exercise A4.15.** Suppose that $S \subset X$ and $x \in X$.
    (i) Show that $x \in \overline{S}$ if and only if there exists a sequence (x_n) in S with $x_n \to x$.
    (ii) Show that $x \in S^\circ$ if and only if there exists $r > 0$ such that $B(x, r) \subset S$.
    (iii) Give a condition involving convergent sequences which holds if and only if $x \in \partial S$ (and prove the equivalence).

- **Exercise A4.16.** Suppose that $S \subset X$. Show that S is closed if and only if $S = \overline{S}$ and that S is open if and only if $S = S^\circ$.

  Note that if $S \subset X$ then the restriction of d to $S \times S$ is a metric on S. Thus any subset of a metric space "is" itself a metric space, and in particular properties of metric spaces (completeness, compactness, connectedness, etc.) can also be attributed to subsets of metric spaces.

- **Exercise A4.17.** Suppose that $S \subset X$ and X is complete. Show that S is closed (as a subset of X) if and only if S is complete (when regarded as a metric space itself).

  We repeat that whether a set is open or closed depends on what metric space we are regarding it as a subset of; thus the basic concept is really "open (closed) subset of X", not "open (closed) set". For example, $[0, 1)$ is an open subset of the metric space $[0, 2]$ (with the standard metric), but not an open subset of $\mathbb{R}$, and similarly $(0, 1]$ is a closed subset of $(0, 2)$ but not a closed subset of $\mathbb{R}$. One often says that $[0, 1)$ is "*relatively open* in $[0, 2]$" and that $(0, 1]$ is "*relatively closed* in $(0, 2)$". Note that the word "relatively" is often omitted here.

- **Exercise A4.18.** Suppose that $S \subset E \subset X$. Show that S is relatively open (closed) in E if and only if there exists an open (closed) subset O of X with $S = O \cap E$.

The function $f : X \to Y$ is *continuous* at $x \in X$ if $f(x_n) \to f(x)$ in Y for every sequence (x_n) in X with $x_n \to x$. If f is continuous at x for every $x \in X$ then f is *continuous*.

- **Exercise A4.19.** Show that $f : X \to Y$ is continuous at $x \in X$ if and only if for every $\epsilon > 0$ there exists $\delta > 0$ such that $\rho(f(x), f(y)) < \epsilon$ for all $y \in X$ with $d(x, y) < \delta$.

The function $f : X \to Y$ is *uniformly continuous* if for every $\epsilon > 0$ there exists $\delta > 0$ such that $\rho(f(x), f(y)) < \epsilon$ for all $x, y \in X$ with $d(x, y) < \delta$.

If you're not reading very carefully it may appear that Exercise A4.19 shows that every continuous function is uniformly continuous. This is not so. The difference is that δ in Exercise A4.19 depends on both ϵ and x, while in the definition of uniform continuity δ depends only on ϵ. That is, if f is uniformly continuous then "the same δ works for every x"; one may choose δ "uniformly" with respect to x.

- **Exercise A4.20.** Give an example of a continuous $f : \mathbb{R} \to \mathbb{R}$ which is not uniformly continuous; explain.

There is a similar distinction between (pointwise) convergence and uniform convergence for sequences of functions. Suppose that $f_n : X \to Y$ ($n = 1, 2, \ldots$) and $f : X \to Y$. We say that $f_n \to f$ (pointwise) if $f_n(x) \to f(x)$ for every $x \in X$; this is the same as saying that for every $x \in X$ and every $\epsilon > 0$ there exists N such that $\rho(f_n(x), f(x)) < \epsilon$ for all $n > N$. On the other hand, we say that $f_n \to f$ *uniformly* if for every $\epsilon > 0$ there exists N such that $\rho(f_n(x), f(x)) < \epsilon$ for all $n > N$ and all $x \in X$; if $f_n \to f$ pointwise then N depends on ϵ and on $x \in X$, while if $f_n \to f$ uniformly then the same N works for all $x \in X$. The next two exercises show why the distinction is important:

- **Exercise A4.21.** Suppose that $f_n : X \to Y$ ($n = 1, 2, \ldots$), each f_n is continuous, and $f_n \to f$ uniformly. Show that f is continuous.

- **Exercise A4.22.** Give an example of $f_n : \mathbb{R} \to \mathbb{R}$ ($n = 1, 2, \ldots$) and $f : \mathbb{R} \to \mathbb{R}$ such that each f_n is continuous, $f_n \to f$ pointwise, but f is not continuous.

We say that X is *sequentially compact* if every sequence in X has a convergent subsequence. We say that X is *compact* if every open cover of X

has a finite subcover; this means that if $\mathcal{F}$ is a family of open subsets of X and
$$X = \bigcup_{S \in \mathcal{F}} S$$
then there exist $S_1, \ldots, S_n \in \mathcal{F}$ such that
$$X = \bigcup_{j=1}^{n} S_j.$$

We recall the Heine-Borel Theorem: A subset of $\mathbb{R}^n$ is compact if and only if it is closed and bounded. (Note that the same statement does not hold in a general metric space!)

An important result here is that *a metric space is compact if and only if it is sequentially compact.* You should note that neither implication holds for general topological spaces in place of metric spaces; see Appendix 6. The proof is tricky enough that I concede and write it out instead of making it a series of exercises. One direction is fairly easy:

Proposition A4.0. *If X is compact then X is sequentially compact.*

Proof. Suppose that X is compact but $(y_j) \subset X$ has no convergent subsequence. If $x \in X$ then (y_j) has no subsequence converging to x; it is not hard to see that this implies the existence of $r(x) > 0$ such that
$$B'(x, r(x)) \cap \{ y_j : j = 1, 2, \ldots \} = \emptyset,$$
where $B'(x, r) = B(x, r) \setminus \{x\}$.

It is clear that
$$X = \bigcup_{x \in X} B(x, r(x));$$
since X is compact, this shows that there exist $x_1, \ldots, x_n \in X$ with
$$X = \bigcup_{k=1}^{n} B(x_k, r(x_k)).$$
Since $y_j \notin B'(x, r(x))$, it follows that
$$\{ y_j : j = 1, 2, \ldots \} \subset \{x_1, \ldots, x_n\}.$$
So there must exist k such that $y_j = x_k$ for infinitely many values of j, which shows that (y_j) has a convergent subsequence after all. $\square$

The other direction requires a little preparation:

Lemma A4.1. *Suppose that X is sequentially compact and $\mathcal{F}$ is an open cover of X. Then there exists $\delta > 0$ such that for every $x \in X$ there exists $S \in \mathcal{F}$ with $B(x, \delta) \subset S$.*

The δ in the lemma is known as a *Lebesgue number* for the open cover $\mathcal{F}$. (Note that the conclusion of the lemma is sometimes misread. It does *not* say that for every $S \in \mathcal{F}$ and every $x \in S$ we have $B(x, \delta) \subset S$; that is typically false.)

Proof. Suppose that $\mathcal{F}$ is an open cover of X but there does not exist a Lebesgue number for $\mathcal{F}$. Then there exist $x_n \in X$ ($n = 1, 2, \ldots$) such that $B(x_n, 1/n)$ is contained in no element of $\mathcal{F}$. Say $x_{n_j} \to x$. Now, since $\mathcal{F}$ is an open cover of X, there exists $S \in \mathcal{F}$ with $x \in S$, and so there exists $r > 0$ with $B(x, r) \subset S$.

But if j is large enough then we have $d(x_{n_j}, x) < r/2$ and $1/n_j < r/2$; the triangle inequality then shows that $B(x_{n_j}, 1/n_j) \subset S$, a contradiction. $\square$

Theorem A4.2. *X is compact if and only if X is sequentially compact.*

Proof. One direction was proved in Proposition A4.0 above. For the other direction, suppose that X is sequentially compact and suppose that $\mathcal{F}$ is an open cover of X; we need to show that $\mathcal{F}$ has a finite subcover.

Let $\delta > 0$ be a Lebesgue number for $\mathcal{F}$, as in the previous lemma. Since every $B(x, \delta)$ is contained in some element of $\mathcal{F}$, it is enough to show that there exist $x_1, \ldots, x_n \in X$ with

$$X = \bigcup_{j=1}^{n} B(x_j, \delta).$$

Suppose this is false. Then we may recursively define a sequence $(x_n) \subset X$ such that

$$x_{n+1} \notin \bigcup_{j=1}^{n} B(x_j, \delta) \quad (n = 1, 2, \ldots).$$

But then (x_n) cannot have a convergent subsequence, since $d(x_n, x_m) \geq \delta$ whenever $n \neq m$. $\square$

Note. Lemma A4.1 is important for many reasons in addition to its use above. But it is not usually stated in terms of sequential compactness, since as we now know this is equivalent to compactness. This is to say that the version of the lemma you should store in your brain for future reference is this:

Lemma A4.1, rephrased. *Suppose that X is a compact metric space and $\mathcal{F}$ is an open cover of X. Then there exists $\delta > 0$ such that for every $x \in X$ there exists $S \in \mathcal{F}$ with $B(x, \delta) \subset S$.*

There is a characterization of compact metric spaces that is often very useful, in terms of a concept known as *total boundedness:* We say that the metric space X is *totally bounded* if for every $\epsilon > 0$ there exist finitely many points $x_1, \ldots, x_n \in X$ such that

$$X = \bigcup_{j=1}^{n} B(x_j, \epsilon).$$

It is clear that this has something to do with compactness: $\mathcal{F} = \{\, B(x, \epsilon) : x \in X \,\}$ is an open cover of X, and saying that X is totally bounded just says that this particular open cover has a finite subcover. What is interesting is that this condition, plus completeness, suffices to characterize compactness:

Theorem A4.3. *The metric space X is compact if and only if it is complete and totally bounded.*

Proof. Suppose first that X is compact. As mentioned above, the fact that X is totally bounded is a trivial consequence of the definitions, and it is easy to see that X is complete: If (x_n) is a Cauchy sequence in X then the fact that X is sequentially compact (cf. Theorem A4.2) shows that (x_n) has a convergent subsequence, and it is easy to show that a Cauchy sequence with a convergent subsequence is itself convergent (if this is not clear you should take it as an exercise; the proof begins "Let $\epsilon > 0$ ... ").

Suppose now that X is complete and totally bounded. We intend to show that X is sequentially compact, so we choose a sequence (x_n) in X; now, since X is complete, we need only show that (x_n) has a Cauchy subsequence.

We can cover X by finitely many balls of radius 1; since there are infinitely many positive integers, one of these balls must contain x_n for infinitely many values of n. That is, there exists $y_1 \in X$ such that S_1 is infinite, where

$$S_1 = \{\, n : x_n \in B(y_1, 1) \,\}.$$

And now we can cover X by finitely many balls of radius $1/2$; since S_1 is infinite, one of these balls must contain x_n for infinitely many $n \in S_1$. That is, there exists $y_2 \in X$ such that S_2 is infinite, where

$$S_2 = \{\, n \in S_1 : x_n \in B(y_2, 1/2) \,\}.$$

Appendix 4. Metric Spaces 469

Repeating this construction we obtain a sequence (S_k) of infinite sets of integers and a sequence $(y_k) \subset X$, such that $S_{k+1} \subset S_k$ and $x_n \in B(y_k, 1/k)$ for every $n \in S_k$. In particular we have

$$d(x_n, x_m) < 2/k \quad (n, m \in S_k).$$

Now let $n_1 \in S_1$. Since S_2 is infinite, there exists $n_2 \in S_2$ with $n_2 > n_1$, and in fact, since all the S_k are infinite, we obtain a strictly increasing sequence (n_k) with $n_k \in S_k$. Now the fact that the S_k are nested shows that $n_j \in S_k$ whenever $j \geq k$, and this implies that

$$d(x_{n_j}, x_{n_k}) < 2/N \quad (j, k \geq N).$$

Hence the subsequence (x_{n_k}) is Cauchy. $\square$

A few examples of why we care about the concept of compactness follow; note that some of the next few exercises may or may not be easier using sequential compactness in place of compactness:

- **Exercise A4.23.** Suppose that X is compact and $f : X \to Y$ is continuous.

    (i) Show that f is uniformly continuous.

    (ii) Show that the image of f, i.e. the set $f(X) = \{ f(x) : x \in X \}$, is compact.

- **Exercise A4.24.** Suppose that X is compact and $f : X \to \mathbb{R}$ is continuous. Show that f is bounded (i.e., there exists K such that $|f(x)| \leq K$ for all $x \in X$).

- **Exercise A4.25.** Suppose that X is compact and $f, g : X \to Y$ are continuous. Define $h : X \to \mathbb{R}$ by $h(x) = \rho(f(x), g(x))$. Show that h is uniformly continuous.

- **Exercise A4.26.** Suppose that X is compact and $f : X \to Y$ is a continuous bijection. Then $f^{-1} : Y \to X$ is continuous. (Since f is a bijection, you need only show that $f(O)$ is open for all open $O \subset X$; again, since X is a bijection, this is the same as showing that $f(K)$ is closed for all closed sets $K \subset X$.)

In general we will let $C(X, Y)$ denote the class of all continuous functions from X to Y. (Since this is a book on analysis, we set $C(X) = C(X, \mathbb{C})$.)

- **Exercise A4.27.** Suppose that X is compact. For $f, g \in C(X, Y)$ define
$$d_u(f, g) = \sup_{x \in X} \rho(f(x), g(x)).$$

  (i) Show that $d_u(f, g) < \infty$ for all $f, g \in C(X, Y)$.

  (ii) Show that d_u is a metric on $C(X, Y)$.

  (iii) Show that $f_n \to f$ in $C(X, Y)$ if and only if $f_n \to f$ uniformly.

  (iv) Show that if Y is complete then so is $C(X, Y)$.

We say that X is *connected* if X and $\emptyset$ are the only subsets of X that are both open and closed.

- **Exercise A4.28.** If X is connected and $f : X \to Y$ is continuous then the image $f(X)$ is connected.

- **Exercise A4.29.** Suppose that $x_0 \in X$, $X = \bigcup_{j \in J} S_j$, each S_j is connected and $x_0 \in S_j$ for every $j \in J$. Then X is connected.

- **Exercise A4.30.** Suppose that $S \subset X$ is connected, S_x is connected and $x \in S_x$ for every $x \in S$, and $X = \bigcup_{x \in S} S_x$. Then X is connected.

- **Exercise A4.31.** The interval $[0, 1]$ is connected.

  Hint: Suppose that $S \subset [0, 1]$ is both open and closed. Since the same is true of the complement $[0, 1] \setminus S$, we may assume that $0 \in S$ and we have to show that $S = [0, 1]$.

  Since S is open, there exists $x > 0$ such that $[0, x) \subset S$, which is to say that $\{ x \in (0, 1] : [0, x) \subset S \}$ is nonempty. Let $a = \sup\{ x \in (0, 1] : [0, x) \subset S \}$. Show that $a < 1$ is impossible (because S is open); conclude that $a = 1$, and hence that $S = [0, 1]$ (because S is closed).

  We say that X is *path-connected* if for all $x, y \in X$ there exists a continuous map $\gamma : [0, 1] \to X$ with $\gamma(0) = x$ and $\gamma(1) = y$ (i.e. "there exists a path from x to y".)

- **Exercise A4.32.** If X is path-connected then X is connected (use the previous three exercises).

The converse of Exercise A4.32 is false (ask someone to show you "the topologist's sine curve", for example.) But the converse is true for open subsets of $\mathbb{R}^n$. We need another concept for the proof: We say that X is *locally path-connected* if for every $x \in X$ and every $r > 0$ there exists a path-connected open set O such that $x \in O \subset B(x, r)$.

- **Exercise A4.33.** If X is connected and locally path-connected then X is path-connected.

  Hint: Fix $x \in X$ and let $S = \{y \in X :$ there exists a path from x to $y\}$. Show that S is nonempty, open and closed (show that the complement is open!) and conclude that $S = X$.

- **Exercise A4.34.** Any convex subset of $\mathbb{R}^n$ is path-connected.

- **Exercise A4.35.** An open subset of $\mathbb{R}^n$ is connected if and only if it is path-connected.

  We should note that the same argument shows that an open subset of $\mathbb{R}^n$ is connected if and only if it is *polygonally connected*, meaning that any two points can be joined by a curve consisting of a finite union of line segments; it follows that an open subset of $\mathbb{R}^n$ is connected if and only if any two points can be joined by a piecewise continuously differentiable curve. (In fact any two points of an open connected subset of $\mathbb{R}^n$ can be joined by a continuously differentiable curve; the proof of this is essentially the same but one of the details is slightly trickier.)

  Finally we need to say a little bit about the connected components of a metric space. Suppose that $x \in X$ and let $\mathcal{F}$ be the family of all connected subsets of X that contain x. Then Exercise A4.29 above shows that

  $$C_x = \bigcup_{S \in \mathcal{F}} S$$

  is connected. Hence $C_x \in \mathcal{F}$, which is to say that C_x is the largest connected subset of X containing x. We call C_x a *component* of X (or "connected component"). The components of X are precisely the maximal connected subsets of X; the next exercise shows that they form a partition of X.

- **Exercise A4.36.** Suppose that $x, y \in X$, and let C_x and C_y be the components of X containing x and y, respectively. Show that either $C_x = C_y$ or $C_x \cap C_y = \emptyset$.

  We will need the fact that the components of an open subset of $\mathbb{R}^n$ are open; this follows from the next exercise. We say that X is *locally connected* if for every $x \in X$ and every $r > 0$ there exists a connected open set O such that $x \in O \subset B(x, r)$.

- **Exercise A4.37.** Suppose that X is locally connected. Show that the components of X are open.

Appendix 5

Convexity

In this appendix we will assume that X is a real vector space (note that this includes the case $X = \mathbb{C}$).

If $x, y \in X$ the line segment with endpoints x and y is

$$[x, y] = \{\, (1-t)x + ty : 0 \leq t \leq 1 \,\}.$$

The set $S \subset X$ is *convex* if $[x, y] \subset S$ for all $x, y \in S$.

- **Exercise A5.1.** The intersection of any family of convex subsets of X is convex.

So in particular if $S \subset X$ then the intersection of all convex subsets of X containing S is convex; this is clearly the smallest convex subset of X containing S and is called the *convex hull* of S.

A *convex combination* of the elements of S is a vector of the form

$$x = \sum_{j=1}^{n} t_j x_j,$$

where n is a positive integer, each x_j is an element of S, and the t_j satisfy $t_j \geq 0$ and $\sum_{j=1}^{n} t_j = 1$.

- **Exercise A5.2.** If S is convex then any convex combination of elements of S is an element of S. (Use induction on n.)

- **Exercise A5.3.** If $S \subset X$ then the convex hull of S is equal to the set of all convex combinations of elements of X. (Say C is the set of all convex combinations of elements of S. Show that C is a convex set containing S, and then use the previous exercise to show that it is the smallest convex set containing S.)

Appendix 6

Four Counterexamples

Here we give a few counterexamples that didn't quite fit into the main text: Two examples illustrating a fact mentioned in Appendix 4, an example showing that E need not be a fundamental domain for G in Lemma 18.2, and finally an example showing that the conclusion of Theorem 19.0.3 fails for non-simply connected covering spaces

For the first three examples we assume the reader knows a little bit more about general point-set topology and set theory than in the rest of the book — the results are not used elsewhere in the text.

To show that sequential compactness does not imply compactness in a general topological space, let Ω be an uncountable well-ordered set, with smallest element 0 and no largest element, such that every initial segment $[0, x)$ is countable (for example Ω could be the set of countable ordinals). Give Ω the order topology.

Then Ω is certainly not compact, because the set of all initial segments $[0, x)$ is an open cover that obviously has no finite subcover. On the other hand Ω *is* sequentially compact; for example any sequence in Ω has a monotone subsequence (the proof of Theorem 11.3 in Ross [**RO**] suffices to show this) and any monotone sequence must be convergent.

To show that compactness does not imply sequential compactness in a topological space let P be the set of all subsets of the natural numbers and let $X = [0, 1]^P$, with the product topology. The Tychonoff Theorem shows that X is compact. But X is not sequentially compact:

Define a sequence $(x_n) \subset X$ by

$$x_n(S) = \begin{cases} 1 & (n \in S), \\ 0 & (n \notin S), \end{cases} \quad (S \in P).$$

Then (x_n) has no convergent subsequence. Indeed, suppose that (x_{n_j}) is a subsequence and let $S = \{\, n_j : j \text{ is even}\,\}$. Then

$$x_{n_j}(S) = \begin{cases} 1 & (j \text{ even}), \\ 0 & (j \text{ odd}), \end{cases}$$

so that $(x_{n_j}(S))$ does not converge; hence (x_{n_j}) does not converge.

Although it seemed very unlikely that the hypotheses of Lemma 18.2 should imply that E is a fundamental domain for G it took me a little while to come up with a counterexample. In retrospect the construction is obvious: Consider the ping-pong procedure used in the proof of Theorem 18.9. As noted elsewhere in Chapter 18, if there does in fact exist $\phi \in \Gamma(2)$ with $\phi(z) \in Q$ then it must be the case that that construction yields ϕ. So we simply concoct an example in the setting of Lemma 18.2 where that procedure never terminates, as follows:

Suppose that G is a free group on two generators a and b. Let X be the set consisting of e, the identity of G, all nontrivial reduced words in a and b, and also all "infinite reduced words" in a and b, where an infinite reduced word is allowed to extend infinitely far to the left, but must terminate on the right (so for example $w = \ldots ab^4 ab^3 ab^2 ab$ is an example of an "infinite reduced word" in a and b).

Define an action of G on X by "concatenate and reduce". For example, if $w = \ldots ab^4 ab^3 ab^2 ab \in X$ and $g = b^{-1} a^{-1} b^2 \in G$ then the action of g on w is given by

$$wg = \ldots ab^4 ab^3 ab^2 abb^{-1} a^{-1} b^2 = \ldots ab^5 ab^4 ab^3 ab^4.$$

A moment's thought shows that wg is always well defined (for a formal proof you can find a book on algebra and modify the proof that any word is equal to a unique reduced word).

Let A consist of e together with all words that terminate in a power of b, and let B consist of e together with all words that terminate in a power of a. It is not hard to see that all the hypotheses of Lemma 18.2 are satisfied. But $E = \{e\}$, so that $G(E)$ consists of just e together with the finite reduced words.

It is not at all important for our purposes, but, since I was surprised at how long it took me to determine whether Theorem 19.0.3 remained true without assuming that X is simply connected, I thought I would include a discussion of that. It turns out that a covering space need not be nearly as homogeneous as I had assumed: My mental picture of what a covering space looks like made it clear that the theorem should still be true whether X is simply connected or not, but I couldn't quite prove it after some effort.

Appendix 6. Four Counterexamples

I couldn't find any mention of the question in a few books; I finally asked Benny Evans and he showed me the standard counterexample:

We will take X to be a union of 4 circles:

$$X = T_0 \cup T_2 \cup T_4 \cup T_6,$$

where

$$T_j = \{\, z \in \mathbb{C} : |z - j| = 1 \,\}.$$

We take Y to be a union of two circles:

$$Y = T_0 \cup T_2.$$

We define $p : X \to Y$ in pieces:

$$p(z) = \begin{cases} z, & z \in T_0, \\ 2 - (z-2)^2, & z \in T_2, \\ (z-4)^2, & z \in T_4, \\ 2 + (z-6), & z \in T_6. \end{cases}$$

The idea is that p wraps T_0 once around T_0, it wraps T_2 twice around T_2, it wraps T_4 twice around T_0, and it wraps T_6 once around T_2. So it is a covering map; we leave it to you to verify carefully that it satisfies the definition. The possibly confusing question is why the point $1 \in Y$ has a regular neighborhood; Figure A6.1 may clarify this:

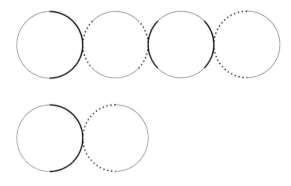

Figure A6.1

We have $p(1) = 1 = p(3)$, but there is no homeomorphism of X taking 1 to 3, so the conclusion of Theorem 19.0.3 fails.

Appendix 7

The Cauchy-Riemann Equations Revisited

We suppose throughout this appendix that D is an open subset of $\mathbb{C}$ and $f: D \to \mathbb{C}$ has continuous partial derivatives in D; we set $u = u(x,y) = \operatorname{Re}(f)$ and $v = v(x,y) = \operatorname{Im}(f)$.

The Cauchy-Riemann equations are

$$u_x = v_y, \ u_y = -v_x.$$

We know that $f \in H(D)$ if and only if the Cauchy-Riemann equations are satisfied at every point of D. We also know that in this case we have

$$f' = u_x + iv_x;$$

this can be put into at least three other forms, using the Cauchy-Riemann equations.

There is another way to phrase all this; the alternate notation makes less sense, at least at first, but is much more suggestive. We define two operators $\partial/\partial z$ and $\partial/\partial \bar{z}$ by

$$\frac{\partial f}{\partial z} = \frac{1}{2}\left(\frac{\partial f}{\partial x} - i\frac{\partial f}{\partial y}\right), \quad \frac{\partial f}{\partial \bar{z}} = \frac{1}{2}\left(\frac{\partial f}{\partial x} + i\frac{\partial f}{\partial y}\right).$$

Hint on how to keep straight which one has the minus sign: You can remember that it appears to be backwards, or you can think of it as rewritten

$$\frac{\partial f}{\partial z} = \frac{1}{2}\left(\frac{\partial f}{\partial x} + \frac{\partial f}{\partial (iy)}\right), \quad \frac{\partial f}{\partial \bar{z}} = \frac{1}{2}\left(\frac{\partial f}{\partial x} + \frac{\partial f}{\partial (-iy)}\right).$$

There are at least two things to be said for this notation: (i) it makes the Cauchy-Riemann equations and the corresponding expression for f' look right; (ii) although the two operators $\frac{\partial}{\partial z}$ and $\frac{\partial}{\partial \bar{z}}$ are not really partial derivatives with respect to two independent variables (z and $\bar{z}$ are certainly not independent!) you can often pretend that they are and hence just "see" what $\frac{\partial f}{\partial z}$ and $\frac{\partial f}{\partial \bar{z}}$ should be. That is:

(i) f satisfies the Cauchy-Riemann equations (and hence $f \in H(D)$) if and only if
$$\frac{\partial f}{\partial \bar{z}} = 0;$$
in this case we have
$$f' = \frac{\partial f}{\partial z}.$$

That looks right, right?

(ii) Now if you have f written as, say, a polynomial in z and $\bar{z}$ then you can tell what $\frac{\partial f}{\partial z}$ and $\frac{\partial f}{\partial \bar{z}}$ are just by looking. For example
$$\frac{\partial (z^2 \bar{z}^3)}{\partial z} = 2z\bar{z}^3, \quad \frac{\partial (z^2 \bar{z}^3)}{\partial \bar{z}} = 3z^2 \bar{z}^2.$$

(If you want to prove that this always works for polynomials all you have to do is verify that
$$\frac{\partial z}{\partial z} = 1, \quad \frac{\partial z}{\partial \bar{z}} = 0, \quad \frac{\partial \bar{z}}{\partial z} = 0 \text{ and } \frac{\partial \bar{z}}{\partial \bar{z}} = 1,$$
and then note that the two operators satisfy the product rule.)

A person sometimes says that all this shows that f is holomorphic if and only if "it doesn't have any $\bar{z}$'s in it". I know I sometimes say that, anyway, even though it is too vague to be literally true (for example I don't see any $\bar{z}$'s in $\log(|z|)$, but it is certainly not holomorphic, being real-valued). We give a precise version of "f is holomorphic if and only if it doesn't have any $\bar{z}$'s in it" in the theorem below.

We say that a function of two complex variables $F(z, w)$ is holomorphic if it is holomorphic in each variable separately. (One of the amazing facts about complex analysis is that this condition actually implies the function is continuous; it then follows easily that it is infinitely smooth. Never mind that for now ...) If F is a holomorphic function of two variables we will denote the (complex) partial derivatives with respect to the first and second variables by F_1 and F_2 respectively.

Appendix 7. The Cauchy-Riemann Equations Revisited

Theorem. *Suppose that F is a holomorphic function of two variables and $f(z) = F(z, \bar{z})$. Then*

$$\frac{\partial f}{\partial z}(z) = F_1(z, \bar{z}), \quad \frac{\partial f}{\partial \bar{z}}(z) = F_2(z, \bar{z}),$$

and hence:

f is holomorphic if and only if $F_2 = 0$, in which case $f'(z) = F_1(z, \bar{z})$.

Note. Most of the things we've stated in this appendix are easy advanced-calculus exercises. The theorem above is also easy *if* we assume that F is continuously differentiable (in the real sense); this actually follows from the hypotheses as given, but that fact is far from trivial.

References

This is not meant to be a list of suggestions for further reading, a list of classic works in the field, or any such thing; it is simply a list of the works that happen to be cited in the text.

[D] Dixon, John D., *A brief proof of Cauchy's integral theorem*, Proc. Amer. Math. Soc. **1971**, no. 29, 625–626.

[DO] Doob, Joseph L., *Classical Potential Theory and its Probabilistic Counterpart*, Classics in Mathematics, Springer-Verlag, Berlin, 2001, Reprint of the 1984 edition.

[DU] Durrett, Richard, *Brownian Motion and Martingales in Analysis*, Wadsworth Mathematics Series, Wadsworth International Group, Belmont, CA, 1984.

[E] Estermann, T., *Complex Numbers and Functions*, The Athlone Press, London: University of London, 1962.

[F] Folland, Gerald B., *Real Analysis: Modern Techniques and their Application*, Pure and Applied Mathematics, John Wiley & Sons Inc., New York, 1999.

[FB] Freedman, David, *Brownian Motion and Diffusion*, Springer-Verlag, New York, 1983.

[G] Garnett, John B., *Applications of Harmonic Measure*, University of Arkansas Lecture Notes in the Mathematical Sciences, 8, John Wiley & Sons Inc., New York, 1986, A Wiley-Interscience Publication.

[H] Helms, Lester L., *Introduction to Potential Theory*, Robert E. Krieger Publishing Co., Huntington, N.Y., 1975, Reprint of the 1969 edition.

[J]   Julia, G., *Leçons sur les fonctions uniformes à point singulier essentiel isolé. Rédigées par P. Flamant. (Collection de Monographies sur la théorie des fonctions)*, Paris: Gauthier-Villars, 1924.

[M]   Markushevich, A. I., *Theory of Functions of a Complex Variable. Vol. I, II, III*, Chelsea Publishing Co., New York, 1977, Translated and edited by Richard A. Silverman.

[MA] Massey, William S., *Algebraic Topology: an Introduction*, Springer-Verlag, New York, 1977, Reprint of the 1967 edition.

[R]   Rudin, Walter, *Real and Complex Analysis*, McGraw-Hill Book Co., New York, 1987.

[RO] Ross, Kenneth A., *Elementary Analysis: the Theory of Calculus*, Springer-Verlag, New York, 1980, Undergraduate Texts in Mathematics.

[S]   Salem, Raphaël, *Algebraic Numbers and Fourier Analysis*, D. C. Heath and Co., Boston, Mass., 1963.

Index of Notations

$\mathbb{C}$	3	$SL_2(\mathbb{R})$	220
O ("big-oh")	4	$\mathcal{R}_A$	231
o ("little-oh")	4	E_n	247
$D(z,r), \overline{D}(z,r)$	9	(f, D)	280
$\dot{+}, \dot{-}$	20	$\mathcal{O}_a, \mathcal{O}(D)$	281
$[a,b]$	21	$[f]_a$	281
$H(V)$	30	$\Gamma, \Gamma(2)$ (modular groups)	319
∞	37, 113	σ, τ (generators for $\Gamma(2)$)	320
$Z(f)$	40	j (involution in Π^+)	321
$D'(z,r)$	43	$\mathrm{Aut}(\Omega, p)$	339
$A(a,r,R)$	46	Γ (Gamma function)	399
Ind	56	$\mathbb{R}$	446
$\gamma_0 \sim \gamma_1$	69	$\bar{z}$	449
Res	71	Im	449
$\mathbb{C}_\infty$	113	Re	449
$\mathrm{Aut}(V)$	118	$|z|$	449
$\mathcal{M}$	120	$\arg(z)$	452
$GL_2(\mathbb{C})$	121	cos	455
Π^+	129	exp	455
$\mathbb{D}$	130	sin	455
ϕ_a	132	e^z	457
P_r (Poisson kernel)	173	$C(X,Y), C(X)$	469
$P[f]$ (Poisson integral)	175	$[x,y]$	473
$\mathrm{Exit}(D,z)$	202	$\partial/\partial z, \partial/\partial \bar{z}$	479
$\mathrm{Pr}(A)$ (probability of A)	205		

Index

Abel summability, 187, 369
Abel's Theorem, 367–372
abelian theorem, 369
absolute convergence, 452
absolutely and uniformly, 10
affine map, 119
almost surely, 200
analytic
   continuation, 277–305
   function, 29
annulus, 46, 66
approximate identity, 174
argument, 452–453
Argument Principle, 80
Arzelà-Ascoli Theorem, 153–155
automorphism group, 118

ball (in a metric space), 462
big oh, 4
biholomorphic, 118
boundary, 463
branch of the logarithm, 54–56
Brownian motion, 199–215

Carathéodory's Theorem, 257–266
Cauchy
   Integral Formula, 26, 51–71, 89, 238, 435–440
   Integral Formula for derivatives, 34, 75
   -Riemann equations, 3–6, 169, 479–481
   sequence, 462
Cauchy's
   Estimates, 35
   Theorem, 17, 19, 22, 26, 30, 51–71, 243, 435–440
Cayley transform, 129–130

chain, 20
circuit, 194
closed set, 451, 463
closure, 463
compact, 465
   sequentially, 465
complete
   analytic function, 287–288
   metric space, 462
complex number, 445–453
component, 471
concave function, 143
conformal
   equivalence, 117
   mapping, 117
conjugate
   harmonic, 171, 441–442
   of a complex number, 449
connected, 470
   component, 471
   path-, 470
   polygonally, 471
continuation (of a function element along a path), 282
continuous function, 451, 465
contour integral, 17
convergence
   uniform, 452, 465
   uniform, on compact sets, 33
convergent sequence, 462
convex, 21
   combination, 473
   hull, 21, 473
   set, 473
convexity, 106
   source of all wisdom, 98

487

cosine, 455–459
cotangent (infinite sum for), 91
counting zeroes, 79
covering
    map, 290, 337–354
    space, 58
curve, 16
cycle, 20

deleted disk, 43
diameter, diam, 463
differentiable function, 3–6
Dirichlet
    domain, 204
    problem, 171–182
diverges to zero, 94
Dominated Convergence Theorem, 146, 313

elementary
    factors, 246–247
    neighborhood, 290, 337
elliptic automorphism, 268
entire function, 37
equicontinuous, 152
essential singularity, 44
Euler, L., 87
Euler's formula for the sine, 87–102, 254–255
Evans, Benny, 477
exhaustion, 150
exponential function, 455–459
extended plane, 113–115

field of quotients, 252
fixed points (of automorphisms), 268
Fourier
    coefficients, 185–187
    series, 185–187
free group, 324
freely generated, 324
Fubini's Theorem, 64
function element, 280
fundamental domain, 322, 325
Fundamental Theorem of Algebra, 37

Gamma function, 399–420
Gauss, K.F., 404
general linear group, 121
germ of a holomorphic function, 281
graph, directed, 194–198
Green's function, 191–193

Hadamard Factorization Theorem, 255
harmonic
    conjugate, 171, 441–442
    function, 167–224, 441–442

Harnack's Inequality, Harnack's Theorem, 220–224
Hausdorff-Young Inequality, 390–397
Heine-Borel Theorem, 466
holomorphic
    function, 30
    interpolation, 252
homotopic
    curves, 69–70
    null-, 286, 294
homotopy, 285
    path-, 285
Hurwitz's Theorem, 85, 158, 224
hyperbolic automorphism, 268

impudence, 390
In principio, 3
index, 56
inequalities, source of, 98
infinite products, 92–94
infinity, 37
integral, 17
integral domain, 252
Integral Tests, 99
interior, 463
invariant Schwarz Lemma, 135–139
inverse functions, 105–109
isolated singularity, 43

Kable, Anthony, 129

Landau notation, 4
Laurent series, 46, 66
Lebesgue number, 467
lifting, 338
limit, 462
    point, 463
Lindelöf's Theorem, 262
line integral, 17
linear-fractional transformation, 117–135
Liouville's Theorem, 37
little oh, 4
Littlewood, J. E., 357, 390
locally
    connected, 471
    path-connected, 470
logarithm, 54–56
    branch of, 54–56
logarithmic derivative, 95

M-test, 452
manifold, 304–305
Maximum Modulus Theorem, 41–43, 191, 385
Mean Value Property, 28, 172, 173, 181
    weak, 179–182
meromorphic function, 113, 252

Index

metric space, 461–471
Mittag-Leffler Theorem, 241
ML inequality, 18
Möbius transformation, 118
modular
  function, 319–335
  group, 319–335
modulus, 449
Monodromy Theorem, 286, 288–297
Montel's Theorem, 155–157, 164–165
Morera's Theorem, 30
multiplicity, 79

Noell, Alan, 218
normal family, 157
null-homotopic, 280, 294

open mapping, 83
Open Mapping Theorem, 83–84, 109–111
open set, 451, 462
order
  of a pole, 45
  of a zero, 39

parabolic automorphism, 268
Parseval, 42, 75, 391
path-connected, 470
path-homotopic, 285
Phragmén-Lindelöf Theorem, 385
ping, 325
ping pong lemma, 324
pointwise convergence, 465
Poisson
  integral, 171–187
  kernel, 173
polar coordinates, 452
pole, 44
polygonally connected, 471
pong, 325
power series, 9–13, 28
Prime Number Theorem, 417
principal part, 46
proper map, 119
pseudo-hyperbolic distance, 135–139
punctured disk, 43

quasi-metric, 143–149

radius of convergence, 10
rational function, 229–231
real-analytic, 29, 37
regular domain, 204
relatively
  closed, 464
  open, 464
removable singularity, 44
residue, 71

Residue Theorem, 71–74
Riemann
  Hypothesis, 417
  Mapping Theorem, 141, 159–163
  sphere, 113–115
  surface, 304–305
Riesz, M., 390
Rouché's Theorem, 81, 85
Rudin, Walter, xi
Rudin-Shapiro Polynomials, 397
Runge's Theorem, 229–241, 238

Schwarz Lemma, 130–139
  invariant, 135–139
  fancy version, 135–139
Schwarz Reflection Principle, 181, 215–220
sequence, 451
sequentially compact, 465
series, 451
sheaf of germs of holomorphic functions, 281
simple
  boundary point, 257
  pole, 45
  zero, 39
Sicut erat in principio, 479
simply connected, 26, 70, 225, 338
sine, 455–459
smooth curve, 17
source of all wisdom, 98
stereographic projection, 113
summability method, 369
summation by parts, 367

tauberian theorem, 369
Three-Lines Theorem, 385
topology, 462
totally bounded, 468
triangle inequality, 450

uniform convergence, 452, 465
  on compact sets, 33
Uniformization Theorem, 423
uniformly continuous function, 465
unrestricted continuation, 288–297

Weierstrass, 455
  elementary factors, 246–247
  Factorization Theorem, 245–251
    for entire functions, 249
    in an open set, 250
well defined, 53
winding number, 56

zeroes, counting, 79

Titles in This Series

97 **David C. Ullrich,** Complex made simple, 2008

96 **N. V. Krylov,** Lectures on elliptic and parabolic equations in Sobolev spaces, 2008

95 **Leon A. Takhtajan,** Quantum mechanics for mathematicians, 2008

94 **James E. Humphreys,** Representations of semisimple Lie algebras in the BGG category $\mathcal{O}$, 2008

93 **Peter W. Michor,** Topics in differential geometry, 2008

92 **I. Martin Isaacs,** Finite group theory, 2008

91 **Louis Halle Rowen,** Graduate algebra: Noncommutative view, 2008

90 **Larry J. Gerstein,** Basic quadratic forms, 2008

89 **Anthony Bonato,** A course on the web graph, 2008

88 **Nathanial P. Brown and Narutaka Ozawa,** C*-algebras and finite-dimensional approximations, 2008

87 **Srikanth B. Iyengar, Graham J. Leuschke, Anton Leykin, Claudia Miller, Ezra Miller, Anurag K. Singh, and Uli Walther,** Twenty-four hours of local cohomology, 2007

86 **Yulij Ilyashenko and Sergei Yakovenko,** Lectures on analytic differential equations, 2007

85 **John M. Alongi and Gail S. Nelson,** Recurrence and topology, 2007

84 **Charalambos D. Aliprantis and Rabee Tourky,** Cones and duality, 2007

83 **Wolfgang Ebeling,** Functions of several complex variables and their singularities (translated by Philip G. Spain), 2007

82 **Serge Alinhac and Patrick Gérard,** Pseudo-differential operators and the Nash–Moser theorem (translated by Stephen S. Wilson), 2007

81 **V. V. Prasolov,** Elements of homology theory, 2007

80 **Davar Khoshnevisan,** Probability, 2007

79 **William Stein,** Modular forms, a computational approach (with an appendix by Paul E. Gunnells), 2007

78 **Harry Dym,** Linear algebra in action, 2007

77 **Bennett Chow, Peng Lu, and Lei Ni,** Hamilton's Ricci flow, 2006

76 **Michael E. Taylor,** Measure theory and integration, 2006

75 **Peter D. Miller,** Applied asymptotic analysis, 2006

74 **V. V. Prasolov,** Elements of combinatorial and differential topology, 2006

73 **Louis Halle Rowen,** Graduate algebra: Commutative view, 2006

72 **R. J. Williams,** Introduction the the mathematics of finance, 2006

71 **S. P. Novikov and I. A. Taimanov,** Modern geometric structures and fields, 2006

70 **Seán Dineen,** Probability theory in finance, 2005

69 **Sebastián Montiel and Antonio Ros,** Curves and surfaces, 2005

68 **Luis Caffarelli and Sandro Salsa,** A geometric approach to free boundary problems, 2005

67 **T.Y. Lam,** Introduction to quadratic forms over fields, 2004

66 **Yuli Eidelman, Vitali Milman, and Antonis Tsolomitis,** Functional analysis, An introduction, 2004

65 **S. Ramanan,** Global calculus, 2004

64 **A. A. Kirillov,** Lectures on the orbit method, 2004

63 **Steven Dale Cutkosky,** Resolution of singularities, 2004

62 **T. W. Körner,** A companion to analysis: A second first and first second course in analysis, 2004

TITLES IN THIS SERIES

61 **Thomas A. Ivey and J. M. Landsberg,** Cartan for beginners: Differential geometry via moving frames and exterior differential systems, 2003
60 **Alberto Candel and Lawrence Conlon,** Foliations II, 2003
59 **Steven H. Weintraub,** Representation theory of finite groups: algebra and arithmetic, 2003
58 **Cédric Villani,** Topics in optimal transportation, 2003
57 **Robert Plato,** Concise numerical mathematics, 2003
56 **E. B. Vinberg,** A course in algebra, 2003
55 **C. Herbert Clemens,** A scrapbook of complex curve theory, second edition, 2003
54 **Alexander Barvinok,** A course in convexity, 2002
53 **Henryk Iwaniec,** Spectral methods of automorphic forms, 2002
52 **Ilka Agricola and Thomas Friedrich,** Global analysis: Differential forms in analysis, geometry and physics, 2002
51 **Y. A. Abramovich and C. D. Aliprantis,** Problems in operator theory, 2002
50 **Y. A. Abramovich and C. D. Aliprantis,** An invitation to operator theory, 2002
49 **John R. Harper,** Secondary cohomology operations, 2002
48 **Y. Eliashberg and N. Mishachev,** Introduction to the h-principle, 2002
47 **A. Yu. Kitaev, A. H. Shen, and M. N. Vyalyi,** Classical and quantum computation, 2002
46 **Joseph L. Taylor,** Several complex variables with connections to algebraic geometry and Lie groups, 2002
45 **Inder K. Rana,** An introduction to measure and integration, second edition, 2002
44 **Jim Agler and John E. McCarthy,** Pick interpolation and Hilbert function spaces, 2002
43 **N. V. Krylov,** Introduction to the theory of random processes, 2002
42 **Jin Hong and Seok-Jin Kang,** Introduction to quantum groups and crystal bases, 2002
41 **Georgi V. Smirnov,** Introduction to the theory of differential inclusions, 2002
40 **Robert E. Greene and Steven G. Krantz,** Function theory of one complex variable, third edition, 2006
39 **Larry C. Grove,** Classical groups and geometric algebra, 2002
38 **Elton P. Hsu,** Stochastic analysis on manifolds, 2002
37 **Hershel M. Farkas and Irwin Kra,** Theta constants, Riemann surfaces and the modular group, 2001
36 **Martin Schechter,** Principles of functional analysis, second edition, 2002
35 **James F. Davis and Paul Kirk,** Lecture notes in algebraic topology, 2001
34 **Sigurdur Helgason,** Differential geometry, Lie groups, and symmetric spaces, 2001
33 **Dmitri Burago, Yuri Burago, and Sergei Ivanov,** A course in metric geometry, 2001
32 **Robert G. Bartle,** A modern theory of integration, 2001
31 **Ralf Korn and Elke Korn,** Option pricing and portfolio optimization: Modern methods of financial mathematics, 2001
30 **J. C. McConnell and J. C. Robson,** Noncommutative Noetherian rings, 2001
29 **Javier Duoandikoetxea,** Fourier analysis, 2001
28 **Liviu I. Nicolaescu,** Notes on Seiberg-Witten theory, 2000
27 **Thierry Aubin,** A course in differential geometry, 2001
26 **Rolf Berndt,** An introduction to symplectic geometry, 2001

For a complete list of titles in this series, visit the AMS Bookstore at **www.ams.org/bookstore/**.